Encyclopaedia of
MEASURE THEORY

Encyclopaedia of
MEASURE THEORY

Volume 1

R.K. Pandey

ANMOL PUBLICATIONS PVT. LTD.
NEW DELHI - 110 002 (INDIA)

ANMOL PUBLICATIONS PVT. LTD.
H.O.: 4374/4B, Ansari Road, Daryaganj,
New Delhi-110 002 (India)
Ph.: 23278000, 23261597
B.O.: No. 1015, Ist Main Road, BSK IIIrd Stage
IIIrd Phase, IIIrd Block,
Bangalore - 560 085 (India)
Visit us at: www.anmolpublications.com

Encyclopaedia of Measure Theory

First Edition, 2009
ISBN 978-81-261-3921-7 (Set)

PRINTED IN INDIA

Printed at Mehra Offset Press, Delhi.

Contents

Preface

Measure Theory has become increasingly essential to *Geometry* as well as *Numerous and Varied Physical Applications*. Over the past thirty years, this theory has contributed to major advances in many classical branches of *Mathematics*. This Encyclopaedia is an introduction to Measure Theory. It is designed for graduate students in a variety of fields (*Mathematics, Statistics, Economics, Management, Finance, Computer Science, and Engineering*) who require a working knowledge of Measure Theory that is Mathematically Precise, but without excessive Technicalities. The text provides complete proofs of all the essential introductory results. Nevertheless, the treatment is focused and accessible, with the Measure Theory and Mathematical details presented in terms of intuitive Probabilistic concepts, rather than as separate, imposing subjects. The text strikes an appropriate balance, rigorously developing *Probability Theory* while avoiding unnecessary detail.

Author

List of Symbols

$\pm$ $\not\subset$ $\subset$ $\supset$ $\subseteq$ $\in$ $\notin$ ∂

$\neq$ $\equiv$ $\approx$ $\angle$ ∇ Σ Π Ω

ϕ α β χ ε γ η λ

μ ν $\perp$ π θ ρ σ τ

$\Re$ ω ξ ψ ζ Υ $\exists$ $\int$

$\iint$ $\iiint$ $\oint$ $\times$ $\sqrt{\ }$ $\therefore$ $\because$ Δ

$\geq$ $\leq$ $\Rightarrow$ $\Downarrow$ $\Leftrightarrow$ ∞ $\propto$ $\mathbb{R}$

$\mathbb{Z}$ $\mathbb{N}$ $\mathbb{C}$ $\mathbb{Q}$

Chapter 1

Introduction to Measure Theory

This introductory section is directed primarily to those readers who have some familiarity with undergraduate-level probability theory, and who may be unclear as to why it is necessary to introduce measure theory and other mathematical difficulties in order to study probability theory in a rigorous manner.

We attempt to illustrate the limitations of undergraduate-level probability theory in two ways: the restrictions on the kinds of random variables it allows, and the question of what sets can have probabilities defined on them.

VARIOUS KINDS OF RANDOM VARIABLES

The reader familiar with undergraduate-level probability will be comfortable with a statement like, "Let X be a random variable which has the Poisson distribution."

The reader will know that this means that X takes as its value a "random" non-negative integer, such that the integer $k \geq 0$ is chosen with probability $P(X = k) = e^{-5}5^k / k!$.

The expected value of, say, X^2, can then be computed as $E(X^2) = \sum_{k=0}^{\infty} k^2 e^{-5} 5^k / k!$. X is an example of a discrete random variable.

Similarly, the reader will be familiar with a statement like, "Let Y be a random variable which has the Normal (0, 1) distribution." This means that the probability that Y lies between two real numbers $a < b$ is given by the integral $P(a \leq Y \leq b) = \int_a^b \frac{1}{\sqrt{2\pi}} e^{-y^2/2} dy$. (On the other hand, $P(Y = y) = 0$ for any particular real number y.) The expected value of, say, Y^2, can then be computed as $E(Y^2) = \int_{-\infty}^{\infty} y^2 \frac{1}{\sqrt{2\pi}} e^{-y^2/2} dy$. Y is an example of an absolutely continuous random variable.

But now suppose we introduce a new random variable Z, as follows. We let X and Y be as above, and then flip an (independent) fair coin. If the coin comes up heads we set $Z = X$, while if it comes up tails we set $Z = Y$.

In symbols, $P(Z = X) = P(Z = Y) = 1/2$. Then what sort of random variable is Z? It is not discrete, since it can take on an uncountable number of different values. But it is not absolutely continuous, since for certain values z (specifically, when z is a non-negative integer) we have $P(Z = z) > 0$. So how can we study the random variable Z? How could we compute, say, the expected value of Z^2?

The correct response to this question, of course, is that the division of random variables into discrete versus absolutely continuous is artificial.

Instead, measure theory allows us to give a common definition of expected value, which applies equally well to discrete random variables (like X above), to continuous random variables (like Y above), to combinations of them (like Z above), and to other kinds of random variables not yet imagined.

THE UNIFORM DISTRIBUTION AND NON-MEASURABLE SETS

In undergraduate-level probability, continuous random variables are often studied in detail. However, a closer examination suggests that perhaps such random variables are not completely understood after all.

To take the simplest case, suppose that X is a random variable which has the uniform distribution on the unit interval [0, 1]. In symbols, X 'Uniform [0, 1]. What precisely does this mean?

Well, certainly this means that $P(0 \leq X \leq 1) = 1$ It also means that $P(0 \leq X \leq 1/2) = 1/2$, that $P(3/4 \leq X \leq 7/8) = 1/8$, etc., and in general that $P(a \leq X \leq b) = b - a$ whenever $0 \leq a \leq b \leq 1$, with the same formula holding if $\leq$ is replaced by $<$. We can write this as

$$P([a, b]) = P((a, b]) = P([a, b)) = P((a, b)) = b - a, 0 \leq a \leq b \leq 1.$$

In words, the probability that X lies in any interval contained in [0, 1] is simply the length of the interval. (We include in this the degenerate case when $a = b$, so that $P(\{a\}) = 0$ for the singleton set $\{a\}$; in words, the probability that X is equal to any particular number a is zero.) Similarly, this means that

$$P(1/4 \leq X \leq 1/2 \text{ or } 2/3 \leq X \leq 5/6)$$

$$= P(1/4 \leq X \leq 1/2) + P(2/3 \leq X \leq 5/6) = 1/4 + 1/6 = 5/12,$$

and in general that if A and B are disjoint subsets of [0, 1] (for example, if $A = [1/4, 1/2]$ and $B = [2/3, 5/6]$), then

$$P(A \cup B) = P(A) + P(B).$$

Equation is called finite additivity.

Indeed, to allow for countable operations (such as limits, which are extremely important in probability theory), we would like to extend to the case of a countably infinite number of disjoint subsets: if $A_1, A_2, A_3, \ldots$ are disjoint subsets of [0, 1], then

$$P(A_1 \cup A_2 \cup A_3 \cup \ldots) = P(A_1) + P(A_2) + P(A_3) + \ldots.$$

Equation is called countable additivity.

Note that we do not extend equation to uncountable additivity. Indeed, if we did, then we would expect that $P([0, 1]) = \sum_{x \in [0,1]} P(\{x\})$, which is clearly false since the left-hand side equals 1 while the right-hand side equals 0. (There is no contradiction to since the interval [0, 1] is not countable.) It is for this reason that we restrict attention to countable operations. (For a review of countable and uncountable sets. Also, recall that for non-negative uncountable collections $(r_\alpha)_{\alpha \in I}$, $\sum_{x \in I} r_\alpha$ is defined to be the supremum of $\sum_{x \in J} r_\alpha$ over finite $J \subseteq I$.)

Similarly, to reflect the fact that X is "uniform" on the interval [0, 1], the probability that X lies in some subset should be unaffected by "shifting" (with wrap-around) the subset by a fixed amount. That is, if for each subset $A \subseteq [0, 1]$ we define the r-shift of A by

$$A \oplus r \equiv \{a + r;\, a \in A,\, a + r \leq 1\} \cup \{a + r - 1;\, a \in A,\, a + r > 1\},$$

then we should have

$$P(A \oplus r) = P(A),\ 0 \leq r \leq 1.$$

So far so good. But now suppose we ask, what is the probability that X is rational? What is the probability that X^n is rational for some positive integer n? What is the probability that X is algebraic, i.e. the solution to some polynomial equation with integer coefficients? Can we compute these things?

More fundamentally, are all probabilities such as these necessarily even defined? That is, does $P(A)$ (i.e., the probability that X lies in the subset A) even make sense for every possible subset $A \subseteq [0, 1]$?

It turns out that the answer to this last question is no, as the following proposition shows. The proof requires equivalence relations, but can be skipped if desired since the result is not used elsewhere in this book.

Proposition: There does not exist a definition of $P(A)$, defined for all subsets $A \subseteq [0, 1]$.

Proof: (optional). Suppose, to the contrary, that $P(A)$ could be so defined for each subset $A \subseteq [0, 1]$. We will derive a contradiction to this.

Define an equivalence relation on [0, 1] by: $x \sim y$ if and only if the difference $y - x$ is rational. This relation partitions the interval [0, 1] into a disjoint union of equivalence classes. Let H be a subset of [0, 1] consisting of precisely one element from each equivalence class. For definiteness, assume that $0 \notin H$ (say, if $0 \in H$, then replace it by 1/2).

Now, since H contains an element of each equivalence class, we see that each point in (0, 1] is contained in the union $\bigcup_{\substack{r \in [0,1) \\ r \text{ rational}}} (H \oplus r)$ of shifts of H. Furthermore, since H contains just one point from each equivalence class, we see that these sets $H \oplus r$, for rational $r \in [0,1)$, are all disjoint.

But then, by countable additivity, we have

$$P((0,1]) = \sum_{\substack{r \in [0,1) \\ r \text{ rational}}} P(H \oplus r)$$

Shift-invariance implies that $P(H \oplus r) = P(H)$, whence

$$1 = P((0,1]) = \sum_{\substack{r \in [0,1) \\ r \text{ rational}}} P(H).$$

This leads to the desired contradiction: A countably infinite sum of the same quantity repeated can only equal 0, or ∞, or –∞, but it can never equal 1.

This proposition says that if we want our probabilities to satisfy reasonable* properties, then we cannot define them for all possible subsets of [0, 1]. Rather, we must restrict their definition to certain "measurable" sets.

Remark: The existence of problematic sets like H above turns out to be equivalent to the Axiom of Choice. In particular, we can never define such sets explicitly – only implicitly via the Axiom of Choice as in the above proof.

INTRODUCTION TO MEASURE THEORY

The Method of Exhaustion is a technique to find out the area of a shape (for example, circle) by approximating (lower bound) it by inscribing it using a polygon. The idea is that as the number of sides of the polygon tends to infinity the area of the polygon will get arbitrarily close to the area of the circle.

This method was used by Archimedes to find out the value of π. Thus, by exhausting the values of the area of the polygon, the area of the circle can be obtained. In fact, the Riemann integral is an application of the method of exhaustion.

Measure theory is concerned with generalizing the notions of area on arbitrary sets of Euclidean spaces and notions of length of subsets of ℝ. Essentially, it is a common ground for analysis of real functions and set theory. For the notion of area, there are certain properties like non-negativity and additivity that hold true. Similarly, in measure theory there is a stronger assumption of countable additivity. Measure theory involves study of σ-algebras (abstract mathematical structures), measures, measurable functions and integrals. Integration in the context of measure theory involves analogous sums and is based on functions constant on sets of some -algebra and not on functions constant on intervals (as is done in the traditional manner).

BASICS OF MEASURE THEORY

Before looking at measure theory more formally, I will review some of the basic concepts including some ideas from abstract algebra that define the underlying mathematical structures on which measures are defined.

Limit of a sequence

A sequence $x_1, x_2, \ldots x_n$ is said to converge to the point or has limit x if for every $\varepsilon > 0$ there is a natural number N_ε such that the neighborhood of $O(x, \varepsilon)$ (a circle with center x and radius ε) has all the points x_n with $n > N_\varepsilon$. Intuitively, it means that the elements of the sequence eventually become arbitrarily close to the point x.

Cauchy Sequence

A sequence $x_1, x_2, \ldots x_n$ is said to be a Cauchy sequence if for every $\varepsilon > 0$ there exists an N_ε such that $d(x_{n'}, x_{n''}) <$ for all $n', n'' \geq N_\varepsilon$.

Intuitively, it means that the elements of the sequences eventually become close to each other. However, Cauchy sequences are not the same as convergent sequences (having a limit) except in certain cases (for example in $\mathbb{R}$). Some of the important properties of Cauchy sequences are:

1. Any Cauchy sequence is bounded.
2. Any Cauchy sequence in $\mathbb{R}$ is convergent.

Algebraic Structure

In pure mathematics, an algebraic structure is a collection of one or more sets that are operated upon by one or more operations.

These operations satisfy certain axioms. An algebraic structure consists of all the possible models that are characterized by the axioms. Algebraic structures are categorized by the number of sets and binary/unary operations they require. Following are some of the relevant structures:

1. *Groups* - A group $(S, *)$ is a set S under a binary operation $*: S \times S \rightarrow S$ that satisfies certain axioms like associativity, closure, e.g. Integers under addition

which also happens to be an abelian group. An abelian groups is a group in which the order of operation does not matter, i.e. $a * b = b * a$.

2. *Rings* - A ring is a set S under two binary operations: addiction and multiplication. A ring under addition is an abelian group. e.g. integers, polynomials. A commutative ring is a ring in which the members satisfy the commutative law.
3. *Fields* - A field is a structure under which the operations of addition, multiplication, division and subtraction are performed. A field is a commutative ring.

σ-algebra and Borel algebra

σ-algebra S is a collection of subsets of a set S that is closed under countable set operations i.e. the complement of a member (subset), the union or intersection of countably many members are also members. Formally, an algebra S of subsets of a set S is a σ-algebra if S contains the limit of every monotone sequence of its sets. The pair (S, S) is then a measurable space and the sets in S are said to be measurable. e.g. 2^S.

Borel algebra of a set S is the minimal σ-algebra that contains all the open sets (or closed sets) on the real line. The elements of Borel algebra are called Borel Sets. σ-algebras (in particular, Borel algebras) allow us to concentrate on sets that are (in some way) easy to use. In other words, these mathematical structures hold certain important properties that allow us to define the concept of a measure on seemingly arbitrary sets.

Countably Additive

Let be a real function, and S be a σ-algebra. The function ϕ is countably additive or σ-additive when

$$\phi(S) = \sum_{i=1}^{\infty} \phi(S_i) \text{ where } S \in S \bigcup_{i=1}^{\infty} S_i; S_i \cap S_j = 0$$

It means that for finite or countably infinite sequence of disjoint sets, the length of the union of these sets is equal to the sum of the lengths of these sets.

The function ϕ is countably sub-additive when

$$\phi(S) \leq \sum_{i=1}^{\infty} \phi(S_i) \text{ where } S \in S;\ S \subseteq \bigcup_{i=1}^{\infty} S_i$$

Metric Space

A metric space is a set wherein the concept of length or distance between elements is defined. It is denoted by (S, d) where S is a set and d is the metric (a single-valued, non-negative distance function) such that $d: S \times S \rightarrow R$. It satisfies the properties of positivity, symmetry, identity and triangle inequality.

If every Cauchy sequence in a metric space (S, d) converges to an element in the space, then the space (S, d) is said to be complete, e.g. The open interval (0, 1) is not complete whereas the closed interval [0, 1] is complete. Every compact (closed and bounded) metric space is complete.

Every normed space (distance function set to the absolute difference between two elements of the space) is also a metric space. As a side note, a complete normed space is siad to be a Banach space.

Measure Space and Measure

If there is a countably additive set function μ defined on σ-algebra S of subsets of S, then the triplet (S, S, μ) is a measure space. An example of measure space is an Euclidean space with Lebesgue measure. The sets of S are called measurable sets and the function μ is called a measure and has the following properties:

1. μ is non-negative, i.e. $\mu(X) \geq 0$ for all $X \in S$.
2. μ is countably additive (as defined in section 2.5).
3. $\mu(\phi) = 0$.
4. μ obeys monotonicity: $\mu(X) \leq \mu(Y)$ for all X, $Y \in S$ and $X \subset Y$.

The concept of measure has connections with integration over arbitrary sets. Moreover, if $\mu(S) = 1$, then the measure space is called a probability space and μ is a probability measure. The sets are called events. One says that a property holds almost everywhere if the set for which the property does not hold is a null set or a set with measure 0. In probability theory, analogous to almost everywhere, almost certain or almost sure means except for an event of probability measure 0. This means that the event has zero probability of not occurring although it is still possible that they might not occur. There are different kinds of measures - Borel measure, Lebesgue measure, Haar measure, Dirac measure to name a few.

The measure space (S, S, μ) is complete if every subset of a set of measure 0 is measurable and has measure 0. Formally, for all $S \subset S$, if $\mu(S) = 0$ then for all subsets $X \subset S$, $\mu(X) = 0$. If not, then a space can be made complete. The Borel measure is not complete and hence the complete Lebesgue measure is preferred. Every Borel set is also Lebesgue measurable.

Lebesgue Measure

The definition of measure as given above is of a generalized form. However, in practice it is often the case that only those sets that are Lebesgue measurable need to be considered. To get to that, let's define an outer measure $\mu^*(X)$ and an inner measure $\mu^*(X)$ on a set $X \subseteq S$.

The outer measure of a set $X \subseteq S$ is defined as:

$$\mu^*(X) = inf\left\{\sum_n m(B_n) : m \text{ is a measure on interval } B_n \text{ and } X \subseteq \bigcup_n B_n\right\}$$

The outer measure as defined above is again of generalized form. Consider an interval B with endpoints a and b such that $a < b$ instead of the set A, then the outer measure for the interval is defined as $b - a$.

The inner measure is defined as:

$$\mu^*(X) = m(S) - \mu^*(S \setminus X)$$

If the following is true,

$$\mu^*(X) = \mu^*(X)$$

then $\mu(X)$ (which denotes the value of $\mu^*(X)$ or $\mu_*(X)$) is said to be a Lebesgue measure on the set X.

Lebesgue Integral

A measure of a measure space (S, S, μ) is determined by its values $\mu(s_i)$ where s is a finite or infinite sequence in the countable space S. Here, $S = 2^S$. If f is a function from this countable space into $\mathbb{R}$ and is μ-integrable over the set S (i.e. the sequence evaluated under f is absolutely convergent), then the Lebesgue integral of f on S is defined as a sum of the sequence evaluated under f as shown below:

$$\int_S f(s)d\mu = \sum f(s.)\mu(s.)$$

If $\mu(S) = 1$ i.e. the μ-algebra S is a probability space and μ is a probability measure, then in probability theory, Equation is referred to as the expectation of f and is denoted by $E(f)$.

A generalized definition of the integral of f when there exists a sequence of functions $\{f_n\}$ μ-integrable over S and uniformly convergent to f is given by:

$$\int_S f(s)d\mu = \lim_{n\to\infty} \int_S f_n(s)d\mu$$

The definition of an integral in this form is in contrast with that of the Riemann integral which is only applicable to intervals and unions of intervals. For Riemann integral (following the idea of method of exhaustion), the space over which f is defined is divided into smaller sub-intervals and represented in the following way:

$$\sum_n f(\eta_n)(x_{n+1} - x_n)$$

where the value of $f(x)$ is substituted by the function evaluated at an arbitrarily chosen point η within the interval $[x_n, x_{n+1}]$. However, this can be done only when $f(x)$ is continuous or when it does not have many points of discontinuity. The Riemann integral of f is obtained by taking the limit of n to infinity in the above sum.

In order to distinguish between the Lebesgue and Reimann integrals consider the values that the function f can take to be on the x-axis (called the domain of the function) and the values of the function evaluated at the chosen points to be on the y-axis (called the range of the function).

The Lebesgue integral defines the sub-intervals along the range of the function and the Riemann integral defines them along the domain of the function. Thus, the Riemann integral requires the values of the domain to be very close to each other (function should be continuous). This is a limitation since it restricts the class of functions on which the integral can be defined. Thus the Lebesgue integral is preferred because it is applicable to arbitrary classes of functions. Another reason is that if there is a sequence of Riemann integrable functions $\{f_n\}$, then it is not true that the function f defined as $f = \lim_{n\to\infty} f_n$ (absolutely convergent) is Riemann integrable.

Now, I present two theorems from Kolmogorov and Fomin that explicitly state the conditions under which the limit of an integral is equal to the integral of the limit as given in the definition in Equation. The need to take the limit inside the integral or to consider piecewise integration of a convergent sequence of functions arises in serveral applications, most notable being Fourier series.

Dominated Convergence Theorem

Theorem: If $\{f_n\}$ is a sequence of measurable functions from a measure space (S, S, μ) into $\mathbb{R}$ and the sequence converges to f and

$$|f_n(s)| \leq g(s)$$

almost everywhere, for all n, where g is a non-negative function integrable over S, then the limit function f is integrable over S and

$$\int_S f(s)d\mu = \lim_{n\to\infty} \int_S f_n(s)d\mu$$

Monotone Convergence Theorem

Theorem: If $\{f_n\}$ is a sequence of measurable functions from a measure space (S, S, μ) into $\mathbb{R}$ such that $f_1 \leq f_2 \leq \ldots \leq f_n \leq \ldots$ and the sequence converges to f, then the limit function f is integrable over S and

$$\int_S f(s)d\mu = \lim_{n\to\infty} \int_S f_n(s)d\mu$$

The theorem is so called because of the use of the integrable function g for stating the condition of integrability of piecewise functions and the theorem is so called because it assumes that the sequence of piecewise functions is monotonically increasing.

Measurable Function

It has been proven that it is impossible to assign a measure to all subsets of $\mathbb{R}$ without preserving some of the properties like additivity. Hence, as seen in the previous section I will concentrate only on certain measurable sets. Consequently, I will generalize the concept of a function adapted to the class of measurable sets and hence to their corresponding σ-algebras. Such functions are called measurable functions. They are defined as follows: Let $(S, 2S')$ be a countable measurable space and consider functions from a measurable space (S, S) into S'. Such a function f is measurable if the following is true for every set $X \subseteq S'$:

$$f^{-1}[X] = \{s \in S \mid f(s) \in X\}$$

This equation is also called the preimage (not an inverse function) of set X under the measurable function f. If (S, S) is equipped with a probability measure, a measurable function is a random variable in the context of probability theory.

Final Comment

Measure theory is a fascinating area of study in its own right and at the same time plays an important role in functional analysis, advanced probability theory and statistics. In this note, I have only laid down a framework that gives an intuitive insight into the world of measures and enables us to define integrals, probability distributions and random variables. For more rigorous proofs, convergence properties and for exploring the use of Lebesgue integral in the context of square integrable functions and Hilbert spaces, the reader is referred to Kolmogorov and Fomin.

DEFINITIONS

POWER SET

A power set of a set S is the set of all possible subsets of S. It is denoted by 2^S or P^S. If there are n elements in the set S, then $|P^S| = 2^n$. 2^S defines the set of all functions from S to $\{0, 1\}$. A power set forms an abelian group when considered with the operation of symmetric difference and a commutative ring when considered with the operations of symmetric difference and intersection.

CONVEX SET, CONVEX CLOSURE, SIMPLEX

A set X in the linear space S is said to be convex if given any two arbitrary points x and y from the set, the segment joining them also belongs to X. The convex closure of the set X is the smallest closed convex set containing X. The convex closure can also be obtained as the intersection of all the closed convex set which contain the set X.

A set of points $x_1, x_2, \ldots x_{n+1}$ in a normed linear space is said to be in general position if no $k+1 (k < n)$ of these points lie in a subspace of dimension less than k (if it does, then it is the degenerate case). The convex closure of such a set of points $x_1, x_2, \ldots x_{n+1}$ that are in general position is called an n-dimensional simplex. In other words, it is the minimal convex set. The n-dimensional simplex with $n+1$ points is also represented in the following form:

$$x = \sum_{k=1}^{n+1} \alpha_k x_k ; a_k \geq 0 \text{ and } \sum_{k=1}^{n+1} \alpha_k = 1$$

INEQUALITIES

Let X be a measurable space with dimension n, $a_1, a_2, \ldots a_n$ and $b_1, b_2, \ldots b_n$ are real or complex numbers; $1 \leq p, q \leq \infty$ and $\frac{1}{p} + \frac{1}{q} = 1$.

Cauchy-Schwarz Inequality

$$\left(\sum_{k=1}^{n} a_k b_k \right)^2 \leq \sum_{k=1}^{n} a_k^2 \sum_{k=1}^{n} b_k^2$$

The Cauchy-Schwarz inequality is used to prove triangle inequality.

Holder's Inequality

$$\sum_{k=1}^{n} | a_k b_k | \leq \left(\sum_{k=1}^{n} | a_k |^p \right)^{\frac{1}{p}} \left(\sum_{k=1}^{n} | b_k | \right)^{\frac{1}{q}}$$

Holders inequality is a generalized case of Cauchy-Schwarz inequality and is used to prove the generalization of triangle inequality and the Minkowskis inequality.

Minkowski's Inequality

$$\left(\sum_{k=1}^{n} | a_k + b_k |^p \right)^{\frac{1}{p}} \leq \left(\sum_{k=1}^{n} | a_k |^p \right)^{\frac{1}{p}} \left(\sum_{k=1}^{n} | b_k |^q \right)^{\frac{1}{q}}$$

Weak and Strong Convergence

A sequence $\{x_n\}$ in a normed linear space L is said to converge weakly to an element x when the following conditions are true:

1. The norms of the elements $\{x_n\}$ are uniformly bounded: $\| x_n \| \leq M$.
2. $f(x_n) \to f(x)$ for all $f \in L$

The sequence $\{x_n\}$ is said to converge strongly when its norm converges to x such that $\| x_n - x \| \to 0$ as $n \to \infty$. Obviously, for finite dimensional spaces, strong convergence and weak convergence is the same. We have analogous definitions for convergence of linear functionals.

LANGUAGE OF MEASURE THEORY

Definition: Let X be a set and χ a collection of subsets of X. Then χ is a σ-algebra if

1. $\phi, X \in \chi$;
2. χ is closed under complementation (i.e., $X_0 \in \chi \Rightarrow X_0^C \in \chi$;
3. χ is closed under countable union and countable intersection (i.e., $\{X_n\}_{n=1}^{\infty} \in \chi \Rightarrow \bigcup_{n=1}^{\infty} X_n \in \chi$ and $\bigcap_{n=1}^{\infty} X_n$)

Definition: Let X be a set and χ be a σ-algebra. Then (X, χ) is a measurable space.

Definition: Let (X, χ) be a measurable space. A measure is a function $\mu: \chi \to \mathbb{R}$ such that:

1. $\mu(\phi) = 0$;
2. $\mu(X_0) \geq 0, \forall X_0 \in X_0 \in \chi$;
3. If $\{X_n\}_{n=1}^{\infty}$ is a countable, disjoint collection of subsets of χ, then $\mu\left(\bigcup_{n=1}^{\infty} X_n\right) = \sum_{n=1}^{\infty} \mu(X_n)$.

Definition: (X, χ, μ) be a measure space. If $\mu(X) = 1$, then μ is a probability measure and (X, χ, μ) is a probability space.

Definition: Given a measurable space (X, χ), a function $f: X \to \mathbb{R}$ is measurable with respect to X if $\{x \in X: f(x) \leq k\} \in \chi$, $\forall k \in \mathbb{R}$.

Definition:Let (X, χ) be a measurable space. A transition function is a function $Q: X \times \chi \to [0, 1]$ such that:

1. For each $x \in X$, $Q(X,.)$ is a probability measure on (X, χ);
2. for each $X_0 \in \chi$, $Q(., X_0)$ is a measurable function.

For any probability measure μ on the probability space $\Lambda(X, \chi)$, define the operator T^* by

$$(T^*\mu)(X_0) = \int_X Q(x, X_0)\mu(dx), \forall X_0 \in \chi.$$

Since for each $X_0 \in \chi$, $Q(., X_0)$ is bounded (in [0, 1] and measurable, then $T^*\mu$ is well defined and takes probability measures into themselves.

Define the operator T on the space of bounded measurable functions $B(X)$ by

$$(Tf)(x) = \int_X f(x')Q(x, dx'), \forall x \in X$$

which takes bounded, measurable functions $f: X \rightarrow \mathbb{R}$ into themselves

Definition: A transition function Q on (X, χ) has the Feller property if the operator T in (2) maps the space of bounded, continuous functions on X, namely $C(X)$, into itself (i.e., $T: C(X) \rightarrow C(X)$).

Definition: A transition function Q on (X, χ) is monotone if the operator T has the property that for every nondecreasing function $f: X \rightarrow \mathbb{R}$, the function Tf is also nondecreasing.

Definition: A measure μ^* is an invariant distribution with respect to the transition function Q if $\forall X_0 \in \chi, \mu^*(X_0) = \int_X Q(x, X_0)\mu^*(dx)$. That is, μ^* is invariant if it is a fixed point of the operator T^*.

Chapter 2

Area and Measure

The modern notion of measure, developed in the late Nineteenth and early Twentieth centuries, is an extension of the notion of area, developed by the Greeks. The area of a planar rectangle is defined to be the product of its length and width. By elementary geometric arguments, the area of a triangle, and hence the area of any polygon, can be determined. To determine the area of a circle, the greeks constructed sequnces of inscribed and circumscribed regular polygons, with the number of sides tending to infinity, to obtain inner and outer approximations to the circle. This gives a sequence of lower and upper estimates of the area fo the circle, and the area of circle is defined to be the common limit as the number of sides tends to infinity.

This procedure, known as the Method of exhaustion, was formulated by Euxodus of Cnidos, and was developed sysematically by Archimedes. Archimedes developed systematic algorithms for calculating not only the areas, but the volumes of wide variety of geometric figures, and laid the foundation of integration theory that are still relevant today. In fact, the Riemann integral is really an application of the method of exhaustion and the principles of Archimedes.

What properites do we want of a notion of 'Area'? Suppose A denotes the area of a set E. The most obvious properties to require of the set function A is that it be non-negative and 'additive'. That is, $A\left(E \cup F\right) = A(E) + A(F)$, provided that $E \cap F = \emptyset$. In addition, in the case of Euclidean geometry, we require A to be invariant under the group of Euclidean motions, that is, trnaslations, rotations, and reflections; but this is not an essential property of a measure.

In the theory of measures, the assumption of additivity is replaced by a slightly stronger assumption, that of countable additivity:

$$\mu\left(\cup_j^\infty E_j\right) = \sum_{j=1}^{\infty} \mu(E_j), \text{ if } E_j \cap E_k = \emptyset, j \neq k$$

The sequence of sets $\{E_j\}$ is said to disjoint.

Definition: Let X be a set and let $M \subset P(X)$ be a collection of subsets of X. *A* set function μ defined on M is said to be a measure if μ is non-negative and countably additive.

We point out immediately that the assumption of countable additivity, together with the axiom of choice, leads in general to the existence of non measurable set – that is, that μ is not defined everywhere on $P(X)$. We demonstrate the existence of non-measurable sets for any measure μ on the unit interval [0, 1] which is invariant under translations.

Define an equivalence relation on [0, 1] by $x \sim y$ if $x - y$ is rational. This partitious the unit interval into equivalence classes. By the Axiom of Choice, there is a set N consisting of one element from each equivalence class. For each rational number r, let $N_r = [N + r]$ mod1, that is, the translate of N by r with warapped around around 1. (Or, do the construction on the unit circle). Now clearly $[0, 1] = U_r N_r$ and the sets N_r are disjoint. Hence by countable additivity, and translation invariance $\mu([0, 1]) = \sum_r \mu(N_r) = \sum_r \mu(N)$. This sum must be infinity or zero,d epending on whesther μ(N) is positive or zero. This μ cannot be defined on N.

We now turn our attention to the collection M, which are called measurable sets. In view of our requirement of countable additivity of μ it is natural to require that M be closed under countable disjoint unions. In addition, it is natural to require that the complement of any measurable set also be measurable. Then if A and B are measurable, $A \cap B = (A^\circ \cup B^\circ)^\circ$ is measurable.

Definition: *A* α*-algebra M is a collection of subsets of which is closed under countable unions and complementations.*

The verification of the following proposition is left to the reader.

Proposition: *Any countable union of sets may be replaced by a countable disjoint union of sets. The intersection of any family of* σ*-alegbras is again a* σ*-algebra. If* $\xi \subset P(X)$ *there is a smallest* σ*-algebra containing* ξ.

The smallest σ-algebra containing a collection of sets ξ is called the σ-algebra generated by ξ. For example, if X is a topological space, the open sets on X generate a σ-algebra, called the Borel σ-algebra, which is denoted by B_X. The Borel algebra of sets of R is generated by the open intervals, closed intervals, half-open intervals, or by rays $(a, \infty), (-\infty, a)$, or by half-open rays, such as $(-\infty, a]$.

A measure space is denoted by X, M, μ), where X is the space of points, M is the σ-algebra of measurable sets, and μ is the measure, defined on M. A measure on a topological space for which the measurable sets is the Borel algebra B_X is called a Borel measure. Borel measure play a pre-eminent role in measure theory on R^n.

Borel measure on the line are consetructed by specifying the measure of each open or half-open interval on the line. If $F(x)$ is a monotone increasing, right continous function on the line, then the measure of the half-open interval $(a, b]$ is defined by $\mu_0((a, b]) = F(b) - F(a)$. the extension of μ_0 ito B_R will be developed below.

A number of important examples arise in this manner. For example:

1. Lebesgue measure on the real line, which extends the notion of length, is obtained by taking $F(x) = x$;
2. In probability theory, the distribution function of a random variable X, defined by $F(x) = P_r(X \leq x)$, generates a probability measure, and $P_r(a < X \leq b) = F(b) - F(a) = \mu_F((a, b])$.
3. The Dirac measure is obtained by taking F equal to the Heaviside step function: $H(x) = 0$ for and $H(x) =$ for $x \geq 1$. The Dirac measure assigns the measure 1 to any set containing the origin, and 0 to all other sets.
4. The Cantor function F generates a measure μ_F which assigns the measure 1 to the Cantor set, which has Lebesgue measure 0; and 0 the the union of open middle thirds, which has total length 1.

Theorem: *Let (X, M, μ) be a measure space. Then μ is*

1. Monotone: If $E \subset F$ and both measurable, then $\mu(E) \leq \mu(F)$.
2. Countably subadditive: For $E_j \in M, \mu(\cup_j E_j) \leq \sum_j \mu(E_j)$;
3. Continuous from below: If $E_j \subset E_{j+1} \subset$ is an increasing sequence of measurable sets, then

$$\mu\left(U_{j=1}^{\infty} E_j\right) = \lim_{j \to \infty} \mu(E_j);$$

4. μ is continuous from above: If...... $\supset E_j \supset E_{j+1} \supset$ and $\mu(E_1) < \infty$, then

$$\mu(\cap_{j=1}^{\infty} E_1) = \lim_{j \to \infty} \mu(E_j)$$

Proof: *If $E \subset F$ then $F = E \cup (F/E)$ so $\mu(F) = \mu(E) + \mu(F/E) \geq \mu(E)$.*

If $\{E_j\}$ is a sequence of sets with with non-empty intersection, then for all n,

$$\mu(\cup_{j=1}^{n} E_j) = \mu(\cup_{j=1}^{n} F_j) = \sum_{j=1}^{n} \mu(F_j) \leq \sum_{j=1}^{n} \mu(E_j)$$

where $F_1 = E_1$, $F_k = E_k \backslash (\cup_{j=1}^{k-1} E_k)$

Then the F_j are disjoint and $F_j \subset E_j$; letting $n \to \infty$ we obtained the second statement in the theorem.

To prove continuity from below, let $E_0 = \emptyset$ and note that

$$\mu(\bigcup_{j=1}^{\infty} E_j) = \sum_{j=1}^{\infty} \mu(E_j / E_{j-1}) = \lim_{n\to\infty} \sum_{j=1}^{n} \mu(F_j) \le \sum_{j=1}^{n} \mu(E_j)$$

Finally, to prove continuity from above, put $F_j = E_{1\setminus} E_j$. Then the sequence $\{F_j\}$ is increasing and

$$E_1 = \left(\bigcup_{j=1}^{\infty} F_j\right) \cup \left(\bigcap_{j=1}^{\infty} E_j\right)$$

Therefore by the previous result,

$$\begin{aligned} \mu(E_1) &= \mu(\bigcup_{j=1}^{\infty} F_j) + \mu(\bigcap_{j=1}^{\infty} = E_j) \\ &= \lim_{j\to\infty} \mu(F_j) + \mu(\bigcap_{j=1}^{\infty} E_j) \\ &= \lim_{j\to\infty} \mu(E_1) - \mu(E_j) + \mu(\bigcap_{j=1}^{\infty} E_j) \end{aligned}$$

Since $\mu(E_1) < \infty$ it follows that

$$\mu(\bigcap_{j=1}^{\infty} E_j) = \lim_{j\to\infty} \mu(E_j)$$

A null set of a measure space (X, M, μ) is a set N for which $\mu(N) = 0$. By subadditivity, countable unions of null sets are null sets. P is a proposition which is true except on a set of measure zero, we say that P holds almost everywhere (a.e.). If that a-algebra M includes all null sets, we say that the measure μ is complete. Any measure can be extended to be complete.

Caratheodary's Theorem

In this section we describe a general constructive procedure for obtaining complete measures. Recall that the area enclosed by a circle is obtained as the limit of the areas of a sequence of circumscribed polygons. In fact, the area of a disk is by definition the infimum of the areas of all circumscribed polygon. This leads to the general notion of outer measure, as follows.

Definition: An outr measure on a non-empty set X is a set function μ defined on $P(x)$ which is non-negative, monotone and countably subadditive.

In practice, an outer measure is obtained by starting with an elementary concept of measure, called a pre-measure, which is deifned on a restricted class of sets $\varepsilon \subset P(X)$ and approximating general subsets X from the outside. We assume that ε is an algebra, that is, that it is clsed under complementation and finit unions. For example, Lebesgue measure in the plane is constructed beginning with the notion of area of a rectangle, extending it to polygons, and defining the 'outer area' of a plane figure by taking the infimum of the areas of all circumscribed polygons.

Proposition: Let $\varepsilon \subset P(X)$ be an algebra of sets and ρ a non negative function defined on, ε such that $\rho(\emptyset) = 0$. For any $A \subset X$ define

$$\mu^*(A) = \inf\left\{\sum_{j=1}^{\infty} \rho(E_j) : A \subset \bigcup_{j=1}^{\infty} E_j, E_j \in \varepsilon\right\}$$

This μ^* is an outer measure.

Proof: By its definition μ^* vanishes on the empty set. If $A \subset B$ then any covering of B is also a covering of A; hence the infimum for A is taken over a larger collection of coverings, and therefore smaller. Hence μ^* is montone.

It remains to domonstrate countable additivity. Let $A = U_j A_j \subset X$, and let $\varepsilon > 0$. For each j there is a covering of A_j by sets $E_{jk} \in \varepsilon$ such that

$$\sum_{k=1}^{\infty} \rho(E_{jk}) < \mu^*(A_j) + 2^{-j}\varepsilon$$

hence

$$\mu^*(A) < \sum_{j,k=1}^{\infty} \rho(E_{jk}) < \sum_{j=1}^{\infty} \mu^*(A_j) + \varepsilon$$

Since ε is arbitrary, $\mu^*(A) \leq \sum_j \mu^*(A_j)$ and the countable suadditivity of μ is established.

A measr is obtaind by restricting the outer measure to a subclass of sets,called measurable sets, on which it is countable additive.

Definitin: Let μ^* be an outer measure on a set X, A subset $A \subset X$ is said to be μ^*-measurable if

$$\mu^*(E) = \mu^*(E \cap A) + \mu^*(E \cap A^\circ), E \subset X$$

By subadditivity $\mu^*(E)$ $\mu^*(E) \leq \mu^*(E \cap A) + \mu^*(E \cap A^\circ)$ for all A and E; hence A measurable iff

$$\mu^*(E) \geq \mu^*(E \cap A) + \mu^*(E \cap A^\circ), \ E \subset X$$

This definition of measurability which is due to the Twentieth century Greek mathematician Carathedory.

Theorem: If μ is an outer measure on *X*, then the collection *M* of measurable sets is a σ – algebra, and the restriction of μ* to *M* is a complete measure.

Proof: First note that *M* is clsed under complementation, since the definition of measurability is symmetric in *A* and *A*°.

If *A, B M* and $E \subset X$, then

$$\mu^*(E) = \mu^*(E \cap A) + \mu^*(E \cap A^\circ) = \mu^*(E \cap A \cap B) +$$

$$\mu^*(E \cap A \cap B^\circ) + \mu^*(E \cap A^\circ \cap B) + \mu^*(E \cap A^\circ \cap B^\circ)$$

Since

$$A \cup B = (A \cap B)(A \cap B^\circ) \cup (A \cap B^\circ)$$

we have, by subadditivity.

$$\mu^*(E \cap A \cap B) + \mu^*(E \cap A \cap B^\circ) + \mu^*(E \cap A^\circ \cap B) + \mu^*(E \cap (A \cap B))$$

Therefore

$$\mu^*(E) \geq \mu^*(E \cap (A \cup B)) + \mu^*(E \cap (A \cup B)^\circ)$$

and $A \cup B \in M$. This shows that *M* is an algebra. *A* and *B* are disjoint sets in *M* then

$$\mu^*(A \cup B) = \mu^*((A \cup B) \cap A) + \mu^*((A \cup B) \cap A^\circ) = \mu^*(A) + \mu^*(B)$$

so μ* is finitely additive on *M*.

Now suppose that $\{A_j\}$ is a sequence of disjoint measurable sets, and let

$B_n = \bigcup_1^n A_j$ with $B = \bigcup_1^\infty A_j$. Then for any $E \subset X$

$$\mu^*(E \cap B_n) = \mu^*(E \cap B_n A_n) + \mu^*(E \cap B_n \cap A^\circ{}_n)$$

$$= \mu^*(E \cap A_n) + \mu^*(E \cap B_{n-1}) = ...$$

$$= \sum_{j=1}^{n} \mu^*(E \cap A_j)$$

Therefore,

$$\mu^*(E) = \mu^*(E \cap B_n) + \mu^*(E \cap B_n^0) \geq \sum_{j=1}^{n} \mu^*(E \cap A_j) + \mu^*(E \cap B^\circ)$$

Letting $n \to \infty$ and using the countable subadditivity of μ^*, we obtain

$$\mu^*(E) \geq \sum \mu^*(E \cap A_j) + \mu^*(E \cap B^c) \geq \mu^*(E \cap B) + \mu^*(E \cap B^c \geq \mu^*(E).$$

Thus $B \in M$; and taking $E = B$ we obtain also that μ^* is countably additive on M.

Finally, if $\mu^*(A) = 0$, then

$$\mu^*(E) \leq \mu^*(E \cap A) + \mu^*(E \cap A^c) = \mu^*(E \cap A) \leq \mu^*(E)$$

hence A is measurable and μ^* restricted to M is a complete measure.

The Caratheodory extension theorem is a primary method of constructing measures. Typically, the set function ρ of Proposition is taken to be a pre-measure, a non-negative set function thatvanishes on the empty set, and is countable additive on the algebra ε in the sense that

$$\rho(\cup_j A_j) = \sum_j \rho(A_j),$$

whenever $\cup_j A_j \in \varepsilon$

For example, Lebesgue measure on the line is an extension of the pre-measuredefined by $\rho((a,b]) = b - a$; ε is the algebra of finite disjoint unions of half-open intervals $(a, b]$.

Theorem: Let μ be the Caratheodory extension of a pre-measure ρ defined on algebra ε. Then every set in ε is measurable, and $\mu(A) = \rho(A)$ for all $A \in \varepsilon$. If ρ is σ-finite then the extension is unique.

In the case of Lebesgue measure on R^n, the algebra ε is the collection of finite unions of multi-intervals,

$$R = \{(x_1 \ldots\ldots, x_n) : a_j \leq x \leq b_j\}, \ \rho(R) = \prod_{j=1}^{n} (b_j - a_j)$$

BOREL MEASURE

Let X be a metric space and ρ its metric. The distance between two sets A and B in X is defined to be

$$\rho(A, B) = \inf\{\rho(x, y) : x \in A, y \in B\}$$

A metric outer measure μ^* on a metric space X is one which satisfies the additional postulate that

$$\mu^*(A \cup B) = \mu^*(A) + \mu^*(B), \text{ whenever } \rho(A, B) > 0$$

The importance of this class of measures is that The Borel sets of X are then μ^*-measurable [1, 2, 4]

An important class of Borel measures on the real line are the Lebesgue Stieltjes measures. These are extensions of the pre-measure $\rho F((a,b]) = F(b) - F(a)$, where F is a non-decreasing, right continuous function on the real line, or on an interval. In the following, ε denotes the algebra of sets which can be expressed as finite disjoint unions of half open intervals $\{(a_j, b_j]\}$, with $a_j < b_j$.

Lemna: The set function ρ*F* is a pre-measure on ε.

The proof of this lemna is somewhat more technical then one might expect. The difficulty comes in that any half-open interval can be decomposed ina non-unique fashion:

$$(a,b) = \bigcup_{j=1}^{n} (c_{j-1}, c_j), \quad c_0 = a, c_n = b$$

Theorem: *Let μ be a Lebesgue-Stieltjes measure and let M_μ be its deomain of measurable sets.Then for any $E \in M_\mu$ we have*

$$\mu(E) = \begin{cases} \inf\{\mu(U) : E \subset U, U \text{ open}\} \\ \sup\{\mu(K) : K \subset E, K \text{ compact}\} \end{cases}$$

Lebesgue measure *m* on the line is the extension of the pre-measure ρ_F, where $F(x) = x$. The Borel algebra is a sub-class of Lebesgue measurable sets. In the theory of Borel sets, $\Im_\sigma$ denotes the class of sets which are countable union of closed sets; while $\aleph_\delta$ denotes the class of countable intersections of open sets. Any Lebesgue measurable set *E* can be written in the form

$$E = \begin{cases} V/N_1 & V \in \aleph_\delta \quad \mu(N_1) = 0; \\ H \cup N_2 & H \in \Im_\sigma \quad \mu(N_2) = 0 \end{cases}$$

A second Lebesgue-Stieltjes measure of interest is that generated by the Cantor function, *c(x)*. This function is defined to be 1/2 on the open interval (1/3, 2/3);1/4on(1/9,2/9); 3/4 on (7/9, 8/9), etc; and then extending this to the entire unit interval [0, 1] by continuity. The Cantor function is continuous and non-decreaseing. Let μ_c be the Lebesgue-Stieltjes measure generated by c.

The Cantor set $C \subset [0,1]$ is obtained by removing the open middle thirds of the remaining intervals, and so forth, successively removing the open middle thirds of the remaining subintervals at each stage. The Cantor measure of *C* is one, while the Lebesgue measure of the complement C^c is one. The Cantor measure and Lebesgue measure are examples of mutually singular measures; they have their support on complementary subset of [0,1].

PRELIMINARIES

Let $\{a_n\}$ be a sequence of real numbers. The limit superior and limit inferior of the sequence are defined as

$$\limsup_n a_n = \inf_{k\geq 1} \sup_{n\geq k} a_n \quad \limsup_n a_n = \sup_{k\geq 1} \inf_{n\geq k} a_n$$

The following is left as an exercise:

$$\liminf_n a_n \leq \limsup_n a_n,$$

Let $\{E_n\}$ be a countable collection of sets. We define

$$\bar{E} = \limsum_n E_n \bigcap_{k=1}^{\infty}\bigcap_{n=k}^{\infty} E_n = \{x : x \in E_n \text{ for infinitely many } n\}$$

$$\underline{E} = \liminf_n E_n = \bigcup_{k=1}^{\infty}\bigcap_{n=k}^{\infty} E_n = \{x : x \in E_n \text{ for all but finitely many } n\}.$$

It is a simple exercise to show that $\underline{E} = \lim\inf f_n E_n \subset \lim\sup_n E_n = \vec{E}$. If $\underline{E} = \vec{E}$ we say the limit of the seuqence of sets exists and is equal to this common set.

A partial ordering of a set X is relation "$\leq$" which satisfies the following axioms; Given $x, y, z \in X$,

1. $x \leq y, y \leq z$ implies $x \leq z$
2. $x \leq y, y \leq x$ implies $x = y$
3. $x \leq x$

   If, in addition,

4. For all $x, y \in X$ either $x \leq y$, or $y \leq x$ then the relations $\leq$ is called a linear or total ordering. For example, the real line is linearly ordered; the space of continuous real-valued functions on a topological space X, denoted by $C(X)$, is partially ordered by the relation: $f \leq g$ if $g(x) - f(x) \geq 0$ for all $x \in X$; the collection of all subsets of a set X, called the power set of X and denoted by $P(x)$ is partially ordered by the inclusion relation.

A maximal element of a partially ordered set X is an element M such that if $x \in X$ and $x \geq M$, then $x = M$. A minimal element is defined similarly. A maximal or minimal element may or may not exist, and in general is not unique.

The Hausdorff Maximal Principle states that very partially ordered set has a maximal linearly ordered subset. Zorn's Lemma states that if X is a partially ordered set for which every linearly ordered subset has an upper bound, then X has a maximal element. The Cartesian product of a collection of set $\{X_\alpha\}_{\alpha \in A}$, denoted by $\prod_\alpha X\alpha$ is defined to be the set of all maps $f : A \to U_\alpha X_\alpha$ such that $f(\alpha) \in X_\alpha$. The Axiom of Choice states that if $\{X_\alpha\}$ is non-empty. That is, it is possible to construct a set with one lement from each of theX_α.

The Hausdorff Maximal principle, Zorn's Lemma and the Axiom of Choice are all equivalent.

MEASURABLE FUNCTIONS

A function $f : \mathbb{R}^n \to \mathbb{R}$ is Lebesgue measurable if $f^{-1}(U)$ is a Lebesgue measurable set for every open set U. More generally, given any measure μ we say real valued function f is measurable with respect to μ if $f^{-1}(U) \in M$ whenever U is open. M being the σ-algebra of μ-measurable sets. In general, we shall simply say f is measurable without reference to the measure μ. If f is complex valued, then f is measurable iff its real and imaginary parts are measurable.

If $f : X \mapsto Y$ then f^{-1} is defined by

$$f^{-1} : P(Y) \mapsto P(X), \quad f^{-1}(E) = \{x : x \in X, f(x) \in E \subset Y\}$$

It is a simple exercise to show that the inverse map preserves complements, unions, and intersections, that is

$$f^{-1}(E \cup F) = f^{-1}(E) \cup f^{-1}(F) \; f^{-1}(E \cap F) = f^{-1}(E) \cap f^{-1}(F),$$

$$f^{-1}(E^e) = f^{-1}(E)^c,$$

It follows that a real-valued function is measurable if $f^{-1}(U)$ is measurable whenever U is an open or closed intervals, or an open ray (a, ∞), etc.

A mapping $f : X \mapsto Y$, where X and Y are metric spaces, is Borel measurable if $f^{-1}(U)$ is a Borel subset of X for every open set $U \subset Y$. By the observations above.

Propositions: *Continuous functions* are Borel measurable. f is Borel *measurable* iff $f^{-1}(B) \in B_X$ whenever $B \in B_Y$. The composition of two Borel measurable functions on $\mathbb{R}$ to itself is Borel measurable.

On the other hand, the inverse image of a Lebesgue measurbale set is not necessarily Lebesgue measurable, and so Lebesgue measurability is not closed under composition.

Give two functions f and g we define.

$$f \vee g = \max\{f, g\}, f \wedge g = \min\{f, g\};$$

$$f^+ = f \vee 0, f^- = (-f) \vee 0$$

Proposition: If f and g are measurable, then so are $f+g$, fg, $f\vee g$, $f\wedge g$, $f^{\pm}$, *and* $|f\}$. The sums, products, and upper and lower envelops of any finit numbers of measurable functions is measurable.

Proof: For any real a we have

$$\{x: f+g>a\} = \bigcup_{t\in\mathbb{Q}} \{x: f(x)>t\}\ (x: g(x)>a-t\};$$

hence $f+g$ is measurable. It is immediate that measurability is invariant under scalar multiplication; hence $f-g$ is also measurable.

Consider the function f^2. If $a<0$ then $\{x: f^2>a\} = X$; while for $a\geq 0$,

$$\{x: f^2>a\} = \{x: f(x)>\sqrt{a}\}\cup\{x: f(x)<-\sqrt{a}\}$$

Hence f^2 is measurable. The identity.

$$fg = \frac{(f+g)^2-f^2-g^2}{2}$$

then shows that fg is measurable.

We have

$$\{f^+>a\} = \begin{cases} x & a<0; \\ \{x: f(x)>a\} & x\geq 0; \end{cases}$$

hence f^+ is measurable. The same argument shows f^- is measurable; and therefore also $|f| = f^+ + f^-$ is measurable. The identities

$$f\vee g = \frac{f-g+|f-g|}{2}, f\wedge g = \frac{f+g-|f-g|}{2}$$

then show that $f\vee g$ and $f\wedge g$ are measurable. The result extends to finite combinations of measurable functions by induction.

Theorem: *If* $\{f_j\}$ *is a sequence of measurable functions, then the limits* $\lim\sup_j f_j$, $\lim\inf_j f_j$, *as well as* $\lim_f f_j$, *when it exists, are all measurable.*

Proof: Let $g_k(x)\ \sup_{j>k} f_j(x)$. Then

$$\{x: g_k(x)>a\} = \bigcup_{j\geq k}\{x: f_j(x)>a\};$$

hence $\{g_k\}$ is a sequence of measurable functions. Putting $f(x) = \inf_k g_k(x)$, we see that

$$\{x: f(x)>a\} = \bigcap_{k=1}^{\infty}\{x: g_k(x)>a\},$$

hence $\overline{f}(x) = \lim\sup_j f_j(x)$ is measurable. A similiar proof shows that lim in $f_j f_j$ is measurable. The limit limj f_j (x)(exists for all x for which the limits inferior and superior of the sequence $\{f_j(x)\}$ coincide; hence the lim f_j is measurable.

INTEGRATION OF NON-NEGATIVE FUNCTIONS

A *characteristic* or *indicator* functions is one which takes on only the values 0 and 1; it is generally denoted by $_{XE}(x)$ where E is the set on which the function is 1. It is useful to note that

$$X_{A\cap B} = X_{AXB}, X_{A\cup B} = X_A + X_B \text{ iff } A \cap B = \phi$$

A simple function is one which takes on only a finite number of values in the extended real numbers. It is thus a finite linear combination of characteristics functions

$$\varphi = \sum_{j=1}^{n} c_j XE_j \,,\ E_j = \{x : (x) = c_j\}$$

It is clear that a simple function is measurable iff the sets E_j are measurable. Moreover, the class of simple functions is closed under multiplication and addition.

Theorem: Let f be a measurable real-valued function on a space X. Then there is a squence of simple functions φ_n such that $\varphi_n \to f$ uniformly on any set on which f is bounded.

Proof: Decompose $f = f_+ - f_-$, where $f_+ = \max\{f, 0\}\{$ and $f_- = \max\{-f, 0\}$. Thus we may assume that $f \geq 0$. For $n \geq 0$ and $0 \leq k \leq 2^{2n} - 1$, let

$$E_{n,k} = f^{-1}((k2^{-n}, (k+1)2^{-n})), F_n = f^{-1}((2^n, \infty]),$$

Then the sequence of simple functions

$$\varphi_n = \sum_{k=0}^{2^{2n}-1} k2^{-n} \chi E_{n.k} + 2^n \chi F_n$$

satisfy

$$0 \leq \ldots\ldots \leq f_n \leq f_{n+1} \ldots\ldots \leq f,\ 0 \leq f - \varphi_n \leq 2^{-n},\ x \in F_n^c$$

The convergence is uniform on sets where f is bounded.

Let (X, M, μ) be a measure space. The integral of a measurable simple function φ is defiend to be

$$\int \varphi d\mu = \sum_{j=1}^{n} c_j \mu(E_j)$$

We see the convention taht $0.\infty = 0$

It is clear that the integral of non-negative simple functions is non-negative. Also, since finite linear combinations and finite products of simple functions are simple, the integral so defined is a linear functional on the vector space of simple functions.

Lemma: Give a non-negative, measurable simple function φ and $A \in M$, define $\int_A \varphi d\mu = \int \varphi \chi_A d\mu$. Then the set function $v(A)$ defined by

$$v(A) = \int_A \varphi d\mu$$

is a (countably additive) measure.

Proof: If A is the countable union of disjoint sets in M then

$$v(A) = \int \varphi \chi_A d\mu = \int \sum_{k=1}^{n} c_k \chi E_k \chi_A d\mu = \int \sum_{k=1}^{n} c_k \chi_{A \cap E_k} d\mu$$

$$= \sum_{k=1}^{n} c_k \mu(E_k \cap A) = \sum_{k=1}^{n} \sum_{j=1}^{\infty} c_k \mu(E_k \cap A_j) = \sum_{j=1}^{\infty} c_k \mu(E_k \cap A_j)$$

$$= \sum_{j=1}^{\infty} \int_{A_2} \varphi d\mu = \sum_{j=1}^{\infty} v(A_j)$$

We now define the integral of any non-negative measurable f by

$$\int f d\mu = \sup_{\varphi \le f} \left\{ \int \varphi d\mu : \varphi \text{ simple} \right\}$$

The reader may check that this definition is consistent with that of the integral of a simple function. If $f \le g$ then $\int f d\mu$ is obtained as the supremum over a simaller set of simple functions thn that for g; hence $\int f d\mu \le \int g d\mu$.

The following theorem is fundamental to the development of the theory of the Lebesgue integral.

Monotone Convergence Theorem

Theorem: If $\{f_n\}$ is an increasing sequence on non-negative functions then

$$\int \lim_n f_n d\mu = \lim_n \int f_n d\mu$$

Proof: Since $\{f_n\}$ is a nondecreasing sequence of measurable functions, $\lim_n f_n = f$

exists and is measurable. It may take on the value infinity. Moreover, the integrals $\int f_n d\mu$ form a nondecreasing sequence of nunbers; and $f_n \le f$ implies that $\int f_n \le \int f$, hence

To prove equality, it suffices to show that for any fixed $\alpha, 0 < \alpha < 1$,

$$\alpha \int f d\mu \le \lim_n \int f_n d\mu$$

Give α and any simple functions φ, let $E_n = \{x : f_n \ge \alpha\varphi\}$. Then

$$\int f_n d\mu \ge \int_{E_n} f_n \ge \int_{E_n} \alpha\varphi d\mu = \alpha v(E_n)$$

By Lemma v is a countable additive measure. Since $\{E_n\}$ is an increasing sequence of sets whose union is X, it follows by Theorem that $\lim_n v(E_n) = v(X) = \int \varphi d\mu$. Thus

$$\lim_n \int f_n d\mu \ge \alpha \int \varphi d\mu$$

Taking the supremum over all simple functions $\varphi \le f$ we obtain

$$\alpha \int f d\mu = \alpha \sup_{\varphi \le 1} \int \varphi d\mu \le \lim_n \int f_n d\mu$$

An immediate consequence of the Monotone Convergence Theorem is the following characterization of the integral of a general non-negative measurable functions.

Corollay: *Let f be a non-negative measurable function, and let φ_n be a monotone increasing sequence of measurable functions converging to f. Then*

$$\int f d\mu = \lim_n \int f \varphi_n d\mu$$

Using Corollary we can prove the additivity of the integral on non-negative functions.

Propositions: *The integral f dμ is finitely additive on non-negative measurable functions.*

Proof: Let f and g be non-negative measurable functions and let $\varphi_n \uparrow f, \varphi_n \uparrow g$, where φ_n and ψ_n are sequence of non-negative simple functions. Then

$$\int f + g d\mu = \int \lim_n (\varphi_n + \psi_n) d\mu$$

$$= \lim_n \int \varphi_n d\mu + \lim_n \int \psi_n d\mu = \int f d\mu + \int g d\mu$$

By the same argument, one can show that $\int cf d\mu = c \int f d\mu$, for any constant c.

Applying the monotone convergence theorem to the partial sums of a series of non-ngeative functions, we obtain

Corollary: If $f_n \geq 0$, then

$$\int \sum f_n d\mu = \sum_n \int f_n d\mu$$

where both sides may be infinite.

Proposition: *If* $f \geq 0$ *then* $\int f = 0$ iff $f = 0$ a.e.

Proof: The statement is immediate for simple functions. For general f, if $f = 0$ a.e. and $\varphi \leq f$ is simple, then $\varphi = 0$ a.e. hence $\int f = \sup_{\varphi \leq f} \int \varphi = 0$. Conversely, let

$$\{x : f(x) > 0\} = \bigcup_{n \geq 1} E_n, \ E_n = \{x : f(x) \geq \frac{1}{n}\}$$

If $f > 0$ on a set of positive measure, then $\mu(E_n) > 0$ for some n; hence

$$\int f d\mu \geq \frac{1}{n} \mu(E_n) > 0$$

An immediate corollary of this proposition is the following strengthening of the monotone convergence theorem.

Corollary: If 0 $0 \leq \ldots\ldots \leq f_n \leq f_{n+1} \leq \ldots\ldots$ and $f_n \uparrow f$ a.e. then $\lim_n \int f_n d\mu = \int f d\mu$

The following lemma, known as Fatou's lemma, can also be used as the starting point in building the convergence theorems of Lebesgue integration theory.

Fatou's Lemma

Lemma: For any non-negative sequence of measuable functions $\{f_n\}$ we have

$$\int \liminf_n f_n d\mu \leq \lim_n d\mu f \int f_n d\mu$$

Proof: We have $\inf_{n \geq k} f_n \leq f_j$ for $j \geq k$ hence $f \inf_{n \geq k} f_n d\mu \leq \int f_j d\mu$ for $j \geq k$ and so

$$\int \inf_{n \geq k} f_n d\mu \leq \inf_{j \geq k} \int f_j d\mu$$

Since both sides are increasing in km, we have, in the limit as $k \to \infty$,

$$\lim_{k\to\infty}\int \inf_{n\geq k} f_n d\mu \leq \lim_{k\to\infty}\inf_{j\geq k}\int f_j d\mu = \sup_{k\geq 1}\inf_{j\geq k}$$

$$\int f_j d\mu = \liminf_n \int f_n d\mu$$

Since the sequence $\{\inf_{n\geq k} f_n\}$ is non-negative and increasing in k, we have, by the Monotone Convergence Theorem

$$\int \liminf_n f_n d\mu = \int \sup_{k\geq 1}\inf_{n\geq k} f_n d\mu = \int \lim_{k\to\infty}\inf_{n\geq k} f_n d\mu = \lim_{k\to\infty}\int \inf_{n\geq k} f_n d\mu$$

THE DOMINATED CONVERGENCE THEOREM

Let (X, M, μ) be a measure space. A real valued function f is said to be integrable if f is measurable with respect to μ and $f \mid f \mid d\mu < +\infty$. Since $| f_{\pm} | \leq | f |$, hence $\int f_{\pm} d\mu < +\infty$ and we can define

$$\int f d\mu = \int f_+ d\mu - \int f_- d\mu$$

The class of integrable functions on (X, M, μ) is denoted by $L_1 (X, d\mu)$ or more briefly L^1 when the measure space is clear. The class L^1 is closed under addition and scalar multiplication, hence is a vector space. Moreover, if M contains an infinite number of sets of positive measure, then $L_1 (X, d\mu)$ is an infinite dimensional vector space. A norm is defined on this vector space by

$$\| f \|_1 = \int | f | d\mu$$

It is easily verified that $\|.\|_1$ satisfies the triangle inequality and $\|cf\|_1 = |c|\|f\|_1$.Thus L^1 is a metric space with the metric $\rho(f, g) = \| f - g \|_1$; and it is natural to ask the question as to whether L^1 is a comlete metric space. The answer to this question is yes; and not only that, but the completness is a direct consequence of the introduction of the notion of measure theory nad measurable functions. If one begins with only the continuous functions in L^1, and seeks the completion of this metric space, then one has to include the measurable functions as well.

Lebesgue Dominated Convergence Theorem

Theorem: Let $\{f_n\}$ be a sequence of functions in L^1 such that $f_n \to f$ a.e. and there exists a non-negative functions $g \in L^1$ such taht $|f| f_n | \leq g$ a.e. for all n. Then $f \in L^1$ and $\int f d\mu = \lim_n \int f_n d\mu$.

Before beginning the proof, let us give an example that shows the necessity of the assumption that the sequence f_n be dominated by a fixed function $g \in L^1$. Consider the fundamental solution of the heat equation on the line:

$$G(x,t) = \frac{e^{-x^2/4t}}{\sqrt{4\pi t}}$$

As $t \downarrow 0$ the function narrows and its maximum value tends to infinity. It is a simple matter to show that

$$\int_{-\infty}^{\infty} G(x,t)dx = 1, \qquad t > 0$$

Yet that $G(x, t)$ tends to 0 for all $x \neq 0$. Thus its pointwise limit is 0 almost everywhere, and for any sequence $t_n \downarrow 0$.

$$0 = \int_{-\infty}^{\infty} \lim_{n\to\infty} G(x,t_n)dx < \lim_{n\to\infty} \int G(x,t_n)dx = 1$$

Thus the conclusion of the Dominated Convergence theorem fails while that of Fatou's lemma holds with a strict inequality.

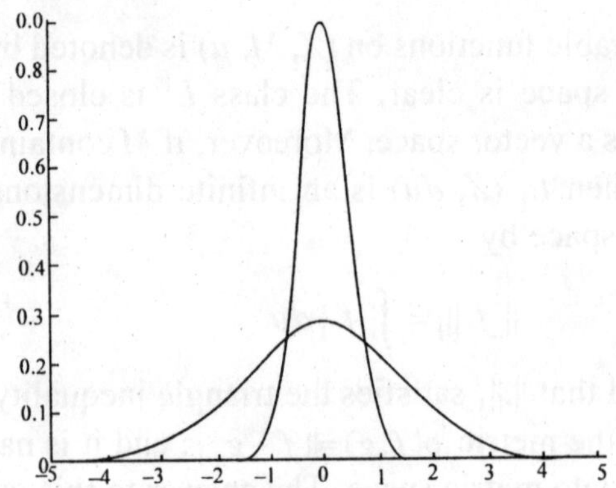

Fig. The Fundamental Solution of the Heat Equation Evaluated at $t = .1$ and $t = 1$.

Proof: Applying Fatou's lemma to the sequence of functions $g + f g + f_n \geq 0$ we obtain

$$\int g + f d\mu = \int (g + \liminf_n f_n) d\mu \leq \int g d\mu + \liminf_n \int f_n d\mu$$

hence $\int f d\mu \leq \liminf_n \int f_n d\mu$. Applying the same argument to the sequence $g - f_n \geq 0$, we obtain.

$$\int -f d\mu \leq \liminf_n \int -f_n d\mu = -\limsup_n \int f_n d\mu$$

Combining both inequalities, we have

$$\int f d\mu \leq \liminf_n \int f_n d\mu \leq \limsup_n \int f_n d\mu \leq \int f d\mu$$

Corollary: Let $\{f_j\}$ be a sequence of integrable functions such that the series $\sum_j \int |f_j| d\mu < +\infty$. Then $\sum_j f_j$ converges a.e. to a function $f \in L^1$ and

$$\int \sum_{j=1}^{\infty} f_j d\mu = \sum_{j=1}^{\infty} \int f_j d\mu$$

Proof: By the monotone convergence theorem.

$$\int \sum_j |f_j| d\mu = \sum_j \int f_j d\mu$$

hence the function $g = \sum_j |f_j| \in L^1$. Thus g is finite a.e. and for each x for which $g(x)$ is finite, the series $\sum_j f_j(x)$ converges absolutely. The partial sums of this series are dominated by g, by the conclusion of the Corollary follows by the dominated convergence theorem.

Theorem: The simple functions are dense in L^1. The class $C_0(\mathbb{R})$ is dense in $L^1(\mathbb{R}, B_{\mathbb{R}}, \mu)$, where μ is any Borel measure on $\mathbb{R}$.

Proof: The simple functiuons are dense in $L^1(d\mu)$ by Theorem and the dominated convergence theorem. To show the continuous functions are dense in $L^1(\mathbb{R}, B_{\mathbb{R}})$ it is sufficient to show that any simple function can be approximated by a continuous function in the L^1 norm. If $I = (a, b)$ is an open interval, we may approximate χ_I by a sequence of continuous, picewise linear functions $C_{n,E}$ which take the value 1 on I, 0 on $(-\infty, a-n^{-1}) \cup (b+n^{-1}, \infty)$. It is a simple calculation to show that $\| \chi_E - C_{n,E} \|_1 = n^{-1}$. Next let E be a Borel measurable set in $\mathbb{R}$, with $\mu(E) = \| \chi_E \|_1 < \infty$. We show that for any $\varepsilon > 0$, there is a finite union of open intervals A such that

$$\mu(E) = \inf \left\{ \sum_{j=1}^{\infty} \mu(I_j) : E \subset \bigcup_{j=1}^{\infty} (a_j, b_j) \right\}$$

Thus, given any $\varepsilon > 0$ there is a collection of disjoint open intervals $I_j = (a_j, b_j)$ such that $\mu(U \setminus E) \frac{\varepsilon}{2}$, $U = \bigcup_{j=1}^{\infty} I_j$

Now chose N so that $\sum_{N+1}^{\infty} \mu(I_j) < \varepsilon/2$, and put $A = \bigcup_{j=1}^{N} I_j, B = U \setminus A$. Then $E \setminus A$

$$= E \cap A^{\circ} \subset (A \cup B) \cap A^{\circ} \subset B \text{; hence}$$

$$\mu(E \setminus A) < \mu(B) < \varepsilon/2 \text{ and,}$$

$$\mu(A \Delta E) = \mu(A \setminus E) + \mu(E \setminus A) < \mu(U \setminus E) + \mu(B) < \varepsilon$$

THE RIEMANN INTEGRALS

The dominated convergence theorem is a fundamental tool. For example, we may use it to prove that the Lebesgue integral is an extension of the Riemann integral; thsat is, the two integrals give the same result for a Riemann-integrable function. Let $[a, b]$ be a bounded interval. A partition of $[a, b]$ is a set of points $\mathrm{P} = \{a = x_0 < x_1 < .....x_n = b\}$. Give a bounded real-valued functions f on $[a, b]$, the upper and lower Riemann sums of f relative to a partition P are given by

$$L(\mathrm{P}, f) = \sum_{j=1}^{n} m_j \Delta x_j, U(\mathrm{P}, f) = \sum_{j=1}^{n} M_j \Delta x_j$$

where $\Delta x_j = x_j - x_{j-1}$ and m_j, M_j are the infimum and supremum of f on $[x_{j-1}, x_j]$. Clearly $L(P, f) \leq U(P, f)$ for any partition P.

Definition: A bounded real-valkued function f defined on a bounded interval $[a, b]$ is said to be Riemann integrable if $\sup_P L(P, f) = \inf_p U(P, f)$. In that case the Riemann integral of f on $[a, b]$ is defined to be

$$\int_a^b f(x)dx = \sup_P L(P, f) = \inf_P U(P, f)$$

The concept of the Riemann integral is cleary an application of the method of exhaustion, proposed originally by Euxodus of Cnidus and developed extensively byt Archimedes as a tool for computing areas and volumes. It was not until Isaac Barrow, thge first Luscasion chair of mathematics, and his student and successor, Sir Isaac Newton that the connection between the method of exhaustion and the notion of anti-derivative was fully established. This is the true content of the Fundamental Theorem of Calculus.

Theorem: Let f be a bounded real-valued function on $[a, b]$. If f is Riemann integrable, then f is Lebesgue measurable and therefore integrable on $[a, b]$; moreover the Lebesgue and Riemann integrals agree. A function is Riemann integrable iff it is continuous a.e.

Proof: Choose a sequence of partitions P_n such that each partition is a refinement of the previous one and for which $\max_j \Delta x_j$ tends to zero as $n \to \infty$. Then the lower sums $\pounds(f, P_n)$ are increasing and the upper sums are decreasing. f is Riemann integrable iff these two sequence converge to the same limit.

The lower and upper sums of a function f are the Lebesgue integrals of the simple functions

$$l(P,f)=\sum_{j=1}^{n} m_j \chi_{(x_{j-1},x_j)},\quad u(P,f)=\sum_{j=1}^{n} M_j \chi_{(x_{j-1},x_j)}$$

Let $h_n = l(P_n, f)$ and $H_n = u(P_n, f)$. Then $h_n \le f \le H_n$, and in the limit $h \le f \le H$, where $h = \lim_n h_n$ and $H = \lim_n H_n$

If f is Riemann integrable, then the sequence of upper and lower sums converge to the same limit. It follows that

$$\lim_n \int_{[a,b]} h_n dm = \int_a^b f dx = \lim_n \int_{[a,b]} H_n dm$$

By the Lebesgue dominated convergence theorem,

$$\lim_n \int_{[a,b]} h_n dm = \int_{[a,b]} h dm = \int_{[a,b]} H dm = \lim_n \int_{[a,b]} H_n dm$$

Since $H \ge h$ this implies that $h = H$ a.e. and hence that $H = f = h$ a.e. It follows that f is Lebesgue measurable and that

$$\int_{[a,b]} H dm = \int_{[a,b]} f dm = \int_a^b f dx$$

The proof that the discontinuities of a Riemann integrable function form a null set is left to the reader.

Theorem: Let $f(x, t)$ be a mapping form $X \times [a, b]$, where (X, M, μ) is a measure space and suppose that f is differentiable in with respect to t and that for each $x \in X, |f_t(x,t)| \le g(x)$ for all $a \le t \le b$, where $g \in L^1(X, d\mu)$. Let $F(t) = \int_X f(x,t) d\mu$ Then F is differentiable on $a < t < b$ and $F'(t) = \int_X f_t(x,t) d\mu$

Proof: For $t \in (a,b)$ we apply the Lebesgue dominated convergence theorem to the sequence

$$h_n(x) = \frac{f(x,t_n) - f(x,t)}{t_n - t}, t_n \to t$$

By the mean value theorem, the difference quotients h_n are bounded by $g(x) = \sup_t |f_t(x, t)|$.

CONVERGENCE THEOREMS

The L^1 integral of each term of the following sequence is 1,

$$\frac{1}{n}\chi_{(0,n)},\ \chi_{(n,n+1)},\ n\chi_{(0,n^{-1})}$$

yet the sequence themseelves tend to 0 uniformly, pointwise, and a.e. respectively. In each of thgese cases the sequence are not dominated by a fixed integrable function g.

On the other hand, there are sequence which tend to 0 in the L^1 norm but which do not tend to zero pointwise. For example, the sequence

$$f_1 = \chi_{(0,1)}, f_2 = \chi_{(0,1,/2)}, f_3 = \chi_{(1/2,1)},$$

$f_4 = \chi_{(0,1/4)}, f_5 = \chi_{(1/4,1/2)}$ etc. and in general $f_n = \chi_{(j2^{-k},(j+1)2^{-k})}$, where $n = 2^k + j$, with $0\,0 \le j < 2^k$. For this sequence, $\| f_n \|_1 \to 0$ as $n \to 0$ but $f_n(x)$ does not converge for any x. This shows that pointwise convergence and convergence in norm are in general not equivalent. We shall prove here that if $\| f_n - f \|_1 \to 0$, then some subsequence of f_n converge to f a.e. In order to do this we introduce the concept of convergence in measure.

Definition: A sequence of functions f_n is said to converge in measure to f if for every $\varepsilon > 0$.

$$\lim_n \mu(\{x :| f_n(x) - f(x) |\ge \varepsilon\}) = 0$$

A sequence is said to be Cauchy in measure if for every $\varepsilon > 0$,

$$\mu(\{x :| f_n(x) - f_m(x) |\ge \varepsilon\}) \to 0,\ m, n \to \infty$$

For example, the first, third, and fourth sequence above converge to 0 in Lebesgue measure on the line, while the second sequence does not. Convergence in norm implies convergence in measure. We need only prove this for a sequence of functions tending to in the L^1 norm. Suppose ||fn||1 $\| f_n \|_1 \to 0$ and, given $\varepsilon > 0$, put $E_{n,\varepsilon} = \{x :| f_n(x) |> \varepsilon\}$. Then for any fixed $\varepsilon > 0$.

$$\int | f_n | d\mu \ge \int | f_n | d\mu \ge \varepsilon\mu(E_n, \varepsilon) \to 0$$

Thus $f_n \to 0$ in measure.

The first and third examples above show that convergence in measure does not imply convergence in norm.

Many authors use the nomenclature $f_n \to f$ [meas], a.e., [unif], or [mean], to signify convergence in measure, pointwise convergence, uniform convergence, or convergence in the L^1 norm.

Theorem: If $\{f_n\}$ is a Cauchy sequence in measure, then there is a measurable function f such that $f_n \to f$ [meas]; and there is a subsequence f_{n_k} such taht $f_{n_k} \to f$ a.e. Moreover, f is uniquely determined a.e.

Proof: We may choose a subsequence n_k such that for $g_k = f_{n_k}$,

Putting $F_k = \bigcup_{j \geq k} E_j$, we have

$$\mu(F_k) \leq \sum_{j \geq j}^{\infty} 2^{-j} = 2^{1-k}$$

for $i \geq j \geq k$, and $x \in F_k^c$. This shows that $\{g_j(x)\}_{j \geq k}$ is a Cauchy sequence for $x \in F_k^c$. This shows that $\{g_j(x)\}_{j \geq k}$ is a Cauchy sequence ofr $x \in F_k^c$.

Let $F = \bigcap_1^{\infty} F_{ki}$; then $\mu(F) = 0$ and $\{g_j(x)\}$ is a Cauchy seuqence on F^c. Let $f = \lim g_k$ on F^c and 0 on F. Then f is measurable, $g_k \to f$ a.e; and letting $i \to \infty$, in (3, 1) we obtain

$$| f(x) - g_j(x) | \leq 2^{1-j}, \quad x \in F_k^c; j \geq k$$

Hence $\{x : | f(x) - g_j(x) | > 2^{1-j}\} \subset F_k$ for $j \geq k$. Since $\mu(F_k) \to 0$ as $k \to \infty$, this shows that $g_j \to f$ [meas].

But then $f_n \to f$ [meas] since for any $\varepsilon > 0$.

$$\mu\{(x : | f(x) - f_n(x) | \geq \varepsilon\}) \leq \mu(\{x : | f(x) - g_j(x) | \geq \varepsilon / 2\})$$

$$+ \mu(\{x : | g_j(x) - f_n(x) | \geq \varepsilon / 2\}) \to 0, \; j, n \to \infty$$

The proof that the limit in measure of a Cauchy sequence in measure is unique is left as an exercise.

Corollary: *If $f_n \to f$ in L^1 then there is a subsequence $\{f\{f_{n_k}\}$ such that $f_{n_k} \to f$ (a.e.)*

The sequence $\{x_{[n,n+1]}\}$ converges pointwise to 0 everywhere, but does not converge by 0 in measure. On a finite measure space, however, pointwise convrgence implies what is called almost uniform convergence. If μ (X) $< +\infty$, a sequence $[f_n]$ is said to converge almost uniformly to f on X if given any $\varepsilon > 0$ there is a set E such that $\mu(E) < \varepsilon$ and $f_n \to f$ [unif] on E^c. It is easily seen that almost uniform convergence implies convergence in measure.

Egorov's Theorem

Theorem: On sets of finite measure, convergence almost everywhere implies almost uniform convergence.

Proof: It suffices to show that if $f_n \to 0$ everywhere on E, and $\mu\,(E) < \infty$, then there exists a subset F $F \subset E$ such that $\mu(E/F) < \varepsilon$ and $f_n \to 0$ uniformly on F. Let $E_{mn} = \{x : |f_j(x)| < n^{-1}, j \geq m$. For fixed n, the sequence $\{E_{mn}\}$ is increasing and $E = \bigcup_1^\infty E_{mn}$. Therefore, by continuity from below, $\lim_m \mu(E_{mn}) = \mu(E)$. Hence given $\varepsilon > 0$ and for $n \in N$, there exists an integer m_n such that $\mu(E / Em_n n) < \varepsilon 2^{-n}$.

Let $F = \bigcap_{n \geq 1} E_{m_n n}$. If $x \in F$ then for all $n \in \mathbb{N}$ there is an m_n such that $|f_j(x)| < n^{-1}$ for all $j \geq m_n$, hence $f_j \to 0$ uniformly on F. On the other hand.

$$\mu(E/F) = \mu\left(\bigcup_{n\geq 1} E \setminus E_{m_n n}\right) \leq \sum_{n=1}^{\infty} \mu(E \setminus E_{m_n n}) < \varepsilon$$

Chapter 3

Abstract Measure Theory

MEASURE THEORY IN FINANCIAL MATHEMATICS

A modern probability theory is measure-theoretic since such approach brings rigour. For a pure mathematician it is a sufficient argument but not for a financial engineer. who has to solve concrete business problems with the least efforts. Well, it is possible to learn some financial mathematics with a calculusbased probability theory and without functional analysis at all. However, avoiding a "complicated" mathematics is akin hiding a head in the sand instead of meeting the challenge face to face. Just to mention that the price of any financial derivative (contingent claim) is nothing else but the expectation of its payoff under the *riskneutral probability measure*.

In order to study the properties of the Wiener process we need to understand an - convergence; the functional space is an excellent example of an interplay between the measure theory and the functional analysis. Considering the relation between an (economically plausible) no-arbitrage assumption and an existence of the equivalent risk neutral measure, one needs *Hahn-Banach theorem*. Finally, many *computational* problems imply solutions of Partial Differential Equations (PDEs), for which the functional analysis is essential.

Both measure theory and functional analysis are deservedly considered complicated and pretty abstract. But the problem lies mostly in the teaching style: in order introduce as much stuff as possible within a limited time the lecturers present an abstract theory, which is rigorous, concise ... and totally unobvious since the historical origins are omitted. I will, however, try to select the topics such that they are closely related to the financial mathematics and/ or help to elucidate the "main thread".

EDIFYING PROBLEMS

Lebesgue measure on $\mathbb{R}^1$ is a generalization of the notion of length. Note that sometimes *any* measure is considered to generalize a length but (as can be proved) the Lebesgue measure is an only one, which is *translation invariant* (an obvious requirement to a "length").

As the first step we consider a measure on an interval (0, 1) rather than on the whole real line because the "length" of $\mathbb{R}^1$ is obviously unbounded; this fact would bring some technical problems. However, the "boundness" of the underlying space does not matter for a general [abstract] measure theory, which allows to introduce not only the [translation invariant] Lebesgue measure on Euclidian space $\mathbb{R}^n$ but also a lot of other measures on many other spaces. After getting familiar with the Lebesgue measure on (0, 1) it is much easier to understand the generalized ideas of the measure theory. The choice of the interval bounds is of course arbitrary, we could have considered any [finite] interval (a, b) as a starting point.

Note: In elementary analysis one learns that (0, 1) is an open *subset* of $\mathbb{R}^1$. But if we consider (0, 1) *by itself*, it is assumed to be *both* open and closed. This assumption can be motivated as follows: whatever point of we consider, it belongs to (0, 1) together with some neighbourhood and thus (0, 1) is open. On the other hand all converging sequences have their limits in (0, 1) and thus it is closed. In this context the sequences "converging" to 0 or to 1 do *not* have limits since these points do not exist in underlying space. Analogously, if we considered an interval [0, 1] *as an underlying subspace*, it also would be both open and closed.

Lebesgue measure was developed as a tool, which helped to generalize the Riemann integral[2]. An old, good Riemann integral turned out to be not so good, when one started to require rigour and generality.

Problem 1: Non-Integrable Functions

The Dirichlet function, defined as

$$D(x) = \begin{cases} 1, & x \text{ is rational} \\ 0, & x \text{ is irrational} \end{cases}$$

is not Riemann integrable.

Problem 2: Interchange of an Integral and a Limit

Consider a sequence of function, defined on the domain as follows:

$$f_k(x) = \begin{cases} k, & 0 < x < \dfrac{1}{k} \\ 0, & \dfrac{1}{k} \le x < 1 \end{cases} \quad k \in \mathbb{N}$$

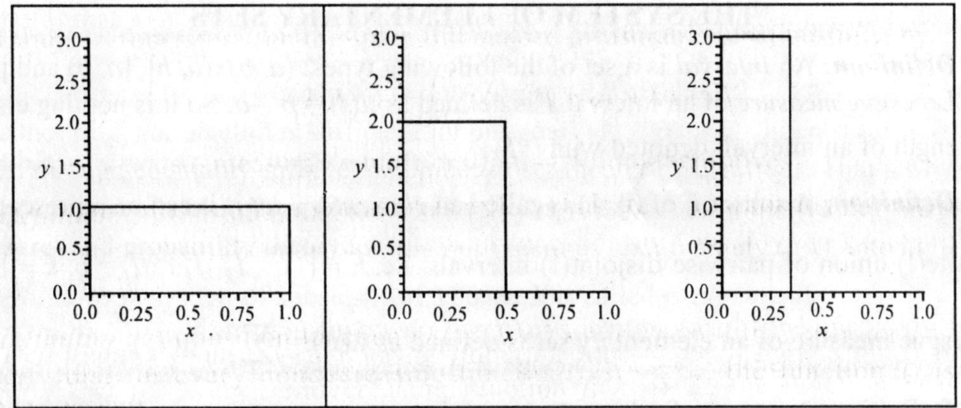

Fig. $f_x(x)$, k = 1, 2, 3

One has $f_k(x) \xrightarrow{\text{point wise}} k(x) \equiv 0$ but for any k it holds that $\int_0^1 f_k dx = k.\frac{1}{k} = 1$

Problem 3: Calculation of a Multiple Integral

Consider a function

$$f(x, y) := \frac{x^2 - y^2}{(x^2 + y^2)^2} \quad (x, y) \in \text{I} := (0, 1) \times (0, 1)$$

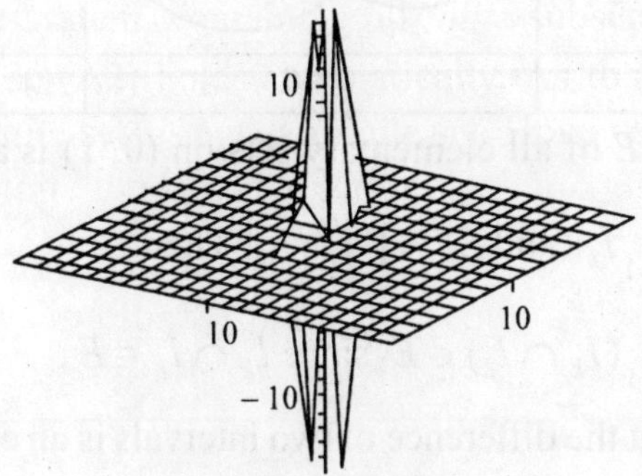

Fig. 3D-Plot of f(x, y)

Then

$$\int_0^1 f(x, y)dy = \frac{1}{1 + x^2} \Rightarrow \int_0^1 \left[\int_0^1 f(x, y)dy \right] dx = \frac{\pi}{4} \text{ but } \int_0^1 \left[\int_0^1 f(x, y)dy \right] dy = -\frac{\pi}{4}.$$

To check this result you can either let integrals.com help you or note that

$$\frac{x^2 - y^2}{(x^2 + y^2)^2} = \frac{x^2 + y^2}{(x^2 + y^2)} - \frac{2y^2}{(x^2 + y^2)^2} = \frac{1}{(x^2 + y^2)} - \frac{2y^2}{(x^2 + y^2)} = \left(\frac{y}{x^2 + y^2} \right)_y$$

THE SYSTEM OF ELEMENTARY SETS

Definition: An *interval* is a set of the following types: (a, b), $(a, b]$, $[a, b)$ and $[a, b]$. The *Lebesgue measure* of an interval T is defined as $\mu(I) := b - a$. So it is nothing else but the length of an interval, denoted with $\ell(I)$.

Definition: A subset F of $(0, 1)$ is called an *elementary set* if it can be represented as a finite(!) union of pairwise disjoint(!) intervals, i.e. $E = \bigcup_{k=1}^{n} I_k \;\; I_k \cap I_l = \phi,\ k \neq l$. The Lebesgue measure of an elementary set is defined as $\mu(E) := \sum_{k=1}^{n} \mu(I_k)$

Definition: A system of subsets $\mathfrak{R}$ of an [abstract nonempty] space X is called *ring* if:

(R1) $A, B \in \mathfrak{R} \Rightarrow A \cup B \in \mathfrak{R}$

(R2) $A, B \in \mathfrak{R} \Rightarrow A \setminus B \in \mathfrak{R}$

Remark: A ring is closed under $\cap$-operation since $A \cap B = A \setminus (A \setminus B)$

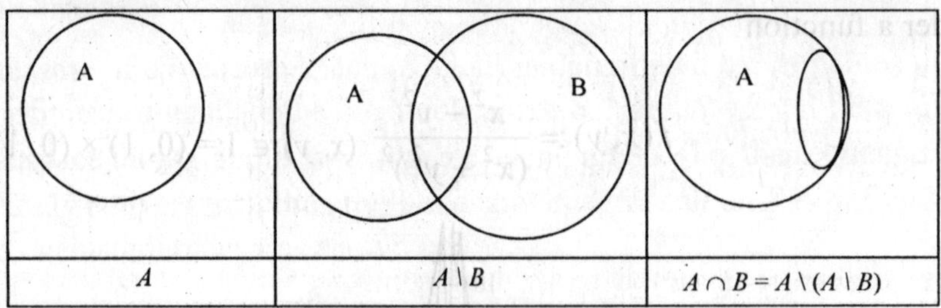

Example: A system E of all elementary sets on $(0, 1)$ is a ring.

Proof: Let $E_1 := \bigcup_{k=1}^{m} I_k$ and $E_2 := \bigcup_{l=1}^{m} I_l$. Then

$E_1 \cap E_2 = \bigcup_{k=1}^{n} \bigcup_{l=1}^{m} (I_k \cap I_l) \in E$ since $I_k \cap I_l \in E$.

One easily checks that the difference of two intervals is an elementary set. Now consider the interval $I := (0, 1)$.

It holds: $I \setminus E_1 = I \setminus \bigcup_{k-1}^{n} I_k = \bigcap_{k-1}^{n} (I \setminus I_k) \in E$. Similarly $I \setminus E_2 \in E$. Due to this and since $E_1 \subset I$ and $E_2 \subset I$ one has $E_1 \cup E_2 = I \setminus [(I \setminus E_1) \cap (I \setminus E_2)] \in E$ $(I \setminus [I \setminus E_1] \cap (I \setminus E_2)]$ is read: an interval I without points that are neither in E_1 nor in E_2).

Finally $E_2 \setminus E_1 = E_2 \cap (I \setminus E_1) \in E$.

Definition: A system of subsets A of a space is X called *algebra* if:

(A1) $A, B \in A \Rightarrow A \cup B \in A$

(A2) $B \in A \Rightarrow X \setminus B \in A$

It holds that $X \in A$ (and hence by (A2)) $\phi \in A$) since $X = B \cup (X \setminus B). \forall\ B \in A$.

Comparing (R2) and (A2) one sees that if $X \in \mathfrak{R}$ then $\mathfrak{R}$ is algebra.

Neither ring, nor algebra is sufficient for our goals since a lot of "interesting" sets are obtained through countably-infinite(!) number of set-theoretic operations. That's why we consider a notion of a σ-*algebra* (sometimes called σ-*field*), which plays the key role both in measure- and probability theory.

Definition: A system of subsets A of a space X is called σ-*algebra* if:

(sA1) $\quad$ $A1, A2, \ldots \in A \Rightarrow \bigcup_{k=1}^{\infty} A_k \in A$

(sA2) $\quad$ $B \in A \Rightarrow X \setminus B \in A$

Example:

1. A system (ϕ, X) is a σ-algebra. A powerset of X (i.e. a set of all subsets of a set X) denoted as 2^X or $P(X)$ is a σ-algebra.

   Note that the number of elements in a σ-algebra can be uncountable but (sA1) only guaranties that a σ-algebra is closed under a countably(!) infinite union.

   Every σ-algebra is an algebra (why?) but the converse is not true as the following example shows:

2. Let $X = \mathbb{R}^1$ and is a system, containing all *finite* subsets of X (including ϕ) and their complements; so (A2) holds automatically. As to (A1) it holds too since a *finite* union of the finite sets is again a finite set. However, (sA2) fails since if we take a *countably* infinite union of the sets, containing just one point, we obtain a set which is not in X.

Theorem: A σ-algebra is closed under $\bigcap^{\infty}$ (countably infinite intersection),\ (complement) and Δ (symmetric difference) operations.

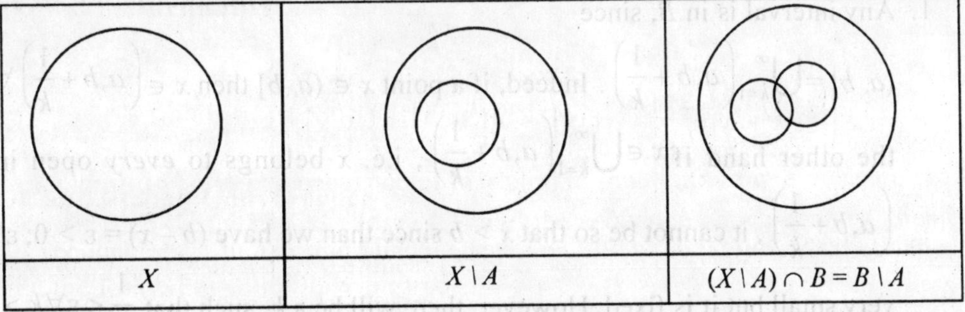

Proof: (i) $\bigcap^{\infty}$-closedness follows from De Morgan's law, which states that $\bigcup_{k=1}^{\infty} A_k = X \setminus \bigcup_{k=1}^{\infty} (X \setminus A_k)$. Indeed $\bigcup_{k=1}^{\infty} (X \setminus A_k)$ contains all elements of X except for those which belong to *all* A_k's. Thus $X \setminus \bigcup_{k=1}^{\infty} (X \setminus A_k)$ consists of the elements, which do belong to all A_k's but this is exactly $\bigcup_{k=1}^{\infty} A_k$.

(ii) Let $A, B \in A$. It holds $B \setminus A = (X \setminus A) \cap B$ thus the\-closedness follows from (i) and (sA2).

(iii) Follows from (ii) and (sA1) since $A \Delta B = (A \setminus B) \cup (B \setminus A)$.

Remark: Analogously to the transition from an algebra to a σ-algebra one can introduce a *σ-ring* so that (R1) becomes

$$\text{(sR1) } A_1, A_2, \ldots\ldots \in \Re \Rightarrow \bigcup_{k=1}^{\infty} A_k \in \Re$$

In (sA1) $\overset{\infty}{\cup}$ can be replaced with $\overset{\infty}{\cap}$. In (sR1) it cannot.

Theorem shows that a σ-algebra is very comfortable to work with: we can be sure that we will not transgress its bounds, applying any feasible set-theoretic operation to a *countably*(!) infinite number of operands. However, a *non-countable* intersection or union of the elements of a σ-algebra does not *necessarily* belongs to it (although it may). We learnt two examples of sigma-algebras, which are thougn "trivial" but important. Even more "interesting" is

Example: (Borel σ-algebra on (0, 1)) *Borel σ-algebra* on a [topological] space (0, 1), denoted by B ((0, 1)) or just by B if the underlying space is unambiguous, is the σ-algebra generated by all open intervals that belong to (0, 1).

It means that in order to "construct" B we must take all open intervals. Either all closed intervals, since a closed interval is a complement of its open counterpart. So if the closed intervals were not in B, (sA2) would fail. Finally, we have to include in B *any* set, which can be obtained by means of *not-more-than-countablyinfinite* number of set-theoretic operations on open and closed intervals. The elements of B are called *Borel sets*. Of course we cannot highlight *all* Borel sets but here are some distinguished examples:

1. Any interval is in B, since

   $(a, b] = \bigcup_{k=1}^{\infty} \left(a, b + \frac{1}{k} \right)$. Indeed, if a point $x \in (a, b]$ then $x \in \left(a, b + \frac{1}{k} \right) \forall k$. On the other hand if $x \in \bigcup_{k=1}^{\infty} \left(a, b + \frac{1}{k} \right)$, i.e. x belongs to *every* open interval $\left(a, b + \frac{1}{k} \right)$, it cannot be so that $x > b$ since than we have $(b - x) = \varepsilon > 0$; ε can be very small but it is fixed. However, there will bé a k_0 such that $\frac{1}{k} < \varepsilon \forall k > k_0$ but then $x \notin \left(a, b + \frac{1}{k} \right) \forall k > k_0$.

2. It immediately follows that all elementary sets are in *B*.
3. All single points are in *B* since $a = \bigcup_{k=1}^{\infty}\left[a, a+\frac{1}{k}\right)$.
4. All countably-infinite sets and in particular $\mathbb{Q} \cap (0, 1)$, i.e. a set of rational numbers within the interval (0, 1), are in *B*.
5. All open sets are in *B* because every open set on (0, 1) can be represented as at most countable union of open intervals (indeed, even as a disjoint union).

It is pretty difficult to give an example of a non-Borel set, although such sets do exist.

Remark: Borel σ-algebra *B*(*X*) can be defined on an arbitrary *topological* space *X*, i.e. a space on which a notion of an open set is [somehow] defined. An Euclidian space $\mathbb{R}^n$ is a topological space. Any *metric* and *normed* space (soon we will learn some of them) is a topological space, in which the topology is *induced* by the metric.

On the other hand it often hardly makes sense to speak about Borel σ-algebra on a *probability space* (or a set of events), since it is not [necessarily] endowed with a topology. For example, suppose we toss a coin two times.

Then the probability space is $\Omega = \{TT, TH, HT, HH\}$ where *T* stands for tail, and *H* for head. Even if we define (somehow) the system of open sets on Ω, it hardly brings us anything (in such a simple case).

But still the idea of the Borel σ-algebra is essential in probability theory, since events are mapped in $\mathbb{R}^1$ (in the simplest cases), which is a topological space. So we couple $B(\mathbb{R}^1)$ and the sigma-algebra of events … but let us postpone it to the chapter on stochastic processes.

The last but not the least important system of sets we consider is a *semiring.*

Semirings are important, for instance, for the measures on *product spaces*, e.g. $\mathbb{R}^n$ is an [*n*-time] *Cartesian product* of $\mathbb{R}^1$.

Definition: A system of subsets *S* of a space *X* is called a *semiring* if:

(smR1) $\phi \in S$

(smR2) $A, B \in S \Rightarrow A \cap B \in S$

(smR3) $A, B \in S, B \subset A$ then $\exists\ C_1, C_2, \ldots\ C_n \in S$ such that $A \setminus B = \bigcup_{k=1}^{n} C_k$

Example:

1. A system *H* of all *half-open* intervals [*a*, *b*) in $\mathbb{R}^1$ (including an "empty interval") is a semiring.
2. A system of all intervals is a semiring.

Proof: (smR1) holds since we include an "empty interval", i.e. an interval which contains no points in H. Consider two half-open intervals and $[a, b)$. If they $[c, d)$. If they are disjoint than $[a, b) \cap [c, d) = \phi$. If not, then:

1. Either one of them includes another, let without loss of generality $[a, b) \subset [c, d)$, so $[a, b) \cap [c, d) = [a, b)$ and $[c, d) \setminus [a, b) = [c, a) \cup [b, a)$.
2. or (letting w.l.o.g. $a < c$) $[a, b) \cap [c, d) = [c, b)$ and $[c, d) \setminus [a, b) = [b, d)$
3. Exercise.

We introduced the Borel -algebra "constructively": we specified the "basic" system of sets (i.e. all open intervals) and the operations to be one in order to obtain $B((0, 1))$. One says that the system of all open sets *generates* Borel σ-algebra or, in other words, the Borel σ-algebra is *the smallest* σ-algebra, which contains all open sets.

Often such "constructive" specification is not possible and that's why we need a technical formal

Definition: Consider an arbitrary [nonempty] space X and an arbitrary [nonempty] system of its subsets $\wp$. A σ-algebra (resp. an algebra, a ring) *generated by* $\wp$ is defined as an intersection of all σ-algebras (resp. algebras, rings) containing $\wp$, i.e.

$$\bigcap_{A_i \sigma-\text{algebra}, A_i \supset \wp} A_i$$

Remark: To show the correctness of the Definition one still needs to prove that $A(\wp)$ is indeed a σ-algebra. Moreover, one has to show that there is no σ-algebra, A^* such that $A^* \supset \wp$ and $A^* \subset A(\wp)$ ("$\subset$" means here that $A(\wp)$ *strictly* includes A^*).

So:

1. Consider a countably infinite sequence of sets $A_1, A_2,...$ all elements of which belongs to *every* A_i. Thus $A_1, A_2, ... \in \bigcap_i A_i$. Note that the index i can run through more than countably-infinite set, the statement still holds true.

Since A_i is a σ-algebra $\forall i, \bigcup_{n=1}^{\infty} A_n$, belongs to every A_i and thus to $\bigcap_i A_i$. Analogously one proves that $X \setminus A_n \in \bigcap_i A_i$.

2. Assume that A^* is a σ-algebra containing $\wp$ and $A^* \subseteq A(\wp)$. But it immediately follows from that $A^* \supseteq A(\wp)$, hence $A^* = A(\wp)$.

Also note that Definition would not pass for the semirings. Indeed, for the space X:= $\{1, 2, 3, 4\}$ we have that $S_1 := \{\phi, X, \{1\}, \{2, 3\}, \{4\}\}$ and S2:= $\{\phi, X, \{1\}, \{2\}, \{3, 4\}\}$ are semirings but is not $S_1 \cap S_2 = \{\phi, X, \{1\}\}$ is not.

Example:

1. B ((0, 1)) is a σ-algebra generated by all open (and all closed) sets on (0, 1). It follows from the definition of B ((0, 1)).
2. The system E of all elementary sets on (0, 1) is a ring, generated by the semiring I, i.e. the system of all intervals on (0, 1)
3. One can prove that the σ-algebra generated by I (denoted σ(I)) is B ((0, 1)). Note that B ((0, 1)) is also generated by E.

Assigning the Length to the Possibly Largest System of Sets

Before going on let us make a short revision. So far we started with the natural idea of length for the intervals. Then we naturally extended it to the system of all elementary sets E, which is a ring. Indeed E is even algebra, since $(0, 1) =: X \in E$ but had we taken $X := \mathbb{R}^1$ it would not be the case.

Now our goal is to assign the "length" to the possibly largest system L of subsets of (0, 1). L is called the *system of Lebesgue measurable sets*. Of course L must include E (and thus I). Moreover, it will be nice if L turns out to be a σ-algebra (it will). A σ-algebra is desired since it is closed under all feasible countably-infinite set-theoretic operations so dealing with measurable sets we can be sure that the result is still a measurable set.

We learnt an interesting example of the Borel -algebra, which is generated by E and thus is a good candidate. However, it turns out that L is larger than B, i.e. there are sets that are Lebesgue measurable but do not belong to the Borel σ-algebra.

The main idea is to approximate a set to be "measured" by not-more-thancountably infinite number of sets, for which the "measure"=length is already defined.

Definition: Consider an arbitrary set $A \subset (0, 1)$. An *outer measure* of A is defined as

$$\lambda^*(A) := \left\{ \sum_{k=1}^{\infty} \ell(I_k) \mid I_k \in I, A \subset \bigcup_{k=1}^{\infty} I_k \right\}$$

that is as the at-most-countably infinite covering of A with intervals with the least total length. Note that we can find such covering for *any* subset A of (0, 1), since.

$A \subset (0, 1)$, $(0, 1) \in I$ and $\lambda^*(A) \le \ell((0,1)) = 1$. Also note that the intervals in the union $\bigcup_{k=1}^{\infty} I_k$ are not necessarily disjoint.

The quantity

$$\lambda'(A) := \left\{ \sup \sum_{0, else}^{\infty} \ell(I_k) \mid I_k \in I, A \supset \sum_{k=1}^{\infty} I_k \right\} \text{ if } \bigcup_{k=1}^{\infty} I_k \ne \phi$$

is a *bad* candidate for the *inner measure* since then a pretty "simple" set $A := (0, 1)\setminus\mathbb{Q}$ $\mathbb{Q}$ (stands for the set of rational numbers) would be nonmeasurable. Indeed, $\lambda^*(A) := 1$ but $\lambda'(A) := 0$ because every interval contains at least one rational point but there is no rational point in, so does not contain any interval.

That's why the *inner measure* of is defined as

$$\lambda_*(A) := \ell((0,1)) - \lambda^*((0,1) \setminus A) = 1 - \lambda^*((0,1) \setminus A)$$

In other words one approximates the complement of A from outside; recall (sA2) from the Definition and you will see that we just made the first step towards a sigma-algebra.

If $\lambda^*(A) = \lambda^*(A)$ then the set A is called *Lebesgue measurable* and we define its Lebesgue measure

$$\mu(A) := \lambda^*(A) = \lambda^*(A)$$

Once again, recall that at the beginning of this chapter the Lebesgue measure was claimed to generalize an idea of the length. It means that should not contradict to the Definitions.

And it really does not! Right now I informally show that the Lebesgue measure of any interval I from I coincides with its length. Consider and arbitrary interval $I \in I$ with endpoints a and b, $0 \leq a \leq b \leq 1$. The least coverage of I in the sense of is itself[6], so the length of I is equal to its outer measure $\lambda^*(I)$. Respectively, the least coverage of its complement, i.e. of $(0, 1)\setminus I$ is $(0, a] \cup [b, 1)$ so accordingly

$$\lambda^*(I) = 1 - a - (1 - b) = b - a = {}^*(I).$$

To go further, we need to prove some auxiliary statements; they are useful to show that is L a σ-algebra as well as for a formal proof of the fact that for every set from E (and thus from I) its length and its Lebesgue measure coincide.

All these statements, which we are going to consider now, are technical and most of them are "obvious". However, some are worth paying attention because when we start dealing with an abstract measure (which is essential for the theory of stochastic processes) you will see a lot of analogies. But if you are short of time, you may (without vital losses) skip the proofs marked with tech.

Lemma: (Additivity in I) Let I, $I_1, I_2, \ldots, I_N \in I$ with $I_k \cap I_l = \phi$ for $k \neq 1$ and $I = \bigcup_{k=1}^{N} I_k$. Then $\ell(I) = \sum_{k=1}^{N} \ell(I_k)$.

Proof: Since $I_1, I_2, \ldots, I_N$ are disjoint and their union constitutes an interval, one has $a1 \leq b1 = a_2 \leq b_2 \ldots b_{N-1} = a_N \leq b_N$. Thus the length of $\bigcup_{k=1}^{N} I_k$ is $(b_1 - a_1) + (b_2 - a_2) + \ldots + (b_N - a_N) = b_N - a_1$, which is the length of I.

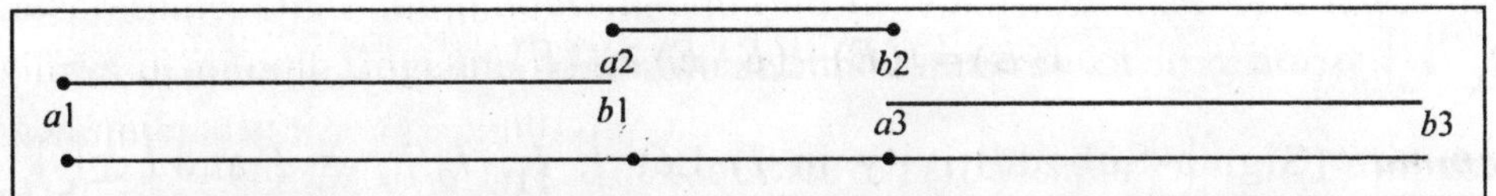

Fig. An Interval as a Union of Three Disjoint Intervals

Remark: Had we considered as a starting point the n-dimensional cube $(0, 1)^n$ instead of the interval (0, 1), the proof of Lemma would be much more complicated.

Lemma: Let $E \in E$ and consider two representations[7] of in the sense of the Definition

$E = \bigcup_{k=1}^{N} I_k$ and $E = \bigcup_{l=1}^{M} J_l$. Then $\bigcup_{k=1}^{N} \ell(I_k) = \bigcup_{l=1}^{M} \ell(J_l)$.

Proof: It holds that $\forall k I_k = E \cap I_k = \left(\bigcup_{l=1}^{M} J_l\right) \cap I_k = \bigcup_{l=1}^{M} (J_l \cap I_k)$.

Note that the union on the right-hand side is actually a union of disjoint sets. By Lemma $\ell(I_k) = \bigcup_{l=1}^{M} \ell(J_l \cap I_k)$. Since $\bigcup_{k=1}^{N} I_k$ is a disjoint union too, from the same Lemma follows that $\sum_{k=1}^{N} \ell(I_k) = \sum_{k=1}^{N} \sum_{l=1}^{M} \ell(J_l \cap_k) = \sum_{l=1}^{M} \sum_{k=1}^{N} \ell(J_l \cap I_k) = \sum_{l=1}^{M} \ell(J_k)$

Remark: Lemma shows that for any elementary set $E \in E$, $E = \sum_{k=1}^{N} I_k$ its length is well-defined:

$\ell(E) := \sum_{k=1}^{N} \ell(I_k)$, where $\bigcup_{k=1}^{N} I_k$ is an *arbitrary* representation of E in the sense of the Definition.

Lemma: (Additivity in E) Let $E_1, E_2, \ldots, E_N \in E$ and $E_k \cap E_l = \phi$ for $k \neq l$. Then $\ell\left(\bigcup_{k=1}^{N} E_k\right) = \bigcup_{k=1}^{N} \ell(E_k)$.

Proof: Since E is a ring and thus $\cup$-closed, it is sufficient to give a proof for the case $N = 2$ and then apply induction. Let $E_1 = \bigcup_{k=1}^{M} I_k$ and $E_2 = \bigcup_{l=1}^{P} J_l$, "$\bigcup$" denotes a disjoint union. Since E_1 and E_2 are disjoint, it holds that $E_1 \cup E_2 = \left(\bigcup_{k=1}^{M} I_k\right) \cup \left(\bigcup_{l=1}^{P} J_l\right)$. Now assign $\hat{I}_i := I_k, i, k = 1, 2, \ldots, M$ and $\hat{I}_i := J_l, i = (M+1), (M+2), \ldots, (M+P)$, $l = 1, 2, \ldots, P$. Then $E_1 \cup E_2 = \bigcup_{i=1}^{M+P} \hat{I}_i$ thus

$$\ell(E_1 \cup E_2 = \sum_{i=1}^{M+P} \ell(\hat{I}_i) \sum_{k=1}^{M} \ell(I_k) + \sum_{l=1}^{P} \ell(J_l) = \ell(E_1) + \ell(E_2)$$

Lemma: (Monotonicity of length on E) Let E, $\hat{E} \in E$ and $E \subset \hat{E}$. Then $\ell(E) \leq \ell(\hat{E})$.

Proof. We can represent $\hat{E}$ as a disjoint union $E \cup (\hat{E} \setminus E)$. By Lemma

$$\ell(\hat{E}) = \ell(E) + (\hat{E} \setminus E) \geq \ell(E)$$

Lemma: (Sigma-Subadditivity in I) Let $I, I_1, I_2, \ldots \in I$ and $I \subset \bigcup_{k=1}^{\infty} I_k$. Then $\ell(I) \leq \sum_{k=1}^{\infty} \ell(I_k)$.

Proof: Take arbitrary $\varepsilon > 0$. There exist a closed (and hence compact) interval $J \subset I$ and open intervals $J_k \supset I_k$ such that $\ell(I) = \frac{\varepsilon}{2} \leq \ell(J) \leq \ell(I)(*)$

and $\ell(I_k) \leq \ell(J_k) \leq \ell(I_k) + \frac{\varepsilon}{2^{k+1}}, k \in \mathbb{N}$. It holds that $J \subset \bigcup_{k=1}^{\infty} J_k$ and by *Heine-Borel* theorem $J \subset \bigcup_{k=1}^{N} J_k$ for sufficiently large (but finite) N.

Now define the following sets (and try to sketch the idea graphically):

$E_1 = J_1$

$E_2 = J_1 \setminus E_1$

$E_3 = J_3 = (E_1 \cup E_2)$

....

$E_N = J_N \setminus (E_1 \cup E_2 \ldots \cup \ldots E_N)$

These sets are elementary and disjoint (by construction).

Moreover, $\bigcup_{k=1}^{N} E_k = \bigcup_{k=1}^{N} J_k$. By Lemmas (Additivity and Monotonicity in E) we obtain

$$\ell(J) \leq \ell\left(\bigcup_{k=1}^{N} E_k\right) = \bigcup_{k=1}^{N} \ell(E_k) \leq \bigcup_{k=1}^{N} \ell(J_k)$$

From (*) it follows that

$$\ell(I) - \frac{\varepsilon}{2} \leq \ell(J) \leq \ell(I) \leq \bigcup_{k=1}^{N} \ell(J_k) \leq \bigcup_{k=1}^{N} \ell(I_k) + \bigcup_{k=1}^{N} \frac{\varepsilon}{2^{k+1}}$$

That is $\ell(I) \leq \sum_{k=1}^{N} \ell(I_k) + \varepsilon$. Since ε is arbitrary, the statement holds.

Lemma: (Sigma-Additivity in I) Let $I, I_1, I_2, \ldots \in I, I_k \cap I_l = \phi$ for $l \neq k$ and $I = \bigcup_{k=1}^{\infty} I_k$. Then $\ell(I) = \bigcup_{k=1}^{\infty} \ell(I_k)$.

Proof: For all $N \in \mathbb{N}$ in holds that $\bigcup_{k=1}^{N} I_k \in E$ and $\bigcup_{k=1}^{N} I_k \subset E$.

Thus it follows from the additivity and monotonicity of length that $\bigcup_{k=1}^{N}\ell(I_k)$ = $\ell\left(\bigcup_{k=1}^{N} I_k\right) \le \ell(I)$ for all $N \in \mathbb{N}$. The sequence of partial sums $\bigcup_{k=1}^{N}\ell(I_k)$ monotonically grows with N and is bounded. Hence letting $N \to \infty$ we have $\bigcup_{k=1}^{\infty}\ell(I_k) \le \ell(I)$. The converse inequality follows from sigmasubadditivity of length.

Remark: The σ-additivity is a key property of any [abstract] measure. In case of the underlying space $X = (0, 1)$ the σ-additivity follows from the additivity (due to Heine-Borel theorem) but generally it is *not* the case. We will see an example of additive but not sigma-additive function when we consider the abstract measure theory.

Lemma: (Sigma-Subadditivity in E) Let $E, E_1, E_2, \ldots \in E$ and $E \in \bigcup_{k=1}^{\infty} E_k$. Then $\ell(E) \le \sum_{k=1}^{\infty}\ell(E_k)$.

Proof: By definition of an elementary set, there exist the following representations of E and $E_1, E_2, \ldots$:

$$E = \bigcup_{l=1}^{N} I_l,\ E_k = \bigcup_{m=1}^{M(k)} I_m^k \quad k \in \mathbb{N}$$

Then $I_l = E \cap I_l \subset \bigcup_{k=1}^{\infty} E_k \cap I_l = \left(\bigcup_{k=1}^{\infty}\bigcup_{m=1}^{M(k)} I_m^k\right) \cap I_l = \bigcup_{k=1}^{\infty}\bigcup_{m=1}^{M(k)}\left(I_m^k \cap I_l\right).$

Since $I_m^k \cap I_l \in I$ we may apply Lemma (Subadditivity in I) and obtain

$$\ell(I_l) \le \sum_{k=1}^{\infty}\sum_{m=1}^{M(k)} \ell\left(I_m^k \cap I_l\right) \underline{\underline{(*)}} \sum_{k=1}^{\infty} \ell\left(E_k \cap I_l\right)$$

where in (*) we used Remark.

So by additivity and monotonicity in E the following holds

$$\ell(E) \le \sum_{l=1}^{N}\sum_{k=1}^{\infty} \ell\left(E_k \cap I_l\right) = \sum_{k=1}^{\infty}\sum_{l=1}^{N} \ell\left(E_k \cap I_l\right) \sum_{k=1}^{\infty} \ell(E_k \cap E) \le \sum_{k=1}^{\infty} \ell(E_k)$$

Lemma: (Sigma-Additivity in E) Let $E_1, E_2, \ldots \in \varepsilon$, $E_k \cap E_1 = \phi$ for $l \ne k$ and $E = \bigcup_{k=1}^{\infty} E_k$. Then $\ell(E) = \sum_{k=1}^{\infty}\ell(E_k)$.

Lemma: (Equality of the Length ℓ and the outer measure λ^* in I) For any interval $I \in I$ it holds that $\lambda^*(I) = \ell(I)$. In other words one writes $\lambda^*|_I = l$ and says that the *projection* of the outer measure on I coincides with the length.

Proof: Because I covers itself and its outer measure $\lambda^*(I)$ is an infimum of all coverings, it holds that $\lambda^*(I) \le \ell(I)$.

On the other hand if I is covered with some $\bigcup_{k=1}^{\infty} I_k$ then by subadditivity $\ell(I) \leq \sum_{k=1}^{\infty} \ell(I_k)$. Since it holds for all such coverings, it also holds that $\ell(I) \leq$ in f $\left\{\sum_{k=1}^{\infty} \ell(I_k)\right\} = \lambda^*(I)$

Lemma: (Equality of the length ℓ and the outer measure λ^* in E) For any elementary $E \in E$ set it holds that $\lambda^*(E) = \ell(E)$ that is $\lambda^*|_E = \ell$.

Proof: Analogous to that of Lemma with the same tricks as in the proof of Lemma

Lemma: (Equality of the length ℓ and the Lebesgue measure μ in E) For any elementary set $E \in E$ it holds that $\mu(E) = \ell(E)$ that is $\mu|_E = \ell$

Proof: According to Lemma $\lambda^*(E) = \ell(E)$ for all $E \in E$. Consider $E^C := (0, 1) \setminus E$. Since $(0, 1) \in E$ and E is a ring we obtain that $E^C \in E$. Thus by Lemma $\ell(E^C) = \lambda^*(E^C)$.

Finally by the definition of the inner measure $\lambda^*(E) = 1 - \lambda^*(E^C) = 1 - \ell(E^C) = \ell(E) = \lambda^*(E)$ so $\mu(E) = \ell(E)$.

The first part of auxiliary statements is over. In particular, we formally proved monotonicity and additivity of the length, which are its obvious properties. Either we proved its σ-additivity, which is less obvious.

Now we consider the second part, which will lead us to the sigma-algebra of Lebesgue-measurable sets. From now on I will frequently write A^c (^{c} stands for compliment) instead of $(0, 1) \setminus A$.

Lemma: For $A \subset (0, 1)$ define

$$\tilde{\lambda}(A) := \inf\left\{\sum_{k=1}^{\infty} \mu(I_k) \mid I_k open, I_k \in I, A \subset \bigcup_{k=1}^{\infty} I_k\right\}$$

Then $\tilde{\lambda}(A) = \lambda^*(A)$. In other words, it does not matter, whether we approximate from outside by a union of *arbitrary* intervals or just by a union of *open* intervals.

Proof. The relation $\tilde{\lambda}(A) = \lambda^*(A)$ is obvious (open intervals are a subclass of I). To show is the opposite inequality.

Take an $\varepsilon > 0$. Since $\lambda^*(A)$ is defined as an infimum of the lengths of all coverings, there exists a covering $\bigcup_{k=1}^{\infty} I_k \supset A$ such that $\lambda^*(A) \leq \sum_{k=1}^{\infty} \mu(I_k) < \lambda^*(A) + \frac{\varepsilon}{2}$ (a).

For all I_k, $k \in \mathbb{N}$ there exists an open interval J_k such that $I_k \subset J_k$ and $\mu(I_k) \leq \mu(J_k) < \mu(I_k) + \frac{\varepsilon}{2^{k+1}}$ (b).

It follows that $\tilde{\lambda}(A) \le \sum_{k=1}^{\infty} \mu(J_k) \overset{(b)}{<} \sum_{k=1}^{\infty} \mu(I_k) + \frac{\varepsilon}{2} \overset{(a)}{<} \lambda^*(A) + \varepsilon$. Since ε is arbitrarily small it holds that $\tilde{\lambda}(A) \le \lambda^*(A)$.

Remark: One can also show that if we define an outer measure of A as $\bar{\lambda}(A) := \inf\left\{\sum_{k=1}^{\infty} \mu(E_k) \mid E_k \in E, A \subset \bigcup_{k=1}^{\infty} E_k\right\}$ then $\bar{\lambda}(A) = \lambda^*(A)$.

Lemma: If a set $A \subset (0, 1)$ is measurable then its complement $A^C := (0, 1)\backslash A$ is measurable too.

Moreover, $\mu(A) + \mu(A^C) = \mu((0, 1) = 1$.

Proof: Due to the measurability of A we have $\lambda^*(A) = \lambda_*(A) = 1 - \lambda^*(A^C)$.

Further $\lambda_*(A^C) = 1 - \lambda^*(A) = 1 - (1 - \lambda^*(A^C)) = \lambda^*(A^C)$.

Finally $\mu(A) + \mu(A^C) = (1 - \lambda^*(A^C)) + \lambda^*(A^C) = 1$

Lemma was yet another simple but important step to the sigmaalgebra of the measurable sets. But before we commit another step and prove the $\overset{\infty}{\cup}$-closedness of the measurable sets, we have to do a lot of work, so be patient.

Next lemma will show the sigma-*sub* additivity of the outer(!) measure in $P((0, 1)$ i.e. for *all* subsets of (0, 1). However, do not expect the sigma-additivity (or even additivity) for an outer measure. For the Lebesgue measure it will be the case (on the sigma-algebra L which is strictly(!) smaller than $P((0, 1))$. But not for the outer measure.

Lemma: (σ-Subadditivity of the outer measure)

If $A \subset \bigcup_{k=1}^{\infty} B_k \subset (0,1)$ then $\lambda^*(A) \le \sum_{k=1}^{\infty} \lambda^*(B_k)$.

Proof: By definition of the outer measure we can find for every B_k a covering

$$\bigcup_{n(k)=1}^{\infty} I_n^k \text{ such that } \sum_{n(k)=1}^{\infty} \ell(I_n^k) < \lambda^*(B_k) + \frac{\varepsilon}{2^k}.$$

It holds that $A \subset \bigcup_{k=1}^{\infty} B_k = \bigcup_{k=1}^{\infty}\bigcup_{n=1}^{\infty} I_n^k$ i.e., $\{I_n^k\}_{k \in \mathbb{N}, n \in \mathbb{N}}$ is a countable covering of A.

It is somewhat unusual that we take two $\cup$'s but it does not change anything, since as is well known, I_n^k can be enumerated so that we need only one index running through $\mathbb{N}$; recall how you proved the countability of the set of all rational numbers in your first Analysis course.

Thus $\lambda^*(A) \le \sum_{k=1}^{\infty}\sum_{n=1}^{\infty} \ell(I_n^k) < \sum_{k=1}^{\infty}\left(\lambda^*(B_k) + \frac{\varepsilon}{2^k}\right) = \sum_{k=1}^{\infty} \lambda^*(B_k) + \varepsilon$.

And since $\varepsilon > 0$ is arbitrary, we yield the claim.

Lemma: (Monotonicity of the outer measure)

If $A \subset B \subset (0, 1)$ then $\lambda^*(A) \le \lambda^*(B)$

Proof: First we note that the outer measure of an empty set is equal to zero, that is $\lambda^*(\phi) = 0$. Indeed, $\lambda \subset (0, \varepsilon)$ where $\varepsilon > 0$ is arbitrary, so $\lambda^*(\phi) \le 0$. And since the outer measure cannot be negative (why?) we obtain that $\lambda^*(\phi) = 0$.

Using a technical trick, we write $A \subset (B \cup \phi \cup \phi....)$ thus by Lemma.

$$\lambda^*(A) \le \lambda^*(B) + \lambda^*(\phi) + \lambda^*(\phi) + ... = \lambda^*(B).$$

Lemma: (Outer Measure of a disjoint countable union of intervals)

Let for $A = \bigcup_{k=1}^{\infty} I_k$ $I_k \in I, I_l \cap I_k = \phi$ for $k \ne 1$.

Then $\lambda^*(A) \sum_{k=1}^{\infty} \ell(I_k) = \sum_{k=1}^{\infty} \lambda^*(I_k) = \sum_{k=1}^{\infty} \mu(I_k)$.

Proof. First we recall that $\lambda^*(E) = \mu(E) = \ell(E)$ for any $E \in E \supset I$.

Further by subadditivity $\lambda^*(A) \le \sum_{k=1}^{\infty} \lambda^*(I_k) = \sum_{k=1}^{\infty} \ell(I_k)$. To obtain the converse inequality, note that $\bigcup_{k=1}^{N} I_k \subset \bigcup_{k=1}^{\infty} I_k$ for any [finite] $N \in \mathbb{N}$ since we consider a disjoint union. Thus by monotonicity and the fact that $\bigcup_{k=1}^{N} I_k \in E$ we obtain the following

$$\lambda^*\left(\bigcup_{k=1}^{N} I_k\right) \le \lambda^*\left(\bigcup_{k=1}^{\infty} I_k\right) = \lambda^*(A) \text{ i.e. } \bigcup_{k=1}^{N} \lambda^*(I_k) \le \lambda^*\left(\bigcup_{k=1}^{\infty} I_k\right) = \lambda^*(A)$$

Since inequality on the right-hand side holds for all $N \in \mathbb{N}$, it also holds if we let $N \to \infty$, that is

$$\lim_{N\to\infty} \sum_{k=1}^{N} \lambda^*(I_k) = \sum_{k=1}^{\infty} \lambda^*(I_k) \le \lambda^*(A)$$

Lemma: (Measurability of open sets) An open set $O \subset (0, 1)$ is measurable.

Proof. Since $O \subset (0, 1)$ is open, it can be represented as a finite or countably infinite disjoint(!) union of *open* intervals $I_k \in I$. In the first case $O \in E$ and thus measurable. So we consider the case $O = \bigcup_{k=1}^{\infty} \bar{I}_k$, where $\bar{I}_k = (a_k, b_k)$ and $0 \le a_1 \le b_1 \le a_2 \le b_2 \le a_3 < ... \le 1$.

By Lemmas $\lambda^*(O)=\sum_{k=1}^{\infty}\lambda^*(\tilde{I}_k)=\sum_{k=1}^{\infty}\ell(\tilde{I}_k)$ On the other hand $O^C=(0,\ 1)\backslash\left(\bigcup_{k=1}^{\infty}\tilde{I}_k\right)$ is a countable disjoint union of [closed] intervals $J_k=[a_k, a_{k+1}]\in$ I. If $a_1>0$ we must also consider and $J_0=(0, a_1]$ analogously for the right end-point. Some of J_k might consist of a single point but it does not matter.

Hence $\lambda^*(O^C)=\sum_{k=0}^{\infty}\lambda^*(J_k)=\sum_{k=0}^{\infty}\ell(J_k)$.

Since $\left(\bigcup_{k=1}^{\infty}\tilde{I}_k\right)\cup\left(\lambda^*(O^C)=\sum_{k=0}^{\infty}\lambda^*(J_k)=\sum_{k=0}^{\infty}J_k\right)=(0,1)$ it follows by σ-additivity in I that

$1=\lambda^*((0,1)=\sum_{k=1}^{\infty}\ell(\tilde{I}_k)+\sum_{k=0}^{\infty}\ell(J_k)\sum_{k=1}^{\infty}\lambda^*(\tilde{I}_k)+=\sum_{k=0}^{\infty}\lambda^*(J_k)$ and finally $\lambda^*(O^C)=\sum_{k=0}^{\infty}\lambda^*(J_k)=1-\sum_{k=1}^{\infty}\lambda^*(\tilde{I}_k)$

Lemma: Consider two open sets O_1 and O_2 such that $O_1\cup O_2\supset(0,\ 1)$.

Then $\mu(O_1\cap O_2)\leq\mu(O_1)+\mu(O_2)-\mu((0,\ 1)=\mu(O_1)+\mu(O_2)-1$.

(Note that $O_1\cap O_2$ is measurable since it is open – as a finite(!) intersection of open sets).

***Proof*.** If both O_1 and O_2 consist of a finite number of [open] intervals, i.e. $O_1, O_2\in E$ and thus $O_1\cap O_2\in E$ then by additivity in E $\mu((0,\ 1))=1=\mu(O_1\backslash(O_1\cap O_2))+\mu(O_2\backslash(O_1\cap O_2))+(O_1\cap O_2)$ whereas $\mu(O_1)+\mu(O_2)=\mu(O_1\backslash(O_1O_2))+\mu(O_2\backslash O_1\cap O_2))+2\mu(O_1\cap O_2)$.

In the general case (when $O_1=\bigcup_k\tilde{I}_k$ and $O_2=\bigcup_l\tilde{J}_l$ or are the countable disjoint unions of open intervals) Heine-Borel theorem helps. Let $0<\varepsilon<0.5$ and consider a closed interval $[\varepsilon,\ 1-\varepsilon]$. By Heine-Borel theorem there exists a finite covering of $[\varepsilon,\ 1-\varepsilon]$ and this covering consists of the opens sets $\widehat{O_1}=\bigcup_{k=1}^{N}\tilde{I}_k$ and $\widehat{O_2}=\bigcup_{l=1}^{M}\tilde{J}_l$ Define $R_1:=\left\{\bigcup_k\tilde{I}_k\mid\tilde{I}_k\subset O_1\text{ but }\tilde{I}_k\not\subset\widehat{O_1}\right\}$ and $R_1:=\left\{\bigcup_l\tilde{J}_l\mid\tilde{J}_l\subset O_1\text{ but }\tilde{J}_l\not\subset\widehat{O_2}\right\}$. Then R_1 and R_2 are open and for sufficiently large N, M we have $\mu(R_1)<\varepsilon$ and $\mu(R_2)<\varepsilon$.

Since $O_1\cap O_2\subset\left(\widehat{O_1}\cap\widehat{O_2}\right)\cup R_1\cup R_2$ it holds

$$\mu(O_1\cap O_2)\leq\mu\left(\widehat{O_1}\cap\widehat{O_2}\right)+\mu(R_1)+\mu(R_2)(*)$$

where in (*) I used subadditivity of the outer measure and measurability of open sets.

On the other hand since $M, N<\infty$ and $[\varepsilon,\ 1-\varepsilon]\subset\widehat{O_1}\cup\widehat{O_2}$ we obtain

$\mu\left(\widehat{O_1}\right)+\mu\left(\widehat{O_2}\right)-\mu\left(\widehat{O_1}\cap\widehat{O_2}\right)>(1-\varepsilon)-\varepsilon=1-2\varepsilon$ (compare to).

Moreover, $\mu\left(\widehat{O_1}\right)\leq\mu(O_1)$ and $\left(\widehat{O_2}\right)\leq\mu(O_2)$ thus applying (*) we finally obtain

$$\mu(O_1\cap O_2)\leq\mu\left(\widehat{O_1}\cap\widehat{O_2}\right)+2\varepsilon<\left[\mu\left(\widehat{O_1}\right)+\mu\left(\widehat{O_2}\right)+2\varepsilon-1\right]+2\varepsilon$$

But because can be taken arbitrarily small, we yield the claim.

Lemma: (Superadditivity[8] of the inner measure) If $A=\bigcup_{k=1}^{\infty}A_k$ and $A_j\cap A_l=\phi, j\neq l$ then $\lambda_*(A)\geq\sum_{k=1}^{\infty}\lambda_*(A_k)$.

Proof. As a first step consider the case $A=A_1\cup A_2$. Take two open sets O_1 and O_2 such that $\lambda^*(A_1^c)<\mu(O_1)<\lambda^*(A_1^c)+\varepsilon$ and $\lambda^*(A_2^c)<\mu(O_2)<\lambda^*(A_2^c)+\varepsilon$ where $\varepsilon>0$ is arbitrary. Since $A_1\cap A_2=\phi$, it holds that $A_1^c\cup A_2^c\supset(0,1)$ thus $O_1\cup O_2\supset(0,1)$. By Lemma $\mu(O_1\cap O_2)\leq\mu(O_1)+\mu(O_2)-1$.

But $(O_1\cap O_2)\supset\left(A_1^c\cap A_2^c\right)=(A_1\cup A_2)^C=A^C$. Thus

$\lambda^*(A^C)\leq\mu(O_1)+\mu(O_2)-1<\lambda^*\left(A_1^c\right)+\lambda^*\left(A_2^c\right)-1+2\varepsilon$. Letting $\varepsilon\to\infty$ we obtain $\lambda^*(A^C)\leq\mu\left(A_1^c\right)+\lambda^*\left(A_2^c\right)-1$ which results in

$$1-\lambda^*(A^C)\geq\left[1-\lambda^*\left(A_1^c\right)\right]+\left[1-\lambda^*\left(A_2^c\right)\right] \text{ i.e. } \lambda_*(A)\geq\lambda^*(A_1)+\lambda_*(A_2).$$

But induction we prove the statement for the case $A=\bigcup_{k=1}^{N}A_k$, where N is finite.

Let, finally, $A=\bigcup_{k=1}^{\infty}A_k$. Consider $S_N:=\bigcup_{k=1}^{N}A_k$.

It holds $\lambda_*(S_N)\geq\sum_{K=1}^{N}\lambda_*(A_K)$. Further $B_N\subset A$ thus $S_N^c\supset A^c$ and by monotonicity of the outer measure $\lambda^*(S_N^c)\geq\lambda^*(A^c)$ and hence

$$\lambda^*(S_N)=1\lambda-\lambda^*(S_N^c)\leq1-\lambda^*(A^c)=\lambda_*(A)$$

so
$$\lambda^*(A)\geq\sum_{k=1}^{N}\lambda_*(A_k).$$

So the sum at the right-hand side is bounded and thus for $N\to\infty$ $\lambda_*(A)\geq\sum_{k=1}^{\infty}\lambda_*(A_k)$

Now we are in position to prove the main theorems for the measurable sets. We start with *Vallée Poussin's measurability criterion*, which sometimes is deservedly accepted as the definition of the measurability. Informally it says that a measurable set can be however closely approximated by the elementary sets.

Theorem: Completeness of the Lebesgue Measure

1. The outer measure of any set $A \subset (0,1)$ is not less that its inner measure, i.e. $\lambda_*(A) \leq \lambda^*(A)$.
2. If $\lambda^*(A) = 0$ then A is Lebesgue measurable and $\mu(A) = 0$.

Proof: Due to Lemma there exist two open sets and $o_1 = \bigcup_k I_k$ $O_2 = \bigcup_k I_k$ are at most countably infinite) such that

$$(*)\begin{cases} O_1 \supset A, & \mu(O_1) \leq \lambda^*(A) + \varepsilon \\ O_2 \supset A^c, & \mu(O_2) \leq \lambda^*(A^c) + \varepsilon \end{cases}$$

Consider an interval $[\varepsilon, 1-\varepsilon] \subset O_1 \cup O_2$, which is compact and thus by Heine-Borel theorem can be covered by a finite union of intervals, which constitute O_1 and O_2. We denote this finite union of intervals by E and note that $E \in E$.

So $[\varepsilon, 1-\varepsilon] \subset E$ and thus by the outer measure monotonicity $\lambda^*(E) = \pi(E) \geq 1 - 2\varepsilon$ and $\mu(E) \leq \mu(O_1) + \mu(O_2)$.

Consequently $\mu(O_1) + \mu(O_2) \leq 1 - 2\varepsilon$ and by $(*) \lambda^*(A) + \lambda^*(A^c) \geq 1 - 4\varepsilon$ i.e. $(*) \lambda^*(A) + \lambda^*(A^c) \geq 1 - 4\varepsilon$ and letting $\in \to 0$ we obtain that $1 - \lambda^*(A^c) = \lambda_*(A) \leq \lambda^*(A)$.

Since both inner and outer measures are nonnegative (why?) we yield the second part of the theorem.

Remark: The measure completeness is very important property in the probability theory. Informally it means that a subset ("sub-event") of an "impossible" event (i.e. an outcome, which probability measure is zero) is also "impossible". Later on we will see an example of an incomplete measure.

Vallée-Poussin's Measurability Criterion

Theorem: A set $A \subset (0, 1)$ is measurable if and only if for any $\varepsilon > 0$ there exists an elementary set $E \in E$ such that $\lambda^*(A \,\Delta\, E) < \varepsilon$, where as you remember $(A \,\Delta\, E) = (A \setminus E) \cup (E \setminus A) = (A \cup E) \setminus (A \cap E)$

Proof. $\Leftarrow$ Let $A \subset (0, 1)$ be measurable. Then $\lambda^*(A) = \lambda_*(A)$ or equivalently

$$\lambda^*(A) + \lambda^*(A^C) = 1 \ (*).$$

By definition of the outer measure there exist open sets $O_1 \supset A$ and $O_2 \supset A^C$ such that $\mu^*(O_1) < \lambda^*(A) + \varepsilon/2$ and $\mu(O_2) < \lambda^*(A^C) + \varepsilon/2$ thus from (*) it follows that $\mu(O_1) + (O_2)$

$< 1 + \varepsilon$. Since $(0, 1) \subset O_1 \cup O_2$ the following holds: $\mu(O_1) + \mu(O_2) \geq 1 + \mu(O_1 \cap O_2)$ thus $\mu(O_1 \cap O_2) < \varepsilon$. Let $O_1 = \bigcup_{k=1}^{\infty} \tilde{I}_k$, $\tilde{I}_k \in I, \tilde{I}_k$ open. Choose sufficiently big N such that $\sum_{k=N+1}^{\infty} \mu(\tilde{I}_k) < \varepsilon/2$ and define $E := \bigcup_{k=1}^{N} \tilde{I}_k$. It holds $A \setminus B \subset O_1 \setminus E$ hence $\lambda^*(A \setminus E)\ \lambda^*(O_1 \setminus E) = \mu(O_1 \setminus E) < \varepsilon/2$, where in equality in the middle I used the fact that $O_1 \setminus E$ is open (why is it?) and then applied Lemmas. Further $E \setminus A = E \cap A^C \subset E \cap O_2 \subset O_1 \cap O_2$ and consequently $\lambda^*(E \setminus A) \leq \lambda^*(O_1 \cap O_2) = \mu(O_1 \cap O_2) < \varepsilon/2$. Finally by the monotonicity of the outer measure $\lambda^*(A \Delta E) \leq \lambda^*(A \setminus E) + \lambda^*(E \setminus A) < \varepsilon/2 + \varepsilon/2 = \varepsilon$.

$\Rightarrow$ Let $E \in E$ and $\lambda^*(A \Delta E) < \varepsilon$. It holds (check!) $A \subset E \cup (A \Delta E)$ hence $\lambda^*(A) \leq \lambda^*(E) = \mu(E) + \lambda^*(A \Delta E) < \mu(E) + \varepsilon$.

Further $E^C \cup (A \Delta E) = E^C \cup (A \cup E) \setminus (A \cap E) \supset A^C$ because if $x \in A^C$ then $x \notin (A \cap E)$ and on the other hand $E^C \cup (A \cup E) = (0, 1)$.

Thus $\lambda^*(A^C) \leq \lambda^*(E^C) + \lambda^*(A \Delta E) = \mu(E^C) + \lambda^*(A \Delta E) < \mu(E^C) + \varepsilon$.

Respectively, $1 - \lambda(A^C) = \lambda^*(A) > 1 - \mu(E^C) + \varepsilon$.

Putting all together: $0 \leq \lambda^*(A) - \lambda_*(A) < 2\varepsilon$ and since ε is arbitrary, it follows that $\lambda^*(A) = \lambda_*(A)$.

Remark: Let us illustrate an idea of the measurability criterion graphically. The problem is, however, is that we unlikely can draw an expressive picture for the case when the underlying space is an interval (0, 1). But things would be better in case of an open square $(0, 1) \times (0\ 1)$.

Well, if the system E of the elementary sets on (0, 1) consists of the sets, which can be represented as a finite disjoint union of the intervals from I, for the case of $(0, 1) \times (0, 1)$it, as you guess, consists of the sets, which are a finite disjoint union of *rectangles* from $R := I \times I$.

At the Figure it is shown how a circle can be measured in the sense of Lebesgue/Vallée-Poussin. First we "tightly" cover a circle G with a finite(!) number of disjoint(!) rectangles; we denote such covering by R_1. Then take the symmetric difference $C \Delta R_1$ of the circle and its covering (coloured in black). Note that at this stage we do not require that the circle must be covered completely, i.e. that $C \subset R_1$, all we want is to make the symmetric difference small; in the sense that it can be covered by a set R_2, depicted by red rectangles.

For R_2 we do require a complete covering, i.e. that $C \Delta R_1 \subset R_2$. Also note that although R_2 consists of rectangles from R it is not necessarily elementary, since these rectangles need not to be disjoint and their number can be infinite.

If we can do the job in such a way that for an arbitrary (but given in advance and afterthat fixed) $\varepsilon > 0$ we find R_1 and R_2 such that an area of R_2 is less than ε then C is measurable.

In other words it means that is however closely approximated by.

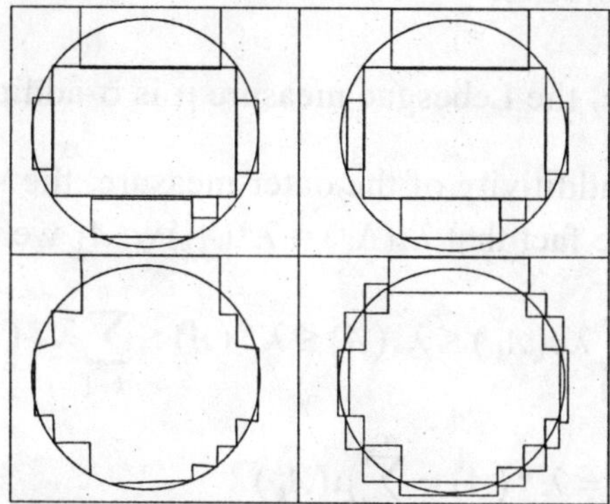

Fig. Geometric Interpretation of the Measurability (Measuring a Circle)

Measurability criterion is useful for the proof of the following theorems, which the essence of our efforts.

Theorem: Union, intersection, difference and symmetric difference of finitely(!) many measurable sets are measurable.

Proof: Let A_1 and A_2 be measurable. Then there exist two elementary sets $E_1, E_2 \in E$ such that $\lambda^*(A_1 \Delta E_1) < \varepsilon/2$ and $\lambda^*(A_2 \Delta E_2) < \varepsilon/2$.

It holds $(A_1 \cup A_2) \Delta (E_1 \cup E_2) \subset (A_1 \Delta E_1) \cup (A_2 \Delta E_2)$.

Indeed, if x belongs to the left-hand side then

$x \in A_1 \cup A_2 \cup E_1 \cup E_2 = (A_1 \cup E_1) \cup (A_2 \cup E_2)$ and

$x \notin (E_1 \cup A_2) \cap (E_1 \cup E_2) = (A_1 \cap E_1) \cup (A_1 \cap E_2) \cup (A_2 \cap E_2) \cup (A_2 \cap E_2)$

thus in particular $x \notin (A_1 \cap B_1)$ and $x \notin (A_2 \cap E_2)$.

On the other hand

$$(A_1 \Delta E_1) \cup (A_2 \Delta E_2) = (A_1 \cup E_1) \backslash (A_1 \cap E_1) \cup (A_2 \cup E_2) \backslash (A_2 \cap E_2)$$

So $\lambda^*[(A_1 \cup A_2) \Delta (E_1 \cup E_2)] < \lambda^*(A_1 \Delta E_1) + \lambda^*(A_2 \Delta E_2) < \varepsilon$.

Since $E_1 \cup E_2 \in E$ we are allowed to apply the measurability criterion.

The proof of the measurability of $A_1 \cap A_2$ is reduced to the equality

$$(A_1 \cap A_2)^c = A_1^c \cup A_2^c$$

Further we note that A1\ A2 $= A_1 \cap A_2^c$ and $A_1 \Delta A_2 = (A_1 \backslash A_2) \cup (A_2 \backslash A_1)$.

And finally extend the proof by induction to the arbitrary but finite number of the measurable sets

Theorem: If the sets $A_1, A_2,...$ (countably infinite sequence) are measurable and disjoint then is measurable and moreover

$\mu(A) = \sum_{k=1}^{\infty} \mu(A_k)$ (i.e. the Lebesgue measure μ is σ-additive)

Proof: From the subadditivity of the outer measure, the superadditivity of the inner measure, Theorem and the fact that $\lambda_*(A_k) = \lambda^*(A_k) \forall A_k$ we obtain:

$$\sum_{k=1}^{\infty} \mu(A_k) = \sum_{k=1}^{\infty} \lambda_*(A_k) \le \lambda_*(A) \le \lambda^*(A) \le \sum_{k=1}^{\infty} \lambda^*(A_k) = \sum_{k=1}^{\infty} \mu(A_k)$$

that is $\lambda_*(A) = \mu(A) = \lambda^*(A) = \sum_{k=1}^{\infty} \mu(A_k)$

Theorem: A countably infinite union of the measurable set (which are not necessarily disjoint) is measurable. The system of the Lebesgue-measurable sets on is a σ-algebra.

Proof. Let $A_1, A_2,...$ be measurable and $A = \bigcup_{k=1}^{\infty} A_k$. Define $\widehat{A_1} := A_1$ and $\widehat{A_k} := A_k \setminus \bigcup_{i=1}^{k-1} A_i$. Then $\widehat{A_j} \cap \widehat{A_l} = \phi$ for $j \neq l$ and $A = \bigcup_{k=1}^{\infty} \widehat{A_k}$.

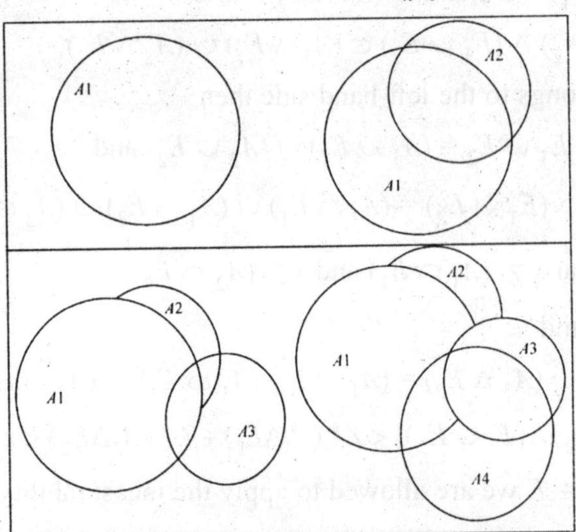

Fig. Building $\widehat{A_1}, \widehat{A_2}, \widehat{A_3}$ and $\widehat{A_4}$ (areas in grey).

By the previous theorem $\widehat{A_k}$'s are measurable. Since they are disjoint and $A = \bigcup_{k=1}^{\infty} \widehat{A_k}$, we obtain that is measurable.

So far we proved that a complement of a measurable set is measurable and that a countable union of measurable sets is measurable. The system of the Lebesgue-measurable sets is a σ-algebra.

In particular it means that a countably infinite intersection of the measurable sets is measurable too.

Upward Continuity of the Lebesgue Measure

Theorem: Vallée-Poussin's Measurability CriterionVallée-Poussin's Measurability Criterion Consider an increasing sequence of measurable sets $A_1 \subset A_2 \subset ...$ and let $A := \bigcup_{k=1}^{\infty} A_k$. Then

$\mu(A) = \lim_{k\to\infty} \mu(A_k)$ or in other words $A_k \nearrow A \Rightarrow \mu(A_k) \nearrow \mu(A)$

Proof: We are going to use a simplified version of a trick, depicted at the Figure. Consider $B_1 := A_1$, $B_2 := A_2 \setminus A_1, ..., B_N := A_N \setminus A_{N-1}, ...$

(Note that $A_N \setminus \bigcup_{k=1}^{N-1} A_k = A_N = A_{N-1}$ because we consider an *increasing* sequence of sets).

Then for $B_i \cap B_j = \phi$ for $i \neq j, A = \bigcup_{k=1}^{\infty} B_k$, and $A_N = \bigcup_{k=1}^{N} B_k$.

It holds by sigma-additivity of measure that $\mu(A_N) = \mu\left(\bigcup_{k=1}^{N} B_k\right) = \sum_{k=1}^{N} \mu(B_k)$

and by monotonicity of the outer measure $\mu(A) = \mu\left(\bigcup_{k=1}^{\infty} B_k\right) \geq \sum_{k=1}^{N} \mu(B_k)$; as you remember, for a measurable set its λ* and its μ are equal.

Since the inequality holds for all $N \in \mathbb{N}$, we let $N \to \infty$ and conclude that

$$\mu(A) \geq \bigcup_{k=1}^{\infty} \mu(B_k) = \lim_{N\to\infty} \sum_{k=1}^{N} \mu(B_k) \lim_{N\to\infty} \mu(A_N).$$

Converse inequality follows from the fact that $A \subset \bigcup_{k=1}^{\infty} B_k$.

Short glance at Lebesgue Measure on the Real line + Bonus

From to (0, 1) to $\mathbb{R}^1$ 4

In order to introduce the Lebesgue measure on $\mathbb{R}^1$ we could have tried to proceed analogously. Of course we would do it with caution since we would have dealt with sets of infinite measure. In particular, the outer measure causes no problem but for inner measure we would have confronted an uncertainty: the "length" of $\mathbb{R}^1$ is infinite thus for any bounded set A we would obtain $\lambda*(A) < \infty$ (well-defined) but $\lambda_*(A) = \infty - \infty$ (generally uncertain). Fortunately, there is a better way: as you remember, $\mathbb{R}^1$ can be represented as a countable union of intervals. So we can consider a sequence of intervals $(I_m)_{m\in\mathbb{Z}}$,

where $I_m := \{m < x \le m+1\}$.

Then $\mathbb{R}^1 = \bigcup_{m=-\infty}^{\infty} I_m$. We say that a set $A \subset \mathbb{R}^1$ is measurable if $A_m := A \subset I_m$ is measurable for every m. We define the measure of A as

$$\mu(A) := \sum_m \mu(A_m)$$

and the infinite sum on the right-hand side can either converge to some finite limit or diverge to $+\infty$.

One says that the measure $\mu(.)$ is *σ-finite* when [even if $\mu(A) = \infty$] there exists a sequence A_m (m countable) such that $A \subset \bigcup_m A_m$ and $\mu(A_m) < \infty$ for all m. You readily see that the Lebesgue measure on $\mathbb{R}^1$ is sigma-finite.

The main results we proved for the Lebesgue measure on (0, 1) hold for the case of $\mathbb{R}^1$ too. Particularly, the Lebesgue measure on the real line is σ-additive and the system $L(\mathbb{R}^1)$ of all measurable sets is a σ-algebra. However, there are some theorems, which hold for a finite measure but fail for an infinite one. Right now we will see an example.

Downward Continuity of Measure

Theorem: For (0, 1) but *not* for $\mathbb{R}^1$) Consider an decreasing sequence of measurable sets $A_1 \supset A_2 \supset \ldots$ and let $A := \bigcap_{k=1}^{\infty} A_k$ and (!) $\mu(A_1) < \infty$. Then

$$\mu(A) = \lim_{k\to\infty} \mu(A_k) \text{ or in other words } A_k \searrow A \Rightarrow \mu(A_k) \searrow \mu(A)$$

Proof. Consider $B := A_1 \setminus A$ and $B_k := A_1 \setminus A_k$ for $k \in \mathbb{N}$.

It holds $B_1 \subset B_2 \subset \ldots$ and $\bigcup_{k=1}^{\infty} B_k = B$.

Since $B \equiv (B\setminus B_k) \cup B_k$ (disjoint union) we obtain $\mu(B) = \mu(B\setminus B_k) + \mu(B_k)$.

Because $\mu(B) < \infty$ an equation $\mu(B\setminus B_k) = \mu(B) - \mu(B_k)$ always makes sense (i.e. we will not confront with something like $\mu(B\setminus B_k) = \infty - \infty$.

By theorem $\lim_{k\to\infty} \mu(B_k) = \mu(B)$ and thus $\lim_{k\to\infty} \mu(B \setminus B_k) = 0$.

But $B \setminus B_k = (A_1 \setminus A) \setminus (A_1 \setminus A_k) = A_k \setminus A$

So $0 = \lim_{k\to\infty} \mu(A_k \setminus A) = \lim_{k\to\infty} \mu(A_k) - \mu(A)$ i.e. $\lim_{k\to\infty} \mu(A_k) = \mu(A)$

Remark: The condition $\mu(A_1) < \infty$ (from which, by monotonicity, immediately follows that $\mu(A_k) < \infty, k \in \mathbb{N}$) cannot be omitted!

Consider a sequence of sets $A_N := (N, \infty), N \in \mathbb{N}$. Then $\bigcap_{N=1}^{\infty} A_N = \phi$; indeed, whatever point x we take, there will be N_0 such that $x \notin (N, \infty) \forall N > N_0$.

But $\mu(A_N) = +\infty$ for all(!) N hence $\lim_{N\to\infty} \mu(A_N) = +\infty \neq 0$.

CANTOR'S DISCONTINUUM A

We already know that if a set has a zero outer measure, it is Lebesguemeasurable and its Lebesgue measure is zero, too[10]. This, in particular, means that every countable set is Lebesgue-measurable and has measure zero. Indeed, let $A = \bigcup_{k=1}^{\infty} x_k, x_k$, is a point in $\mathbb{R}^1, k \in \mathbb{N}$. Take arbitrary $\varepsilon > 0$ and cover A as follows: $x_1 \in (x_1 - \varepsilon/4, x_1 + \varepsilon/4) := I_1$, $x_2 \in (x_2 - \varepsilon/8, x_2 + \varepsilon/8) := I_2$, etc.

Then $\lambda^*(A) \le \sum_{k=1}^{\infty} \ell(I_k) = \sum_{k=1}^{\infty} \frac{\varepsilon}{2^k} = \frac{\varepsilon/2}{1/2} = \varepsilon$, which is however small.

Thus one asks, whether the converse is true, i.e. whether a set with [outer] measure zero is countable. The answer is negative, as the following example shows.

Example: (Cantor set) Consider an interval $[0, 1] \subset \mathbb{R}^1$ and remove its middle open part $G_1 = I_1^1 := \left(\frac{1}{3}, \frac{2}{3}\right)$.

What remains is $C_1 = J_1^1 \cup J_2^1 := \left[0, \frac{1}{3}\right] \cup \left[\frac{2}{3}, 1\right]$.

As the second step we remove $G_2 = I_1^2 \cup I_2^2 := \left(\frac{1}{9}, \frac{2}{9}\right) \cup \left(\frac{7}{9}, \frac{8}{9}\right)$ from C_1.

$C_2 = J_1^2 \cup J_2^2 \cup J_3^2 \cup J_4^2 := \left[0, \frac{1}{9}\right] \cup \left[\frac{2}{9}, \frac{1}{3}\right] \cup \left[\frac{2}{3}, \frac{7}{9}\right] \cup \left[\frac{8}{9}, 1\right]$ remains[11].

We repeat the process and *after* the k-th step it is $C_k = \bigcup_{l=1}^{2^k} J_l^k$ what remains.

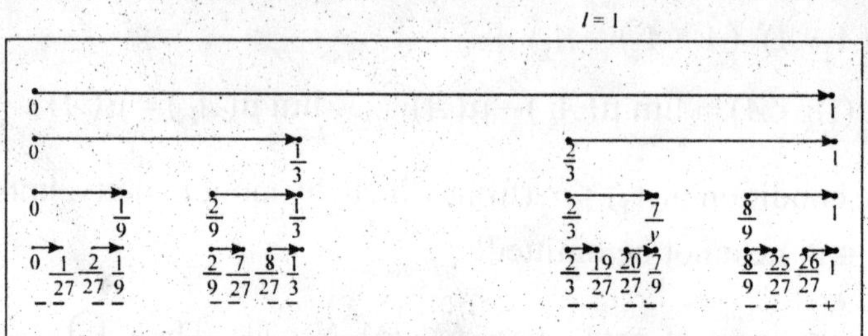

Fig. Constructing Cantor's Discontinuum

Respectively, $G_k = \bigcup_{l=1}^{2^k} I_l^k$ will be removed *at* the *k*-th step. And $\ell\left(I_l^k\right) = \frac{1}{3^k}$.

So for $N \in \mathbb{N}$ we have $C_N = [0,1] \setminus \bigcup_{k=1}^{N} G_k$. Letting $N \to \infty$ we obtain the *Cantor set*, defined as $C := [0,1] \setminus \bigcup_{k=1}^{\infty} G_k$, where $G_k = \bigcup_{l=1}^{2^{k-1}} I_l^k$. Here we emphasize that the unions are disjoint because it helps us to calculate the measure of C (which is closed and thus measurable).

We have $\mu\left(\bigcup_{k=1}^{\infty} G_k\right) = \frac{1}{3} + 2\frac{1}{3^2} + 2^2\frac{1}{3^3} + \ldots + 2^n\frac{1}{3^{n+1}} + \ldots - \frac{1}{3}\sum_{k=0}^{\infty}\left(\frac{2}{3}\right)^k = 1$

hence $\mu(C) = 1 - \mu\left(\bigcup_{k=1}^{\infty} G_k\right) = 0$.

It remains to show that the cardinality of C is continuum. Let point $x \in C$.

Then $x \in J_1^{\nu_1}, x \in J_2^{\nu_2}, \ldots,$ where ν_n runs through $1, \ldots, 2^{n-1}$.

Conversely, a sequence $J_1^{\nu_1} \supset J_2^{\nu_2} \supset \ldots$ converges to a single point (recall the *nested interval lemma* from your first Analysis course).

Thus there is a bijection between a point $x \in C$ and a countably-infinite system of indices (ν_1, ν_2,...) in which every element (except the 1st) runs at least through two values independently(!) of other elements. But it means that C is uncountable.

Cantor set is a classical example of a "bad" set. It is mainly used to construct various (counter)examples. Particularly, it allows to prove an existence of a set [with zero Lebesgue measure], which does not belong to the Borel σ-algebra. However, we can consider this topic only after we get familiar with *measurable* functions.

Chapter 4

Product Measures

In mathematics, given two measurable spaces and measures on them, one can obtain the product measurable space and the product measure on that space. Conceptually, this is similar to defining the Cartesian product of sets and the product topology of two topological spaces.

Given two measures μ and v on X and Y, we can form a new measure $\mu \times v$ on the product space. There are two main things we want to establish: Firstly, that this measure makes sense – it can be defined and there is truly only one choice for how we would to do this; secondly, Fubini's theorem: integrating over $\mu \times v$ is the same as integrating by μ and then v or by v and then μ.

It will be helpful in the course of the proofs to use a couple of the basic convergence theorems. You would probably have seen them before, at least in some form, but there is no harm in quickly recalling the proofs.

All the spaces in this section will be standard Borel. All the measures will be non-negative Borel probability measures. The results hold in the more general context of σ-finite Borel measures, but it is usually a routine exercise to obtain the general result from the specific ones we give below.

Monotone Convergence Theorem

Theorem: Let (f_n) be a sequence of measurable functions on a standard Borel probability space (X, μ). Assume the functions are non-negative and non-decreasing:

$$\forall_n \forall_x 0 \le f_n(x) \le f_{n+1}(x).$$

Assume the functions converge pointwise to some function f:

$$f_n(x) \to f(x)$$

as $n \to \infty$.

Then f is measurable and

$$\int f_n d\mu \to \int f d\mu .$$

Proof: It is a routine consequence of the measurable sets forming a σ-algebra that f will be measurable: For I the interval (a, ∞) we have

$$f^{-1}[I] = \bigcup_{n\in\mathbb{N}} f_n^{-1}[I],$$

and then one just verifies that the Borel sets are generated as a σ-algebra by intervals of this form.

Case 1: $\int f_n d\mu = \infty$. Then we can find a sequence of disjoint measurable sets, (E_i), and positive reals a_i, such that

$$\sum \mu(E_i).a_i = \infty$$

while $f(x) > a_i$ on each E_i. Then for each i and $x \in E_i$ there will be some n_x such that $f_{n_x}(x) > a_i$; and from this it follows that there will be some m_i such that $\{x \in E_i: n_x \leq m_i\}$ has measure at least $\mu(E_i)/2$. Then if we let N_i be the maximum of $m_j, j \leq i$, then

$$\int f_{N_i} d\mu > \frac{1}{2}\sum_{j\leq i} \mu(E_j).a_j .$$

Case 2: $\int f_n d\mu < \infty$.

Fix ε > 0. At each n let A_n be the set of x for which $|f(x) - f_n(x)|$ is less than or equal to ε/ 2. These are increasing sets with

$$\bigcup_n A_n = X,$$

and so eventually we can find some N with

$$\int_{X-A_N} f(x) d\mu < \varepsilon/2,$$

and then $\int f d\mu - \int f_N d\mu < \varepsilon$.

Dominated Convergence Theorem

Theorem: Let (f_n) be a sequence of measurable functions on a standard Borel probability space (X, μ) converging pointwise to some function f. Let g be a non-negative integrable function on X with

$$|f_n(x)| \leq g(x)$$

for all $n \in \mathbb{N}, x \in X$. Then

$$\int f_n d\mu \to \int f d\mu$$

Proof: As in the last argument, we let let A_n be the set of x for which $|f(x) - f_n(x)|$ is less than or equal to $\varepsilon/2$. Eventually we come to some N with

$$\int_{X \setminus A_N} g(x) d\mu < \varepsilon/4$$

$$\therefore \left| \int_{X \setminus A_N} f - f_N d\mu \right| \leq 2 \int_{X \setminus A_N} g d\mu < \varepsilon/2$$

and then $\left| \int f d\mu - \int f_N d\mu \right| \leq \left| \int_{A_N} (f - f_N) d\mu \right| + \int_{X \setminus A_N} g d\mu < \varepsilon$

On to product spaces. There will be a little bit of notation to clean up.

Notation: If $A \subset X \times Y$ and $x \in X$, then

$$A_x = \{y: (x, y) \in A\}.$$

If $y \in Y$ then

$$A^y = \{x: (x, y) \in A\}.$$

Definition: If X and Y are standard Borel spaces, then we equip $X \times Y$ with the σ-algebra generated by the rectangles of the form

$$A \times B \subset X \times Y$$

for A Borel in X and B Borel in Y.

Lemma: $X \times Y$ in this σ-algebra is a standard Borel space.

Proof Fix compatible Polish topologies for X and Y. Recall from an earlier exercise that $X \times Y$ is then Polish in the product topology. We finish by observing that every rectangle of the form $A \times B$ as above appears in the σ-algebra generated by the open sets in $X \times Y$.

Lemma: For X, Y as above, and $B \subset X \times Y$ Borel,

$$B_x$$

and

$$B_y$$

are Borel for all $x \in X, y \in Y$.

Proof: Observe that the set of B's with these slices Borel forms a σ-algebra containing the open sets.

Lemma: Let (X, μ) and (Y, ν) be standard Borel probability spaces. If $C \subset X \times Y$ is Borel then

$$x \mapsto \nu(C_x)$$

is measurable.

Proof: Actually this takes some work to prove, and I am going to skip through some of the details. (In fact, an even stronger result is true, namely that $x \mapsto \nu(C_x)$ is Borel, but that takes more trouble and would be beyond what we need.)

This is clear for rectangles of the form $A \times B$, A, B Borel subsets of X, Y. We obtain the result for finite unions of rectangles by using that the finite sums of measurable functions are measurable and since every such union can be represented as a finite union of disjoint rectangles. From there we can obtain the result for C a countably infinite unions of rectangles by appealing to the monotone convergence theorem. Note that the intersections of rectangles are rectangles, and hence the intersection of two sets formed as the countable unions of rectangles is again a countable union of rectangles.

Claim: If $C \subset X \times Y$ is Borel, then for any $\varepsilon > 0$ we can find unions $D^0 = \bigcup_{i \in \mathbb{N}} A_i^0 \times B_i^0$ and $D^1 = \bigcup_{i \in \mathbb{N}} A_i^1 \times B_i^1$ with

$$C \subset \mathrm{D}^0,$$

$$X \setminus C \subset D^1,$$

and $R\,(x \mapsto \nu(D^0 \cup D^1)x)d\mu < \varepsilon$.

Proof of Claim: Clearly the claim is true for C itself a rectangle of the form $A \times B$, since we can then take $D^0 = A \times B$, $D^1 = (X \setminus A) \times Y \cap X \times (\mathbf{Y} \setminus B)$. Clearly if the claim is true for C then it is true for $X \setminus C$. The main battle is to show closure under countable intersections.

For this purpose, suppose the claim is true for $C_1, C_2, C_3, \ldots$. At each n choose D_n^0, D_n^1 as in claim, with $C_n \subset D_n^0, X \setminus C_n \subset D_n^1$

$$\int \left(x \mapsto \nu\left(D_n^0 \cap D_n^1\right)_x\right) d\mu < 2^{-n}\varepsilon.$$

Then

$$\int \left(x \mapsto \nu\left(\bigcup_{n \in \mathbb{N}} D_n^0 \cap \left(\bigcap_{n \in \mathbb{N}} D_n^1\right)_x\right)\right) d\mu < \varepsilon$$

By the monotone convergence theorem we eventually get large N with

$$\int \left(x \mapsto \nu\left(\bigcup_{n \in \mathbb{N}} D_n^0 \cap \left(\bigcap_{n \in \mathbb{N}} D_n^1\right)_x\right)\right) d\mu < \varepsilon$$

Since $\bigcap_{n\leq N} D_n^1$ can be expressed as a countable union of rectangles, we have established the claim for $\bigcup_{n\in\mathbb{N}} C_n$.

Since the Borel sets in $X \times Y$ are the smallest σ-algebra containing the rectangles of the form $A \times B$ for $A \subset X$, $B \subset Y$ both Borel, we are done.

We can then obtain for any Borel C two sequence of decreasing sets as above, $\left(D_n^0\right)_n, \left(D_n^1\right)_n$, with $C \subset D_n^0$, $X \setminus C \subset D_n^1$,

$$\int x \mapsto v\left(\left(D_n^0\right)_x \cap \left(D_n^1\right)_x\right) d\mu <\!\!\rightarrow 0$$

and hence for μ-almost every x

$$v\left(\left(D_n^0\right)_x \cap \left(D_n^1\right)_x\right) \rightarrow 0.$$

Since

$$1 = v\left(\left(D_n^0\right)_x \cup \left(D_n^1\right)_x\right) = v\left(\left(D_n^0\right)_x\right) - v\left(\left(D_n^0\right)_x \cap \left(D_n^1\right)_x\right) + v\left(\left(D_n^1\right)_x\right)$$

we obtain for a.e. x

$$v\left(\left(D_n^0\right)_x\right) - \left(1 - v\left(\left(D_n^1\right)_x\right)\right) \rightarrow 0.$$

Since $x \mapsto v(C_x)$ is trapped between the measurable functions $x \mapsto \left(D_n^0\right)_x$ and $x \mapsto 1 - v\left(D_n^1\right)_x$ we obtain that it as well will be measurable.

Lemma: Let (X, μ) and (Y, v) be standard Borel probability spaces. If we define m on $X \times Y$ by

$$m(B) = \int_X (x \rightarrow v(B_x))\mu$$

then m is a Borel probability measure on X.

Proof: The main issue is to show σ-additivity. So suppose (B_i) is a sequence of disjoint Borel subsets of $X \times Y$. But then at each finite N we have

$$m\left(\bigcup_{i<N} B_i\right) = \int_X \left(x \mapsto v\left(\left(\bigcup_{i<N} B_i\right)_x\right)\right)\mu$$

$$= \sum_{i<N} \int (x \mapsto v((B_i)_x)\mu = \sum_{i<N} m(B_i)$$

Then the case for $\bigcup_{i \in \mathbb{N}} B_i$ and the observation that

$$v\left(\left(\bigcup_{i \in \mathbb{N}} B_i\right)_x\right) = \lim_{n \to \infty} v\left(\left(\bigcup_{i<n} B_i\right)_x\right)$$

Lemma: Let (X, μ) and (Y, v) be standard Borel probability spaces. Then for any Borel $B \subset X \times Y$

$$\int_X (x \mapsto v(B_x))\mu = \int_Y (y \mapsto \mu(B^y))v.$$

Proof Let us first define m as in 4.6 and then likewise m^* by

$$m^*(B) = \int_Y (y \mapsto \mu(B_y))v.$$

This likewise will give us a measure, and we easily check that m and m^* agree on the measurable rectangles.

Since every set arising in the algebra generated by the rectangles can be written as a finite union of disjoint rectangles, we quickly obtain that these two measures agree on this algebra.

From this it follows that $m = m^*$.

Definition: We let $\mu \times v$ be the measure m obtained by either

$$m(B) = \int_X (\mapsto v(B_x))\mu$$

or

$$m(B) = \int_Y (y \mapsto \mu(B^y))v.$$

Summarising the above:

Proposition: Let (X, μ) and (Y, v) be standard Borel probability spaces. Then the probability measure $\mu \times v$ on $X \times Y$ has the following properties for A, B, C Borel sets:-

1. $\mu \times v(A \times B) = \mu(A) \times v(B)$;
2. $\mu \times v(C) = R_X(x \mapsto v(C_x))d\mu$;
3. $\mu \times v(C) = R_Y(y \mapsto \mu(C_y))dv$.

Fubini's Theorem:

Theorem: Let (X, μ) and (Y, v) be standard Borel probability spaces and let

$$f: X \times Y \to \mathbb{R}$$

be a measurable non-negative function. Then

$$\int_{X\times Y} fd(\mu\times\nu)=\int_X\left(x\mapsto\left(\int_Y y\mapsto f(x,y)d\nu\right)d\mu\right)=\int_Y\left(y\mapsto\left(\int_X x\mapsto f(x,y)d\mu\right)d\nu\right)$$

Proof: We will traverse through a sequence of special cases, finally obtaining the result for general f.

First of all if f has the form

$$a.\ \chi_C,$$

for $a\in\mathbb{R},\chi_C$ the characteristic function of some measurable set. Similarly in the case that f is a finite sum of such functions, a simple function, we obtain the result by the additivity of integration. Finally in the general case we can obtain a sequence of simple functions, (f_n), with

$$0\le f_i(x)\le f_{i+1}(x)\le f(x)$$

and $f_n(x)\to f(x)$ at every x.

Usually one would just put this as

$$\int_{X\times Y} fd(\mu\times\nu)=\int_X\int_Y fd\nu d\mu=\int_Y\int_X fd\mu d\nu\,.$$

We are not quite ready to state the measure disintegration theorem, but it is possible to give a sense of its content. Suppose we have measure on $X\times Y$. Call it m. Suppose μ is a measure on X. Now there is no guarantee that there will be a ν on Y for which $m=\mu\times\nu$. Just doesn't necessarily hold. Even under the additional assumption of

$$m(B\times Y)=\nu(B)$$

for all Borel $B\subset X$.

However something else is true. We can find a range of measures, ν_x on Y, for each $x\in X$ such that m equals the integral, as it were, along μ of the various ν_x's. More precisely, for $A\subset X\times Y$,

$$m(A)=\int_X(x\mapsto\nu_x(A_x))d\mu\,.$$

There are some subtleties here being swept under the rug by our notation. To even make sense of this we need to know that $x\mapsto\nu(A_x)$ will always be measurable, and this in turn needs us to have some idea of what it would mean for the function $x\mapsto\nu_x$ to be Borel or measurable. It won't be until we see the Riesz representation theorem that this can be made precise. A slightly more sophisticated version of the lemma applies to measure preserving functions. Suppose (X,μ), (Z,λ) are standard Borel probability spaces and

$$f\colon X\to Z$$

is measure preserving function. As a matter of notation, let $Y_z=\{x\in X\colon f(x)=z\}$ for any $z \in Z$. Then our more sophisticated theorem states this: We can find a (suitably measurable) assignment $z \mapsto \mu_z$, where each μ_z is a measure on Y_z, such that

$$\mu(A) = \int_Z (z \mapsto \mu_z(A \cap Y_z)) d\lambda.$$

THE RIESZ REPRESENTATION THEOREM

Definition: Let (K, d) be a compact metric space. $C(K)$ consists of all continuous functions

$$f: K \to \mathbb{R}.$$

For $f \in C(K)$ we let $\|f\|$ be

$$\sup_{x \in K} |f(x)|.$$

Given $r \in \mathbb{R}$ and $f \in C(K)$ we define rf by pointwise multiplication,

$$(rf)(x) = r((f(x)),$$

and similarly for $f, g \in C(K)$ we define $f + g$ by pointwise addition,

$$(f + g)(x) = f(x) + g(x).$$

There is a whole train of remarks set into motion these simple definitions.

First of all, any continuous function on a compact metric space is bounded, so we are indeed assured that $\|f\|$ will always be a real number – the sup does not attain $+\infty$. Given the norm we can define $\rho(f, g) = \|f - g\|$. It is not hard to verify this satisfies the triangle inequality, and then that $\rho(.,.)$ defines a metric.

Secondly, any continuous function on a compact metric space is uniformly continuous; in other words

$$\forall \varepsilon > 0 \exists \delta > 0 \forall x, y \in K (d(x, y) < \delta \Rightarrow | f(x) - f(y) | < \varepsilon)$$

Then given a sequence $(f_n)_n$ in $C(K)$ which is Cauchy with respect to ρ we can easily check that $(f_n(x))_n$ is Cauchy for any x and we can define $f: K \to \mathbb{R}$ by simply letting $f(x)$ be the limit of $(f_n(x))_n$. By considering their being Cauchy with respect to ρ we have $(f_n)_n$ converges to f not just pointwise but also uniformly:

$$\forall \varepsilon > 0 \exists N \forall x \in K \in \forall n > N (| f(x) - f_n(x) | \leq \varepsilon)$$

(Take N to be large enough such $\forall n, m > N \forall x \in K (| f_n(x) - f_m(x) | < \varepsilon.)$ It is a standard result of undergraduate analysis[1] that the pointwise limit of a uniformly convergent sequence of uniformly continuous functions is again uniformly continuous, and thus $f \in C(K)$.

In conclusion then we have that $C(K)$ is a Banach space.

And given a Banach space we can sensibly ask for its dual. The Riesz representation theorem states that the positive elements of $C(K)$ with norm 1 may be identified with the collection of probability measures on K.

We will prove this. It is a long proof.

Definition: For K a compact metric space, $C(K)^*$, the dual space, is the collection of functions

$$\Lambda : C(K) \to \mathbb{R}$$

that are continuous and linear: $\Lambda(f+g) = \Lambda(f) + \Lambda(g)$, $\Lambda(rf) = r\Lambda(f)$. We then let $\|\Lambda\|$ be

$$\sup_{\{f \in C(K): \|f\|=1\}} |\Lambda(f)|.$$

(It is a routine exercise to verify $\|\Lambda\|$ finite from continuous.)

We say that $f \in C(K)$ is positive if $f(x) \geq 0$ all $x \in K$. We then write $f \leq g$ if $g - f$ is positive. For $r \in \mathbb{R}$ we write $f \leq r$ (respectively $r \leq f$) if $f(x) \leq r$ (respectively $r \leq f(x)$) all $x \in K$. For $r \in$ R we abuse

notation and also use r to indicate the continuous function

$$r: K \in \mathbb{R},$$

$$x \mapsto rx.$$

We say that $\Lambda \in C(K)^*$ is positive if $\Lambda(f) \geq 0$ whenever f is positive.

Definition: Let K be a compact metric space and $f \in C(K)$. The support of f is the closure of the set of $x \in K$ with $f(x) \neq 0$. For $C \subset K$ closed,

$$C \prec f$$

indicates $0 \leq f \leq 1$ and $f(x) = 1$ for all $x \in K$. For $V \subset K$ open,

$$f \prec V$$

indicates

$$\overline{\{x : f(x) \neq 0\}} \subset V;$$

in other words, V includes the support of f.

Lemma: If K is a compact metric space and $C \subset V$ with C closed and V open, then there is a $f \in C(K)$ with $0 \leq f \leq 1$ and

$$C \prec f \prec V.$$

Proof: Assume $d(C, K \setminus V) = \varepsilon > 0$. Let

$$g(x) = \frac{1}{\varepsilon} d(x, K \setminus V)$$

and

$$f(x) = \min(g(x), 1).$$

Lemma: If K is a compact metric space and $V_1, \ldots, V_n$ are open sets with

$$\bigcup_{i \leq n} V_i \supset C,$$

C closed, then there are continuous functions $h_1, \ldots, h_n$ with

$$0 \leq h_i \leq 1,$$

$$h_i \prec V_i,$$

at each i and

$$h_1(x) + h_2(x) + \ldots + h_n(x) = 1$$

at each $x \in C$.

Proof For each $x \in C$ we can find an open neighborhood U_x such that

$$\overline{U_x} \subset V_i$$

for some i. By compactness we may cover C with finitely many of these sets of the form U_x; call them $O_1, O_2, \ldots, O_\ell$. At $i \leq n$ we let C_i be the union of all $\overline{O_j}$'s with $\overline{O_j} \subset V_i$; this set is a finite union of closed sets and hence closed. We may find continuous g_i with $C_i \prec g_i \prec V_i$. The C_i's cover C and hence every $x \in C$ has some i with $g_i(x) = 1$, and so

$$(1 - g_1)(1 - g_2) \ldots (1 - g_n)(x) = 0.$$

Thus if we let

$$h_1 = g_1,$$

$$h_2 = (1 - g_1)g_2,$$

$$h_3 = (1 - g_1)(1 - g_2)g_3,$$

through to

$$h_n = (1 - g_1)(1 - g_2) \ldots (1 - g_{n-1})g_n$$

then at each j

$$h_1 + h_2 + \ldots + h_j = 1 - (1 - g_1)(1 - g_2) \ldots (1 - g_j)$$

yields

$$h_1 + h_2 + \ldots + h_j + h_j + 1 = 1 - ((1 - g_1)(1 - g_2) \ldots$$

$$(1-g_j)-(1-g_1)(1-g_2)\dots(1-g_j)\,g_{j+1})$$
$$=1-(1-g_1)(1-g_2)\dots(1-g_j)(1-g_{j+1}),$$

until at last

$$h_1+h_2+\dots+h_n=1-(1-g_1)(1-g_2)\dots(1-g_n),$$

which by the above constantly assumes the value 1 on C.

Theorem: Let K be a compact metric space and suppose $\Lambda\in C(K)^*$ is positive with $\|\Lambda\|=1$. Then there is a Borel probability measure μ on K with

$$\Lambda(f)=\int f(x)d\mu(x)$$

for any $f\in C(K)$.

Proof: For V open define $\mu(V)$ to to be

$$\sup_{f\prec V}\Lambda(f)$$

For $B\subset K$ Borel we let $\mu(B)$ be

$$\inf_{V\supset B, V \text{ open}}\mu(V)$$

We let Σ be the collection of all Borel B for which

$$\mu(B)=\sup_{C\subset B, C}\text{ closed }\mu(C).$$

Immediately we have every closed set in C. Our main battle will be to show that Σ includes every Borel set. This in turn is staggered into a sequence of skirmishes.

Claim 1: If V_1, V_2 are open, then

$$\mu(V_1\cup V_2)\le\mu(V_1)+\mu(V_2).$$

Proof of Claim: Let $g\prec V_1\cup V_2$. Applying 5.2 to the support of g we can find h_1, h_2 with

$$0\le h_1, h_2\le 1,$$
$$h_i\prec V_i,$$
$$h_1(x)+h_2(x)=1$$

all x in the support of g. Letting $g_i=h_ig$ we have

$$g=g_1+g_2,$$
$$g_i\prec V_i,$$

and then

$$\Lambda(g)=\Lambda(g_1)+\Lambda(g_2)\le\mu(V_1)+\mu(V_2).$$

(Claim)

Clearly then we also obtain $\mu(V_1 \cup V_2 \cup ... \cup V_n) \le \mu(V_1) + \mu(V_2) + ... + \mu(V_n)$ whenever $V_1, V_2, ... V_n$ is a sequence of open sets.

Claim 2: If $(B_i)_i$ is a sequence of Borel sets, then

$$\mu\left(\bigcup_{i=1}^{\infty} B_i\right) \le \sum_{i=1}^{\infty} \mu(B_i)$$

Proof of Claim: Fix $\varepsilon > 0$. We take open sets with $(V_i)_i$ with

$$B_i \subset V_i,$$

$$\mu(B_i) > \mu(V_i) - 2^{-i}\varepsilon.$$

Letting $V = \bigcup_i V_i$, we have $V \supset \bigcup_i B_i$. If $f \prec V$ then by compactness

$$f \prec \bigcup_{i<N} V_i,$$

some N, whence by the last lemma

$$\Lambda(f) \le \mu\left(\bigcup_{i<N} V_i\right) \le \sum_{i<N} \mu(V_i) \le \sum_{i=1}^{\infty} \mu(V_i) \le \varepsilon + \sum_i \mu(B_i)$$

Claim 3: If C closed then

$$\mu(C) = \inf_{C \prec f} \Lambda(f).$$

Proof of Claim: Given an open V with $V \supset C$ and $\mu(V) < \mu(C) + \varepsilon$, 5.1 gives us some f with $C \prec f \prec V$, and so

$$\Lambda(f) \le \mu(V) < \mu(C) + \varepsilon$$

Conversely, given $C \prec f$ and $\varepsilon > 0$ we may find open $U \supset C$ with

$$f(x) > 1 - \varepsilon$$

all $x \in U$. Then for any $f_0 \prec U$ we have $f_0(x) - f(x) < \varepsilon$ at every x, and hence

$$\Lambda(f_0) < \Lambda(f) - \varepsilon.$$

Claim 4: Every open set is in Σ.

Proof of Claim: Let V be open and $\varepsilon > 0$. Let $f \prec V$ with

$$(f) > \mu(V) - \varepsilon.$$

Then letting C be the support of f we obtain

$$\mu(C) \ge \Lambda(f) > \mu(V) - \varepsilon.$$

Claim 5: If C_1, C_2 are disjoint and closed, then

$$\mu(C_1 \cup C_2) = \mu(C_1) + \mu(C_2).$$

Proof of Claim: $\leq$ follows from claim 2.

For $\geq$ we apply the criterion of claim 3 and suppose

$$C_1 \cup C_2 \prec g.$$

we get f_1, f_2 with

$$f_1 + f_2 = 1,$$
$$f_1 \mid C_2 = 0,$$
$$f_2 \mid C_1 = 0.$$

$C_i \prec f_i g$ and $g = f_1 g + f_2 g$.

Claim 5 then extends to finite unions of disjoint closed sets.

Claim 6: If $(B_i)_i$ are disjoint sets in Σ and $B = \bigcup_i Bi$, then

$$\mu(B) = \sum_i \mu(B_i)$$

and $B \in \Sigma$.

Proof of Claim: (i) $\mu(B) \geq \sum_i \mu(B)$: Let $\varepsilon > 0$. Take closed $C_i \subset B_i$ with

$$\mu(C_i) > \mu(B_i) - \varepsilon 2^{-i}.$$

Claim 5 gives

$$\sum_{i<N} \mu(C_i) = \mu\left(\bigcup_{i<N} C_i\right)$$

for every finite N. Thus

$$\left(\sum_{i<N} \mu(B_i)\right) - \varepsilon < \sum_{i<N} \mu(C_i) = \mu\left(\bigcup_{i<N} C_i\right) \leq \mu(B).$$

(ii) $\mu(B) \leq \sum_i \mu(B_i)$: By claim 2.

(iii) $B \in \Sigma$: Fix $\varepsilon > 0$. Use the first part of the proof to find N with

$$\sum_{i<N} \mu(B_i) > \mu(B) - \varepsilon / 2.$$

Then find $C_i \subset B_i$ closed with

$$\mu(C_i) > \mu(B_i) - 2^{-i-1}\,\varepsilon.$$

Then

$$\mu\left(\bigcup_{i<N} C_i\right) = \sum_{i<N} \mu(C_i) > \left(\sum_{i<N} \mu(B_i)\right) - \varepsilon/2 > \mu(B) - \varepsilon.$$

Claim 7: If $B \in \Sigma$ then there is a closed C and open V with

$$C \subset B \subset V$$

and $\mu(V \setminus C) < \varepsilon$.

Proof of Claim: Find C, V with $\mu(B) < \mu(C) + \varepsilon/2$ and $\mu(B) > \mu(V) - \varepsilon/2$. Claim 6 yields

$$\mu(V) = \mu(V \setminus C) + \mu(C),$$

thereby giving the required bound.

Claim 8: If $B_1, B_2 \in \Sigma$, then $B_1 \setminus B_2$ is in Σ.

Proof of Claim: Fix $\varepsilon > 0$ and use claim 7 to choose $C_i \subset B_i \subset V_i$ with $\mu(V_i - C_i) < \varepsilon/2$.

$$B_1 \setminus B_2 \subset (V_1 \setminus C_1) \cup (C_1 \setminus V_2) \cup (V_2 \setminus C_2).$$

Since these last three sets are disjoint, claim 2 yields

$$\mu(B_1 \setminus B_2) < \mu(C_1 \setminus V_2) + \varepsilon.$$

Claim 9: Σ is a Boolean algebra.

Proof of Claim: Given A, B in Σ we can repeatedly claim 8 to

$$A \cap B = A \setminus (B \setminus A)$$

and claim 6 to

$$A \cup B = A \cup (B \setminus A).$$

Claim 10: Σ is a σ-algebra.

Proof of Claim: Given $(B_i)_i$ all in Σ, let

$$C_i = B_i \setminus \bigcup_{j<i} B_j$$

These are all in Σ by claim 9, and then by claim 6

$$\bigcup_i B_i = \bigcup_i C_i$$

is as well.

Σ is a σ-algebra incuding the closed sets, and thus every Borel set.

We have marched through vale and stormed through precarious mountain parts, but there is one last garrison of resistance. We still need to show the equality of integration by μ and application of Λ.

Claim 11: If C is closed and f positive with $f(x) \geq r$ on C, then

$$\Lambda(f) \geq r\,\mu(C).$$

Proof of Claim: Rescaling by and applying linearity, we may assume $r = 1$. Then it follows from claim 3.

Claim 12: For $f \in C(K)$ with $f \geq 0$,

$$\int f d\mu \leq \Lambda(f).$$

Proof of Claim: Choose $\varepsilon > 0$ and $y_0 < y_1 <\ldots y_n$ with each $y_{i+1} - y_i < \varepsilon/2$ and the range of f included in the open interval (y_0, y_n). Let E_i be the set on which $f(x)$ lies in the semi open interval $(y_{i-1}, y_i]$. Let $V_i \supset E_i$ be open and $C_i \subset E_i$ closed with

$$\mu(V_i \setminus E_i) < \frac{\varepsilon}{2ny_n},$$

and hence

$$\sum \mu(V_i \setminus E_i) y_n < \frac{\varepsilon}{2}.$$

Note that

$$\sum_{i<n} y_{i+1}\mu(V_{i+1}) \geq \int f d\mu.$$

Let

$$U_i = V_i \setminus \bigcup_{j \neq i} C_j.$$

The U_i's cover K so we can apply 5.2 to obtain $h_1, h_2, \ldots, h_n$ with

$$h_1 + h_2 + \ldots + h_n = 1,$$

each h_j having support in U_j. Since $U_i \cup C_j = 0$ for $i \neq j$, each h_j assumes the constant value 1 on the corresponding C_j. Letting $f_i = h_i f$ we have

$$f = \sum_{i \leq n} f_i.$$

Applying this and claim 11 we obtain

$$\Lambda(f) = \sum_{i<n} \Lambda(f_{i+1}) \geq \sum_{i<n} y_i \mu(c_{i+1}).$$

Thus

$$\int f d\mu = \Lambda(f) \leq \left(\sum_{i<n} \mu(V_{i+1} \setminus C_{i+1}) y_{i+1} \right) + \left(\sum_{i<n} \mu(C_{i+1})(y_{i+1} - y_i) \right)$$

$$< y_n \sum_{i<n} \mu(V_{i+1} \setminus C_{i+1}) + \left(\sum_{i<n} \mu(C_{i+1}) \right) \frac{\varepsilon}{2} < \varepsilon.$$

Claim 13: For any $f \in C(K)$

$$\int f d\mu \leq \Lambda(f).$$

Proof of Claim: For r a constant

$$\Lambda(r) = \int r d\mu = r.$$

Thus if we choose r a sufficiently large positive real to ensure $f + r \geq 0$ we can apply the last claim to obtain

$$\Lambda(f + r) \geq \int (f + r) d\mu$$

Since $\Lambda(f + r) = \Lambda(f) + \Lambda(r) = \Lambda(f) + r$ and since $\int (f + r) d\mu = \int f d\mu = \int f d\mu + r$ we are done.

Our victory is almost complete. Replacing f by $-f$ we can as well obtain

$$\int f d\mu \geq \Lambda(f).$$

And it is done and done.

Corollary: If $\Lambda \in C(K)^*$ is positive, then there is a finite Borel measure μ with

$$\Lambda(f) = \int f d\mu .$$

Proof: Apply the last theorem to

$$f \mapsto \frac{\Lambda(f)}{\Lambda(1)}$$

Notation: For K a compact metric space, let $P(K)$ be the probability measures on K equipped with the topology generated by the basic open sets

$$\{\mu: s_1 < \mu(f_1) < r_1, s_2 < \mu(f_2) < r_2, \ldots, n < \mu(f_n) < r_n\}$$

for $f_1, f_2, \ldots, f_n \in C(K)$.

Lemma: Let $\Lambda \in C(K)^*$. Then there is a positive $\Phi \in C(K)^*$ with

$$\Phi(f) \geq |\Lambda(f)|$$

for all $f \geq 0$.

Proof: For $f \geq 0$ set $\Phi^+(f)$ to be

$$\sup\{|\Lambda(g)| : |g| \leq f\}$$

Claim: Φ^+ is linear on its domain, in the sense that for $f_1, f_2, r \geq 0$

$$\Phi^+(f_1 + f_2) = \Phi^+(f_1) + \Phi^+(f_2)$$

and

$$\Phi^+(rf_1) = r\ \Phi^+(f_1).$$

Proof of Claim: The preservation with respect to multiplication by positive scalars is pretty immediate. The main issue is verifying the sups involved in finite additivity.

To see $\Phi^+(f_1 + f_2) \geq \Phi^+(f_1) + \Phi^+(f_2)$, consider some g_1, g_2 with $|g_i| \geq f_i$. After possibly replacing g_i by $-g_i$ we can assume $\Lambda(g_i) \geq 0$, when $|\Lambda(g_1 + g_2)| = |\Lambda(g_1)| + |\Lambda(g_2)|$. Conversely if $|g| \leq f_1 + f_2$, we can again assume without loss of generality that $\Lambda(g) \geq 0$ and write

$$g = g^+ - g^-,$$

where $g^+, g^- \geq 0$, $|g| = |g^+| + |g^-|$. Let $g_1^+ = \min(g^+, f_1)$, $g_1^- = \min(g^-, f_1)$, and then $g_2^+ = g^+ - g_1^+$, $g_2^- = g^- - g_1^-$. So we have found g_1, g_2 with $|g_i| \leq f_i$, $g_1 + g_2 = g$. (Claim)

Then given any $f \in C(K)$ we can let f^+ be defined by $f^+(x) = f(x)$ if $f(x) \geq 0$ and $f^+(x) = 0$ if $f(x) < 0$. Similarly we can let f^- be defined by $f^-(x) = -f(x)$ if $f(x) \leq 0$ and $f^+(x) = 0$ if $f(x) > 0$. Therefore we have represented f as the difference of two continuous functions:

$$f(x) = f^+(x) - f^-(x).$$

We let $\Phi(f) = \Phi^+(f^+) - \Phi^+(f^-)$.

Claim: If $f = g_1 - g_2$ with $g_1, g_2 \geq 0$, then $\Phi(f) = \Phi^+(g_1) - \Phi^+(g_2)$.

Proof of Claim: Let $g(x) = \min(g_1(x), g_2(x))$. Then $f^+(x) = g_1(x) - g(x)$ and $f^-(x) = g_2(x) - g(x)$.

Thus by linearity of Φ^+ we have

$$\begin{aligned}\Phi(f) &= \Phi^+(f^+) - \Phi^+(f^-)\\ &= \Phi^+(g_1) - \Phi^+(g) - (\Phi^+(g_2) - \Phi^+(g)) = \Phi^+(g_1) - \Phi^+(g_2).\end{aligned}$$

From this we have we can routinely verify linearity: For instance,

$$\Phi(f_1 + f_2) = \Phi^+(f^+) - \Phi^+(f^-)$$

$$\Phi^+(f_1^+ + f_2^+) - \Phi^+(f_1^- + f_2^-) = \Phi^+(f_1^+) + \Phi^+(f_2^+) - \Phi^+(f_1^-) - \Phi^+(f_2^-)$$

$$= \Phi(f_1) + \Phi(f_2).$$

Φ provides a positive element of the dual, and the very way in which it has been defined gives $\Phi(f) \geq |\Lambda(f)|$ for all $f \geq 0$.

Theorem: If $\Lambda \in C(K)^*$ then there is a signed measure μ with

$$\Lambda(f) = \int f d\mu$$

$$\Phi(f) = \int f d\mu$$

Note then that $\int |f| d\mu < r$ entails $\Lambda(f) < r$ and so Λ defines a bounded linear function on the continuous functions in $L^1(\mu)$. The continuous functions are dense in $L^1(\mu)$ and so Λ extends uniquely to the dual of $L_1(\mu)$.

The dual of L^1 is L^∞ and so we can find $\phi \in L^\infty(\mu)$ with

$$\Lambda(f) = \int f.\phi d\mu.$$

v defined by

$$v(B) = \int_B \phi d\mu$$

is as required.

There is also a version of this result for the complex dual of the continuous functions from K to $\mathbb{C}$, $C(K, \mathbb{C})$: For every $\Lambda \in C(K, \mathbb{C})^*$ there is a finite complex valued measure μ with $(f) = \int f d\mu$. This can be easily derived from the previous results.

Give $\Lambda \in C(K, \mathbb{C})^*$ and $f: K \to \mathbb{R}$ continuous, we can write

$$\Lambda(f) = \Lambda_0(f) + i\Lambda_1(f),$$

where $\Lambda_0(f), \Lambda_1(f) \in \mathbb{R}$. It is easily seen that Λ_0, Λ_1 are linear and so we obtained signed, real valued measures μ_0, μ_1 with

$$\Lambda(f) = \int f d\mu_0 + i \int f \mu_1.$$

All real valued *f*. By linearity of integration and Λ, the same formula holds for all complex valued continuous functions. Thus:

Theorem: Let *K* be a compact metric space. Let Λ be a continuous linear function from $C(K, \mathbb{C})$ (the continuous complex valued functions on *K*) to $\mathbb{C}$. Then there is a complex valued Borel measure μ with

$$\Lambda(f) = \int f d\mu$$

for all continuous $f: K \to \mathbb{C}$.

STONE-WEIERSTRASS

There is an extended discussion of this fundamental result.

Definition: A subset *A* of a vector *V* is said to be convex if for all $a, b \in A$ and $\alpha \in [0, 1]$ we have

$$\alpha a + (1 - \alpha)b \in A.$$

$c \in A$ is then said to be an extreme point of *A* if whenever $\alpha \in (0, 1)$, $a, b \in A$ with

$$\alpha a + (1 - \alpha)b = c$$

we have $a = b = c$.

Definition: Let *B* be a Banach space with dual space B^*. The weak star topology on B^* is the one generated by the basic open sets

$$\{\varphi \in B^* : \varphi(z_0) \in (a_0, b_0), \varphi(z_1) \in (a_1, b_1), \ldots, \varphi(z_n) \in (a_n, b_n)\}$$

for $z_0, \ldots, z_n \in B$, $a_0, \ldots, a_n$, $b_0, \ldots, b_n \in \mathbb{R}$. We say that a set is weak star open or weak star closed when it is open or, respectively, closed in this topology.

As a warning on notation, many people use "weak''" rather than "weak star" to denote this topology.

Lemma: Let *B* be a Banach space and $A \subset B^*$ convex and and weak star open. Let ϕ be in the weak star closure of *A*. Then for any $\psi \in A$ and $t \in (0, 1)$ we have

$$t\phi + (1 - t)\psi \in A.$$

Proof: Let *V* be a convex, weak star open neighborhood of the identity with $\psi + V \in A$. Note that

$$t^{-1}(1 - t)V =_{df} \{t^{-1}(1 - t)\phi : \phi \in V\}$$

is again an open neighborhood of 0, and so

$$\phi - t^{-1}(1 - t)V$$

is an open neighborhood of ϕ, and thus we can find some $\theta \in A$ lying in this set, which amounts to

$$\phi - \theta \in t^{-1}(1 - t)V.$$

This in turn yields

$$t\phi \in t\theta + (1 - t)V$$

$$\therefore \qquad t\phi + (1 - t)\psi \in t\theta + (1 - t)\psi(+ V).$$

Choosing some $\zeta \in (1 - t)\psi(+ V)$ with

$$t\phi + (1 - t)\psi = t\phi + (1 - t)\psi(+ \zeta),$$

we have $\psi + \zeta \in A$ since $\psi + V \subset A$, and then $t\phi + (1 - t)\psi$ is a convex linear combination of elements in A, as required.

Krein-Milman

Theorem: Let B be a Banach space and let $A \subset B^*$ satisfy:

1. A non-empty;
2. A convex;
3. A compact in the weak star topology.

Then A contains an extreme point.

Proof: Let u be the collection of all convex proper subsets of A which are relatively open in the weak star topology.

The compactness of A ensures that u is closed under directed unions, and so we can apply Zorn's lemma to get a maximal element, V. It suffices to show $A \setminus V$ consists of a singleton.

Let $\overline{V}$ be the weak star closure of V.

Claim: $V \neq V$.

Proof of Claim: Choose any $\psi \in V, \phi \notin V$. Let $S = \{s \in [0, 1]: s\psi + (1 - s)\phi \in V\}$. This set is weak star open in V since $s \mapsto s\psi + (1 - s)\psi \in V$ is weak star continuous. Let s_0 be the infinum of $[0, 1] \setminus S$. Then $s_0\psi + (1 - s_0)\phi$ is in $\overline{V} \setminus V$.

Note that if $\psi \in V$, $\alpha \in (0, 1)$ then the function

$$T: \phi \mapsto \alpha\psi + (1 - \alpha)\phi$$

$$A \to A$$

is continuous, affine, injective, and has $T[\overline{V}] \subset V$. Hence $T^{-1}[V]$ is a convex open subset of A including $\overline{V}$, and thus by maximality of V and the last claim it must equal A.

Claim: If W is a convex subset of K, then $V \cup W$ is a convex subset of A.

Proof of Claim: Since given any $\psi \in V$, $\phi \in W$, and $\alpha \in (0, 1)$, the argument above showed that

$$\alpha\psi + (1-\alpha)\phi \in V.$$

Now for a contradiction, assume ϕ_0, ϕ_1 are distinct points in $A \setminus V$. Let W_0 be a convex open set containing the first point but not the second. Then $V \cup W_0$ is a convex set, weak star open set providing a counterexample to the maximality of V.

The version of Krein-Milman presented above is rather weak. First of all, the proper conclusion of the result is not just that A contains an extreme point, but moreover the convex closure of the extreme points is dense in A. Secondly, the result holds in greater generality: We only really used that B^* is a topological vector space with a basis consisting of convex sets.

Definition: For K a compact metric space and μ a signed measure, we appeal to the Jordan decomposition of 3 above to find orthogonal, finite (positive) measures μ^+, μ^- with

$$\mu = \mu^+ - \mu^-.$$

We then let $|\mu| = \mu^+ + \mu^-$.

Definition: $A \subset C(K)$ is an algebra if it is closed under addition ($f, g \in A \Rightarrow (x \mapsto f(x) + g(x)) \in A$), multiplication ($f, g \in A \Rightarrow (x \mapsto f(x)g(x)) \in A$), and scalar multiplication ($f \in A, r \in \mathbb{R} \Rightarrow (x \mapsto rf(x)) \in A$).

An algebra A is said to separate points if for all $x_0 \neq x_1$ in K there is some $f \in A$ with

$$f(x_0) \neq f(x_1).$$

Stone-Weierstrass

Theorem: Let $A \subset C(K)$ be an algebra which is closed in the sup norm, separates points, and contains the constant functions. Then $A = C(K)$.

Proof: Consider the subset A of $C(K)^*$ consisting of Λ with norm less than one having

$$\Lambda(f) = 0$$

all $f \in A$. By Alaoglu's theorem, this is a compact set in the weak topology. It is clearly convex. In light of the Hahn-Banach theorem we will be done if we show that A only consists of 0.

Note that if $A \neq \{0\}$ then all of its extreme points would have to have norm one. So, instead, for a contradiction, apply theorem to obtain some extreme $\Lambda \in A$, $\|\Lambda\| = 1$ and to get a signed measure μ with

$$\Lambda(f) = \int f d\mu$$

all $f \in C(K)$. By the exercise above we have $|\mu|(K) = 1$. Let C be the support of $|\mu|$:

$$C =_{df} k \setminus \bigcup \{U : U \text{ open}, |\mu|(U) = 0\}.$$

Thus C is the minimal closed set with $\mu(E) = 0$ all Borel $E \subset K \setminus C$.

Claim: C does not consist of a single point.

Proof of Claim: Otherwise suppose $C = \{x_0\}$. Then for any $g \in C(K)$

$$\Lambda(g) = \Lambda(c_0),$$

where $c_0 = g(x_0)$. But c_0 is in A since it is a constant function, and after all we have $\Lambda(g) = 0$ all $g \in C(K)$.

Let $x_0 \neq x_1$ be in C. Since A separates points we obtain some $f_0 \in A$ with

$$f_0(x_0) \neq f_0(x_1).$$

Using the constant functions in A along with closure under addition we can get $f_1 \in A$ with

$$f_1(x_0) = 0,$$
$$f_1(x_1) \neq 0.$$

Letting $f_2 = (f_1)^2$ we obtain a positive element of A with the same property. Letting

$$f_3 = \frac{f_2}{\| f_2 \|}$$

we obtain $f_3 \in A$ with

$$f_3 \colon K \to [0, 1],$$
$$f_3(x_0) = 0,$$
$$f_3(x_1) \to 0.$$

We now define a new measure, $f_3\mu$, with

$$\int f d(f_3\mu) = \int (f.f_3)\mu.$$

$|f_3\mu| = f_3|\mu|$ since $f_3 \geq 0$. Letting

$$\alpha = \| f_3\mu \|$$

we have $\alpha > 0$ since $f_3(x_1) \neq 0$ and $\alpha < 1$ since $f_3(x_0) = 0$. We have constrained f_3 so it only takes values between 0 and 1, and hence $1 - f_3 \geq 0$, and so

$$\| (1-f_3)\mu \| = |(1-f_3)\mu|(K) = \int (1-f_3)|\mu| = 1-\alpha.$$

Then

$$\mu = \alpha \frac{f_3\mu}{\| f_3\mu \|} + (1-\alpha)\frac{(1-f_3)\mu}{\| 1-f_3\mu \|}.$$

From Λ being an extreme point we have $\mu = f_3\mu = (1-f_3)\mu$, but our assumptions on x_0, x_1 give us open neighborhoods U_0, U_1 of x_0, x_1 with

$$|\mu|(U_0), |\mu|(U_1) \neq 0$$

and some constant c such that for all $z \in U_0, z' \in U_1$

$$f_3(z) < c < f_3(z'),$$

with a contradiction.

A similar argument holds for algebras of complex valued functions with the sup norm

$$d(f, g) = \sup_{x \in K} |f(x) - g(x)|,$$

given $f, g: K \to \mathbb{C}$. There are however two key differences.

The first of these differences is minor. In the proof above we needed to use the Riesz representation theorem to summon in to being the indicated measure. The second difference is more telling. Trailing through the proof above we reached the function f_1 which separated the points x_0, x_1 as indicated, and from there we passed to $f_2 \geq 0$ which performed a similar task. In the complex value case we cannot simply take $f_2 = (f_1)^2$ since this will not necessarily be real valued. Instead we must make an additional assumption on the algebra. We need to assume it is closed under complex conjugation – which is to say whenever $f \in A$ we also have in A the function

$$\overline{f} : x \to \overline{f(x)},$$

mapping x to the complex conjugate of $f(x)$. With this key adjustment we can let

$$f_2 = f_1 \cdot \overline{f_1}$$

and continue the proof as above.

Theorem: Let K be a compact metric space and let $\mathbb{C}(K, \mathbb{C})$ be the complex valued continuous functions on K. Let $A \subset \mathbb{C}(K, \mathbb{C})$ be such that:

1. A is an algebra, in the set that $f_1, f_2 \in A$, $\alpha \in \mathbb{C}$ yield

$$f_1 + f_2 \in A,$$
$$\alpha f_1 \in A;$$

2. A contains the constant complex valued functions;
3. A separates points;
4. A is closed under complex conjugation.

Then A is dense in $C(K, \mathbb{C})$.

Chapter 5

Construction of Lebesgue Measure

In the Lebesgue measure, named after Henri Lebesgue, is the standard way of assigning a length, area or volume to subsets of Euclidean space. It is used throughout real analysis, in particular to define Lebesgue integration. Sets which can be assigned a volume are called Lebesgue measurable; the volume or measure of the Lebesgue measurable set A is denoted by $\lambda(A)$. A Lebesgue measure of infinity sign is possible, but even so, assuming the axiom of choice, not all subsets of R^n are Lebesgue measurable. The "strange" behavior of non-measurable sets gives rise to such statements as the Banach-Tarski paradox, a consequence of the axiom of choice.

Lebesgue measure is often denoted *dx*, but this should not be confused with the distinct notion of a volume form. We begin our construction of lebesgue mesure by checking that the notion of length we are used to really does give a measure on our usual semi -ring of half -open intervals,p.in order to do this we need to briefly discuss compactness

COMPACTNESS OF CLOSED INTERVALS

In terms of squences,a metric space is compact if every sequence in X has a subsequnce which converges in x. By the bolzano-weiestrass theorem,then every closed and bounded interval [a, b] in R is compact however there is a topological version of compactness which we need here concerning coverring using open sets lets a < b $\in$ R. suppose we have a collection of open sets in R, $[u_{y:}\ y \in$ r] s.t.

$$[a,b] \subseteq \bigcup_{y \in r} U_r$$

Then it is always true that there are

$$[y_1, y_2,,,,,, y_n \in \Gamma$$

$$[a,b] \subseteq \bigcup_{j=1}^{n} \Upsilon j$$

Lemma:

1. If $(a,b] \subseteq \bigcup_{k=1}^{n} (a_k, b_k]$,

$$\sum_{k=1}^{n} (b_k - a_k) \geq b - a.$$

2. If $(a,b] \supseteq \bigcup_{k=1}^{n} (a_{k,} b_k]$ then

$$\sum_{k=1}^{n} (b_k - a_k) \leq b - a.$$

3. $(a,b] = \bigcup_{k=1}^{n} (a_k, b_k]$

$$(b - a] = \sum_{k=1}^{n} (b_k - a_k).$$

Proof: (There is fairly easy direct proof avalilable but we use the Riemann intergral)

We can assume that all the intervals

$$(a_{k,}\ b_k], (a,b]$$

are contained in some closed interval $[-m, n]$. then

$$[-b - a) = \int_{-m}^{m} X_{(a,b]}(t)dt$$

and

$$(b_k - a_k) = \int_{-m}^{m} X_{(b_k, a_k]}(t)dt$$

1. If

$$(a,b] \subseteq \bigcup_{k=1}^{n} a_k, b_k]$$

then

$$X(a,b] \leq \sum_{k=1}^{n} X_{(a_k, b_k]}$$

at every point,so

$$b-a=\int_{-m}^{m}\chi_{(a,b)}(t)dt$$

$$\leq \int_{-m}^{m}\sum_{k=1}^{n}\chi_{(a_k,b_k)}(t)dt$$

$$=\sum_{k=1}^{n}(b_k-a_k).$$

if

$$(a,b]\geq\bigcup_{k=1}^{n}(a_k,b_k)$$

then

$$X_{(a,b)}(t)\geq\sum_{k=1}^{n}\chi_{(a_k,b_k)}(t)\ \forall t.$$

Integrating as in (i),

$$(b-a)\geq\sum_{k-1}^{n}(b_k-a_k).$$

3. If

$$(a,b)=\bigcup_{k=1}^{n}(a_k-b_k],$$

then (i) and (ii) apply and

$$(b-a)=\sum_{k=1}^{n}(b_k-a_k).$$

Recall: closed intervals $[a, b]$ are compact. In particular, we have the following.

Proposition: Let $[a, b]\subseteq\mathbb{R}$ be a closed interval,and suppose that]

$$(a_k, b_k)\ (k = 1, 2, 3,....)$$

are open intervals s.t.

$$[a,b]\subseteq\bigcup_{k=1}^{\infty}(a_k,b_k).$$

then $\exists m\in$ N$s.t$

$$[a,b]\subseteq\bigcup_{k=1}^{m}(a_kb_k).$$

Returing to half-open intervals, we need the follwing.

Corollary: If

$$[a,b] \subseteq \bigcup_{k=1}^{\infty} (a_k b_k).$$

then

$$(b-a) \leq \sum_{k=1}^{\infty} (b_k - a_k)$$

By porposition there exist $m \in N$ such that $[a,b] \subseteq \bigcup_{k=1}^{m} (a_{k,} b_k)$. But then $(a,b]$ $\subseteq \bigcup_{k=1}^{m} (a_{k,} b_k],$.

Theorem: Define $\mu : p \to [0,\infty]$ by $\mu((a,b]) = b-a.$ then μ measure on P.

Proof: Certainly $\mu(\phi) = 0$. Now suppose that

$$(a,b) = \bigcup_{n=1}^{\infty} (a_n, b_n].$$

We must show that

$$b - a = \sum_{n=1}^{\infty} (b_n - a_n)$$

Certainly we may assume $b > a$. First note that for all $m \in \mathbb{N}$,

$$\bigcup_{a=1}^{m} a_{n,} \mathrm{b}_n] \subseteq (a,b].$$

and so by 5.1 (ii).

$$\sum_{n=1}^{m} (b_n, a_n) \leq b - a.$$

Since this is true for all $m \in \mathbb{N}$, we have

$$\sum_{n=1}^{\infty} (b_n - a_n) \leq b - a.$$

To conclude the proof we must prove the reverse inequality.let $\varepsilon > 0$. Then, provide that $\varepsilon < b - a$,

$$[a+\varepsilon,b]\subseteq\bigcup_{n=1}^{\infty}\left(a_n,b_n+\frac{\varepsilon}{2^n}\right)$$

so, by corollary

$$(b-(a+\varepsilon))\leq\sum_{n=1}^{\infty}\left(b_n+\frac{\varepsilon}{2^n}-a_n\right).$$

i.e.

$$(b-a-\varepsilon))\leq\sum_{n=1}^{\infty}((b_{n-}a_n))+\varepsilon$$

$$b-a\leq\sum_{n=1}^{\infty}(b_n-a_n)+2\varepsilon$$

Since this is true for all $\varepsilon \in (0, b-a)$, we obtain

$$(b-a)\leq\sum_{n=1}^{\infty}(b_{n-}a_n)$$

As required. A similar result holds in R^n: eg. in R^2 the function on half - open rectangles by

$$v\,(a,\,b]\times(c,\,d])=(b-d)\,(d-c)$$

is also a measure (on p^2).

We now wish to measure the size of more complicated sets. good start is to extend μ to measure on the elementary figures ε (finite unions of sets in P).the fact that this is possibe is special case of a more general theorem. First we define extension.

Definition: Let X be a set and let $\varepsilon_1,\varepsilon_2$ be subsets of $p(X)$with $\varepsilon_1\subseteq\varepsilon_2$, let

$$\mu_{1:}\varepsilon_2\to[0,\infty],\ \ \mu_{2:}\varepsilon_2\to[0,\infty],$$

Be funct ion.then μ_2 is an extension of μ_1 if

$$\mu_2(E)=\mu_1(E)(E\in\varepsilon_1)$$

In this case we say μ_1 is the restriction of μ_2 to ε_1

[This agrees with the usual notion of extension and restriction for function,]

SUMMATION

In the next theorem we will need to be able to charge the oreder of summation in various

Series. Recall, when we proved proposition we saw that if

$$a_{jk} \in [0,\infty], (j=1,2,,3,....,1 \leq k \leq m)$$

then

$$\sum_{j=1}^{\infty}\left(\sum_{k=1}^{m} a_{jk}\right) = \sum_{k=1}^{m}\left(\sum_{j=1}^{\infty} a_{jk}\right).$$

We shall now show how to extend measure from semi - rings, and in particualar from P to ε.

USEFUL FACTS ABOUT SERIES OF ELEMENTS OF [0, ∞]

Suppose we have $a_{jk} \in [0,\infty]$ where $i \in \mathbb{N}$ and $1 \leq k \leq n_i$. Then the set

$$[(i,k_t) : k,i \in N, 1 \leq k \leq n_i]$$

Is contable,so we can enumerate this set as

$$[(i_t k_t): t =,1, 2, 3,.....]$$

$$\sum_{i=1}^{\infty}\left(\sum_{k=1}^{m_j} a_{jk}\right) = \sum_{t=1}^{\infty} a_{j_t} k_t,$$

From this we obtain a crucial fact about measures

Lemma: Let X be a set,let $\varepsilon \subseteq p(X)$ suppose and let be a measure

If A and A satisfies

$$A = \bigcup_{i=k}^{\infty} \bigcup_{k=1}^{n_i} D_{ik},$$

where $n_{1.}\ n_2, ... \in \mathbb{N}$, and the sets D_{ik} are all in ε the then

$$\mu(A) = \sum_{i=1}^{\infty}\sum_{k=1}^{n_i} \mu(D_{ik}).$$

[In some books this is regarded as obvious!]

Proof: Let $(i_k, k_l,), (i_2, k_2), (i_3, k_3),$ be and enumeration of the set

$$[(i,k) : k,i \in \mathbb{N}, 1 \leq k \leq n_i],$$

Then

$$A = \bigcup_{t=1}^{\infty} D_{i,k,}$$

$$\mu(A) = \sum_{t=1}^{\infty} \mu(D_{i,k,})$$

$$= \sum_{t=1}^{\infty} \sum_{k=1}^{n_i} \mu(D_{ik})$$

by (*) with $a_{jk} = \mu(D_{ik})$.

With these fact at our disposal,we can prove our theorem.

Theorem: Let be a semi-ring of subsets of a set X, and $\mu_1 : \varepsilon \to [0,\infty]$ let be a measure. Then is a unique measure $\mu_2 : R(\delta) \to [0,\infty]$ such that is an extension of μ_1.

Proof: For any $A \in R(\delta)$ there are disjoint sets $A_1, \ldots\ldots, A_m$ in δ with

$$A = \bigcup_{j=1}^{m} A_j.$$

The only possible value for

$$= \sum_{j=1}^{m} \mu_1(A_j).$$

Thus if such a measure μ_2 exist it is unque.We now wish to *define,* suppose that

$$\mu_2\left(\bigcup_{j=1}^{m} A_j\right) = \sum_{j=1}^{m} \mu_1(A_j)$$

Whenever $A_1, \ldots\ldots, A_m$ are disjoint sets in ε. To see that μ_2 well defined, suppose that

$$\bigcup_{j=1}^{m} A_j = \bigcup_{k=1}^{n} \mathrm{B}_{k,}$$

with $A_{j,}\ B_k$ all ε in then set $C_{jk=} A_j \cap B_k$, and obtain

$$A_j = \bigcup_{k=1}^{n} C_{jk} \text{ all } j,$$

$$B_k = \bigcup_{j=1}^{m} C_{jk} \text{ all } k$$

Thus

$$\sum_{j=1}^{m} \mu_1(A_j) = \sum_{j=1}^{m} \sum_{k=1}^{n_i} \mu_1(C_{jk})$$

$$= \sum_{k=1}^{n} \sum_{j=1}^{n} \mu_1(C_{jk})$$

$$= \sum_{k=1}^{n} \mu_1(\mathrm{B}_k).$$

Thus μ_2 is well defined. To see that μ_2 is measure, suppose that

$$A \in R(\varepsilon) \text{ and } A = \bigcup_{i=1}^{\infty} B_i.$$

with each $B_i \in R(\varepsilon)$. Then there are pairwise disjoint sets A_j in $\varepsilon(1 \le j \le m)$ with

Also for each i, there are disjoint sets $C_{l1}, C_{i2}, \ldots C_{\mathrm{In}}$ in ε such that

$$B_i = \bigcup_{k=1}^{n} C_{ik}.$$

We have

$$A_j = \bigcup_{j=1}^{\infty} (A_j \cap B_1) = \bigcup_{i=1}^{\infty} \bigcup_{k=1}^{n} (A_{j \cap} C_{ik})$$

and

$$B_1 = \bigcup_{j=1}^{m} \bigcup_{k=1}^{n} (A_{j \cap} C_{ik})$$

Since the $(A_j \cap C_{ik})$ sets are in ε we have

$$\mu_2(\mathrm{B}_i) = \sum_{j=1}^{m} \sum_{k=1}^{n} \mu_1(A_{j \cap} C_{ik})$$

and since μ_1 is a measure settting $D_{ik} = A_j \cap C_{ik}$ in lemma gives

$$\mu_1(A_j) = \sum_{i=1}^{\infty}\sum_{k=1}^{n}\mu_1(A_{j\cap}C_{ik})$$

Thus

$$\mu_2(A)\sum_{j=1}^{\infty}\mu_1(A_j) = \sum_{j=1}^{m}\sum_{i=1}^{\infty}\sum_{k=1}^{n_1}\mu_1(A_j \cap C_{ik})$$

$$= \sum_{i=1}^{\infty}\sum_{j=1}^{m}\sum_{k=1}^{n_1}\mu_1(A_j \cap C_{ik})$$

$$= \sum_{i=1}^{\infty}\mu_2(B_i).$$

This conclude the proof in particular tthe measure on P

$$\mu((a,b]) = b-a$$

extend to a unique measure on the ring of elementry figures in R

Example: Define $\mu_1 : p \to [0,\infty]$ by $\mu_1((a,b]) = b-a$. then has a unique extension $\mu_2 : \varepsilon \to [0,\infty]$ which is measure stisfying

$$\mu_2\left(\bigcup_{k=1}^{n} a_k, b_k]\right) = \sum_{k=1}^{n}(b_k - a_k).$$

(Similarly in higher dimensions.)

For our main extension theorem we will need the notion of an outer measure μ^* and sets which are 'measurable with respect to μ^*'.

Definition: Let X be a set,and let

$$\mu^* : p(X) \to [0,\infty].$$

then μ^* is an outer measure on X if

1. $\mu^*(\phi) = 0,$
2. If $A, \mathrm{B} \in p(X)$ and $A \subseteq B$ then $\mu^*(A) \le \mu^*(B)$(μ* is monotone).
3. If $A \subseteq \bigcup_{n=1}^{\infty} A_n,$ where A, are subsets of X, then $\mu^*(A) \le \sum_{n=1}^{\infty}(u^*(A_n).-a_k).$ (μ^* is countably subadditive.)

Note: An outer measuere on X is defined ofn all subsets of X

Example:

1. Any measure $p(X)$ on is also an outer measure on X
2. Defining

$$\mu^*(E) = \begin{cases} 0(E = \phi), \\ 1(E \neq \phi), \end{cases}$$

defines an outer measure which is not a measure (provided that X has at least two point!)

Definition: Let X be a set and let μ^* be and outer measure on X. A set $A \subseteq X$ is said to be measurable with respect to μ^*(or μ^*-mesurable) if, for every set $E \subseteq X$ the equality

$$\mu^*(E) = \mu^*(E \setminus A) + \mu^*(E \cap A)$$

holds.

Notes:

1. We know $\mu^*(E) \leq \mu^*(E \setminus A)\mu^*(E \cap A)$ by subadditivity,thus A is μ^*-measurable if and only if,for all $E \subseteq X$,

$$\mu^*(E) \geq \mu^*(E \setminus A) + \mu^*(E \cap A).$$

2. Similarly, if $\mu^*(E) = \infty$ the equality is automatic, so the condition need only be checked for those E with $\mu^*(E) < \infty$
3. A is μ^* measurable if and only if $X \setminus A \mu^*$ is measurable,because

$$E \setminus A = E \cap (X \setminus A), \text{ while}$$
$$E \cap A = E \setminus (X \setminus A).$$

we will be interested in using a measure defined on a ring to define an outer measurer on all subnsets of a set. [e.g,we shall show that defining,for $A \subseteq \mathbb{R}$,

$$\mu^*(A) = \inf\left\{\sum_{n=1}^{\infty}(b_n - a_n) : a_n \leq b_n \in R, A \subseteq \bigcup_{n=1}^{\infty} a_{n,} b_n]\right\}$$

μ^* is an outer measure on R.We will show that all the Borel sets are μ^* measrable,and that the restriction of μ^* to the Borel sets is a measure]

We now begin to prove the results we need about outer measures.The first Lemma is fairly weak

Lemma: Let X be set and let $\mu^* : f(X) \to [0, \infty]$ be an outer measure.set

$$f = [A \in p(X) : A \text{ is } \mu^* \text{ measurable}]$$

Then f is a field, and for any E$\subseteq$Xand any disjoint sets $A_1, A_2, \ldots\ldots A_n \in f$,

$$\mu^*\left(\bigcup_{k=1}^{n}(E \cap A_k)\right) = \sum_{k=1}^{n}\mu^*(E \cap A_k).$$

in particular, μ^* is finitely additive on f

Proof: Let $A, B \in f$ and let $E \subseteq X$. Then note that $E \setminus (A \cup B) = (E \setminus A) \setminus B$, and that $E \cap (A \cup B) = ((E \setminus A) \cap B \cup (E \cap A))$.

Thus

$$\mu^*(E) \leq \mu^*(E \setminus A \cup B)) + \mu^*(E \cap (A \cup B))$$

$$= \mu^*((E \setminus A) \setminus B) + \mu^*((((E \setminus A) \cap B) \cup (E \cap A))$$

$$= \mu^*((E \setminus A) \setminus B) + \mu^*((E \setminus A) \cap B) + \mu^*(E \cap A)$$

(by subadditivity)

$$= \mu^*(E \setminus A) + \mu^*(E \cap A)$$

(B is measurable)

$$= \mu^*(E)$$

(A is measurable).

Since u^*(E) appeares at both ends,all the inequaliaties in the meddle are equalites, in particular,

$$u^*(E) = u^*(E \setminus A \cup B)) + u^*(E \cap (A \cup B)).$$

This shows $A \cap B$ is mesurable.

The fact ϕ that is measurable is trivial, and we already know that

$$A \in f \Rightarrow X \setminus A \in f.$$

Thus f is field

For the last part, let $E \subseteq X$ and define

$$v : f \to [0, \infty]$$

$$v(A) = \mu^*(E \cap A).$$

Then we are required to prove that v is finitely additive of f

Note: If $A_1,...., A_n$ $A_1,,,,,,,, A_n$ are disjioint sets in f, then

$$\left.\bigcup_{k=1}^{n}(E \cap A_k) = E \cap \bigcup_{k=1}^{n} A_k.\right]$$

It is enough to show that if $A, B \in f$ and $A \cap B = \phi$ then $v(A \cup B) = v(A) + v(B)$. But

$$v(A \cup B) = \mu^*(E \cap (A \cup B))$$

$$= \mu^*((E \cup (A \cup B)) \cap A) + \mu^*((E \cap (A \cup B)) \setminus A)$$

(since A is measurable)

$$= \mu^*(E \cap A) + \mu^*(E \cap B)$$

(since $A \cap B = \phi$)

$$= v(A) + v(B)$$

as claimed.

The result follows.

we shall see that, in fact, F is a $\sigma-$ field, and restricted to F is a measure. we will then be able to prove the extension results we want.

Lemma: If F is field of subsets of a set of a X, then the follwing are equivalent:

(i) F is a $\sigma-$ field,

(ii) Whenever $A_1, A_2, A_3,...$ are *pairwise disjoint* sets in F then

$$\bigcup_{n=1}^{\infty} A_n \in f.$$

(i) $\Rightarrow$ (ii) is trivial,

(ii) $\Rightarrow$ (i) assume (ii) holds,

Let $B_1, B_2, B_3,... \in f$.

We must show $\bigcup_{n=1}^{\infty} B_n \in f$, to see this set

$$A_1 = B_1, A_{n+1} = \mathrm{B}_{n+1} \setminus \bigcup_{k=1}^{n} B_k\, (n = 1,2,3,).$$

Then $A_1, A_2, A_3,....$ are pairwise disjoint elements of f.

Also

$$\bigcup_{n=1}^{\infty} B_n = \bigcup_{n=1}^{\infty} A_n$$

and this is in f by (ii).

If X is non-empty set and μ^* is amn outer measure on X, then, defining f to be the set μ^* of measurable subsets of X know that f is a field. Also for any $E \subseteq X$ and sets $A_1, A_2, \ldots A_n \in f$ which are pairwise disjoint

we have

$$\mu^*\left(\bigcup_{k=1}^{n} E \cap A_k\right) = \sum_{k=1}^{n} \mu^*(E \cap A_k).$$

[Taking $E = X$, (*) shows μ^* is finitely additive on f.]

Theorem: Let μ^* be an outer measure on a non-empty set X.Let f be the set of μ^* - measurable subsets of X Then

1. f is a field on X,
2. For any set $E \subseteq X$ and pairwise disjoint sets $A_1, A_2, \ldots$ in f,

$$\mu^*\left(\bigcup_{n=1}^{\infty} E \cap A_n\right) = \sum_{n=1}^{\infty} \mu^*(E \cap A_n),$$

3. The restrication of μ^* to f is a measure.

Proof: First note that (iii) follows immediately from (ii) by setting $E = X$, so we need on y prove (i) and (ii). Also we know f is a field,so to prove (i) we need only check countable disjoint unions.

To prove (i) and (ii), let $A_1, A_2, \ldots$ be pairwise disjoint element of f and let $E \subseteq X$

Set

$$A = \bigcup_{n=1}^{\infty} A_n.$$

Note that

$$E \cap A = \bigcup_{n=1}^{\infty} (E \cap A_n).$$

Since μ^* is countably subaditive, we have

$$\mu^*(E \cap A) = \mu^*\left(\bigcup_{n=1}^{\infty} E \cap A_n\right) \leq = \sum_{n=1}^{\infty} \mu^*(E \cap A_n).$$

Set

$$B_n = \bigcup_{k=1}^{n} A_{n.}$$

We have $B_n \in f$ and $B_n \subseteq A$. Thus

$$\mu^*(B_n) \leq \mu^*(A)$$

(by monotocity) and

$$\mu^*(B_n \cap E) \leq \mu^*(E \cap A).$$

By Lemma

$$\mu^*(B \cap E) = \mu^*\left(B_n = \bigcup_{k=1}^{n} E \cap A_k\right)$$

$$= \sum_{K=1}^{n} \mu^*(E \cap A_k).$$

Thus

$$\mu^*(A \cap E) \geq \sum_{K=1}^{n} \mu^*(E \cap A_k).$$

Letting $n \to \infty$, we have

$$\mu^*(A \cap E) \geq \sum_{K=1}^{\infty} \mu^*(E \cap A).$$

$$\mu^*(A \cap E) \geq \sum_{K=1}^{\infty} \mu^*(E \cap A).$$

we thus obtain (1) and (2)

$$\mu^*(A \cap E) \geq \sum_{K=1}^{\infty} \mu^*(E \cap A_k),$$

providing (3)

To show that

$$\bigcup_{n=1}^{\infty} A_n \in f$$

we need to show

$$u^*(E) = u^*(E \cap A) + u^*(E \setminus A).$$

We know that $B_n \in f$, we have

$$\mu^*(E) = \mu^*(E \cap B_n) + \mu^*(E \setminus B_n)$$

$$\geq \mu^*(E \cap B_n) + \mu^*(E \setminus A)$$

(since $E \setminus B_n \supseteq E \setminus A$ and μ^* is monotone)

$$= \left(\sum_{k=1}^{n} \mu^*(E \cap A_k) \right) + \mu^*(E \setminus A).$$

Letting $n \to \infty$ we have

$$\mu^*(E) \geq \left(\sum_{k=1}^{\infty} \mu^*(E \cap A_k) \right) + \mu^*(E \setminus A)$$

$$= \mu^*(E \cap A) + \mu^*(E \setminus A)$$

$$\geq \mu^*(E)$$

Thus equality holds,and

$$\mu^*(E) = \mu^*(E \cap A) + \mu^*(E \setminus A).$$

This is true for all $(E \subseteq X)$ so A is measurable

Let X be a set, and define

$$\mu^*(E) = \begin{cases} 0 E = \phi, \\ 1 otherwise \end{cases}.$$

Then is an outer measure. The theorem applies to tell us the collection of μ^* measurable sets is a σ field on which μ^* is a measure.

Proposition: If $a_{nk} \in [0, \infty]$ for $n, k \in \mathbb{N}$ then whateve orde we addd up the extended real numbers a_{nk} we always get the same answer: in particular if you enumerate $\mathbb{N} \times \mathbb{N}$ as $(n_1, k_1), (n_2, k_2), (n_3 k_3), \ldots$ then

$$\sum_{j=1}^{\infty} a_{n_j} k_j = \sum_{n=1}^{\infty} \left(\sum_{k=1}^{\infty} a_{nk} \right).$$

Remark: all these result about series have elementry proof,but can also be deduced from results in ourt sect ion on integration (later)

Lemma: Let X be a set, let f be a set of subsets *of* X with $\phi \in f$. with suppose that $\mu : f \to [0,\infty]$ is a function, and that $\mu^*(\phi) = 0$

Define

$$\mu^*(A) = \inf\left\{\sum_{j=1}^{\infty}\mu(E_n) : E_1, E_2 ... \in fandA \subseteq \bigcup_{n=1}^{\infty} E_n\right\}.$$

then μ^* is an outer measure on X

Remarks: We will use proposition in this proof.note that μ^*(A)maybe $+\infty$ for two reasons: either there are no sets $E_1, E_2, E_3 ...$ in with

$$A \subseteq \bigcup_{n=1}^{\infty} E_n$$

in which $\mu^*(A) = \inf(\phi) = +\infty$ case or it could be that

$$\mu^*(A) = \inf\{\infty\}.$$

but certainly

$$\mu^*(A) \in [0,\infty] \forall A \subseteq X.$$

Proof: Certainly $\mu^* : p(X) \to [0,\infty]$. To see that $\mu^*(\phi) = 0$ take

$$E_1 = E_2 = ... = \phi \in f.$$

Then

$$\phi \subseteq \bigcup_{n=1}^{\infty} E_n,$$

so

$$\mu^*(\phi) \le \sum_{n=1}^{\infty}\mu(E_n) = 0.$$

Thus $\mu^*(\phi) = 0$.

To check, monotonicity, suppose that set $A \subseteq F$. Set

$$S_n = \left\{\sum_{n=1}^{\infty}\mu(E_n) := E_1 .... \in \varepsilon, A \subseteq \bigcup_{n=1}^{\infty} E_n\right\}.$$

$$S_F = \left\{ \sum_{n=1}^{\infty} \mu(E_n) := E_1 .... \in \varepsilon, F \subseteq \bigcup_{n=1}^{\infty} E_n \right\}.$$

since $A \subseteq F$, we have $S_F \subseteq S_A$

$\therefore$ (S_A) inf$\leq$ inf(S_F),

i.e. $\mu^*(A) \leq \mu^*(F)$.

To prove countable subadditivity suppose that

$$A \subseteq \sum_{n=1}^{\infty} A_n$$

where $A, A_1, A_2, ...$ are in $P(X)$. We must show

$$\mu^*(A) \leq \sum_{n=1}^{\infty} \mu^*(A_n).$$

Two case

1. If $\sum_{n=1}^{\infty} \mu^*(A_n) = \infty$, the result is trivial;

2. Otherwise $\sum_{n=1}^{\infty} \mu^*(A_n) < \infty$, so also each $\mu^*(A_n) < \infty$.

Let $\varepsilon > 0$ Because $\mu^*(A_n)$ is finite, we can choose $E_{nk} \in \varepsilon(k \in \mathbb{N})$ s.t.

$$A \subseteq \sum_{k=1}^{\infty} E_{nk}$$

and

$$\left(\sum_{k=1}^{\infty} \mu(E_{nk}) \right) < \mu^*(A_n) + \frac{\varepsilon}{2^n}$$

[by definition of].

Then

$$A \subseteq \bigcup_{n=1}^{\infty} A_n \subseteq \bigcup_{n=1}^{\infty} \bigcup_{k=1}^{\infty} E_{nk}.$$

N.B.

$$\sum_{n=1}^{\infty} \left(\sum_{k=1}^{\infty} \mu(E_{nk}) \right) \leq \sum_{n=1}^{\infty} \left(\mu^*(A_n) + \frac{\varepsilon}{2^n} \right)$$

$$= \left(\sum_{n=1}^{\infty} \mu^*(A_n) \right) = \varepsilon$$

$$< \infty .$$

Thus, the countable sum $\mu^*(E_{nk})$ of all the is equal to

$$\sum_{n=1}^{\infty}\sum_{k=1}^{\infty} \mu(E_{nk}) \le \left(\sum_{k=1}^{\infty} \mu(E_{nk}) \right) + \varepsilon.$$

Thus

$$\mu^*(A) \le \sum_{k=1}^{\infty} \mu^*(A_n) + \varepsilon$$

since this is true $\forall \varepsilon > 0$, we obtain

$$\mu^*(A) \le \sum_{k=1}^{\infty} \mu^*(A_n).$$

Extension Theorem

Theorem: Let R be a ring of a subsets of a set X and let $\mu : R \to [0, \infty]$ be a measure on R. Then has an extension to a measure μ defined on a σ-field $f \supseteq R$

Proof: Define

$$\mu^*(E) \left\{ \sum_{k=1}^{\infty} \mu(A_n) : A_1, A_2, \ldots \in R, E \subseteq \bigcup_{n=1}^{\infty} A_n \right\}.$$

Then the conditions of the Lemma are satisified and so μ^* is an outer measure on X Let be the set of μ^* measurable subsets of let μ be the resrtriction μ^* to f of then f is a σ – field and μ is a measure on f.

It remains to show that $R \subseteq f$ and that $\mu(A) = \mu(A) \forall A \in R$.

Let $A \in$ **R**, then

$$A \subseteq A \cup \phi \cup \phi \cup \ldots$$

so $\mu^*(A) \le \mu(A)$.

But if $A_1, A_2, \ldots \in R$ and

$$A \subseteq \bigcup_{n=1}^{\infty} A_n$$

Then, since μ is measure on R, $\mu(A) \le \sum_{n=1}^{\infty} \mu(A_n)$. This holds for all such sequences A_n so $\mu(A) \le \mu^*(A)$. Thus $\mu(A) = \mu^*(A)$.

Finally, we show A *is* μ^* measurable. Let $E \subseteq X$. We must show that

$$\mu^*(E) = \mu^*(E \cap A) + \mu^*(E \setminus A)$$

as usual we only need to prove that LHS $\ge$ RHS. Set

$$S_E = \left\{ \sum_{n=1}^{\infty} \mu(A_n) : A_1, A_2, \ldots. \in R, E \subseteq \bigcup_{n=1}^{\infty} A_n \right\}$$

Since $\mu^*(E) = \inf S_E$ it is enough to show that $\mu^*(E \cap A) + \mu^*(E \setminus A)$ is a lower bound for $A_1, A_2, \ldots. \in R$ with S_E. Let $\subseteq \bigcup_{n=1}^{\infty} A_n$. Then

$$E \cap A \subseteq \bigcup_{n=1}^{\infty} (A_n \cap A).$$

$$E \setminus A \subseteq \bigcup_{n=1}^{\infty} (A_n \cap A).$$

so

$$\mu^*(E \cap A) \le \sum_{n=1}^{\infty} \mu(A_n \cap A),$$

$$\mu^*(E \setminus A) \le \sum_{n=1}^{\infty} \mu(A_n \setminus A)$$

Addding give

$$\mu^*(E \cap A) + \mu^*(E \setminus A) \le \sum_{n=1}^{\infty} \mu(A_n \cap A) + \mu(A_n \cap A))$$

$$= \sum_{n=1}^{\infty} \mu(A_n)$$

because μ a measure on R.

This shows $\mu^*(E \cap A) + \mu^*(E \setminus A)$ is a lower bound for S_E. Thus

$$\mu^*(E \cap A) + \mu^*(E \setminus A) \le \mu^*(E).$$

This concludes the proof.

Proof: The sets of extended real numbers

$$S_1 = \left\{ \sum_{n=1}^{\infty} \mu(A_n) : A_1, A_2, ... \in R, E \setminus A \subseteq \bigcup_{n=1}^{\infty} A_n \right\}$$

and

$$S_2 = \left\{ \sum_{n=1}^{\infty} \mu(A_n) : A_1, A_2, ... \in f, E \subseteq \bigcup_{n=1}^{\infty} A_n \right\}$$

are in fact the same.The fact that $S_2 \subseteq S_1$ is obvious. Now Suppose $x \in S_1$. Then there are $A_1, A_2, ... \in R$ with

$$E \subseteq \bigcup_{n=1}^{\infty} A_n \text{ and } x = \sum_{n=1}^{\infty} \mu(A_n).$$

But we may write

$$A_n = \bigcup_{k=1}^{m_n} B_{nk}$$

with B_{nk} in p, Then

$$\mu(A_n) = \sum_{k=1}^{m_n} \mu(B_{nk,}) \text{ all } n,$$

$$\sum_{n=1}^{\infty} \mu(A_n) = \sum_{n=1}^{\infty} \sum_{k=1}^{m_n} \mu(\mathrm{B}_{nk}).$$

Since

$$E \subseteq \bigcup_{n=1}^{\infty} \bigcup_{k=1}^{m_n} B_{nk}.$$

A countable union of sets in *f* deduce that $x \in S_2$. This proves our claim

Corollary: In the particaular case of lebbasgue outer measure λ^*, we find

$$\lambda^*(E)\inf = \left\{ \sum_{n=1}^{\infty} (b_n - a_n) : E \subseteq \bigcup_{n=1}^{\infty} (a_n, b_n] \right\}.$$

Proposition:

(i) for $a, b \in \mathbb{R}$ with a<b

$$\lambda([a, b]) = \lambda([a, b)) = \lambda((a, b)),$$

$$= \lambda((a, b]) = b - a.$$

2. For $E \subseteq \mathbb{R}$

With our measure $\mu : \varepsilon \to [0,\infty]$ defined by

$$\mu\left(\bigcup_{K=1}^{\infty} a_{k,b_k}\right) = \sum_{k=1}^{n} (b_k - a_k)$$

We obtain an outer measure μ^* as usual. This particular outer measure is called *Lebesgue* outer measure on $\mathbb{R}$ and we shall denote it by λ^*. The set, *f*, of subsets of $\mathbb{R}$ which λ^* are measurable, is known as collection of Lebesgue measurable subsets of $\mathbb{R}$. Our theorem tells us that *f* is a σ - field contaning and that restriction of λ^*. to, *f*, is a measure. We shall denote this restriction by and call it *Lebesgue measure*. Note that *f* contains the σ - field genertated by ε ,which is precisely *B*, the collection of all Borel sets in $\mathbb{R}$.

With this notation we have

$$p \subseteq \varepsilon \subseteq B \subseteq p \subseteq p(\mathbb{R}),$$

and we can measuer the size of all Borel sets using λ.

Definition: Let (*X*, *P*) be measurabl space. Then we say that the sets *E* in *P* are *P*-measurahle sets (or measurahle if the σ - field is unambiguous). In paricular, if is a metric space, by default we take $P = \{E \subseteq X \mid E \text{ is Borel set}\}$ In this case the measurable sets are the Borel sets in *X*, so We say such sets are Borel measurable. When we have an outer μ^* on a sets in *X* we already have defined μ^* measurable. This coinncides with the *P* measurable sets when $P = [E \subseteq X \backslash E$ is μ^* - meaurable]. In the particular case of Lebesgue outer measure on $\mathbb{R}$, λ^*, the λ^*-measurable sets are the Lebesgue mesurable sets (or lebesgue sets) in $\mathbb{R}$.

Every Borel set is Lebesgue measurable. The converse is false but tricky. It turn out that the cardinality of *B* is the same as that of $\mathbb{R}$ whereas the cardinality of is true same as that of $P(\mathbb{R})$

[Recall: Two sets *A*, *B* are said to have the same cardinality if there is a bijection *f*: A $\to$ *B*. For every set *X*,*X* and *P*(*X*) have different cardinalites.]

Is every subset of $\mathbb{R}$ Lebesgue measurable? The answer is no, but most sets you meet are,

Proposition: Let ℓ be a semi-ring of subsets of a set *X* and set $R = R(\ell)$.

Suppose $\mu : R \to [0,\infty]$ is measure, and we μ^* form as usual by

$$\mu^*(E) = \inf\left\{\sum_{n=1}^{\infty} \mu(A_n) : A_1, A_2 .... \in R, E \subseteq \bigcup_{n=1}^{\infty} A_n\right\}.$$

Then in fact we also have

$$\mu^*(E) = \inf\left\{\sum_{n=1}^{\infty}\mu(A_n) : A_1, A_2.... \in \ell, E \subseteq \bigcup_{n=1}^{\infty} A_n\right\}.$$

$$\lambda^*(E) = \inf\left\{\sum_{n=1}^{\infty}(b_n = a_n) : E \subseteq \bigcup_{n=1}^{\infty}(a_n, b_n)\right\}$$

$$= \inf\left\{\sum_{n=1}^{\infty}(b_n - a_n) : E \subseteq \bigcup_{n=1}^{\infty}[a_n, b_n]\right\}$$

Lemma: Let μ^*: $P(X) \to [0, \infty]$ be an outer measure on a set X.

(i) If $A \subseteq X$ and $\mu^*(A) = 0$ then A is μ^*-measurable.

(ii) If B is μ^* measurable. and $A \subseteq B$ and $\mu^*(B) = 0$ then A is μ^* measurable.

(iii) If $A \subseteq C \subseteq B$ with A, $B\,\mu^*$ measurable and $\mu^*(B \backslash A\backslash) = 0$ then C is μ^*-A is measurable

Proof:

(i) Given $\mu^*(A) = 0$ let $E \subseteq X$. Then

$$\mu^*(E) \leq (E \setminus A) + (E \cap A),$$

$$\leq \mu^*(E)\mu^*(A),$$

$$= \mu^*(E)$$

Thus equality holds and A is μ^* measurable.

(ii) $A \subseteq B$, and $\mu^*(B)$ then. μ^* (A)\=0. SoA is μ^* -measurable.

(iii) A, $B\,\mu^*$, measurable $A \subseteq C \subseteq B$, $\mu^*(B\backslash A) = 0$.

Then $C \backslash A \subseteq B \backslash A$, so $C \backslash A$ is μ^*-measurable. But $C = (C \backslash A) \cup A$, which is μ^*-measurable.

In particular, given that λ (Cantor set) = 0 we deduce that every subset of the cantort set is Lebesgue measurble. But the Cantor set has the same cardinality as $\mathbb{R}$ (FACT)and so the collection of subsets of the Cantor sets has the same cardinality as $p(\mathbb{R})$, It follows that the collection of Lebesgue sets has the same cardeinality as $p(\mathbb{R})$ It follwos that the collection of subsets of the Cantor set has thesame cardinality as $P()$. It follows that the collection of Lebesgue sets has the same cardinality as $p(\mathbb{R})$[using Schroder-Bernstein theorem]. Not all of ther subsets of the Cantor set are Borel sets (there are too many!). So if you restrict λ to $\mathfrak{R}$ you have an incomple te measure i.e. therte are sts $A \subseteq B$ with

$$B \in \mathfrak{R},\ \lambda(B) = 0 \text{ but } A \in \mathfrak{R}.$$

Notation: For any $E \subseteq R$, we write

$$E + x = [y + x: y \in E] \text{ (for any } x \text{ in } \mathbb{R})$$

Preposition:

(i) If $E = (a, b]$ then

$$\lambda(E) = \lambda(E + x)(x \in \mathbb{R}).$$

(ii) If $E \subseteq \mathbb{R}$

$$\lambda(E + x) = \lambda^*(E)(x \in \mathbb{R}).$$

(iii) If $A \subseteq \mathbb{R}$ is Lebesgue measurable, then so is $A + x(x \in \mathbb{R})$, and

$$\lambda(A + x) = \lambda(A).$$

(iv) If $A \subseteq \mathbb{R}$ is Lebesgue measurable, then so is $-A$ and

$$\lambda(-A) = \lambda(A).$$

Proof:

(i) $E + x = (a + x, b + x]$ and $(b + x) - (a + x) = b - a$.

(ii) This is an easy exercise, based on (i) and the definition of λ^*

(iii) $\lambda^*(E - x) = \lambda^*(E) = \lambda^*(E \cap A) + \lambda^*(E \setminus A)$

$$E \cap (A + x) = ((E - x) \cap A) + x,$$

$$E \setminus (A + x) = ((E - x) \setminus A) + x.$$

$$\lambda^*(E) = \lambda^*(E - x) = \lambda^*((E - x) \cap A) + \lambda^*((E - x) \setminus A)$$

because A is measurable

$$= \lambda^*(E \cap (A + x)) + \lambda^* E \setminus A + x).$$

This is true for alol EX, and so $A + x$, is Lebesgue measurabel. by (ii)

$$\lambda(A + x) = \lambda(A).$$

(iv) This is similar to (i) – (iii) using $-(a, b) = (-b, -a)$, and definition of in λ^* terms of open intervals.

So Lebesgue measure is translation invariant.

A NON-MEASURABLE SET

To find a non-measurable set we will need the axiom of choice.

Example: Working in [0, 1] we define an equivalence relation – by $x - y$ if $x - y \in \mathbb{Q}$. This equivalence relation splits [0, 1] up into equivalence classes. Note for $x \in [0, 1]$, the equivalence class of x is $(\mathbb{Q} + x) \cap [0, 1]$.

We now form a set E by choosing one element from each equivalence class, [This uses the axiom of choice.] We show that E is not Lebesgue measurable.

Set. $S = \mathbb{Q} \cap [-1, 1]$, a countable set. Then

$$[0,1] \subseteq \bigcup_{q \in S} (E + q) \subseteq [-1,2]$$

Because $E \subseteq [0,1]$, and $\forall y \in [0,1]$, $\exists x \in E$ with $(y - x) \in \mathbb{Q}$ since x,y are in [0,1],we have then $(y - x) \in S$, and $y \in E(y - x)$.

The collection of sets $[E + q: q \in S]$ is countable, and pairwise disjoint.

(Reason: if $p \neq q \in S$, then suppose we had

$$x \in (E + p) \cap (E + q).$$

Then we would have $(x - p) \in E$ and $(x - q) \in E$.This is impossible because E has *j* only one member ofr reach equivalenvce class.)

By assumption, E is measurable, and so $E + q$ is measurable $\forall\ q \in S$. We have

$$1 = \lambda([0,1]) \leq \lambda\left(\bigcup_{q \in S}(E+q)\right) \sum_{q \in S} \lambda(E+q) = \sum_{q \in S} \lambda(E).$$

$$\lambda\left(\bigcup_{q \in S}(E+q)\right) \leq \lambda([-1,2]) = 3$$

$$\sum_{q \in S} \lambda(E) \leq 3.$$

This is impossible: $\sum_{q \in S} \lambda(E)$ must be either 0 or ∞.

REGULARITY OF LEBESGUE MEASURE

$$\lambda^*(E) = \inf\left\{\sum_{n=1}^{\infty}(b_n - a_n) : E \subseteq \bigcup_{n=1}^{\infty}(a_n, b_n)\right\}.$$

Lemma: For any $E \subseteq \mathbb{R}, \lambda^*(E) = \{\lambda \inf(U) : U \text{ open}, U \supseteq E\}$.

Proof: Certainly $\lambda^*(E) \leq \inf\{\lambda(U) : U \text{ open}, U \supseteq E\}$.

Suppose $x > \lambda^*(E)$. Then $\exists\ (a_1, b_1)(a_2, b_2)$.... such that

$$E \subseteq \bigcup_{n=1}^{\infty}(a_n, b_n) \text{ and } \sum_{n=1}^{\infty}(b_n - a_n) < x.$$

set $V = \bigcup_{n=1}^{\infty}(a_n, b_n)$. Then $\lambda(V) \leq \sum_{n=1}^{\infty}(b_n - a_n) < x$.

Also $E \subseteq V$ and so we have

$$\inf\{\lambda(U) : U \text{ open } U \supseteq E\}x.$$

This is true $\forall x > \lambda^*(E)$ and so the result follows.

N.B. We do not claim to be able to obtain

$$\lambda^*(U \setminus E) < \varepsilon$$

Theorem: Let E be a Lebesgue measurable subset of $\mathbb{R}$. Then

(i) $\forall \varepsilon > 0$ there is an open set $U \subseteq \mathbb{R}$ with

$$\lambda(U \setminus E) < \varepsilon \text{ and } U \subseteq E$$

(ii) $\forall \varepsilon > 0 \exists$ a closed set $F \subseteq E$ with

$$\lambda(E \setminus F) < \varepsilon.$$

Proof:

(i) Set $E_n = E \cap [-n, n]$ for $n \in \mathbb{N}$. Then E_n is measurable, and $\lambda(E_n) < \infty$. Choose open sets V_n with $E_n \subseteq V_n$ satisfying $\lambda(V_n) < \lambda^*(E_n) + \frac{\varepsilon}{2^n}$. We can do this because $\lambda^*(E_n) < \infty$. Then set $U = \bigcup_{n=1}^{\infty} V_n$. Certainly U is open, and

$$U \setminus E = \bigcup_{n=1}^{\infty}(V_n \setminus E) \subseteq \bigcup_{n=1}^{\infty}(V_n \setminus E_n).$$

Thus

$$\lambda U \setminus E \leq \sum_{n=1}^{\infty} \lambda(V_n \setminus E_n) < \sum_{n=1}^{\infty} \frac{\varepsilon}{2_n} = \varepsilon$$

(because the sets E_n are measurable proving (i).

(ii) To prove (ii), we see (i), and set $A = E^c$ by (i)$\exists$ an open set $U \supseteq A$ with $\lambda(U \setminus A) < \varepsilon$.

$$\text{Set } F = \mathbb{R} \setminus U. \text{ Then } F \subseteq \mathbb{R} \setminus A = E \text{ and}$$

$$(E \setminus F) = (\mathbb{R} \setminus F) \setminus (\mathbb{R} \setminus E) = U \setminus A$$

so

$$\lambda(E \setminus F) = \lambda(U \setminus A) < \varepsilon.$$

The fact that you can approximate lebesgue measurable sets from the inside by closed sets, and from the outside by open sets, is describe by saying that Lebesgue measure λ is regular.

If (X, f) is a measurable space and $E \subseteq F$ then

$$[F \cap E; F \in f] = \{F \in f : F \subseteq E\}$$

is a $\sigma-$ field on E denoted by $f|_E$ (non- standard notation) or f_E.

If $\mu : f \to [0,\infty]$ is ameasure, then $\mu | f_E$ is ameasure on f_E. we will sometime denote this measure by μ_E

For any Lebesgue measurable set $E \subseteq R$,we have $\sigma-$field consisting of all Lebesgue measurable sets contained in E, and we can restrict Lebesgue measure to this. In particular we can work on any interval [a,b].

Chapter 6

Measure and Probability Theory

The evolution of probability theory was based more on intuition rather than mathematical axioms during its early development. In 1933, A. N. Kolmogorov provided an axiomatic basis for probability theory and it is now the universally accepted model. There are certain 'non commutative' versions that have their origins in quantum mechanics, see for instance K. R. Parthasarathy, that are generalizations of the Kolmogorov Model. We shall however use exclusively Kolmogorov's framework.

The basic intuition in probability theory is the notion of randomness. There are experiments whose results are not predicatable and can be determined only after performing it and then observing the outcome. The simplest familiar examples are, the tossing of a fair coin, or the throwing of a balanced die. In the first experiment the result could be either a head or a tail and the throwing of a die could result in a score of any integer from 1 through 6. These are experiments with only a finite number of alternate outcomes. It is not difficult to imagine experiments that have countably or even uncountably many alternatives as possible outcomes.

Abstractly then, there is a space Ω of all possible outcomes and each individual outcome is represented as a point ω in that space Ω. Subsets of Ω are called events and each of them corresponds to a collection of outcomes. If the outcome ω is in the subset A, then the event A is said to have occurred. For example in the case of a die the set $A = \{1, 3, 5\} \subset \Omega$ corresponds to the event 'an odd number shows up'. With this terminology it is clear that union of sets corresponds to 'or', intersection to 'and', and complementation to 'negation'.

One would expect that probabilities should be associated with each outcome and there should be a 'Probability Function' $f(\omega)$ which is the probabilty that ω occurs. In the case of coin tossing we may expect $\Omega = \{H, T\}$ and

$$f(T) = f(H) = \frac{1}{2}.$$

Or in the case of a die

$$f(1) = f(2) = ... = f(6) = \frac{1}{6}.$$

Since 'Probability' is normalized so that certainty corresponds to a Probability of 1, one expects

$$\sum_{\omega \in \Omega} f(\omega) = 1.$$

If Ω is uncountable this is a mess. There is no reasonable way of adding up an uncountable set of numbers each of which is 0. This suggests that it may not be possible to start with probabilities associated with individual outcomes and build a meaningful theory. The next best thing is to start with the notion that probabilities are already defined for events. In such a case, $P(A)$ is defined for a class B of subsets $A \subset \Omega$. The question that arises naturally is what should B be and what properties should $P(\cdot)$ defined on B have? It is natural to demand that the class B of sets for which probabilities are to be defined satisfy the following properties:

The whole space Ω and the empty set Φ are in B. For any two sets A and B in B, the sets $A \cup B$ and $A \cap B$ are again in B. If $A \in B$, then the complement A^c is again in B. Any class of sets satisfying these properties is called a *field*.

Definition: A 'probability' or more precisely 'a finitely additive probability measure' is a nonnegative set function $P(\cdot)$ defined for sets $A \in B$ that satisfies the following properties:

$$P(A) \geq 0 \text{ for all } A \in B,$$

$$P(\Omega) = 1 \text{ and } P(\Phi) = 0.$$

If $A \in B$ and $B \in B$ are disjoint then

$$P(A \cup B) = P(A) + P(B).$$

In particular

$$P(A^c) = 1 - P(A)$$

for all $A \in B$.

A condition which is some what more technical, but important from a mathematical viewpoint is that of countable additivity. The class B, in addition to being a field is assumed to be closed under countable union (or equivalently, countable intersection); i.e. if $A_n \in B$ for every n, then $A = \bigcup_n A_n \in B$. Such a class is called a σ-field. The 'probability' itself is presumed to be defined on a σ-field B.

Definition: A set function P defined on a σ-field is called a 'countably additive probability measure' if in addition to satsfying equations, it satisfies the following countable additivity property: for any sequence of pairwise disjoint sets A_n with $A = \bigcup_n A_n$

$$P(A) = \sum_n P(A_n).$$

Definition: A probability measure P on a field F is said to be countably additive on F, if for any sequence $A_n \in F$ with $A_n \downarrow \Phi$, we have $P(A_n) \downarrow 0$.

Definition: The σ-field in the above exercise is called the σ-field generated by F.

CONSTRUCTION OF MEASURES

The following theorem is important for the construction of countably additive probability measures. A detailed proof of this theorem, as well as other results on measure and integration, can be found or in any one of the many texts on real variables. In an effort to be complete we will sketch the standard proof.

Caratheodory Extension Theorem

Theorem: Any countably additive probabilty measure P on a field F extends uniquely as a countably additive probability measure to the σ-field B generated by F.

Proof: The proof proceeds along the following steps:

Step 1. Define an object P^* called the *outer measure* for all sets A.

$$P^*(A) = \inf_{\cup_j A_j \supset A} \sum_j P(A_j)$$

where the infimum is taken over all countable collections $\{A_j\}$ of sets from F that cover A. Without loss of generality we can assume that $\{A_j\}$ are disjoint. (Replace A_j by $\left(\bigcap_{i=1}^{j-1} A_i^c\right) \cap A_j$).

Step 2. Show that P^* has the following properties:

1. P^* is countably sub-additive, i.e.

$$P^*\left(\bigcup_j A_j\right) \le \sum_j P^* A_j$$

2. For $A \in F$, $P^*(A) \le P(A)$. (Trivial)

3. For $A \in F$, $P^*(A) \ge P(A)$. (Need to use the countable additivity of P on F)

Step 3. Define a set E to be measurable if

$$P^*(A) \ge P^*(A \cap E) + P^*(A \cap E^c)$$

holds for all sets A, and establish the following properties for the class M of measurable sets. The class of measurable sets M is a σ-field and P^* is a countably additive measure on it.

Step 4. Finally show that $M \supset F$. This implies that $M \supset B$ and P^* is an extension of P from F to B.

Uniqueness is quite simple. Let P_1 and P_2 be two countably additive probability measures on a σ-field B that agree on a field $F \subset B$. Let us define $A = \{A: P_1(A) = P_2(A)\}$. Then A is a monotone class i.e., if $A_n \in A$ is increasing (decreasing) then $\bigcup_n A_n \left(\bigcap_n A_n\right) \in A$. Uniqueness will follow from the following fact left as an excercise.

Exercise: The smallest monotone class generated by a field is the same as the σ-field generated by the field.

It now follows that A must contain the σ-field generated by F and that proves uniqueness.

The extension Theorem does not quite solve the problem of constructing countably additive probability measures. It reduces it to constructing them on fields. The following theorem is important in the theory of Lebesgue integrals and is very useful for the construction of countably additive probability measures on the real line. The proof will again be only sketched. The natural σ-field on which to define a probability measure on the line is the Borel σ-field. This is defined as the smallest σ-field containing all intervals and includes in particular all open sets.

Let us consider the class of subsets of the real numbers, $I = \{I_{a,b}: -\infty \leq a < b \leq \infty\}$ where $I_{a,b} = \{x: a < x \leq b\}$ if $b < \infty$, and $I_{a,\infty} = \{x: a < x < \infty\}$. In other words I is the collection of intervals that are left-open and right-closed. The class of sets that are finite disjoint unions of members of I is a field F, if the empty set is added to the class. If we are given a function $F(x)$ on the real line which is nondecreasing, continuous from the right and satisfies

$$\lim_{x \to -\infty} F(x) = 0 \text{ and } \lim_{x \to \infty} F(x) = 1,$$

we can define a finitely additive probability measure P by first defining .

$$P(I_{a,b}) = F(b) - F(a)$$

for intervals and then extending it to F by defining it as the sum for disjoint unions from I. Let us note that the Borel σ-field B on the real line is the σ-field generated by F.

Lebesgue

Theorem: P is countably additive on F if and only if $F(x)$ is a right continuous function of x. Therefore for each right continuous nondecreasing function $F(x)$ with $F(-\infty) = 0$ and

$F(\infty) = 1$ there is a unique probability measure P on the Borel subsets of the line, such that $F(x) = P(I_{-\infty,x})$. Conversely every countably additive probability measure P on the Borel subsets of the line comes from some F. The correspondence between P and F is one-to-one.

Proof: The only difficult part is to establish the countable additivity of P on F from the right continuity of $F(\cdot)$. Let $A_j \in F$ and $A_j \downarrow \Phi$, the empty set.

Let us assume, $P(A_j) \geq \delta > 0$, for all j and then establish a contradiction.

Step 1. We take a large interval $[-\ell, \ell]$ and replace A_j by $Bj = A_j \cap [-\ell, \ell]$. Since $|P(A_j) - P(B_j)| \leq 1 - F(\ell) + F(-\ell)$, we can make the choice of ℓ large enough that $P(B_j) \geq \frac{\delta}{2}$.

In other words we can assumes without loss of generality that $P(A_j) \geq \frac{\delta}{2}$ and $A_j \subset [-\ell, \ell]$ for some fixed $\ell < \infty$.

Step 2. If

$$A_j = \bigcup_{i=1}^{k_j} I_{a_{j,i}, b_{j,i}}$$

use the right continuity of F to replace A_j by B_j which is again a union of left open right closed intervals with the same right end points, but with left end points moved ever so slightly to the right. Achieve this in such a way that

$$P(A_j - B_j) \leq \frac{\delta}{10.2^j}$$

for all j.

Step 3. Define C_j to be the closure of B_j, obtained by adding to it the left end points of the intervals making up B_j. Let $E_j = \bigcap_{i=1}^{j} B_i$ and $D_j = \bigcap_{i=1}^{j} C_i$. Then, (i) the sequence D_j of sets is decreasing, (ii) each D_j is a closed bounded set, (iii) since $A_j \supset D_j$ and $A_j \downarrow \Phi$, it follows that $D_j \downarrow \Phi$. Because $D_j \supset E_j$ and $P(E_j) \geq \frac{\delta}{2} - \sum_j P(A_j - B_j) \geq \frac{4}{10\delta}$, each D_j is nonempty and this violates the finite intersection property that every decreasing sequence of bounded nonempty closed sets on the real line has a nonempty intersection, i.e. has at least one common point.

The rest of the proof is left as an exercise.

The function F is called the distribution function corresponding to the probability measure P.

Example: Suppose $x_1, x_2, \ldots, x_n, \ldots$ is a sequence of points and we have probabilities p_n at these points then for the discrete measure

$$P(A) = \sum_{n: x_n \in A} p_n$$

we have the distribution function

$$F(x) = \sum_{n: x_n \leq x} p_n$$

that only increases by jumps, the jump at x_n being p_n. The points $\{x_n\}$ themselves can be discrete like integers or dense like the rationals.

Example: If $f(x)$ is a nonnegative integrable function with integral 1, i.e. $\int_{-\infty}^{\infty} f(y)dy = 1$ then $F(x) = \int_{-\infty}^{x} f(y)dy$ is a distribution function which is continuous. In this case f is the density of the measure P and can be calculated as $f(x) = F'(x)$.

There are (messy) examples of F that are continuous, but do not come from any density. More on this later.

Exercise: Let us try to construct the Lebesgue measure on the rationals $Q \subset [0, 1]$. We would like to have

$$P[I_{a,b}] = b - a$$

for all rational $0 \leq a \leq b \leq 1$. Show that it is impossible by showing that $P[\{q\}] = 0$ for the set $\{q\}$ containing the single rational q while $P[Q] = P\left[\bigcup_{q \in Q} \{q\}\right] = 1$. Where does the earlier proof break down?

Once we have a countably additive probability measure P on a space (Ω, Σ), we will call the triple (Ω, Σ, P) a probabilty space.

INTEGRATION

An important notion is that of a random variable or a measurable function.

Definition: A random variable or measurable function is map $f: \Omega \to R$, i.e. a real valued function $f(\omega)$ on Ω such that for every Borel set $B \subset R, f^{-1}(B) = \{\omega : f(\omega) \in B\}$ is a measurable subset of Ω, i.e $f^{-1}(B) \in \Sigma$.

A function that is measurable and satisfies $|f(\omega)| \leq M$ all $\omega \in \Omega$ for some finite M is called a bounded measurable function.

The following statements are the essential steps in developing an integration theory. Details can be found in any book on real variables.

1. If $A \in \Sigma$, the indicator function A, defined as

$$1_A(\omega) = \begin{cases} 1 & \text{if } \omega \in A \\ 0 & \text{if } \omega \notin A \end{cases}$$

is bounded and measurable.

2. Sums, products, limits, compositions and reasonable elementary operations like min and max performed on measurable functions lead to measurable functions.

3. If $\{A_j : 1 \le j \le n\}$ is a finite disjoint partition of into measurable sets, the function $f(\omega) = \sum_j c_j 1_{A_j}(\omega)$ is a measurable function and is referred to as a 'simple' function.

4. Any bounded measurable function f is a uniform limit of simple functions. To see this, if f is bounded by M, divide $[-M, M]$ into n subintervals I_j of length $\frac{2M}{n}$ with midpoints c_j. Let

$$A_j = f^{-1}(I_j) = \{\omega : f(\omega) \in I_j\}$$

and

$$f_n = \sum_{j=1}^{n} c_j 1_{A_j}.$$

Clearly f_n is simple, $\sup_\omega |f_n(\omega) - f(\omega)| - \frac{M}{n}$, and we are done.

5. For simple functions $f = \sum c_j 1_{A_j}$ the integral $\int f(\omega)dP$ is defined to be $\sum_j c_j P(A_j)$. It enjoys the following properties:

   (a) If f and g are simple, so is any linear combination $af + bg$ for real constants a and b and

   $$\int (af + bg)dP = a\int fdP + b\int gdP.$$

   (b) If f is simple so is $|f|$ and $\left|\int fdP\right| \le \int |f|\, dP \le \sup_\omega |f(\omega)|$

6. If f_n is a sequence of simple functions converging to f uniformly, then $a_n = \int f_n dP$ is a Cauchy sequence of real numbers and therefore has a limit a as $n \to \infty$. The integral $\int fdP$ of f is defined to be this limit a. One can verify that a depends only on f and not on the sequence f_n chosen to approximate f.

7. Now the integral is defined for all bounded measurable functions and enjoys the following properties.

   (a) If f and g are bounded measurable functions and a, b are real constants then the linear combination $af+bg$ is again a bounded measurable function, and

   $$\int (af+bg)dP = a\int fdP + b\int gdP.$$

   (b) If f is a bounded measurable function so is $|f|$ and $\left|\int fdP\right| \le \int |f|\,dP \le \sup_\omega |f(\omega)|$.

   (c) In fact a slightly stronger inequality is true. For any bounded measurable f,

   $$\int |f|\,dP \le P\big(\{\omega : |f(\omega)| > 0\}\big)\sup_\omega |f(\omega)|$$

   (d) If f is a bounded measurable function and A is a measurable set one defines

   $$\int_A f(\omega)dP = \int 1_A(\omega) f(\omega)dP$$

   and we can write for any measurable set A,

   $$\int fdP = \int_A fdP + \int_{A^c} fdP$$

In addition to uniform convergence there are other weaker notions of convergence.

Definition: A sequence f_n functions is said to converge to a function f everywhere or pointwise if

$$\lim_{n\to\infty} f_n(\omega) = f(\omega)$$

for every $\omega \in \Omega$.

In dealing with sequences of functions on a space that has a measure defined on it, often it does not matter if the sequence fails to converge on a set of points that is insignificant. For example if we are dealing with the Lebesgue measure on the interval [0, 1] and $f_n(x) = x_n$ then $f_n(x) \to 0$ for all x except $x = 1$. A single point, being an interval of length 0 should be insignificant for the Lebesgue measure.

Definition: A sequence f_n of measurable functions is said to converge to a measurable function f almost everywhere or almost surely (usually abbreviated as a.e.) if there exists a measurable set N with $P(N) = 0$ such that

$$\lim_{n\to\infty} f_n(\omega) = f(\omega)$$

for every $\omega \in N^c$.

Note that almost everywhere convergence is always relative to a probability measure.

Another notion of convergence is the following:

Definition: A sequence f_n of measurable functions is said to converge to a measurable function f 'in measure' or 'in probability' if

$$\lim_{n\to\infty} P[\omega : | f_n(\omega) - f(\omega) | \geq \varepsilon] = 0$$

for every $\varepsilon > 0$.

Let us examine these notions in the context of indicator functions of sets $f_n(\omega) = 1_{A_n}(\omega)$. As soon as $A \neq B$, $\sup_\omega | 1_A(\omega) - 1_B(\omega) | = 1$, so that uniform convergence never really takes place. On the other hand one can verify that $1_{A_n}(\omega) \to 1_A(\omega)$ for every ω if and only if the two sets

$$A_n = \bigcap_n \bigcup_{m\geq n} A_m$$

and

$$\liminf_n A_n = \bigcup_n \bigcap_{m\geq n} A_m$$

both coincide with A. Finally $1_{A_n}(\omega) \to 1_A(\omega)$ in measure if and only if

$$\lim_{n\to\infty} P(A_n \Delta A) = 0$$

where for any two sets A and B the symmetric difference $A\Delta B$ is defined as $A\Delta B = (A \cap B^c) \cup (A^c \cap B) = A \cup B \cap (A \cap B)^c$. It is the set of points that belong to either set but not to both. For instance $1_{A_n} \to \omega$ in measure if and only if $P(A_n) \to 0$.

Lemma: If $f_n \to f$ almost everywhere then $f_n \to f$ in measure.

Proof: $f_n \to f$ outside N is equivalent to

$$\bigcap_n \bigcup_{m\geq n} [\omega : | f_m(\omega) - f(\omega) | \geq \varepsilon] \subset N$$

for every $\varepsilon > 0$. In particular by countable additivity

$$P[\omega : | f_n(\omega) - f(\omega) | \geq \varepsilon] \leq P\left[\bigcup_{m\geq n} [\omega : | f_m(\omega) - f(\omega) | \geq \varepsilon]\right] \to 0$$

as $n\to\infty$ and we are done.

Bounded Convergence Theorem

Theorem: If the sequence $\{f_n\}$ of measurable functions is uniformly bounded and if $f_n \to f$ in measure as $n \to \infty$, then $\lim_{n\to\infty} \int f_n dP = \int f dP$.

Proof: Since

$$\left|\int f_n dP - \int f dP\right| = \left|\int (f_n - f) dP\right| \leq \int |f_n - f| \, dP$$

we need only prove that if $f_n \to 0$ in measure and $|f_n| \leq M$ then $\int |f_n| \, dP \to 0$. To see this

$$\int |f_n| \, dP = \int_{|f_n| \leq \varepsilon} |f_n| \, dP + \int_{|f_n| > \varepsilon} |f_n| \, dP \leq \varepsilon + MP[\omega : |f_n(\omega)| > \varepsilon]$$

and taking limits

$$\limsup_{n \to \infty} \int |f_n| \, dP \leq \varepsilon$$

and since $\varepsilon > 0$ is arbitrary we are done.

The bounded convergence theorem is the essence of countable additivity. Let us look at the example of $f_n(x) = x^n$ on $0 \leq x \leq 1$ with Lebesgue measure. Clearly $f_n(x) \to 0$ a.e. and therefore in measure. While the convergence is not uniform, $0 \leq x^n \leq 1$ for all n and x and so the bounded convergence theorem applies. In fact

$$\int_0^1 x^n dx = \frac{1}{n+1} \to 0.$$

However if we replace x^n by nx^n, $f_n(x)$ still goes to 0 a.e., but the sequence is no longer uniformly bounded and the integral does not go to 0.

We now proceed to define integrals of nonnegative measurable functions.

Definition: If f is a nonnegative measurable function we define

$$\int f dP = \left\{\sup \int g dP : g \text{ bounded}, 0 \leq g \leq f\right\}.$$

Fatou's Lemma

Theorem: If for each $n \geq 1$, $f_n \geq 0$ is measurable and $f_n \to f$ in measure as $n \to \infty$ then

$$\int f dP \leq \liminf_{n \to \infty} \int f_n dP.$$

Proof: Suppose g is bounded and satisfies $0 \leq g \leq f$. Then the sequence $h_n = f_n \wedge g$ is uniformly bounded and

$$h_n \to h = f \wedge g = g.$$

Therefore, by the bounded convergence theorem,

$$\int g dP = \liminf_{n \to \infty} \int h_n dP$$

Since $\int h_n dP \le \int f_n dP$ for every n it follows that

$$\int g dP \le \liminf_{n\to\infty} \int f_n dP.$$

As g satisfying $0 \le g \le f$ is arbitrary and we are done.

Monotone Convergence Theorem

Corollary: If for a sequence $\{f_n\}$ of nonnegative functions, we have $f_n \uparrow f$ monotonically then

$$\int f_n dP \to \int f dP \text{ as } n \to \infty.$$

Proof: Obviously $\int f_n dP \le \int f dP$ and the other half follows from Fatou's lemma.

Now we try to define integrals of arbitrary measurable functions. A nonnegative measurable function is said to be *integrable* if $\int f dP < \infty$. A measurable function f is said to be integrable if $|f|$ is integrable and we define $\int f dP = \int f^+ dP - \int f^- dP$ where $f^+ = f \vee 0$ and $f^- = -f \wedge 0$ are the positive and negative parts of f. The integral has the following properties.

1. It is linear. If f and g are integrable so is $af + bg$ for any two real constants and $\int (af + bg) dP = a\int f dP + b\int g dP$.
2. $\left|\int f dP\right| \le \int |f| dP$ for every integrable f.
3. If $f = 0$ except on a set N of measure 0, then f is integrable and $\int f dP = 0$. In particular if $f = g$ almost everywhere then $\int f dP = \int g dP$.

Theorem: If $\phi(x)$ is a convex function of x and $f(\omega)$ and $\phi(f(\omega))$ are integrable then

$$\int \phi(f(\omega)) dP \ge \phi\left(\int f(\omega) dP\right)$$

Proof: We have seen the inequlity already for $\phi(x) = |x|$. The proof is quite simple. We note that any convex function can be represented as the supremum of a collection of affine linear functions.

$$\phi(x) = \sup_{(a,b)\in E} \{ax + b\}$$

It is clear that if $(a, b) \in E$, then $a f(\omega) + b \le (f(\omega))$ and on integration this yields

$am + b \leq E[(f(\omega))]$ where $m = E[f(\omega)]$. Since this is true for every $(a, b) \in E$, in view of the represenattion, our theorem follows.

The Dominated Convergence Theorem

Theorem: If for some sequence $\{f_n\}$ of measurable functions we have $f_n \to f$ in measure and $|f_n(\omega)| \leq g(\omega)$ for all n and ω for some integrable function g, then $\int f_n dP \to \int f dP$ as $n \to \infty$.

Proof: $g + f_n$ and $g - f_n$ are nonnegative and converge in measure to $g + f$ and $g - f$ respectively. By Fatou's lemma

$$\liminf_{n\to\infty} \int (g + f_n) dP \leq \int (g + f) dP$$

Since $\int g dP$ is finite we can subtract it from both sides and get

$$\liminf_{n\to\infty} \int f_n dP \geq \int f dP.$$

Working the same way with $g - f_n$ yields

$$\liminf_{n\to\infty} \int f_n dP \leq \int f dP$$

and we are done.

Exercise: Take the unit interval with the Lebesgue measure and define $f_n(x) = n^\alpha 1_{\left[0, \frac{1}{n}\right]}(x)$. Clearly $f_n(x) \to 0$ for $x \neq 0$. On the other hand $\int f_n(x) dx = n^{\alpha - 1} \to 0$ if and only if $\alpha < 1$. What is $g(x) = \sup_n f_n(x)$ and when is g integrable ?

If $h(\omega) = f(\omega) + i g(\omega)$ is a complex valued measurable function with real and imaginary parts $f(\omega)$ and $g(\omega)$ that are integrable we define

$$\int h(\omega) dP = \int f(\omega) dP + i \int g(\omega) dP$$

TRANSFORMATIONS

A measurable space (Ω, B) is a set Ω together with a σ-field B of subsets of Ω.

Definition: Given two measurable spaces (Ω_1, B_1) and (Ω_2, B_2), a mapping or a transformation from $T: \Omega_1 \to 2$, i.e. a function $\omega_2 = T(\omega_1)$ that assigns for each point $\omega_1 \in \Omega_1$ a point $\omega_2 = T(\omega_1) \in \Omega_2$, is said to be measurable if for every measurable set $A \in B_2$, the inverse image

$$T^{-1}(A) = \{\omega_1 : T(\omega_1) \in A\} \in B_1.$$

Theorem: If f: $\Omega_2 \rightarrow R$ is a real valued measurable function on Ω_2, then $g(\omega_1) = f(T(\omega_1))$ is a measurable real valued function on (Ω_1, B_1). Moreover g is integrable with respect to P if and only if f is integrable with respect to Q, and

$$\int_{\Omega_2} f(\omega_2)dQ = \int_{\Omega_1} g(\omega_1)dP$$

Proof: If $f(\omega_2) = 1_A(\omega_2)$ is the indicator function of a set $A \in B_2$, the claim in equation is in fact the definition of measurability and the induced measure. We see, by linearity, that the claim extends easily from indicator functions to simple functions. By uniform limits, the claim can now be extended to bounded measurable functions. Monotone limits then extend it to nonnegative functions. By considering the positive and negative parts separately we are done.

A measurable trnasformation is just a generalization of the concept of a random variable. We can either think of a random variable as special case of a measurable transformation where the target space is the real line or think of a measurable transformation as a random variable with values in an arbitrary target space.

The induced measure $Q = PT^{-1}$ is called the *distribution* of the random variable F under P. In particular, if T takes real values, Q is a probability distribution on R.

PRODUCT SPACES

Given two sets Ω_1 and Ω_2 the Cartesian product $\Omega = \Omega_1 \times \Omega_2$ is the set of pairs (ω_1, ω_2) with $\omega_1 \in \Omega_1$ and $\omega_2 \in \Omega_2$. If Ω_1 and Ω_2 come with σ-fields B_1 and B_2 respectively, we can define a natural σ-field B on Ω as the σ-field generated by sets (measurable rectangles) of the form $A_1 \times A_2$ with $A_1 \in B_1$ and $A_2 \in B_2$. This σ-field will be called the product σ-field.

Definition: The product -field B is the σ-field generaated by the field F.

Given two probability measures P_1 and P_2 on (Ω_1, B_1) and (Ω_2, B_2) respectively we try to define on the product space (Ω, B) a probability measure P by defining for a measurable rectangle $A = A_1 \times A_2$

$$P(A_1 \times A_2) = P_1(A_1) \times P_2(A_2)$$

and extending it to the field F of finite disjoint unions of measurable rectangles as the obvious sum.

Lemma: The measure P is countably additive on the field F.

Proof: For any set $E \in F$ let us define the section E_{ω_2} as

$$E_{\omega_2} = \{\omega_1 : (\omega_1, \omega_2) \in E\}.$$

Then $P_1(E_{\omega_2})$ is a measurable function of ω_2 (is in fact a simple function) and

$$P(E) = \int_{\Omega_2} P_1(E_{\omega_2}) dP_2.$$

Now let $E_n \in F \downarrow \Phi$, the empty set. Then it is easy to verify that E_{n,ω_2} defined by

$$E_{n,\omega_2} = \{\omega_1 : (\omega_1, \omega_2) \in E_n\}$$

satisfies $E_{n,\omega_2} \downarrow \Phi$ for each $\omega_2 \in \Omega_2$. From the countable additivity of P_1 we conclude that $P_1(E_{n,\omega_2}) \to 0$ for each $\omega_2 \in \Omega_2$ and since, $0 \leq P_1(E_{n,\omega_2}) \leq 1$ for $n \geq 1$, it follows from equation and the bounded convergence theorem that

$$P(E_n) = \int_{\Omega_2} P_1(E_{n,\omega_2}) dP_2 \to 0$$

establishing the countable additivity of P on F.

By an application of the Caratheodory extension theorem we conclude that P extends uniquely as a countably additive measure to the σ-field B (product σ-field) generated by F. We will call this the Product Measure P.

Corollary: For any $A \in B$ if we denote by A_{ω_1} and A_{ω_2} the respective sections

$$A_{\omega_1} = \{\omega_2 : (\omega_1, \omega_2) \in A\}$$

and

$$A_{\omega_2} = \{\omega_1 : (\omega_1, \omega_2) \in A\}$$

then the functions $P_1(A_{\omega_2})$ and $P_2(A_{\omega_1})$ are measurable and

$$P(A) = \int P_1(A_{\omega_2}) dP_2 = \int P_2(A_{\omega_1}) \, dP_1.$$

In particular for a measurable set A, $P(A) = 0$ if and only if for almost all ω_1 with respect to P_1, the sections A_{ω_1} have measure 0 or equivalently for almost all ω_2 with respect to P_2, the sections A_{ω_2} have measure 0.

Proof: The assertion is clearly valid if A is rectangle of the form $A_1 \times A_2$ with $A_1 \in B_1$ and $A_2 \in B_2$. If $A \in F$, then it is a finite disjoint union of such rectangles and the assertion is extended to such a set by simple addition. Clearly, by the monotone convergence theorem, the class of sets for which the assertion is valid is a monotone class and since it contains the field F it also contains the σ-field B generated by the field F.

Warning. It is possible that a set A may not be *measurable* with respect to the product σ-field, but nevertheless the sections A_{ω_1} and A_{ω_2} are all measurable, $P_2(A_{\omega_1})$ and $P_1(A_{\omega_2})$ are measurable functions, but

$$\int P_1(A_{\omega_2})dP_2 \neq \int P_2(A_{\omega_1})\,dP_1.$$

In fact there is a rather nasty example where $P_1(A_{\omega_2})$ is identically 1 whereas $P_2(A_{\omega_1})$ is identically 0. The next result concerns the equality of the double integral, (i.e. the integral with respect to the product measure) and the repeated integrals in any order.

Fubini's Theorem

Theorem: Let $f(\omega) = f(\omega_1, \omega_2)$ be a measurable function of ω on (Ω, B). Then f can be considered as a function of ω_2 for each fixed ω_1 or the other way around. The functions $g_{\omega_1}(\cdot)$ and $h_{\omega_2}(\cdot)$ defined respectively on Ω_2 and Ω_1 by

$$g_{\omega_1}(\omega_2) = h_{\omega_2}(\omega_1) = f(\omega_1, \omega_2)$$

are measurable for each ω_1 and ω_2. If f is integrable then the functions $g_{\omega_1}(\omega_2)$ and $h_{\omega_2}(\omega_1)$ are integrable for almost all ω_1 and ω_2 respectively. Their integrals

$$G(\omega_1) = \int_{\Omega_2} g_{\omega_1}(\omega_2)dP_2$$

and

$$H(\omega_1) = \int_{\Omega_1} h_{\omega_2}(\omega_2)dP_1$$

are measurable, finite almost everywhere and integrable with repect to P_1 and P_2 respectively. Finally

$$\int f(\omega_1, \omega_2)dP = \int g(\omega_1)dP_1 = \int H(\omega_2)dP_2$$

Conversely for a nonnegative measurable function f if either G are H, which are always measurable, has a finite integral so does the other and f is integrable with its integral being equal to either of the repeated integrals, namely integrals of G and H.

Proof: The proof follows the standard pattern. It is a restatement of the earlier corollary if f is the indicator function of a measurable set A. By linearity it is true for simple functions and by passing to uniform limits, it is true for bounded measurable functions f. By monotone limits it is true for nonnegative functions and finally by taking the positive and negative parts seperately it is true for any arbitrary integrable function f.

Warning. The following could happen. f is a measurable function that takes both positive and negative values that is not integrable. Both the repeated integrals exist and are unequal. The example is not hard.

DISTRIBUTIONS AND EXPECTATIONS

Let us recall that a triplet (Ω, B, P) is a Probability Space, if Ω is a set, B is a σ-field of subsets of Ω and P is a (countably additive) probability measure on B. A random variable X is a real valued measurable function on (Ω, B). Given such a function X it induces a probability distribution on the Borel subsets of the line $\alpha = PX^{-1}$. The distribution function $F(x)$ corresponding to is obviously

$$F(x) = ((-\infty, x]) = P[\omega: X(\omega) \leq x].$$

The measure α is called the distibution of X and $F(x)$ is called the distribution function of X. If g is a measurable function of the real variable x, then $Y(\omega) = g(X(\omega))$ is again a random variable and its distribution $\beta = PY^{-1}$ can be obtained as $\beta = \alpha g^{-1}$ from α. The *Expectation* or *mean* of a random variable is defined if it is integrable and

$$E[X] = E^P[X] = \int X(\omega)dP.$$

By the change of variables formula it can be obtained directly from α as

$$E[X] = \int x d\alpha.$$

Here we are taking advantage of the fact that on the real line x is a very special real valued function. The value of the integral in this context is referred to as the expectation or mean of α. Of course it exists if and only if

$$\int |x| d\alpha < \infty$$

and

$$\left|\int x d\alpha\right| \leq \int |x| d\alpha.$$

Similarly

$$E(g(X)) = \int g(X(\omega))dP = \int g(x)d\alpha$$

and anything concerning X can be calculated from. The statement X is a random variable with distribution α has to be interpreted in the sense that somewhere in the background there is a Probability Space and a random variable X on it, which has for its distribution.

Usually only α matters and the underlying (Ω, B, P) never emerges from the background and in a pinch we can always say Ω is the real line, B are the Borel sets, P is nothing but α and the random variable $X(x) = x$.

Some other related quantities are

$$Var(X) = \sigma^2(X) = E[X^2] - [E[X]]^2.$$

$Var(X)$ is called the variance of X.

MEASURE THEORY PROBABILITY

The Riemann Integral the integral that is treated in freshman calculus applies to continuous functions. It can be extended a little beyond the class of continuous functions, but not very far. It can be used to define the lengths, areas, and volumes of sets in R, R^2, and R^3, provided those sets are reasonably nice, in particular not too irregularly shaped. In R^2, the Riemann Integral defines the area under the graph of a function by dividing the x-axis into a collection of small intervals.

On each of these small intervals, two rectangles are erected: one lies entirely inside the area under the graph of the function, while the other rectangle lies entirely outside the graph. The function is Riemann integrable (and its integral equals the area under its graph) if, by making the intervals sufficiently small, it is possible to make the sum of the areas of the outside rectangles arbitrarily close to the sum of the areas of the inside rectangles.

Measure theory provides a way to extend our notions of length, area, volume etc. to a much larger class of sets than can be treated using the Riemann Integral. It also provides a way to extend the Riemann Integral to Lebesgue integrable functions, a much larger class of functions than the continuous functions.

The fundamental conceptual difference between the Riemann and Lebesgue integrals is the way in which the partitioning is done. As noted above, the Riemann Integral partitions the domain of the function into small intervals.

By contrast, the Lebesgue Integral partitions the *range* of the function into small intervals, then considers the set of points in the domain on which the value of the function falls into one of these intervals. Let f: [0, 1] $\rightarrow R$.

Given an interval $[a, b) \subseteq R$, $f^{-1}([a, b))$ may be a very messy set. However, as long as we can assign a "length" or "measure" $\mu(f^{-1}([a, b)))$ to this set, we know that the contribution of this set to the integral of f should be between $a\mu\ (f^{-1}([a, b)))$ and $b\mu\ (f^{-1}([a, b]))$. By making the partition of the range finer and finer, we can determine the integral of the function.

Clearly, the key to extending the Lebesgue Integral to as wide a class of functions as possible is to define the notion of "measure" on as wide a class of sets as possible. In an ideal world, we would be able to define the measure of *every* set; if we could do this, we could then define the Lebesgue integral of *every* function. Unfortunately, as we shall see, it is not possible to define a measure with nice properties on every subset of R.

Measure theory is thus a *second best* exercise. We try to extend the notion of measure from our intuitive notions of length, area and volume to as large a class of *measurable* subsets of R, R^2, and R^3 as possible. In order to be able to make use of measures and integrals, we need to know that the class of measurable sets is closed under certain types of operations. If we can assign a sensible notion of measure to a set, we ought to be able to assign a sensible notion to its complement.

Probability and statistics focus on questions about convergence of sequences of random variables. In order to talk about convergence, we need to be able to assign measures to countable unions and countable intersections of measurable sets. Thus, we would like the collection of measurable sets to be a σ-algebra:

Definition: A *measure space* is a triple (Ω, B, μ), where

1. Ω is a set
2. B is a σ-algebra of subsets of Ω, i.e.
   (a) $B \subset 2^{\Omega}$, i.e. B is a collection of subsets of Ω
   (b) $\phi, \Omega \in B$
   (c) $B_n \in B,\ n \in N \Rightarrow \bigcup_{n \in N} \in B$
   (d) $B \in B \Rightarrow \backslash B \in B$
3. μ is a nonnegative, countably additive set function on B, i.e.
   (a) $\mu: B \to R^+ \cup \{\infty\}$
   (b) $B_n \in B,\ n \in N,\ B_n \cap B_m = \phi$ if $n \neq m \Rightarrow \mu\left(\bigcup_{n \in N} B\right) = \sum_{n \in N} \mu(B_n)$ and $\mu(\phi) = 0$.

Remark: The definition of a σ-algebra is closely related to the properties of open sets in a metric space. Recall that the collection of open sets is closed under (1) arbitrary unions and (2) finite intersections; by contrast, a σ-algebra is closed under (1) countable unions and (2) countable intersections. Notice also that σ-algebras are closed under complements; the complement of an open set is closed, and generally not open, so closure under taking complements is not a property of the collection of open sets. The analogy between the properties of a σ-algebra and the properties of open sets in a metric space will be very useful in developing the Lebesgue integral. Recall that a function $f: X \to Y$ is continuous if and only if $f^{-1}(U)$ is open in X for every open set U in Y.

Recall from the earlier discussion that the Lebesgue integral of a function f is defined by partitioning the range of the fuction f into small intervals, and summing up numbers of the form $a\mu\,(f^{-1}([a, b)))$; thus, we will need to know that $f^{-1}([a, b)) \in B$. We will see in a while that a function f: $(\Omega, B, \mu) \to (\Omega', B')$ is said to be measurable if $f^{-1}(B) \in B$ for every $B' \in B'$. Thus, there is a close analogy between measurable functions and continuous functions.

As you know from calculus, continuous functions on a closed interval can be integrated using the so-called Riemann integral; the Lebesgue integral extends the Riemann integral to all bounded measurable functions (and many unbounded measurable functions).

Remark: Countable additivity implies $\mu(\phi) = 0$ provided there is some set B with $\mu(B) < \infty$; thus, the requirement $\mu(\phi) = 0$ is imposed to rule out the pathological case in which $\mu(B) = \infty$ for all $B \in B$.

Remark: If we have a finite collection $B_1,..., B_k \in B$ with $B_n \cap B_m = \phi$ if $n \neq m$, we can write $B_n = \phi$ for $n > k$, and obtain $\mu(B_1 \cup ... \cup B_k) = \mu\left(\bigcup_{n \in N} B_n\right) = \sum_{n \in N} \mu(B_n) = \sum_{n=1}^{k} \mu(B_k) + \sum_{n=k+1}^{\infty} \mu(\phi) = \sum_{n=1}^{k} \mu(B_k)$, so the measure is additive over finite collections of disjoint measurable sets.

Example: Suppose that we are given a partition $\{\Omega_\lambda : \lambda \in \Lambda\}$ of a set Ω, i.e. $\Omega = \bigcup_{\lambda \in \Lambda} \Omega_\lambda$ and $\lambda \neq \lambda' \Rightarrow \Omega_\lambda \bigcap \Omega_{\lambda'} \phi$. Then we can form a σ-algebra as follows: Let

$$B_\Lambda = \left\{\bigcup_{\lambda \in C} \Omega_\lambda : C \subseteq \Lambda\right\}$$

In other words, B_Λ is the collection of all subsets of Ω which can be formed by taking unions of partition sets. B_Λ is closed under complements, as well as arbitrary (not just countable) unions and intersections. Suppose the partition is finite, i.e. Λ is finite, say $\Lambda = \{1,..., n\}$. Then B_Λ is finite; it has exactly 2^n elements, each corresponding to a subset of Λ. Suppose now that Λ is countably infinite; since every subset $C \subseteq \Lambda$ determines a different set $B \in B_\Lambda$, B_Λ is uncountable. Suppose finally that Λ is uncountable. For concreteness, let $\Omega = R$, $\Lambda = R$, and $\Omega_\lambda = \{\lambda\}$, i.e. each set in the partition consists of a single real number. There are many $\tilde{A}$-algebras containing this partition. As before, we can take

$$B_\Lambda = \left\{\bigcup_{\lambda \in C} \Omega_\lambda : C \subseteq \Lambda\right\} = \{C: C \subseteq R\} = 2^R$$

This is a perfectly good σ-algebra; however, as we indicated above (and as we shall show) it is not possible to define a measure with nice properties on this σ-algebra. The smallest σ-algebra containing the partition is

$$B0 = \{C: C \subseteq R,\ C \text{ countable or } R \setminus C \text{ countable}\}$$

This σ-algebra is too small to allow us to develop probability theory. We want a σ-algebra on R which lies between B_0 and B_Λ.

Definition: The Borel *σ-algebra* on R is the smallest σ-algebra containing all open sets in R. In other words, it is the σ-algebra

$$B = \cap \{C: C \text{ is a } \sigma\text{-algebra}, U \text{ open} \Rightarrow U \in C\}$$

i.e. it is the intersection of the class of all σ-algebras that contain all the open sets in R. A set is called *Borel* if it belongs to B.

Remark: Obviously, if U is an open set in R, then $U \in B$. If C is a closed set in R, then $R \setminus C$ is open, so $R \setminus C \in B$, so $C \in B$; thus, every closed set is a Borel set. Every countable intersection of open sets, and every countable union of closed sets, is Borel.

Every countable set is Borel (exercise), and every set whose complement is countable is Borel. If $\{U_{nm}: n, m \in N\}$ is a collection of open sets, then $\bigcup_n \left(\bigcap_m U_{nm}\right)$ is a Borel set. Thus, Borel sets can be quite complicated.

The most important example of a measure space is the Lebesgue measure space, which comes in two flavors. The first flavor is (R, B, μ), where B is the Borel σ-algebra and μ is Lebesgue measure, the measure defined in the following theorem; the second flavor is (R, C, μ), where C is the σ-algebra defined in the proof sketch of the following theorem.

Theorem: If B is the Borel σ-algebra on R, there is a unique measure μ (called Lebesgue measure) defined on B such that $\mu((a, b)) = b - a$ provided that $b > a$. For every Borel set B,

$$\begin{aligned}\mu(B) &= \sup\{\mu(K): K \text{ compact}, K \subseteq B\}\\ &= \inf\{\mu(U): U \text{ open}, U \supseteq B\}\end{aligned}$$

The proof works by gradually extending μ from the open intervals to the Borel σ-algebra. First, one shows one can extend μ to bounded open sets; it follows that one can extend it to compact sets. Then let $C_n = \{C \subset [-n, n]: \sup\{\mu(K): K \text{ compact}, K \subset C\} = \inf\{\mu(U): U \text{ open}, C \subset U\}\}$, and $C = \{C \subset \mathrm{R}: C \cap [-n, n] \in C_n \text{ for all } n\}$. If $C \in C$, we define $\mu(C) = \sup\{\mu(K): K \text{ compact}, K \subset C\}$. One can verify that C is a σ-algebra containing every open set, hence $C \supset B$.

Definition: A measure space (Ω, B, μ) is *complete* if $B\ \mu\ B, A \subseteq B, \mu(B) = 0 \Rightarrow A \in B$. The *completion* of a measure space (Ω, B, μ) is the measure space $(\Omega, \overline{B}, \overline{\mu})$, where

$$\overline{B} = \{B \subseteq \Omega : \exists_{D, D \in B} C \subseteq B \subseteq D, \mu(D \setminus C) = 0\}$$

and

$$\overline{\mu}(B) = \sup\{\mu(C) : C \in B, C \subseteq B\} = \inf\{\mu(D) : D \in B, B \subseteq D\}$$

for $B \in \overline{B}$.

It is easy to verify that the Lebesgue measure space (R, C, μ) is complete, and is the completion of (R, B, μ), where B is the Borel σ-algebra.

Definition: Suppose μ is a measure on the Lebesgue σ-algebra C on R. μ is *translation-invariant* if, for all $x \in R$ and all $C \in C$, $\mu(C + x) = \mu(C)$.

Theorem: The Lebesgue measure space (R, C, μ) is translation-invariant.

The theorem follows readily from the construction of Lebesgue measure, since translation doesn't change the length of intervals.

Observe if $x \in R$, $\mu(\{x\}) \leq \mu((x - \varepsilon/2, x + \varepsilon/2)) = \varepsilon$ for every positive ε, so $\mu(\{x\}) = 0$.

As we have already indicated, it is not possible to extend Lebesgue measure to every subset of R, at least in the conventional set-theoretic foundations of mathematics.

Theorem: There is a set $D \subset R$ which is not Lebesgue measurable.

Proof: We actually prove the following stronger statement: there is no translation-invariant measure μ defined on all subsets of R such that $0 < \mu([0, 1]) < \infty$.

Let μ be a translation-invariant measure defined on all subsets of R. Define an equivalence relation $\sim$ on R by

$$x \sim y \Leftrightarrow x - y \in Q$$

Let D be formed by choosing exactly one element from each equivalence class, and such that all the chosen elements lie in $[0, 1)$. Thus,

$$\forall_{x \in R} \exists_{d \in D} d - x \in Q$$

$$\forall_{d_1, d_2 \in D} d_1 - d_2 \notin Q$$

Given $x, y \in [0, 1)$, define

$$x +' y = \begin{cases} x + y & \text{if } x + y \in [0,1) \\ x + y - 1 & \text{if } x + y \in [1,2) \end{cases}$$

The operation +' is *addition modulo one*. It is easy to check that, given any $C \subseteq [0, 1)$ and $y \in [0, 1)$,

$$\mu(C +' y) = \mu(C)$$

i.e. μ is translation-invariant with respect to translation using the operation +'. Then

$$[0, 1) = \bigcup_{q \in Q \cap [0,1)} D +' q$$

so

$$\mu([0, 1)) = \sum_{q \in Q \cap [0,1)} \mu(D +' q) = \sum_{q \in Q \cap [0,1)} \mu(D)$$

Then either $\mu(D) = 0$, in which case $\mu([0, 1)) = 0$, so $\mu([0, 1]) = 0$, or $\mu(D) > 0$, in which case $\mu([0, 1)]) \geq \mu([0, 1)) = \infty$.

Definition: A measure space (Ω, B, μ) is a *probability space* if $\mu(\Omega) = 1$.

From now on, we will restrict attention to probability spaces.

Example: By abuse of notation, let C denote the collection of Lebesgue measurable sets which are subsets of $[0, 1]$, and let μ be the restriction of

Lebesgue measure to this σ-algebra. $([0, 1], C, \mu)$ is called the Lebesgue probability space.

Theorem: Let (Ω, B, μ) be a probability space. Suppose $B_1 \supseteq B_2 \supseteq \ldots \supseteq B_n \supseteq \ldots$ with $B_n \in B$ for all n. Then

$$\lim_{n\to\infty} \mu(B_n) = \mu\left(\bigcap_{n\in N} B_n\right)$$

In particular, if $\bigcap_{n\in N} B_n = \phi$, then

$$\lim_{n\to\infty} \mu(B_n) = 0$$

Proof: Let $C_m = B_m \setminus B_{m+1}$. Then $C_n \cap C_m = \phi$ if $n \neq m$, and

$$B_1 = \left(\bigcap_{n\in N} B_n\right) \cup \left(\bigcup_{m\in N} C_m\right)$$

so

$$\sum_{m\in N} \mu(C_m) = \mu(B_1) - \mu\left(\bigcap_{m\in N} B_n\right) \leq \mu(B_1) < \infty$$

so $\sum_{m\in N} \mu(C_m)$ converges and hence

$$\sum_{m=n}^{\infty} \mu(C_m) \to 0 \text{ as } n \to \infty$$

$$\mu(B_n) = \mu\left(\bigcap_{m\in N} B_m\right) + \sum_{m=n}^{\infty} \mu(C_m)$$

$$\to \mu\left(\bigcap_{m\in N} B_m\right)$$

We now turn to the definition of a *random variable*. We think of Ω as the set of all possible states of the world tomorrow. Exactly one state will occur tomorrow. Tomorrow, we will be able to observe some function of the state which occurred. We do not know today which state will occur, and hence what value of the function will occur. However, we do know the probability that any given state will occur, and we know the mapping from possible states into values of the function, so we know the probability that the function will take on any given possible value. Thus, viewed from today, the value the function will take on tomorrow is a random variable; we know its probability distribution, but not its exact value. Tomorrow, we will observe the value which is realized.

Definition: Let (Ω, B, P) be a probability space. A random variable on Ω is a function $X: \Omega \to R \cup \{-\infty, \infty\}$ satisfying

$$P(\{\omega \in \Omega: X(\omega) \in \{-\infty, \infty\}) = 0$$

which is measurable, i.e. for every Borel set $B \subset R$, $X^{-1}(B) \in B$.

Observe the analogy beteen the definition of a measurable function and the characterization of a continuous function in terms of open sets: f is continuous if and only if $f^{-1}(U)$ is open for every open set U.

Lemma: Let (Ω, B, P) be a probability space. A function $X: \Omega \to \mathrm{R} \cup \{-\infty, \infty\}$ satisfying

$$P(\{\omega \in \Omega: X(\omega) \in \{-\infty, \infty\}\}) = 0$$

is a random variable if and only if for every open interval (a, b), $X^{-1}((a, b)) \in B$.

Proof: Since every open interval is a Borel set, if X is a random variable, then $X^{-1}((a, b)) \in B$.

Now suppose that $X^{-1}((a, b)) \in B$ for every open interval (a, b). Consider

$$M = \{B \subset R: X^{-1}(B) \in B\}$$

We claim that M is a σ-algebra. Note that $X^{-1}(\phi) = \phi \in B$, so $\phi \in M$. $X^{-1}(R) = \Omega\backslash X^{-1}(\{-\infty, \infty\})$. Since $P(X^{-1}(\{-\infty, \infty\}))$ is zero (and hence defined), $X^{-1}(\{-\infty, \infty\}) \in B$; since B is a σ-algebra, it is closed under complements, so X–1 (R) Ω B, and RΩ M.

If $B \in M$, $X^{-1}(B) \in B$, so $X^{-1}(R\backslash B) = X^{-1}(R)\backslash X^{-1}(B) \in B$ because $X^{-1}(R) \in B$ and $X^{-1}(B) \in B$. Therefore, $R\backslash B \in M$, so M is closed under complements.

Now suppose $B_n \in M$, $n \in N$. Then $X^{-1}(B_n) \in B$ for all n, so

$$X^{-1}\left(\bigcup_{n\in N} B_n\right) = \bigcup_{n\in N} X^{-1}(B_n) \in B$$

and hence M is closed under countable unions.

Therefore, M is a Ω-algebra. Since every open set is a countable union of open intervals, M contains every open set; since the Borel Ω-algebra is the smallest Ω-algebra containing the open sets in R, M contains every Borel set, so X is measurable, and hence a random variable.

Theorem: If f: $[0, 1] \to R$ is continuous, then f is a random variable on the Lebesgue probability space $([0, 1], C, \mu)$.

Proof: Since f takes values in R, $f^{-1}(\{-\infty, \infty\}) = \phi$. If (a, b) is an open interval, then since f is continuous, $f^{-1}((a, b))$ is an open subset of $[0, 1]$, so $f^{-1}((a, b)) \in C$. By Lemma, f is a random variable.

Example: A random variable f: $[0, 1] \to R$ can be discontinuous everywhere. Consider $f(x) = 1$ if x is rational, 0 if x is irrational. f is discontinuous everywhere. Suppose B is a Borel set. There are four cases to consider:

1. If $0, 1 \notin B\ f^{-1}(B) = \phi \in C$.
2. If $0 \in B,\ 1 \notin B, f^{-1}(B) = [0, 1] \backslash Q \in C$.
3. If $0 \notin B,\ 1 \in B, f^{-1}(B) = [0, 1] \cap Q \in C$.
4. If $0, 1 \in B, f^{-1}(B) = [0, 1] \in C$.

Thus, f is measurable.

In elementary probability theory, random variables are not rigorously defined. Usually, they are described only by specifying their cumulative distribution functions. Continuous and discrete distributions often seem like entirely unconnected notions, and the formulation of mixed distributions (which have both continuous and discrete parts) can be problematic. Measure theory gives us a way to deal simultaneously with continuous, discrete, and mixed distibutions in a unified way.

We shall first define the cumulative distribution function of a random variable, and establish that it satisfies the defining properties of a cumulative distribution functions, as defined in elementary probability theory. We will then show (first in examples, then in a general theorem) that given any function F satisfying the defining properties of a cumulative distribution function, there is in fact a random variable defined on the Lebesgue probability space whose cumulative distribution function is F.

Definition: Given a random variable X: $(\Omega, B, P) \to R \cup \{-\infty, \infty\}$, the *cumulative distribution function* of X is the function F: $R \to [0, 1]$ defined by

$$F(t) = P(\{\omega \in \Omega: X(\omega) \leq t\})$$

Theorem: If X is a random variable, its cumulative distribution function F satisfies the following properties:

1. $\lim_{t \to -\infty} F(t) = 0$

   this is often abbreviated as $F(-\infty) = 0$.

2. $\lim_{t \to -\infty} F(t) = 1$

   this is often abbreviated as $F(\infty) = 1$.

3. F is increasing, i.e.

   $s < t \Rightarrow F(t) \geq F(s)$

4. F is right-continuous, i.e. for all t,
$\lim_{s \searrow t} F(s) = F(t)$

Proof: We prove only right-continuity, leaving the rest as an exercise. Since F is increasing, it is enough to show that (as an exercise, think through carefully why this *is* enough)

$$\lim_{n\to\infty} F\left(t+\frac{1}{n}\right) = F(t)$$

$$F\left(t+\frac{1}{n}\right) - F(t)$$

$$= P\left(\left\{\omega \in \Omega : X(\omega) \le t+\frac{1}{n}\right\}\right) - P(\{\omega \in \Omega : X(\omega) \le t\})$$

$$= P\left(X^{-1}\left(-\infty, t+\frac{1}{n}\right]\right) - P(\{X^{-1}((-\infty, t])\})$$

$$= P\left(x^{-1}\left(t, t+\frac{1}{n}\right]\right)$$

$$\to P\left(\bigcap_{n\in N} X^{-1}\left(t, t+\frac{1}{n}\right]\right)$$

$$= P\left(X^{-1}\left(\bigcap_{n\in N} t, t+\frac{1}{n}\right]\right)$$

$$= P(X^{-1}(\phi))$$

$$= P(\phi) = 0$$

Example: The *uniform distribution* on [0, 1] is the cumulative distribution function

$$F(t) = \begin{cases} 0 & \text{if} \quad t < 0 \\ t & \text{if} \quad t \in [0,1] \\ 1 & \text{if} \quad t > 1 \end{cases}$$

Consider the random variable X defined on the Lebesgue probability space X: ([0, 1], C, μ) $\to R$, $X(t) = t$. Observe that

$$\mu(\{\omega \in [0, 1]: X(\omega) \le t\}) = \mu([0, t]) = t$$

Thus, X has the uniform distribution on [0, 1]. Notice also that F is strictly increasing and hence one-to-one on [0, 1]. Thus, $F|_{[0,1]}$ has an inverse function $\left(F|_{[0,1]}\right)^{-1}$: $[0,1] \to [0,1]$ In fact, $X = \left(F|_{[0,1]}\right)^{-1}$.

Example: The *standard normal distribution* has the cumulative distribution function

$$f(t) = \frac{1}{\sqrt{2\pi}} \int_{-\infty}^{t} e^{-x^2/2} dx$$

Notice that F is strictly increasing, hence one-to-one, and the range of F is (0, 1), so F has an inverse function X: $(0, 1) \to R$; extend X to [0, 1] by defining $X(0) = -\infty$ and $X(1) = \infty$. One can show that the inverse of a strictly increasing, continuous function is strictly increasing and continuous. Since $X|_{(0,1)}$ is continuous, it is measurable on the Lebesgue probability space. Observe that

$$\mu(\{\omega \in [0, 1]: X(\omega) \leq t\}) = \mu X^{-1}((-\infty, t]) = F(t)$$

so X has the standard normal distribution.

Example: Consider the cumulative distribution function

$$F(t) = \begin{cases} 0 & \text{if } t < -1 \\ \frac{1}{2} & \text{if } t \in [-1,1] \\ 1 & \text{if } t \geq 1 \end{cases}$$

F is the cumulative distribution function of a random variable which takes on the values -1 and 1, each with probability $\frac{1}{2}$. We define a random variable on the Lebesgue probability space by

$$X(\omega) = \begin{cases} -\infty & \text{if } \omega = 0 \\ -1 & \text{if } \omega \in \left(0, \frac{1}{2}\right) \\ 1 & \text{if } \omega \in \left[\frac{1}{2}, 1\right) \\ \infty & \text{if } \omega = 1 \end{cases}$$

Clearly, the cumulative distribution function of X is F. Notice that F is weakly but not strictly increasing, so it is not one-to-one and hence does not have an inverse function. However, notice that X satisfies the following weakening of the definition of an inverse:

$$X(\omega) = \inf \{t: F(t) \geq \omega\}$$

Theorem: Let $F: R \to [0, 1]$ be an arbitrary function satisfying conclusions. There is a random variable X defined on the Lebesgue probability space whose cumulative distribution function is F.

Proof: Let

$$X(\omega) = \inf \{t: F(t) \geq \omega\}$$

Since

$$\omega' > \omega$$

$$\Rightarrow \{t: F(t) \geq \omega'\} \subseteq \{t: F(t) \geq \omega\}$$

$$\Rightarrow \inf\{t: F(t) \geq \omega'\} \geq \inf \{t: F(t) \geq \omega\}$$

$$\Rightarrow X(\omega') e'' X(\omega)$$

X is increasing (not necessarily strictly). Since $\lim_{\sigma \to -\infty} F(s) = 0$ and $\lim_{t \to \infty} F(t) = 1$, if X^{-1} (0, 1), then there exist $s, t \in R$ such that $F(s) < \omega < F(t)$; since F is increasing, $-\infty < X < X(\omega) < \infty$. Thus,

$$\mu(\{\omega: X(\omega) \in \{-\infty, \infty\}\}) \leq \mu\{0, 1\}) = 0$$

Since X is increasing, if (a, b) is an open interval in R, $X^{-1}((a, b))$ is an interval (not necessarily open) in [0, 1], and hence $X^{-1}((a, b)) \in C$. By Lemma, X is a random variable.

$$X(\omega) \leq t \Leftrightarrow \inf \{s: F(s) \geq \omega\} \leq t$$

$$\Leftrightarrow \exists_{s \leq t} F(s) \geq \omega$$

$$\Leftrightarrow F(t) \geq \omega$$

so

$$\mu(\{\omega: X(\omega) \leq t\}) = \mu(\{\omega: \omega \leq F(t)\})$$

$$= \mu([0, F(t)])$$

$$= F(t)$$

In probability theory, the choice of a particular probability space is usually considered arbitrary and unimportant; you should choose a probability space which is convenient for the particular problem at hand, in particular one on which you can easily write down random variables with the desired joint distribution. In what follows, we will allow (Ω, B, P) to be an arbitrary probability space, not necessarily the Lebesgue probability space.

Next, we develop a notion of integration.

Definition: A simple function is a function of the form

$$f = \sum_{i=1}^{n} \alpha_i \chi_{B_i}$$

where each B_i is a measurable set in Ω and χ_{B_i} is the characteristic function of B_i:

$$\chi_{B_i}(\omega)=\begin{cases}1 & \text{if} \quad \omega\in B_i\\ 0 & \text{if} \quad \omega\notin B_i\end{cases}$$

We define the integral of f by

$$\int_\Omega f dP=\sum_{i=1}^{n}\alpha_i P(B_i)$$

A given simple function may be expressed in the form of Equation in more than one way; one can show that one gets the same value of the integral from all of these different expressions.

Given a nonnegative random variable f: $(\Omega, B, P)\to R_+\cup\{\infty\}$, we can construct a simple function which closely approximates it from below. Choose constants

$$0=\alpha_1<\alpha_2<\dots<\alpha_n<\alpha_{n+1}=\infty$$

Let $$f_n=\sum_{i=1}^{n}\alpha_i\chi_{f^{-1}}([\alpha_i,\alpha_{i+1}))$$

Thus,

$$f_n(\omega)=\alpha_i\to f(\omega)\in[\alpha_i,\omega_{i+1})$$

Because f is a random variable, $f^{-1}([\alpha_i,\omega_{i+1}))\in B$, so f_n is a simple function.

Definition: If f: $(\Omega, B, P)\to R_+\cup\{\infty\}$ is a random variable, define

$$\int_\Omega f dP=\sup\left\{\int_\Omega g dP: 0\le g\le f, g \text{ simple}\right\}$$

The value of the integral may be ∞. We say that f is integrable if $\int_\Omega f dP<\infty$.

Example: Consider the function $f(x)=\frac{1}{x}$ on the Lebesgue probability space ([0, 1], C, μ). Let

$$f_n=\sum_{i=1}^{n}i\chi_{f^{-1}([i,i+1))}=\sum_{i=1}^{n}i\chi_{\left(\left(\frac{1}{i+1},\frac{1}{i}\right]\right)}$$

f_n is a simple function and $0\le f_n\le f$.

$$\int f_n d\mu=\sum_{i=1}^{n}i\mu\left(\left(\frac{1}{i+1},\frac{1}{i}\right]\right)$$

$$=\sum_{i=1}^{n}i\left(\frac{1}{i}-\frac{1}{i+1}\right)=\sum_{i=1}^{n}\frac{1}{i+1}\to\infty \text{ as } n\to\infty$$

Thus, $\int_{[0,1]} fd\mu = \infty$.

Definition: If f: $(\Omega, B, P) \to R \cup \{-\infty, \infty\}$ is a random variable, define $f_+(\omega) = \max\{f(\omega), 0\}$, $f_-(\omega) = -\min\{f(\omega), 0\}$. Note that $f = f_+ - f_-$ and $|f| = f_+ + f_-$. We say f is *integrable* if f_+ and f_- are both integrable, and we define

$$\int_\Omega fdP = \int_\Omega f_+ dP - \int_\Omega f_- dP$$

The following theorem, which we shall not prove, shows that the Lebesgue and Riemann integrals coincide when the Riemann integral is defined.

Theorem: Suppose that f: $[a, b] \to R$ is Riemann integrable (in particular, this is the case if f is continuous). Then f is Lebesgue integrable and

$$\int_{[a,b]} fd\mu = \int_a^b f(t)dt$$

In other words, the Lebesgue integral of f equals the Riemann integral of f.

Definition: We say that a property holds *almost everywhere* (abbreviated a.e.) or *almost surely* (abbreviated a.s.) if the set of ω for which it does *not* hold is a set of measure zero.

Definition: Suppose f: $(\Omega, B, P) \to R$ is a random variable. If $B \in B$, we define

$$\int_B fd\mu = \int_\Omega f\chi_B dP$$

where χ_B is the characteristic function of B.

Theorem: Suppose f, g: $(\Omega, B, P) \to R$ are integrable, $A, B \in B$, $A \cap B = \phi$, and $c \in R$. Then

1. $\int_B CfdP = c\int_B fdP$
2. $\int_B f + gdP = \int_B fdP + \int_B gdP$
3. $f \leq g$ a.e. on $B \Rightarrow fdP \leq \int_B gdP$
4. $\int_{A\cup B} fdP = \int_A fdP + \int_B fdP$

Definition: Suppose f, g: $(\Omega, B, P) \to R$ are random variables. Say $f \sim g$ if $f(\omega) = g(\omega)$ a.e. Let $[f]$ denote the equivalence class of f under the equivalence relation ~, i.e. $[f] = \{g: (\Omega, B, P) \to R$: g is measurable, $g \sim f\}$. Define

$L^1(\Omega, B, P) = \{[f]: f: \Omega \to R$, f a random variable, f integrable$\}$

$L^2(\Omega, B, P) = [f]: f: \Omega \to R$, f a random variable, f^2 integrable$\}$

Theorem: If (Ω, B, P) is a probability space,

$$L^2(\Omega, B, P) \subseteq L^1(\Omega, B, P)$$

Definition: Given a metric space (X, d), the *completion* of (X, d) is a metric space $(\overline{X}, \overline{d})$ such that $(\overline{X}, \overline{d})$ is complete, $\overline{d}\,|_X = d$, and X is dense in $(\overline{X}, \overline{d})$. We are justified in calling $(\overline{X}, \overline{d})$ *the* completion of (X, d) because if $(\hat{X}, \hat{y})$ is another completion of (X, d), then there is an isomorphism $\phi: (\overline{X}, \overline{d}) \to (\hat{X}, \hat{y})$ such that $\phi(x) = x$ for every $x \in X$, and $\hat{d}(\phi(\overline{x}), \phi(\hat{y})) = \overline{d}(\overline{x}, \overline{y})$ every $\overline{x}, \overline{y} \in \overline{X}$.

Theorem: $L^1(\Omega, B, P)$ and $L^2(\Omega, B, P)$ are Banach spaces under the respective norms

$$\| f \|_1 = \int_\Omega | f(\omega) | \, dP$$

$$\| f \|_2 = \sqrt{\int_\Omega f^2(\omega) dP}$$

For the Lebesgue probability space, $L^1([0, 1], C, \mu)$ and $L^2([0, 1], C, \mu)$ are the completions of the normed space $C([0, 1])$, with respect to the norms $\|.\|_1$ and $\| \cdot \|_2$.

Example: Recall that $C([0, 1])$ is a Banach space with the sup norm $f = \sup\{| f(x) | : x \in [0, 1]\}$. $(C[0, 1], \| \cdot \|_1)$ and $(C[0, 1], \| \cdot \|_2)$ are normed spaces, but they are not complete in these norms, hence are not Banach spaces. To see this, let $f_n(x)$ be the function which is zero for $x \in \left[0, \frac{1}{2} - \frac{1}{n}\right]$, one for $x \in \left[\frac{1}{2} + \frac{1}{n}, 1\right]$ and linear for $x \in \left[\frac{1}{2} - \frac{1}{n}, \frac{1}{2} + \frac{1}{n}\right]$. f_n is not Cauchy with respect to $\| \cdot \|_\infty$. However, f_n *is* Cauchy and has a natural limit with respect to $\| \cdot \|_1$. Indeed, let

$$f(x) = \begin{cases} 0 & \text{if} \quad x < \frac{1}{2} \\ 1 & \text{if} \quad x \geq \frac{1}{2} \end{cases}$$

Then

$$\int_{[0,1]} | f_n - f | \, d\mu \leq \frac{1}{2} \times \frac{2}{n} = \frac{1}{n} \to 0$$

so f_n converges to f in $L^1([0, 1], C, \mu)$, and hence must be Cauchy with respect to $\| \cdot \|_1$. Notice that this limit does not belong to $C([0, 1])$.

Definition: For $f \in L^1(\Omega, B, P)$, define the *mean* or *expectation* of f by

$$E(f) = \int_\Omega f dP$$

For $f \in L^2(\Omega, B, P)$, define the variance of f by

$$\text{Var}(f) \int_{\Omega} (f(\omega) - E(f))^2 \, dP$$

Definition: If $f, g \in L^2(\Omega, B, P)$, define the *inner product* of f and g by

$$f.g = \int_{\Omega} f(\omega) g(\omega) dP$$

The properties of the inner product are closely analogous to those of the dot product of vectors in Euclidean space R^n. In particular, they determine a geometry, including lengths$\left(\| f \|_2 = \sqrt{f.f}\right)$ and angles. The most basic property of the dot product, the Cauchy-Schwarz inequality, extends to the inner product.

Cauchy-Schwarz Inequality

Theorem: If $f, g \in L^2(\Omega, B, P)$,

$$| f.g | \leq \| f \|_2 \| g \|_2$$

Definition: If $f, g \in L^2(\Omega, B, P)$, define the *covariance* of f and g by

$$\text{Covar}(f, g) = E((f - E(f))(g - E(g)))$$

Proposition: If $f, g - L^2(\Omega, B, P)$, then

$$\text{Covar}(f, g) = E(fg) - E(f)\, E(g)$$

Observe from the definition that Covar (f, g) = Covar (g, f). Thus, given $f_1,..., f_n - L^2(\Omega, B, P)$, the covariance matrix C whose (i, j) entry is $c_{ij} = \text{Covar}(f_i, f_j)$ is a symmetric matrix. Hence, there is an orthonormal basis of R^n composed of eigenvectors of C; expressed in this basis, the matrix becomes diagonal. Observe also that

$$\text{Covar}\left(\sum_{i=1}^{n} \alpha_i f_i, \sum_{j=1}^{n} \beta_j f_j,\right) = (\alpha_1, ..., \alpha_n) C \begin{pmatrix} \beta_1 \\ \vdots \\ \beta_n \end{pmatrix}$$

is a quadratic form of the type we studied earlier.

One very useful consequence of the inner product structure on L^2 is the existence of orthogonal projections. Given a linearly independent family

$f_1,..., f_n - L^2(\Omega, B, P)$, let V be the vector space spanned by $f_1,..., f_n$. Then any $g \in L^2$ can be written in a unique way as $g = \pi(g) + \omega$, where $\pi(v) = \sum_{i=1}^{n} \alpha_i f_i \in V$ and $\omega \cdot v = 0$ for all $v \in V$ (in particular, $\omega \cdot f_i = 0$ for each i). $\pi(g)$ is also characterized as the point in V closest to g. The coefficients $\alpha_1,..., \alpha_n$ are the regression coefficients of g on $f_1,..., f_n$. If $f_1,..., f_n$ are orthonormal (i.e. $f_i \cdot f_j = 1$ if $i = j$ and 0 if $i \neq j$), $\alpha_i = g \cdot f_i$ for each i.

One frequently encounters two measures living on a single probability space. For example, Lebesgue measure is one measure on the real line R; any random variable determines another measure on R by its distribution.

Definition: Let μ and v be two measures on the same measurable space (X, A), i.e. (X, A, μ) and (X, A, v) are measure spaces. We say v is *absolutely continuous with respect to* μ if

$$B \in A,\ \mu(B) = 0 \Rightarrow v(B) = 0$$

We say μ is σ-finite if there exist $B_1, B_2, \ldots \in A$ such that $X = \bigcup_{n\in N} B_n$ and $\mu(B_n) < \infty$ for all n.

Radon-Nikodym

Theorem: Suppose that μ and v are two σ-finite measures on the same measurable space (X, A). Then v is absolutely continuous with respect to μ if and only if there exists $f \in L^1(X, A, \mu)$, $f \geq 0$ almost everywhere, such that

$$v(B) = \int_B f d\mu$$

for all $B \in A$.

The Radon-Nikodym theorem tells us that if a random variable Z has a continuous distribution (in other words, the measure v determined by its cumulative distribution function F is absolutely continuous with respect to Lebesgues measure), then Z has a density function f, namely f is the Radon-Nikodym derivative of v with respect to Lebesgue measure.

We now turn to products of probability spaces (X, A, P) and (Y, B, Q). If $A \in A$ and $B \in B$, it is natural to define

$$(P \times Q)(A \times B) = P(A)Q(B)$$

However, most of the subsets of $X \times Y$ we are interested in are not of the form $A \times B$ with $A \in A$ and $B \in B$. For example, if $X = Y = [0, 1]$, the diagonal $\{(x, y): x = y\}$ is *not* of that form. However, it is possible to write the diagonal in the form

$$\bigcap_{m\in N} \bigcup_{n=1}^{i_m} A_{mn} \times B_{mn}$$

with $A_{mn} \in A$ and $B_{mn} \in B$. This suggests we should extend $P \times Q$ to the smallest σ-algebra containing $\{A \times B: A \in A, B \in B\}$.

Definition: Suppose (X, A, P) and (Y, B, Q) are probability spaces. The product probability space is $(X \times Y, A \times B, P \times Q)$, where

1. $A \times' B$ is the smallest σ-algebra containing all sets of the form $A \times B$, where $A \in A$ and $B \in B$

2. $P \times' Q$ is the unique measure on $A \times' B$ that takes the values $P(A)Q(B)$ on sets of the form $A \times B$, where $A \in A$ and $B \in B$.
3. $(X \times Y, A \times B, P \times Q)$ is the completion of $(X \times Y, A \times' B, P \times' Q)$

The existence of the product measure $P \times Q$ is proven by extending the definition of $P \times Q$ for sets of the form $A \times B$ to $A \times B$, in a manner similar to the process by which Lebesgue measure is obtained from lengths of intervals.

Fubini

Theorem: Suppose (X, A, P) and (Y, B, Q) are complete probability spaces. If $f: X \times Y \to R$ is $A \times B$ measurable and integrable, then

1. For almost all $x \in X$, the function defined by $f_x(y) = f(x, y)$ is integrable in Y;
2. For almost all $y \in Y$, the function defined by $f_y(x) = f(x, y)$ is integrable in X;
3. $\int_Y f(x, y)dQ(y)$ is an integrable function of $x \in X$;
4. $\int_X f(x, y)dP(x)$ is an integrable function of $y \in Y$;
5. $\int_X\left(\int_Y f(x,y)dQ(y)\right)dP(x) = \int_Y\left(\int_X f(x,y)dP(x)\right)dQ(y)$

$$\int_{X\times Y} f(x,y)d(P\times Q)(x,y)$$

There are many notions of convergence of functions, and results showing that the integral of a limit of functions is the limit of the integrals.

Definition: Suppose $f_n, f: (\Omega, B, P) \to R$ are random variables.

1. We say f_n converges to f in probability or in measure if

$$\forall_{\varepsilon>0}\exists_N n > N \Rightarrow P\left(\{\omega : | f_n(\omega) | > \varepsilon\}\right) < \varepsilon$$

2. We say f_n converges to f almost everywhere (abbreviated a.e.) or *almost surely* (abbreviated *a.s.*) if

$$P(\{\omega : f_n(\omega) \to f(\omega)\}) = 1$$

Theorem: If f_n converges to f almost surely, then f_n converges to f in probability.

In the lecture, I gave an example of a sequence of functions which converges in probability, but not almost surely. The example is much easier to describe with pictures than with words, so I won't give the details here. The idea is to construct a sequence of functions f_n: $[0, 1] \to R$ so that $\mu(\{x: f_n(x) \neq 0\}) \to 0$; hence f_n converges in probability to the identically zero function f. However, we can move the set on which f_n is not zero around so that, for every $x \in [0, 1]$, $f_n(x) = 1$ for infinitely many n, so almost surely, $f_n(x)$ fails to converge to $f(x)$.

Example: Although convergence almost surely is a strong notion of convergence of functions, it is not sufficient for convergence of the integrals of the functions to the integral of the limit. For example, consider f_n: $[0, 1] \to R$ defined by $f_n = n\chi_{[0, 1/n]}$, which takes the value n on $[0, 1/n]$ and 0 everywhere else. f_n converges almost surely to f, the function which is identically zero, but if P is Lebesgue measure, $[0\ 1] f_n\, dP = n \times \frac{1}{n} = 1$, while $\int_{[0,1]} fdP = 0$. Roughly speaking, f_n is converging to a function which is ∞ with probabilty 0, and in the Lebesgue integral, $0 \times \infty = 0$; some of the mass in f_n is lost in the limit operation. The next result tells us that, for nonnegative functions, this is the only thing that can go wrong, and the limit of the integrals is always at least as big as the integral of the limit function.

Fatou's Lemma

Theorem: If (Ω, B, P) is a probability space, $f_n, f: \Omega \to R^+$ *are* random variables, and f_n converges to f almost surely, then

$$\int_\Omega fdP \le \liminf_{n\to\infty} \int_\Omega f_n dP$$

In the next theorem, the functions f_n converge to f from below; this guarantees that no mass which is present in the f_n can suddenly disappear at the limit.

Monotone Convergence Theorem

Theorem: If (Ω, B, P) is a probability space, f_n: $\Omega \to R^+$ are random variables, f_n converges to f almost surely, and for all n and ω, $f_n(\omega) \le f(\omega)$. Then

$$\int_\Omega fdP = \lim_{n\to\infty} \int_\Omega f_n dP$$

Lebesgue's Dominated Convergence Theorem

Theorem: Suppose (Ω, B, P) is a probability space, g: $\Omega \to R$ is integrable, fn random variables, $|f_n| \le$ g a.s., and $f_n \to f$ a.s. Then f is integrable and

$$\int_\Omega f_n dP \to \int_\Omega fdP$$

Definition: Suppose (Ω, B, P) is a probability space. We say $\{f_n\}$ is uniformly integ-rable if

$$\forall_{\varepsilon>0} \exists_{\delta>0} \forall_{n\in N} P(B) < \delta \Rightarrow \left| \int_B f_n dP \right| < \varepsilon$$

The following theorem is a useful generalization the Dominated Convergence Theorem.

Theorem: Suppose (Ω, B, P) is a probability space. If $\{f_n\}$ is uniformly integrable and $f_n \to f$ a.s., then f is integrable and

$$\int_\Omega f_n dP \to \int_\Omega fdP$$

Chapter 7

Levenshtein Measure Theory

INTRODUCTION

The problem we consider here, called common motifs extraction, is stated as follows: Given a set of $N \geq 2$ subject strings $\{S_i, | i = 1 ... N\}$, each of average length n over a finite alphabet Σ and a threshold D, find common motifs of length P that occur with at most D differences (or simply D-occurrence) in $1 \leq q \leq N$ distinct sequences of the set. We consider this problem with respect to the Levenshtein metric (or simply Edit distance). We consider here the simple unit measure, whereby, the Edit distance between two strings $S_1[1 ... i]$ and $S_2[1 ... j]$, $\delta(i, j)$ is defined as follows:

$$\delta(i, j) = \min[\delta(i-1, j) + 1,\ \delta(i, j-1) + 1,\ \delta(i-1, j-1) + t(i, j)],$$

where $t(i, j)$ is 0, if $S_1(i) = S_2(j)$ else 1. As far as we know, this work will be the first that discusses this problem with respect to Edit distance not only theoretically (as been done before) but also experimentally.

The above problem has found applications in finding the DNA binding sites. There are various methods to extract common motifs. Most of the approach made use of the Hamming distance metric. It is only the work of Rocke and Tompa and Sagot that considered extracting common motifs under Edit distance but did so theoretically only. Various pointers have indicated that extracting common motifs under Edit distance is also an important biological problem. It is this open problem that we consider in this paper.

Our contributions are in two folds: First, we have implemented the interesting but involved algorithm of Sagot for motifs finding for Edit distance. The theoretical running time of this algorithm is $O(N^2 nP^D |\Sigma|^D)$. It has only been implemented before for Hamming distance. Second, we present an alternative algorithmic approach also for Edit distance based on the concept described by Adebiyi et al. The resulting algorithm is simpler and more efficient for moderately large P and D in theory as well as in practice. It runs in $O(N$

$\cdot n + D(N \cdot n)^{1+pow(\varepsilon)})$ expected time for $D \leq 2\log_{|\Sigma|}(N \cdot n)$, where $\varepsilon = D/P$ and $pow(\varepsilon)$ is an increasing, concave function that is 0 when $\varepsilon = 0$ and about 0.9 for DNA and protein sequences. We implemented this algorithm in C and compare its effectiveness using our implementation of Sagot's algorithm on sequences that include B. Subtilis, yeast S. cerevisiae and H. Pylori.

BASIC CONCEPTS AND SAGOT'S ALGORITHM FOR EDIT DISTANCE

Here, we will describe in detail the algorithm of Sagot for Edit distance, since we use some of the concepts later on for the new algorithm, and start with some definitions. A *model* of length P is a pattern over Σ^P for some positive $P \geq 1$ and a *valid model* is a model that present a single representation for all reoccurrences of a motif. A valid model is elsewhere also known as a consensus.

The distance between a valid model and the reoccurrences of the motif that it represents is not more than D. In particular, we refer to a motif and its reoccurrences in the sequences using its valid model.

Sagot's method is basically based on the concept of validating all words/models (they are elsewhere known as *external objects*, and are the candidates for motifs) over Σ^P that occur in $1 \leq q \leq N$ distinct sequences of the set $\{S_i, | \ i = 1... N\}$. The generation or *SPELLING* after Sagot of each character of the models is done simultaneously as we search them in the preprocessed generalized suffix tree of the sequences. *GST* is the suffix tree that represent the suffixes of multiple sequences. The SPELLING of a model stops, if we have reached the length required for models or the models may not be further extended while remaining valid.

If we introduce errors (Hamming or Edit distance), for any model, we may also stop the descent down a path in *GST* as soon as too many insertions, deletions, or substitutions have been accumulated along it. The models of length P, which have a chance to be valid, are those in the D-neighborhood of the words present in the sequences of the set. Let u be a word present in the sequences. Model X is in the D-neighborhood of u, if the distance between X and u is not more than D. For Hamming distance, Sagot proved that the D-neighborhood of a word of a length P contains

$$V(D,P) = \sum_{i=1}^{D} \binom{P}{i} (|\Sigma - 1|)^i \leq P^D \, |\Sigma|^D$$

elements. Recently, Crochemore and Sagot indicated that this also bounds the number of D-neighborhood for Edit distance from above. To introduce the concept of generalized suffix tree, for brevity, we adopt the combinatorial definitions given by Blumer.

Definition: A substring, x, of a string S, occurs in m left (right) contexts if x occurs at least m times, with these occurrences preceded (followed) by at least m distinct characters.

Definition: Given a finite set *s* of strings and *x*, *y* in *s*, *y* is a minimal extension of *x* (in *s*) if *x* is a prefix of *y* and there is no *z* in s shorter than *y* such that *x* is a prefix of *z* and *z* a prefix of *y*.

We can now define a suffix tree for a single string *S* of size *n* as follows,

Definition: Let *s* be the set of all substrings of *S*. The suffix tree (henceforth known as *ST*) for *S* is the directed graph (*V*, *E*), where *V* is the set of substrings of *S* that are either suffixes of *S* or that occur in *S* in at least two right contexts, and *E* is the set of all directed edges (*x*, *y*), *x*, *y* in *s*, such that *y* is a minimal extension of *x*.

In particular, *ST* of a string *S* of size *n* is a tree with $O(n)$ edges and internal vertices and *n* external vertices (the leaves). Each leaf in the suffix tree is associated with an index $i\ (1 \leq i \leq n)$. The edges are labeled with characters so that the concatenation of the labeled edges on the path from the root to the leaf with index *i* is the *i*th suffix of *S*. Here, *ST* is used to represent a single string. This can be extended to represent multiple strings $\{S_i, | i = 1 ... N\}$. The resulting tree is called GST. When two or more input strings have the same suffix, *GST* has several leaves corresponding to that suffix, but each corresponds to a different input string.

Before giving a detailed description of Sagot algorithm for Edit distance, note that SPELLING all occurrences of a model if errors are not allowed leads to a node *x* in the suffix tree, while it leads to more than one node $\{x_1, x_2, ...\}$, once errors are allowed. Her algorithm is based on the following recurrence formula.

Lemma: Let *x* be a node in the suffix tree and err $\leq D$. (*x*, err) is a node-occurrence of $m' = m\alpha$ with $m \in \Sigma^k$ and $\alpha \in \Sigma$ (i.e node *x* is a substring approximate to m' within distance err) if and only if, one of the following conditions is verified:

(match) (father(*x*), err) is a node-occur. of *m* and the label of father(*x*) to *x* is α;

(substitution) (father(*x*), err-1) is a node-occurrence of *m* and the label of the arc from father(*x*) to *x* is $\beta \neq \alpha$;

(deletion) (*x*, err-1) is a node-occurrence of *m*;

(insertion) (father(*x*), err-1) is a node-occurrence of $m\alpha$.

To perform the SPELLING step, some additional information has to be added to the nodes of the *GST*. The first preprocessing task, is to calculate for each node *x*, the number *CSSx*[8] of different sequences of the set $\{S_i, | i = 1 ... N\}$, that the leaves in the subtree rooted at *x* refer to. This has been shown to be solvable in $O(N \cdot n)$ time by Hui. Furthermore, to each *x*, a boolean array $Colors_x$ of dimension *N* is associated, whose entries are used to determine which sequences are common to two or more nodes of the tree. The *i*-th entry of $Color\ s_x$ is 1 if at least one leaf in the subtree rooted at *x* represents a suffix of string S_i and 0 otherwise. $Color\ s_x$ for all nodes *x* may be obtained by a simple traversal of the tree.

The pseudo-code of the Edit distance algorithm, as outlined by Sagot but implemented by us, are presented below. The following auxiliary structures are used:

1. A set Ext_m of symbols by which a model may be extended at the next step;
2. A set Occ_m of node-occurrences of a model *m*. Recall that these correspond to classes of occurrences. Each node-occurrence *x* is represented by a pair (x, x_{err}) where x_{err} is the number of insertions, deletions or mismatches between *m* and the label of the path leading from the root to *x* in the tree;
3. A variable CSS_x as defined above;
4. A boolean array $Color\ s_x$ as also defined above;
5. *minseq* that indicates the minimum of CSS_x for all node-occurrences *x* of the extended model;
6. A variable *maxseq* that indicates the sum of CSS_x for all node-occurrences *x* of the extended model;
7. A $Color_m$, whose *i*-th entry is 1 if *m* occurs in string S_i and 0 otherwise;
8. A function KeepModel(*m*), that either stores all information concerning a valid model of required length for printing later, or immediately prints this information.

In procedure *SPELLMODELS*, the models *m* are spelled as they are searched against the words in the sequences via the paths that encoded them in *GST*.

Note that the Edit operations can be rewritten in any of the following forms: $a \to \varepsilon$ (a deletion), $\varepsilon \to a$ (an insertion), or $a \to b$ (a change or substitution), where $a, b \in \Sigma, a \neq b$, and ε is the empty string. Let the characters at the left-hand side of the binary operator $\to$ come from the substrings of the node occurrences, while that on the right-hand side come from the models. To spell the models, while allowing its re-occurrences to contain gaps, sub-procedure *TREAT* in *SPELLMODELS* of fig. 1 simulates the adding of the last row of a dynamic programming matrix of the model *m* against the nodes of *GST*.

TREAT is called recursively as TREAT(Occ_m, $Occ_{m\alpha}$, $Ext_{m\alpha}$, $Color\ s_{m\alpha}$, *minseq*, *maxseq*, *x*, x_{err}, α, *level* + 1) and works as follows. At *level* = 0, the edit script of deleting a character from the model m_α is carried out if $x_{err} < D$ and parameters *maxseq*, *minseq*, $Color\ s_{m\alpha}$, $Ext_{m\alpha}$ and $Occ_{m\alpha}$ are updated accordingly using the node *x* in concern. At *level* > 0, the nodes *x* of the subtree rooted at *x* are pruned accordingly using their *minerr*, i.e, the minimum error that exist between them and the model considering insertions, deletions and substitutions. And if $min_{err} \leq D$, parameters *maxseq*, *minseq*, $Color\ s_{m\alpha}$, $Ext_{m\alpha}$ and $Occ_{m\alpha}$ are again updated accordingly using the nodes *x* and *x'* in concern.

Note that in all cases, the number of distinct sequences, the model occurs in, is in the range between *minseq* and *maxseq*. For simplicity, we assumed that motifs of a fixed length *P* are been sought for.

The function *SPELLMODELS* is called with the arguments $(0, \lambda, Occ_\lambda = \{(root, 0)\}, Ext_\alpha)$, where

$$Ext_1 = \begin{cases} \Sigma : D > 0 \\ label_b \ \ for\ branhes\ b\ from\ root : otherwise \end{cases}$$

Note that the number of all possible models in line 4 that does not necessarily represent a motif is $| \Sigma |^P$ and the loop of line 8 is executed $O(N \cdot n)$ times. In fact, the number of nodes x, considered in loop of line 8, are further reduced in sub-procedure *TREAT*. The models which might be valid (those representing motifs) are those in the D-neighborhood of the words in the sequences of the set $\{S_i, | i = 1 ... N\}$. Therefore, the nested loops starting in line 4 and line 8 are executed $O(N \cdot n \cdot V(D, P))$ times, where $V(D, P) \leq P^D | \Sigma |^D$. Within these loops, the dominating operation is the operation of line 14 and this can be carried out in $O(N)$ time. Therefore, it follows that for all possible models in Σ^P, the running time of the algorithm $SPELLMODELS(l, P, Occ_m, Ext_m)$ is $O(N^2 n P^D | \Sigma |^D)$.

1. Procedure SPELLMODELS(l, P, Occ_m, Ext_m)
2. If ($l = P$) then KeepModel(m)
3. Else if ($l < P$)
4. For each symbol α in Ext_m
5. $Maxseq = 0$, $minseq = \infty$.
6. $Color\ s_{m\alpha}$ is initialized with no colors
7. $Ext_{m\alpha} = 0$, $Occ_{m\alpha} = 0$
8. For each pair (x, x_{err}) in Occ_m
9. Remove(x, x_{err}) from Occ_m
10. TREAT($Occ_m, Occ_{m\alpha}, Ext_{m\alpha}, Color\ s_{m\alpha}, minseq, maxseq, x, x_{err}, \alpha, level = 0$)
11. If ($maxseq < q$) return (no hope)
12. Else if ($minseq \geq q$)
13. SPELLMODELS($l + 1, m\alpha, Occ_{m\alpha}, Ext_{m\alpha}$)
14. Else if the number of bits at 1 in $Colors_{m\alpha}$ is not less than q
15. SPELLMODELS ($l + 1, m\alpha, Occ_{m\alpha}, Ext_{m\alpha}$)

An Alternative Algorithmic Approach

We develop another algorithmic approach by a careful generation of potential models using the adapted techniques of Adebiyi et al. Any string S of size n is a combination of exact repeated and non-repeated substrings respectively. Two types of exact repeats exist: maximal and non maximal repeats. An exact repeat β is maximal if $a\beta c$ and $b\beta d$ occur in

S for some $a \neq b$ and $c \neq d$, $a, b, c, d \in \Sigma$. Otherwise it is non-maximal. Note that a maximal repeat might be a proper substring of another maximal repeat. Therefore, a supermaximal repeat is a maximal repeat that does not occur as a substring of any other maximal repeat. Models are constructed by partitioning them into repeats-oriented equivalence classes and avoiding those models that are contained in another maximal models or that will extend to the same motif. Here, we use the ideas developed in Adebiyi et al. The number of such models is linear in the total length of the input strings.

Using the foregone details, the algorithm is based on a clever construction of models that are extended into common motifs.

THE ALGORITHM

The algorithm is divided into two steps. First, we extract the models, second, we extend these models to derive common motifs. We now discuss how the models are constructed. We begin with formal definitions for identifying *maximal and supermaximal repeats respectively* in a suffix tree, *ST*.

Definition: For each position i in string S, character $S(i-1)$ is called the left character of i. The left character of a leaf of *ST* is the left character of the suffix position represented by that leaf.

Definition: A node v of *ST* is called left diverse, if at least two leafs in v's subtree have different left characters. By definition, a leaf cannot be left diverse.

Lemma: The string α, labeling the path to a node v of *ST* is a maximal repeat if and only if v is left diverse. A left diverse internal node v represents a supermaximal repeat α, if and only if all of v's children are leaves and each has a distinct left character.

Lemma: Each maximal or non-maximal repeat is contained in some supermaximal repeats.

We have the following observations, the first one is straight forward.

Observation: A model substring must be extendable in all strings in $\{S_i, | i = 1... N\}$. That is, its left index in each string shifted by $P-1$ positions to the right, must be less than or equal to the end index of each string.

Observation: A substring that is completely non-maximal in all strings in $\{S_i, | i = 1... N\}$ cannot be a model, since it is contained in some supermaximal repeats.

Our models now should satisfy the two observations above and are defined using the following definitions. Note that the keywords *maximal* used in the naming of our models, have different meaning from their use with *repeat* as given in Lemma above.

Definition: A repeat model is a repeat (i.e., non-maximal, maximal or supermaximal) in at least two of the subject strings.

Definition: A maximal model is a non repeated pattern occurring in at least q of the subject strings and a maximal repeat when considering its occurrences in at least q of the subject strings.

Let R and M be the sets of repeat and maximal models respectively. In general, we will call R and M models, the set of maximal models. Blumer[3] showed that the minimum expected length of a repeated pattern is $f \log_{|\Sigma|} n$ for $f \leq 1$. In Adebiyi et al, we derived that the expected length of the longest repeat is $2 \log_{|\Sigma|} n + o(\log n)$.

This implies that, the length of each model l defined above satisfied $l = \Theta(\log_{|\Sigma|} n)$. This information will be useful for the efficient extension step of each model.

Step 1: Extracting the Maximal Models

We will use the above definitions to extract models from a given set of $N \geq 2$ subject strings $\{S_i, | i = 1... N\}$, each of average length n using a generalized suffix tree. To do this, we extend the algorithm for finding exact maximal and supermaximal repeats as described by Gusfield, using the ideas of Hui for solving the CSS_x problem. This is encapsulated in procedure FINDMODELS (N, n).

The main auxiliary structures and functions used by Hui include LeafList and lastleaf(x), where x is a leaf in *GST*. LeafList is the list of leaves ordered according to a post-order traversal of *GST*. For each leaf that belongs to a sequence in $\{S_i, | i = 1... N\}$, last leaf(x) is the last leaf which precedes x in LeafList and belongs to the same sequence. If no leaf precedes x in LeafList, *Nil* = last leaf(x). We add the following preprocessing to extract our maximal models.

Let $lca(x, y)$ be the least common ancestor of x and y, where x and y are vertices in *GST*. An internal vertex v is a repeat representative pair-lca (of a sequence in $\{S_i, | i = 1... N\}$, if there exists a leaf x of a sequence S_i, such that $v = lca(y = lastleaf(x), x)$ and lastleaf $(y) = Nil$. Let *RRP count* (v) denotes the number of such situations for each node v. Therefore, to verify the conditional part of Definition for a node u, the problem to solve is equivalent to the calculation of its number of Non-*Nil* lastleaf that refers to distinct sequences in $\{Si, | i = 1... N\}$. This value for a node u, is called *distinct* (u) and is recursively calculated as $distinct\ (u) = \sum_{v \in subtree(u)} RRP\ count\ (v) = RRPcount\ (u) + \sum_{\omega \in W} distinct\ (w)$, where W is the set of children of u.

1. Procedure FINDMODELS (N, n)
2. Build *GST*
3. Create the LeafList
4. Compute the lastleaf() values
5. Compute the RRP count() values

6. Compute the distinct () values
7. For each node x (in bottom-up format)
8. if (!non-maximal$_x$)
9. if (*distinct* ≥ 2 and $CSS_x \geq 2$)
10. Append x to | R
11. else if (distinct = 0 | 1 and true = *TM*)
12. Append x to M

The property expressed in Observation 2 above, can be verified by adding a boolean variable non-maximal*x* to each node *x*.

The boolean variable nonmaximal$_x$ is 0, if the node is at least once maximal in a sequence and 1 if non-maximal in all strings in $\{S_i, \mid i = 1... N\}$. This can be done by a simple combination of the original algorithm of Gusfield and the auxiliary structure and function LeafList and lastleaf. To verify the properties from Definition for a node *u*, distinct (*u*) must be zero or one and what remains is to test its maximality (hence forth *TM*) with respect to its occurrences in at least *q* strings in $\{S_i, \mid i = 1... N\}$. This involves collecting and scanning of the left characters of the suffixes represented by its children. Using the above facts, we can now compactly present FINDMODELS (*N*, *n*).

Lemma: All the models defined in Definitions can be extracted from *GST* in $O(N \cdot n)$ expected time.

Proof: The preprocessing tasks before extracting models include Creating LeafList, computing last leaf (), RRP count() and distinct() values respectively. lastleaf() values are calculated via a single scan of LeafList, therefore, since the set of leaves is not more than $N \cdot n$, creating Leaf List and computing lastleaf values can be done in $O(N \cdot n)$ time. Given an arbitrary pair of leaves, their *lca* can be computed in $O(1)$ time. After initializing all *RRP count*() values by zero, for every leaf *x*, we compute $u = lca\ (x, y = lastleaf\,(x))$ and if $lastleaf\,(y) = Nil$, which means, it has not been counted, we increment *RRP count* (*u*) by 1. Therefore *RRP count*() values can be computed in $O(N \cdot n)$ time for all leaves. What then follows is the computation of all distinct() values by its recursive definition given above via a post-order traversal. This is also achieved in $O(N \cdot n)$ time.

This is also achievable in $O(N{\cdot}n)$ time because the total number of nodes and leaves in a *GST* is bounded by $O(N \cdot n)$.

Step 2: Extending the Maximal Models

The algorithm we use to extend the models is similar to the algorithm used in Adebiyi et al. to extend seeds to find approximate repeats in one long subject string of length *n*. To extend a seed *W*, given *P*, the length of short substrings sought for, that repeat with at most

D-differences, *i.e.* insertions, deletions, and mismatches, we split W into a left and right part W^l and W^r, where $|W^l| = \log_{|\Sigma|} n$. Next, using SAM (Sublinear Algorithm of Myers), we find all locations of all approximate repeats $(W^l)'$ of W^l. Their right indices are stored in set H. Next, we check whether we can extend $(W^l)'$ to the right to get approximate repeats for $W = W^l W^r$.

Therefore, for every index H_i in H corresponding to a word W_i^l with differences D_i, we consider the word W_i^r starting from H_i and having length $P - \log_{|\Sigma|} n$. Let E_W denote the set of possible rightward extensions of W^r to a word of length $P - \log_{|\Sigma|} n$.

Note that $|E_W|$ is the number of occurrences of the seed W in string S, which is, fortunately, at most $|\Sigma|$ for supermaximal repeats. We compute the pairwise alignments of W_i^r with all elements E_j of E_W and, if the difference between any of the two substrings is at most $D_i = D - D_i$, then the extension works and we can add $W_i = W_i^l W_i^r$ to the output list SA for the short approximate repeats. The procedure is called ADDWR (D_i, H_i, P).

To extract common motifs, the situation is different, the subject string is of size $N \cdot n$. We assume in the following exposition that $P = 2 \log_{|\Sigma|} (N \cdot n)$. Note that if $P \leq \log_{|\Sigma|} (N \cdot n)$, we just directly apply SAM and we do not need any extension by subprocedure ADDWR described below. For each model, after we ran SAM on their left part, we sort the set H, so that we can partition them into N subclasses, depending on where they occur in each subject string S_j, $j = 1 \ldots N$. That is, $H_{ij} \in H$, $i = 1 \ldots |H|$ and $j = 1 \ldots N$. Then in subprocedure ADDWR, for every index H_{ij} in H corresponding to a word W_{ij}^l with differences D_i, we consider the words W_{ij}^r starting from H_{ij} of length $P - \log_{|\Sigma|} (N \cdot n)$.

Next, we can no longer assume that $|E_W|$ is at most $|\Sigma|$, since supermaximal, maximal and non-maximal repeats are all used to construct our models. Note that the concept of maximality is still paramount in our models construction, so that $|E_W|$ in expectation, is the number of maximal substring occurrences divided by the number of internal nodes in a generalized suffix tree. Therefore $|E_W|$ is constant in expectation. The only exceptional case is when a model appears as maximal in one sequence but non-maximal in the others. In this case, its logarithmic length can be used to show that in expectation, its number of occurrences is still bounded by the number of maximal substring occurrences. Finally, we compute the pairwise alignment of W_{ij}^r with all elements E_o of E_W and, if the difference between substrings of W_{ij}^r s.t. $j = 1 \ldots t$, $t \geq q$ and one element of E_o is at most $D_i' = D - D_i$, then $W_{ij} = W_{ij}^l W_{ij}^r$ are the locations of a common motif in the subject strings $\{S_j \mid j = 1 \ldots t\}$.

This one element of E_o is stored in the set of Common Motifs, *CM*.

The operation described above is encapsulated in figure 3 below.

1. Procedure FINDSMOTIF(D, P, N, n)
2. $R, M \longleftarrow$ FINDMODELS (N, n)
3. for W_i, $i = 1 \ldots |R \cup M|$ do
4. $W^l, W^r \longleftarrow$ SPLITW (W_i)
5. $H \longleftarrow$ SAM(D^l, W^l)
6. Sort H
7. for each H_{ij}, $i = 1 \ldots |H|, j = 1 \ldots N$ do
8. $CM \longleftarrow$ ADDWR(D^r, H_{ijr}, P)

Theorem: Given a set of $N \geq 2$ subject sequences $\{S_i | i = 1 \ldots N\}$, each of average length n over a finite alphabet Σ and threshold D, all common motifs, each of size $P = O(\log (N \cdot n))$ can be found in $O(N \cdot n + D(N \cdot n)^{1+\text{pow}(\varepsilon)} \log(N \cdot n))$ expected time for $D \leq 2 \log_{|\Sigma|} N \cdot n$, where $\varepsilon = D/P$, $\text{pow}(\varepsilon) = \log_{|\Sigma|} (c+1)/(c-1) + \varepsilon \log_{|\Sigma|} c + \varepsilon$ and $c = \varepsilon^{-1} + \sqrt{1 + \varepsilon^{-2}}$.

Proof: The subprocedure FINDS MODELS includes the building of the generalized suffix tree and finding the models. *GST* can be built in $O(N \cdot n)$ time. Lemma shows that we can also extract the maximal models in $O(N \cdot n)$ time. Therefore the expected running time of FIND MODELS (N, n) is $O(N \cdot n)$.

Note that the number of maximal models is bounded by $O(N \cdot n)$ and the operation perform in SPLITW can be done for each model in constant time. SAM running time for each model is $O(D(N \cdot n)^{pow(\varepsilon)} \log (N \cdot n))$.

It is known that the size of the output H of SAM for each query model is $O(N \cdot n^{pow(\varepsilon)})$. Therefore the time to sort set H is at most $O(N \cdot n^{pow(\varepsilon)} \log (N \cdot n))$, while ADDWR requires for all indices in H at most $O(N \cdot n^{pow(\varepsilon)} \log^2_{|\Sigma|} (N \cdot n))$ time.

Therefore, the total running time for calling SAM and ADDWR for all models is bounded by $O(D(N \cdot n)^{1+pow(\varepsilon)} \log(N \cdot n))$, for all values of N, n, and Σ. The bound stated in the lemma then follows.

We show the correctness of our method using the following proposition.

Proposition: It is true that the above algorithm in Fig. 3 with the above complexity solves the problem of extracting common motifs under the Levenshtein measure.

Proof: In our construction of maximal models, we optimally ensure that all possible valid node-occurrences of common motifs are covered. To specifically establish it, we make use again of the algorithm of Sagot. For simplicity, we assume that motifs of fixed length P are sought for. The basic ingredients in the validating algorithm include CSS_x, *Color* s_x and the simultaneous SPELLING of each characters of the 'external objects', the models, over Σ^P and the recursively *grouping and validating* of its reoccurrences in each sequence using the given basic parameters P, D and q.

In the following, when we refer to repeats, we mean the reoccurrences of a pattern in the same sequence. The nodes in a generalized suffix tree can be 150 E.F. Adebiyi and M. Kaufmann categorized into two:

1. The ones that represent patterns that are repeated but also occur (or not) in other sequences.
2. The ones that represent patterns that are non-repeated but also occur (or not) in other sequences. The set of Repeat and maximal models covers the first and the second group respectively.

The calculation of *CSSx* is performed in line 6, while *Colors*$_x$ is technically performed in lines 3-5. Finally, the grouping and validating processes are performed in line 7-8. These lines also check to make sure that too many insertions, deletions and substitutions have not been accumulated.

The basic reason of using 'external objects' as models in Sagot's algorithm, is because the real motif (in some literature, consensus) itself may not exist in the set of sequences, but it may be represented by one of its occurrences. This implies that the set of occurrences of a motif may share approximately a common pattern. In our algorithm above, the concept of D-neighborhood also used in SAM but in a different way, allows us to find approximate motif reoccurrences that are related via an approximately common pattern instead of an exact pattern.

Remark: Comparison of the running times $O(N^2 n P^D |\Sigma|^D)$ of Sagot algorithm and $O(N \cdot n + D(N \cdot n)^{1+pow(\varepsilon)} \log(N \cdot n))$ of our algorithm shows that our algorithm is a major improvement for moderately large P and D that normally arise in practice. Note further that for small values of D (for example, $D = 1$) as in some cases in practice, the complexity of Sagot's algorithm will therefore scale linearly and thus its complexity is better than our algorithm in this case.

Chapter 8

Real Anaylsis

Real analysis is a branch of mathematical analysis dealing with the set of real numbers. In particular, it deals with the analytic properties of real functions and sequences, including convergence and limits of sequences of real numbers, the calculus of the real numbers, and continuity, smoothness and related properties of real-valued functions.

Real analysis is an area of analysis, which studies concepts such as sequences and their limits, continuity, differentiation, integration and sequences of functions. However, the scope of real analysis is restricted to the real numbers, and this defines the range of tools available. Real analysis is closely related to complex analysis, which studies broadly the same properties of complex numbers. In complex analysis, it is natural to define differentiation via holomorphic functions, which have a number of useful properties, such as repeated differentiability, expressability as power series, and satisfying the Cauchy integral formula. However, in real analysis, it is usually more natural to consider differentiable, smooth, or harmonic functions, which are more widely applicable, but may lack some more powerful properties of holomorphic functions. Also results such as the fundamental theorem of algebra are simpler when expressed in terms of complex numbers.

On the other hand, the real numbers have several important analytic properties of their own. They are totally ordered, and have the least upper bound property, and these properties lead to a number of important results in real analysis, such as the monotone convergence theorem, the intermediate value theorem and the mean value theorem. However, whilst results in real analysis are generally stated for real numbers, they may still be used in other areas of mathematics - such as by considering real and imaginary parts of complex sequences, or by pointwise evaluation of operator sequences. Conversely, techniques from other areas are often used in real analysis - such as evaluation of real integrals by residue calculus.

The foundation of real analysis is the construction of the real numbers from the rational numbers, usually either by Dedekind cuts, or by completion of Cauchy sequences. Key concepts in real analysis are real sequences and their limits, continuity, differentiation, and integration.

Real analysis is also used as a starting point for other areas of analysis, such as complex analysis, functional analysis, and harmonic analysis, as well as motivating the development of topology, and as a tool in other areas, such as applied mathematics. Important results include the Bolzano-Weierstrass and Heine-Borel theorems, the intermediate value theorem and mean value theorem, the fundamental theorem of calculus, and the monotone convergence theorem.

σ – ALGEBRAS

Let X be a set. We will use the notation: $A^c = \{x \in X : x \notin A\}$ and $B - A = A \cap B^c$. (The notation $A \setminus B$ is also commonly used.)

Defination: An alegebra is a collection A of subsets of X such that

1. $\varnothing, X \in A$;
2. If $A \in A$, then $A^c \in A$;
3. If $A_1, ..., A_n \in A$, then $\bigcup_{i=1}^{\infty} A_i$ and $\bigcap_{i=1}^{\infty} A_i$ are in A.

A is a σ – algebra (or σ – field) if in addition in (d) we allow countable unions and intersecrtions only;we do not allow uncountable unions and intersections.

Example:

1. Let and $X = \mathbb{R}$ and A be the collection of all subsets of $\mathbb{R}$.
2. Let $X = \mathbb{R}$ and let

   $A = \{A = \mathbb{R} : A$ is countable of A^c is countable$\}$.

   Parts (a) and (b) of the definition are easy. Suppose $A_1, A_2 .....$ are all in A. If each of the A_i are countable , then $\bigcup_i A_i$ is countable, so in A If $A_{i_0}^c$ is countable for some i_0, then

   $$(\bigcup A_i)^c = \bigcap_i A_i^c \subset A_{i_0}^{\ c}$$

   is countable, and again $\bigcup_i A_i$ is in A. Since $\bigcap A_i = (\bigcup_i A_i^c)^c$, then the countable intersection *f* sets in A is again in A.
3. Let $X = [0, 1]$ and $A = \left\{\phi, X, \left[0, \frac{1}{2}\right], \left(\frac{1}{2}, 1\right]\right\}$.
4. $X = \{1, 2, 3\}$ and $A = \{X, \phi, \{1\}, \{1, 2\}\}$.
5. Let $X = [0, 1,]$ and $B_1, ... A_8$ subsets of of X which are pairwise disjoint and whose union is all of X. Let A the collection of all finite unions of the B_i`sas well as the empty set (so A consists of 2^8 elements.)

Note that if we take and intersection of σ – algebras, we get a σ – algebra; this is just a matter of checking the definition. If we hve a collection C of subsets of X, there is at least one σ – algebra containing C, namely, the one consisting of all subsets of X. We can take the intersection of all σ – algebras that contain C; we denote this intersection by σ (C). If A any σ algebra containing C then $A \supset \sigma(C)$.

If X has some additional structure, say. it is ametric space, then we can talk about open sets If G is the collection of open subsets of X, then we call σ (G) the Botel σ -algebra on X and this is often denoted $\boldsymbol{B}$ we will see later that when X is the real line, that B is not equal to the collection of all subsets of X.

We end this section with the following proposition.

Proposition: If $X = \mathbb{R}$, then the Borel σ -algebra is genratated by each of the following collection of sets:

1. $C_1 = \{(a,b) : a,b \in \mathbb{R}\}$.
2. $C_2 = \{[a,b] : a,b \in \mathbb{R}\}$;
3. $C_3 = \{(a,b] : a,b \in \mathbb{R}\}$;
4. $C_4 = \{(a,\infty) : a \in \mathbb{R}\}$;

Proof: 1. Let G be the collection of open sets. Then is $C_1 \subset G \subset \sigma(G).\sigma(G)$ is the Borel σ algebras and contains C_1 since σ (C_1) is the intersection of all σ algebras containing C_1, then σ (C_1) $\subset \sigma(G)$.

To get the reverse inclusion, if G is open, it is the countable union of open intervals. So $G \in \sigma(C_1)$, and hence $G \subset \sigma(C_1)$. $(G)\sigma$ is the intersection of all σ–algebras containing G;$\sigma(C_1)$ is one such so $\sigma(G) \subset \sigma(C_1)$.

2. If $[a, b] \in C_2$ then $[a,b] = \bigcap^{\infty}{}_{n=1}\left(a - \frac{1}{n}, b + \frac{1}{n}\right) \in \sigma(G)$. so $C_2 \subset \sigma(G)$, and by an argument similar to that in (1), we conclude $(C_2) \subset \sigma(G)$.

   If $[a, b] \in C_1$, choose $n_0 \geq \backslash(b-a)$ and note $(a, b) = \bigcap^{\infty}{}_{n=n_0}\left(a, b + \frac{1}{n}, b + \frac{1}{n}\right) \in$ $\sigma(C_2)$. So the Borel σ algebra, which is equal to $\sigma(C_1)$ by part (1), is contained in $\sigma(C_2)$.

3. The proof here is similarto (2), using $(a,\ b] = \bigcap^{\infty}{}_{n=1}\left(a, b + \frac{1}{n}\right)$ and $(a,\ b) = \bigcup^{\infty}_{n=n_0}\left(a, b - \frac{1}{n}\right]$ provided n_0 taken large enough.

4. The proof of this comes from (3), using that $(a, b\} = (a, \infty) - (b, \infty)$ and $(a, \infty) = \bigcup_{n=1}^{\infty} (a, a+n]$.

MEASURES

Definition: A measure on (X, A) is a function $\mu : A \to [0, \infty]$ such that

(a) $\mu(A) \geq 0$ for all $A \in A$;

(b) $\mu(\phi) = 0$;

(c) $A_i \in A$ if are disjoint, then

$$\mu\left(\bigcup_{i=1}^{\infty} A_i\right) \sum_{i=1}^{\infty} \mu(A_i).$$

Example:

1. A is the collection of all subsets, and $\mu(A)$ is the number of elements in A. This is called counting measure.
2. $X = \mathbb{R}$, A the collection of all subsets, $x_1, x_2, \ldots \in \mathbb{R}$, and $a_1, a_2, \ldots > 0$. Set $\mu(A) = \sum_{\{i : x_i \in A\}} a_i$. A particular case of this is $x_i = 1$ and all the $a_i = 1$. We will see later that this allows us to view infinte series as function on this space.
3. $\delta_x(A) - 1$ if $x \in A$ and 0 otherwise. This measure is called point mass at x

We will contruct Lebsegue measure on $\mathbb{R}$;this is an extension of the notion of length. However , the construction is a bit lenghthy. We will also construct Lebesgue measure on $\mathbb{R}^n$ when $n = 2$, this this is an extension of the notion of area, when $n = 3$, of volume.

Proposition: The following hold:

(a) If $A, B \in A$ with $A \subset B$, then $\mu(A) \leq \mu(B)$.

(b) If $A_i \in A$ and $A = \bigcup_{i=1}^{\infty} A_i$, then $\mu(A) \leq \sum_{i=1}^{\infty} \mu(A_i)$.

(c) If $A_i \in A$ and $A_1 \subset A_2 \subset \ldots$, and $A = \bigcup_{i=1}^{\infty} A_i$, then $\mu(A) = \lim_{n\to\infty} \mu(A_n)$.

(d) If $A_i \in A, A_1 \supset A_2 \supset \ldots, \mu(A_1) < \infty$, and $A = \bigcap_{i=1}^{\infty} A_i$, then we have $\mu(A) = \lim_{n\to\infty} \mu(A_n)$.

Proof: (a) Let $A_1 = A, A_2 = B = A$, and $A_3 = \ldots A_4 = q$. Now use part (c) of the definition of measure.

(b) Let $B_1 = A_1, B_2 = A_2 - B_1, B_3 A_3 - (B_1 \cup B_2)$, and so on. The B_i are disjoint and $\bigcup_{i=1}^{\infty} B_i = \bigcup_{i=1}^{\infty} A_i$. So $\mu(A) = \sum \mu(B_i) \leq \sum \mu(A_i)$.

(c) Define the B_i as in (b). Since $= \mu\bigcup_{i=1}^{n} B_i = \bigcup_{i=1}^{n} A_i$, then

$$\mu(A) = \mu\left(\bigcup_{i=1}^{\infty} A_i\right) = \mu\left(\bigcup_{i=1}^{\infty} B_i\right) = \sum_{i=1}^{\infty} \mu(B_i)$$

$$= \lim_{n\to\infty} \sum_{i=1}^{n} \mu(B_i) = \lim_{n\to\infty} \mu\left(\bigcup_{i=1}^{n} B_i\right) = \lim_{n\to\infty} \mu\left(\bigcup_{i=1}^{n} A_i\right).$$

(d) Apply (c) to the sets $A_1 - A_i$, $i = 1, 2, \ldots\ldots$

Example: To see that $\mu(A_1) < \infty$ is necessary, let X be the positive integers, μ counting measure, and $A_i = \{i, i+1, \ldots\}$. Then the A_i decrease $\mu(A_i) = \infty$ for all i, but $\mu\left(\bigcap_i A_i\right) = 1\mu(\theta) = 0$

Definition: A probababilty or probality measure is a measure such that $\mu(X) = 1$. In this case we usually write (Ω, f, P) instead of (X, A, μ).

CONSTRUCTION OF LEBESGUE MEASURE

Define $m((a, b)) = b - a$. If G is an open set and $G \subset \mathbb{R}$, then $G = \bigcup_{i=1}^{\infty} (a_i b_i)$ with the interval disjoint. Define $m(G) = \sum_{i=1}^{\infty} (b_{i-a_i})$. If $A \subset \mathbb{R}$, define

We will show the following.

1. m^* is not a measure on the collection of all subset of $\mathbb{R}$.
2. m^* is a measure on a srtrictly smaller σ algebra that strictly contains the Borel σ algebre.

We will prove these two facts (and a bit more) in amoment, but, let`s first make some remarks.

A set N is a null set with respect to m^* if $m^*(N) = 0$. Let L be the smallest σ algebra containing B and all the null sets.

More precisely, let N be the collection of all sets that are null sets with respect to m* and let $L = \sigma(B \cup N)$. L is called the Lebesgue σ algebra, and sets in L are called Lebesgue then the collection of all subsets of $\mathbb{R}$. .It is easy to get lost in construction of Lebesgue measure, so let us summarize our steps.

First we prove (*A*).

We then turn to construction of Lebesgue measure. It is more convenient for technical reasons to define.

$$m^*(A) = \inf\left\{\sum_{i=1}^{\infty}(b_i - a_i) : A \subset \bigcup_{i=1}^{\infty}(a_i, b_i]\right\}.$$

There is no real difference between this and (3.1) since it is clear that the differece between an open set and a set of the form $\bigcup_{i=1}^{\infty}(a_i, b_i]$ is countable, and with either definition of m^*, the measure of a point is 0, so the measure of a set consisting of countably many point is 0. However, when we talk about Lebesgue -Stieltjes measure, then there is a real difference.

We define what it means to be an outer measure and prove that m^* is an outer measure. We then define what it means for a set to be m^*-measurable and prove that the collection of m^*-measurable sets is a σ -algebra and that m* restricted to this σ algebra is a measure.

This looks promising, but we do not yet know that enough sets are m^*- measurable.That takes one more step. We show in Proposition that the collection of m^*-measurable sets contains the Borel σ -algebra.

Proposition: m^* is not a measure on collection of all subsets of $\mathbb{R}$.

Proof: Suppose m^* is a measurable. Define $x \sim y$ if $x - y$ is rational. This is an equivalence reltionship on [0.1]. for each equivalence class, pick an element out of that class (by the axiom of choice) call the collection of such points *A*.Given a set *B*, define $B + x = \{y + x : y \in B\}$. Note $m^*(A + q) = m^*(A)$ since this translation invariance holds for intervals, hence for open sets, hence for all sets.moreove, the sets $A + q$ are disjiont for different ,rationlal *q*.

Now

$$[0, 1] \subset \bigcup_{q\in[-2,2]}(A+q),$$

where the sum is only over rational *q*, so $1 \le \sum_{q\in[-22]} m^*(A+q)$, and therefore $m^*(A) >$ 0 But

$$\bigcup_{q\in[-2,2]}(A+q),[-6,6],$$

where again the sum is only over rational *q*,so if m*is a measure , then $12 \ge \sum_{q\in[-2,2]} m^*(A+q)$, which implies $m^*(A) = 0$, a contradition.

Definition: A function n on the collection of all subsets satisfying

(a) $n(\theta) = 0$;

(b) $A \subset B$, then $n(A) \le n(B)$;

(c) $n\left(\bigcup_{i=1}^{\infty} A_i\right) \leq \sum_{i=1}^{\infty} n(A_i)$. is called an outer measure.

Proposition: m^* defined as an outer measure.

Proof: (a) and (b) are obvious. To prove (c), let $\varepsilon > 0$. For each i there exist intervals $I_{i1}, l_{i2}, \ldots$ each of the form $(a_{ij}, a_{ij}]$, such that $A_i \subset \bigcup_{j=1}^{\infty} i_{ij}$ and

$$\sum_{i,j} m(l_{ij}) \leq \sum_i m^*(A_j) + \sum_i \varepsilon/2^i = \sum_i m^*(A_i) + \varepsilon.$$

Since ε is arbitrary, $m^*\left(\bigcup_{i=1}^{\infty} A_i\right) \leq \sum_{i=1}^{\infty} m^*(A_i)$.

Definition: Let m^* be an outer measure. A set $A \subset X$ is m^*-measurable if

$$m^*(E) = m^*(E \cap A) + m^*(E \cap A^C)$$

for all $E \subset X$.

Theorem: If m^* is an outer measure on X, then the collection A of m^* measurable sets is a σ-algebra and the restriction of m^* to A is a measure Moreover, A contains all the null sets.

Proof: By Proposition

$$m^*(E) \leq m^*(E \cap A) + m^*(E \cap A^C)$$

for all $E \subset X$. So to check (3.3) it is enough to show $m^*(E) \geq m^*(E \cap A) + m^*(E \cap A^C)$. This will be trivial in the case $m^*(E) = \infty$.

If $A \in A$, then $A^C \in A$ by symmetry and the definition of A. Suppose $A, B \in A$ and $E \subset X$. Then

$$m^*(E) = m^*(E \cap A) + m^*(E \cap A^C)$$
$$= m^*(E \cap A \cap B) + m^*(E \cap A \cap B^C)) + (m^*(E \cap A^C \cap B) + m^*(E \cap A^C \cap B^C).$$

The first three terms on the right have a sum greater than or equal to $m^*(E \cap (A \cup B))$ because $A \cup B \subset (A \cap B) \cup (A \cap B^C) \cup (A^C \cap B)$. Therefore

$$m^*(E) \geq m^*(E \cap (A \cup B) + m^*(E \cap (A \cup B)^C),$$

which shows $A \cup B \in A$. Therefore A is an algebra.

Let A_i be disjoint sets in A, let $B_n = \bigcup_{i=1}^{n} A_i$. If $E \subset X$,

$$m^*(E \cap B_n) = m^*(E \cap B_n \cap A_n) + m^*(E \cap B_n \cap A_n^c)$$
$$= m^*(E \cap A_n) + m^*(E \cap B_{n-1})$$

Repeating for $m^*(E \cap B_{n-1})$, we obtain

$$m^*(E \cap B_n) \geq \sum_{i=1}^{n} m^*(E \cap A_i).$$

Since $B_n \in A$, then

$$m^*(E) = m^*(E \cap B_n) + m^*(E \cap B_n^c) \geq \sum_{i=1}^{n} m^*(E \cap A_i) + m^*(E \cap B^c).$$

Let $n \to \infty$. Recalling that m* is an outer measure.

$$m^*(E) \geq \sum_{i=1}^{\infty} m^*(E \cap A_i) + m^*(E \cap B^c)$$

$$\geq m^*\left(\bigcup_{i=1}^{\infty}(E \cap A_i)\right) + m^*(E \cap B^c)$$

$$= m^*(E \cap B) + m^*(E \cap B^c) \geq m^*(E).$$

This shows $B \in A$.

If we set $E = B$ in this last equation, we obtain

$$m^*(B) = \sum_{i=1}^{\infty} m^*(A_i),$$

or m^* is contably additive on A.

If $m^*(A) = 0$ and $E \subset X$, then

$$m^*(E \cap A) + m^*(E \cap A^c) = \text{m}^*(E \cap A^C) \leq m^*(E),$$

which shows A contains all null sets.

Define $m\left(\bigcup_{i=1}^{n}(a_i, b_i]\right) = \sum_{i=1}^{n}(b_i - a_i)$ if the $(a_i, b_i]$ are disjoint. Note m is well-defined (a set might be expressible as a union of such intervals in more than one way).

The last step in the construction is the following.

Proposition: Every set in the Borel σ-algebra is m^*-measurable.

Proof: Since the collection of m^*-measurable set is a σ-algebra, it suffices to show that overy interval J of the form $(a, b]$ is m^*-measurable. Let E be any set with $m^*(E) < \infty$; we need to show

$$m^*(E) \geq m^*(E \cap J) + m^*(E \cap J^c).$$

Choose $I_1, I_2, \ldots$ of the form $(a_i, b_i]$ such that $E \subset \bigcup_i I_i$ and

$$m^*(E) \geq \sum_i (b_i - a_i) - \varepsilon.$$

Since $E \subset \bigcup I_i$, we have $m^*(E \cup J) \leq \sum m^*(I_i \cap J)$ and $m^*(E \cap J^c) \leq \sum m^*(I_i \cap J^c)$. Hence we have

$$m^*(E \cap J) + m^*(E \cap J^c) \leq \sum [m^*(I_i \cap J^c)].$$

Now m* $(I_i \cap J)$ is the length of an interval and $m^*(I_i \cap J^c)$ is the length of two interval, so

$$m^*(I_j \cap J) + m^*(I_j \cap J\ ) = m^*(I_i).$$

$$m^*(E \cap J) + m^*(E \cap J^c) \leq \sum_i [m^*(I_i) \leq m^*(E) + \varepsilon.$$

We now drop the asterisks from m^* and call m^*-Lebesgue measure.

EXAMPLES AND RELATED RESULTS

Examples:

1. Recall the cantor set is constructed by taking the interval [0, 1], removing the middle third, removing the meddle thirds of each of the two remaining subintervals, and continuing. The Cantor set is what remains; it is closed, uncountable, and every point is a limit point Moreover, it contains no intervals.

After one stage, the measure of the two intervals is a $2\left(\frac{1}{3}\right)$, after two stages 4(1/9), and after n stages, $(2/3)^n$. Since the Contor set C is the intersection of all these sets, the Lebesgue measure of C is 0.

Suppose we define f_0 to be 1/2 on the interval (1/3, 2/3,), 1/4 on the interval (1/9, 2/9) and(7/9,8/9), and so on. Define $f(x) = \inf\{f_0(y) : y \geq x\}$. Notice $f = f_0$ on the complement, of the Cantor set. f is montone, so it has only jump discountinuties. But if it has a jump cotnuity,there is a rational of the form $k/2^n$ with $k \leq 2^n$ that is not in the range of f. On the other hand, by construction, each of these values is taken by f_0 for some point in the complement of C, and so is taken by f. The only way this can happen is f is called the Cantor- Lebesgue function. We will use it in example later on. For now. We can see that it a function that increases only on Cantor set, which is of Lebesgue measure 0, yet is continuous.

2. Let q_1, q_2....be an enumeration $\varepsilon > 0$, and let I_i be the interval $(q^i - \varepsilon/2^i, q^i + \varepsilon 2^i)$. Then the the measure of I_i is, 2^{i-1} so the measure or $\bigcup_i I_i$ is at most 2ε. (It is not equal to that because there is a lot of overlap.) So the measure of $A = [0, 1] - \bigcup_i I_i$ is larger then $1 - 2\varepsilon$. But A contains no rational numbers.

3. Let us follow the construction of the Contor set, with this difference. Instead of removing the middle third at the first stage, remove the middle fourth, i.e., remove (3/8, 5/8). On each of the two intervals that remains, remove the middle sixteenths. On each of the four interval that remains, remove the middle interval of length 1/64, and so on. The total that we removed is

$$\frac{1}{4}+2\left(\frac{1}{16}\right)+4\left(\frac{1}{64}\right)+....=\frac{1}{2}.$$

The set that remains contains no intervals, is closed ,every point is limit point, is uncountable and has measure 1/2. such a set is called a fat. Cantor set. Of courese, other choices that 1/4, 1/16, etc.are possible.

Let $A \subset [0, 1]$ be a Borel measurabhle set. We will show that Aalmost equal to the countable union of closed sets. (A similar argument to what follows is possible for srts that have infinte measure.)

Proposition: Suppose $A \subset [0, 1]$ is a Borel measurable set.

(a) There exists a set H that is the countable intersection of open sets which contains Aand $m(H-A) = 0$

(b) There exists a set F that is the countable union of closed sets which is contained in A and $(A-F) = 0$.

Proof: (a) For each i, there is an open set G_i that contains A and such that m $(G_i - A) < 2^{-1}$. This follows from the fact that $m(A) = m^*(A)$ and the definition of m*. Then $H_i = \bigcap_{j \le i} G_j$ will contain A, is open, and since it is contain A, is open and since it is contained in i G_i ,then $m(H_i - A) < 2^{-i}$. Let $H = \bigcap_{i=1}^{\infty} H_i$. H need not be open, but it is the intersection of countably many open sets. The set H is Borel set, contains A, and $m(H-A) \le m(H_i - A) < 2^{-1}$ for each i, hence $m(H-A) = 0$

(b) If $A \subset [0, 1]$, let $F_i = [0, 1] - H_i$, where H_i is decreasing sequece of open sets containing A^c such that $m(H_i - A^c) < 2^{-i}$. (The H_i are constructed as in the proof of (a), but in terms of A^c instead of A.) Then F_i is increasing sequence of closed sets, $F_i \subset A$ for each i, and $m(A-F_i) < 2^{-i}$ for each i. Our result follows from letting $F = \bigcup_i F_i$, since $m(A-F) \le m(A-F_i) < 2^{-i}$ for each i, hence $m(A-F) = 0$.

The countable intersection of open sets are sometimes called G_δ sets; the G is for *geoffnet*, the german word for "open" and the δ for Durchschnitt, the German word for "intersection".

The countable unions of closed sets are called se F_σ sets, the F coming from *ferme*, the French word for "closed"and the σ coming form summe the Genrmen word for "union"

Therefore, when trying to undertstand Lebesgue measure, we can look at G_δ or F_σ sets which are not so bad, and at null sets, which can be quite bad but positive measure.

Next we prove the Caratheodory extension theorem. We say that a measure sure μ is σ finite if there exist $E_i, E_2, ...$ such $\mu(E_i) < \infty$ that for all i and $X \subset \bigcup_{i=1}^{\infty} E_i$.

Theorem: Suppose A_0 is an algebra and m restricted to A_0 is a measure.
Define

$$m^*(E) = \inf\left\{\sum_{i=1}^{\infty} m(A_i) : A_i \in A_0, E \subset \bigcup_{i=1}^{\infty} A_i\right\}.$$

Then

(a) $m^*(A) = m(A)$ if $A \in A_0$;

(b) Every set in A_0 is m^*-measurable;

(c) if m is σ– finite, then there is a uniquie extension to the smallest σ– algebra containing A_0.

Proof: We start with (a). Suppose $E \in A_0$. We know $m^*(E)\, m(E) \leq m(E)$ since we can take $A_1 = E$ and $A_2, A_3, ...$ empty in the definition of m^*. If $E \subset \bigcup_{k=1}^{\infty} A_i$ with $A_i \in A_0$, let $B_n = E \cap \left(A_n - \bigcup_{i=1}^{n=1} A_i\right)$. Then the B_n are disjoint, they are each in A_0, and thier union is E. Therefore

$$m(E) = \sum_{i=1}^{\infty} m(B_i) \leq \sum_{i=1}^{\infty} m(A_i).$$

Thus $m^*(E) \leq m^*(E)$.

Next we look at (b). Suppose $A_i \in A_0$. Let $\varepsilon > 0$ and let $E \subset X$ pick $B_i \in A_0$ such that

$$E \subset \bigcup_{i=1}^{\infty} B_i \text{ and } \sum_i m(B_i) \leq m^*(E) + \varepsilon. \text{ Then}$$

$$m^*(E) + \varepsilon \geq \sum_{i=1}^{\infty} m(B_i) = \sum_{i=1}^{\infty} m(B_i \cap A) + \sum_{i=1}^{\infty} m(B_i \cap A^c)$$

$$\leq m^*(E \cap A) + m^*(E \cap A^c).$$

Since ε is arbitrary, $m^*(E) \geq m^*(E \cap A) + m^*(E \cap A^c)$. So A is m^*-measurable.

Finally, suppose we have two extension to the smallest σ -algebra containing A_0; let the other extension be called n. We will show that if E is in this smallest σ - algebra, then $m^*(E) = n(E)$. Since m is σ-finite, we can reduce to the case where m is a finite measure: if $X = \bigcup_i K_i$ with $m(K_i) < \infty$ and we prove uniqueness for the measure m_i defined by $m_i(A) = m(A \cap K_i)$, then uniqueness for m follows. So we suppose $m(X) < \infty$.

Since E must be m^*- measurable.

$$m^*(E) = \inf\left\{\sum_{i=1}^{\infty} m(A_i) : E \subset \bigcup_{i=1}^{\infty} A_i, A_i \in A_0\right\}.$$

But $m = n$ on A_0 so $\sum_i m(A_i) = \sum_i n(A_i)$. Therefore $n(E) \le \sum_i n(A_i)$, which implies $n(E) \le m^*(E)$.

Since we do not know that n is constructed via an outer measure, we must use a different argument to get the reverse inequqality let $\varepsilon > 0$ and choose

$A_i \in A_0$ such that $m^*(E) + \varepsilon \ge \sum_i m(A_i)$ and $E \subset \bigcup_i A_i$. $A = \bigcup_i A_i$ and $B_k = \bigcup_{i=1}^{k} A_i$.

Observe $m^*(E) + \varepsilon \ge m^*(A)$, hence $m^*(A - E) < \varepsilon$. We have

$$m^*(A) = \lim_{k\to\infty} m^*(B_k) = \lim_{k\to\infty} n(B_k) = n(A).$$

Then

$$m^*(E) \le m^*(A) = n(A) = n(E) + n(A - E) \le n(E) + m(A - E) \le n(E) + \varepsilon.$$

Since ε is arbitrary, this completes the proof.

Remarks:

1. Uniqueness implies there is only one possible Lebesgue measure.
2. We will use the Caratheodory extension theorem in the study of product measures. It is also used in the Tiesz representation theorem and in the Daniell-Kolmogorow extension theorem.

We now define Lebesgue-Stieltjes measuere, Let $\alpha : \mathbb{R} \to \mathbb{R}$ be non decreasing and right continuous ($i.e., \alpha +) = \alpha(x)$ for all x, where $\alpha(\alpha+) = \lim_{y \to x, y>} \alpha(y))$. Suppose we define

$$m_\alpha((a,b]) = \alpha(b) - \alpha(a_i])$$

define

$$m_\alpha\left(\bigcup_{i=1}^{\infty}(a_i, b_i)]\right) = \sum_i (\alpha(b_i) - \alpha(a_i))$$

when the intervals $[a_i, b_i]$ are disjiont, and and define

$$m^*_\alpha(A) = \inf \left\{ \sum_i (\alpha(b_i) - \alpha(a_i)) : A \subset \bigcup_i (a_i, b_i)] \right\}$$

Very much as in the previous section we can show that m^*_α is a measure on the Borel $\sigma-$algebra.

The m_α measure of a point $\alpha(x) - \alpha(x-)$, where

$$\alpha(x-) = \lim_{y \Rightarrow x, y<x} \alpha(x).$$

so $m_\alpha(\{x\})$ is equal to size of the jump (if any) of α at x.

Lebesgue measure is the special case of m_α when $\alpha(x) = x$.

Given a measure μ on $\mathbb{R}$ such that $\mu(K) < \infty$ whenever K is compact define $\alpha(x) = \mu((0, x])$ if $x \geq 0$ and $\alpha(x) = -\mu((0, x])$ if $x < 0$. Then α is nondecreasing, right contionuous, and it is not hard to see that $\mu = m_\alpha$.

MEASURABLE FUNCTION

Suppose we have a set X together with a $\sigma-$algebra A.

Definition: $f : X \to \mathbb{R}$ is measurable or A-measurable if $\{x : f(x) > a\} \in A$ for all a $\in \mathbb{R}$.

Example:

1. Suppose f is identically constandt. Then $\{x : f(x) > a\}$ is either empty or the whole space, f is measurable.
2. Suppose $f(x) = 1$ if $x \in A_0$ otherwise. Then $\{x : f(x) > a\}$ is either θ, A, or X. So f is measurable if and only if A is in the $\sigma-$algebra.
3. Suppose X is the real line with the Borel $\sigma-$algebra and $f(x) = x$. Then $\{x : f(x) > a\} = (a, \infty)$, and so f is measurable.

Proposition: The following are equivalent.

(a) $\{x : f(x) > a\} \in A$ for all a;

(b) $\{x : f(x) \leq a\} \in A$ for all a;

(c) $\{x : f(x) < a\} \in A$ for all a;

(d) $\{x : f(x) \geq a\} \in A$ for all a;

Proof: The equivalence of (a) and (b) and (c) and (d) follow from taking complements. The remaining equuivalences follow from the equation

$$\{x : f(x) \geq a\} = \bigcap_{n=1}^{\infty} \{f : (x) > a - 1/n\},$$

$$\{x : f(x) > a\} = \bigcup_{n=1}^{\infty} \{x : f(x) \geq a + 1/n\}.$$

Proposition: If X is a metric space, A contaian all the open sets, and f is continuous, then f is measurable.

Proof: $\{x : f(x) > a\} = f^{-1}(a, \infty)$ is open.

Proposition: If and g are measurable, so are $f + g$, cf, fg max (f, g) and min (f, g).

Proof: If $f(x) + g(x) < \alpha$, then $f(x) < \alpha - g(x)$, and there exists a rational r such that $f(x) < r < \alpha - g(x)$. So

$$\{x : f(x) + g(x) < \alpha\} = \bigcup_{r \text{ rational}} (\{x : f(x) < r\} \cap \{x : g(x) < \alpha - r\})$$

f^2 is measurable since $\{x : f(x)^2 > a) = \{x : f(x) > \sqrt{a}\} \cup \{x : f(x) < -\sqrt{a}\}$. The measurable of fg follows since $fg = \frac{1}{2}\left[(f+g)^2 - f^2 - g^2\right]$.

(x: max $f(x).g(x)) > a\}$ $\{x : f(x) > a\} \cup \{x : g(x) > a\}$. and the argument for min (f, g) is similar.

Proposition: if f_i is measrable for each j_i, then so are sup $i f_i$ $\inf_i f_i$ lim $\sup_{i\to\infty} f_i$ and lim $\inf_{i\to\infty} f_i$.

Proof: The result will follow for lim sup and lim and lim inf once we have the result for the sup and inf by using the definitions. We have $\{x: \sup_i f_i > a\} = \bigcap_{i=i}^{\infty} \{x : f_i(x) > a\}$, and the proof for inf f_i similar.

Definition: We say $f = g$ almost everywhere, written $f = g$ a.e., if $\{x : g(x) \neq g(x)\}$ has measurer zero similarly, we say if $f_i \to f$ a, eg., if the set of x where this fails has measure zero.

We saw in Proposition 5.at all continuous function are Borel measureable. The same is true for monotone function on the real line.

Proposition: If $f : \mathbb{R} \to \mathbb{R}$ is nondecreasing of nonincreasing, then f is Borel measurable.

Proof: Let us suppose f is nondecreasing. The set $A_a = \{ x: f(x) > a\}$ is then either a semi - infinite open interval or semi- infinte closed interval. This can be seen by a picture. To be more careful, given a $\in \mathbb{R}$, let x_0 = sup $\{y: f(y) \leq a\}$. If $f(x_0)$ = a, then $A_0 = ((x_0 \infty),)$ while if $f(x_0) \neq a$, then $A_a = (x_0, \infty)$. In each case A_a is Borel set.

Proposition: Let X be a space , A a σ-algebra on X, and $f::X \rightarrow \mathbb{R}$ a A- measurable function. If A is in the Borel σ-algebrea on $\mathbb{R}$ then $f^1(A) \in A$.

Proof: Let ***B*** be the Borel σ-algebrea on $\mathbb{R}$ and $C = \{A \in B : F^{-1}(A) \in A\}$.

If $A_1, A_2, \ldots\ldots\ldots \in C$, then since $f^{-1}\bigcup_i Ai = \bigcup_i f^{-1}(A_i) \in A$, we have that C is closed under countable unions.Similarly C is closed under countable. intersections and complements, so Cis a σ-algebrea. Since f is measurable, C contains (a, ∞) (for every real a, hence C contains the σ-algebrea generated by these intervals, that is , C contains ***B***

Example: We want to construct a set that is Lebesgue measurable, but not, Borel measurable. Define f as follows:if $x \in (1/2, 2/3)$, let f(x) = 1/3; if $x \in (1/9, 2/9)$, let f(x) = 1/9, and so on. In general , we let f(x) = sup $\{y \in C : y \leq x\}$, where C is the Cantor set. (f is different from the Cantor Lebesgue function;in particular, f is not continuous).Note that jthe range of f is the Cantor set. f is nondecreasing so f^{-1} maps Borel measurable sets to Borel measurable sets.

Let a be the non-measurable set we constructed in proposition. Let B = f(A). Since f(A) $\subset C$ and m(C) = 0, then $f(A)$ is a null set, hence is Lebesgue measurable.on the other hand. $f(A)$ is not Borel measurable, because if it were , then Af^{-1} $f(A)$) would measurable, a contradicton.

INTEGRATION

In this section we introduce jthe Lebesgue intergral.

Definition: *If E $\subset$X. define the characteristic function of E by*

$$X_E(x) = \begin{cases} 1 x \in E; \\ 0 x \notin E. \end{cases}$$

A simple function s is one of the form

$$s(x) = \sum_{i=1}^{n} a_i x_{E_i}(x)$$

for real a_i and measurable sets E_i.

Proposition: Suppose $f \geq 0$ is measurable. Then there exists a sequece of nonnegative measurable simple functions increasing to f.

Proof: Let $E_{ni} = \{x : (i-1)/2^n \leq f(x) < i/2^n\}$ and $F_n = \{x: f(x) \geq n\}$

for $n = 1, 2,...$ and $i = 1, 2,....$ $n2^n$. Then define

$$s_n = \sum_{i=1}^{n2^n} \frac{i-1}{2^n} X_{E_{ni}} + n\chi_{F_n.}$$

It is easy to see that s_n has desired properties.

Defination: If $s = \sum_{i=1}^{n} a_i \chi_{E_i}$ *is a nonegative measurable simple function, define the Lebesgue intergeral of s to be*

$$\int s d\mu = \sum_{i=1}^{n} a_i \mu(E_i).$$

If $f \geq 0$ is a measurable function, define

$$\int f d\mu = \sup\left\{\int s d\mu : 0 \leq s \leq f, s \text{ simple}\right\}.$$

If f is measurable and at least one of the intergrals $\int f^+ d\mu \int f^{-1} d\mu$ is finite where $f^+ \max(f, 0)$ and $\int f^- = -\min(f, 0)$, define

$$\int f d\mu = \int f^+ d\mu - \int f^{-1} d\mu.$$

finally, if $f = u + i\upsilon$ and $\int (|u| + |\upsilon|) d\mu$ is finite, define

$$\int f d\mu = \int u d\mu + i \int \upsilon d\mu.$$

A few remarks are in order. A function s might be written as a simple function in more than one way. For example $\chi_{A \cup B} = \chi_A + \chi_B$ Is A and B are disjoint. It is clear that the definition of $\int s d\mu$ is unaffected by how s is writen. Secondly, if s is a simple function, one has to third a moment to verify that the definition of $\int s d\mu$.

Definition: If $\int |f| d\mu < \infty$, we say f is integrable.

The proof of the next proposition foilwos from the definitions.

Proposition:

(a) If f is measurable, $a \leq f(x) \leq b$ for all x and $\mu(X) < \infty$, then $a\mu(X) \leq \int f d\mu \leq b\mu(X)$;

(b) If $f(x) \leq g(x)$ for all x and f and g are measurable and integrable, then $\int f d\mu \leq \int g d\mu$.

(c) If f is intergtrable, then $\int cfd\mu = c\int fd\mu$ for all real c.

(d) If $\mu(A) = 0$ and f is measurable, then $\int f\chi_A d\mu = 0$.

The integral $\int f\chi_A d\mu$ is often written $\int_A fd\mu$. Other notation for the integral is to omit the μ if it is clear which measure is being used, to write $\int f(x)\mu(dx)$, or to write $\int f(x)d\mu(x)$

Proposition: *If f is integrable,*

$$|\int f| \int |f|.$$

Proof: For the real case, this is easy. $f \leq |f|$, so $\int f \leq \int |f|$. Also $-f \leq |f|$, so $-\int f \leq \int |f|$. Now combine these two facts.

For the complex case, $\int f$ is complex number. If it is 0, the inequality is trivial, If it is not, then $\int f| = r = re^{-i\theta}$ for some r and θ .Then

$$|\int f| = r = e^{-i\theta}\int f = \int e^{-i\theta} f.$$

From the definition of $\int f$ when is complex, we have Re $\left(\int f\right) = \int \text{Re}(f)$. since $|\int f|$ is real, we have

$$|\int f| = \text{Re}\left(\int e^{-i\theta} f\right) = \int (e^{-i\theta} f) \leq \int |f|.$$

We do not yet have that $\int (f+g) = \int f + \int g$.

LIMIT THEOREMS

One of the most important results concerning Lebesgue Integration is the monotone convergence theorem.

Theorem: Suppose f_n is a sequence of nonnegative measurable functions with $f_i(x) \leq f_2(x) \leq ....$ for all x and with $\lim_{n\to} f_n(x) = f(x)$ for all x. Then $\int f_n d\mu \to \int fd\mu$.

Proof: By proposition, $\int f_n$ is an increasing sequence of real numbers. Let L be the limit. Since $f_n \leq f$ for all n thel $L \leq \int f$. We mjust show $L \geq \int f$.

Let $s = \sum_{i=1}^{m} a_i \chi E_i$ be any nonegtive simple function less than f and let $c \in (0, 1)$. Let

$A_n = \{x : f_n(x) \geq cs(x)\}$. Since the $f_n(x)$ increases to $f(x)$ for each x and $c < 1$ then $A_1 \subset A_2 \subset \ldots$, and the union of the A_n is all of X. For each n.

$$\int f_n \geq \int_{An} f_n \geq c \int_{An} s^n$$

$$= c \int_{An} \sum_{i=1}^{m} a_i \chi_{E_i}$$

$$= c \sum_{i=1}^{m} a_i \mu(E_i \cap A_n).$$

If we let $n \to \infty$, by proposition, the right hand coverages to

$$c \sum_{i=1}^{m} a_i \mu(E_i) = c \int s.$$

Therefore $L \geq c \int s$. Since c is arbitrary in the interval (0, 1], then $L \geq \int s$. Taking the supremum over all simple $s \leq f$, we obtain $L \geq \int f$.

Example: let $X = [0, \infty)$ and $f_{n=}(x) = -1/n$ for all x. Then $\int f_n - \infty$, but $f_n \uparrow f$ where $f = 0$ and $\int f = 0$. The problem here is that the f_n are nonegatgive.

Example: Suppose $f_n = n\chi_{(0,1/n).}$ Then $f_n \geq 0, f_n \to 0$ for eacch x but $\int f_n = 1$ does not converge to $\int 0 = 0$. The trouble here is that the f_n do not increase for each x

Once we have the monotone convergence theorem, we can prove that the Lebesgue integral is linear

Theorem: *If f_1 and f_2 are integrable, then*

$$\int (f_1 + f_2) = \int f_1 + \int f_{2.}$$

Proof: First suppose f_1 and f_2 are nonnegative. Take s_n simple and increasing to f_1 and t_n simple and increasing to f_2. Then $s_n + t_n$ increases to $f_1 + f_2$ so the result follows from the monotone convergence theorem and the result for simple functions.

Finally in the general case, write $f_1 = f_1^+ - f_1^-$ and similarly for f_2 and use definition and the result for nonnegative functions.

Proposition: Suppose f_n are nonegative measurable functions. Then

$$\int \sum_{n=1}^{\infty} f_n = \sum_{n=1}^{\infty} \int f_n.$$

Proof: Let $F_N = \sum_{n=1}^{N} f_n$ and write

$$\int \sum_{n=1}^{\infty} f_n = \int \lim_{n \to \infty} \sum_{n=1}^{N} f_n$$

$$\int \lim_{n \to \infty} F_N = \lim_{N \to \infty} \int F_N$$

$$= \lim_{N \to \infty} \sum_{n=1}^{N} \int f_n = \sum_{n=1}^{N} \int f_n,$$

using the monotone convergence theorem and linearity of the integral.

The next theorem is known as Fatou's lemma.

Theorem: Suppose the f_n are nonnegative and measurable. Then

$$\int \liminf_{x \to \infty} f_n \leq \liminf_{x \to \infty} \int f_n.$$

Proof Let $g_n = \inf_{i \geq n} f_i$. then g_n are nonnegative and g_n increases to $\lim\inf f_n$. Clearly $g_n \leq f_i$ for each $i \leq n$, so $\int g_n \leq \int f_i$. Therefore

$$\int g_n \leq \lim_{i \geq n} \int f_i.$$

If we take the limit as $n \to \infty$ on the left hand side we obtain $\int \lim\inf f_n$ by the monotone convergence theorem, while on the right hand side we obtain $\lim\inf_n \int f_n$.

A typical use of Fatou's lemma is the follwwing. Suppose we have $f_n \to f$ and and $\sup_n \int | f_n | \leq K < \infty$. Then $| f_n | \to | f |$, by Fatou's lemma, $\int | f_n | \leq | K$.

Another very important theorem is the domonated convergence theorem/

Theorem: Suppose f_n are measurable functions and $f_n(x) \to f(x)$. suppose there exists an integrable function g such that $| f_n(x) | \leq g(x)$ for all x. Then $\int f_n d\mu \to \int f d\mu$.

Proof: Since $f_n + g \geq 0$, by Fatou's lemma,

$$\int (f + g) \leq \liminf \int (f_n + g).$$

Since g is intergrable,

$$\int f \leq \liminf \int f_n.$$

Similarly, $g - f_n \geq 0$, so

$$\int (g-f) \leq \liminf \int (g-f_n),$$

and hence

$$-\int f \leq \liminf \int (-f_n) = -\limsup \int f_n.$$

Therefore

$$\int f \leq \liminf \int f_n,$$

which with the above proves the theorem.

An example where the limit of the integrals is not the integral of the limit because there is no dominating function g.

If in the monotone convergence theorem or dominted convergence theorem we have only $f_n(x) \to f(x)$ almost everywhere, the conclusion still holds. For if the f_n and f are measurable and $A = \{x: f_n(x) \to f(x)\}$, then $f\chi_A \to f\chi_A$ for each x. And since A^c has measure 0, we see from Proposition that $\int f\chi_A = \int f$, and similarly with f replaced by f_n.

PROPERTIES OF LEBESGUE INTEGRALS

Later on will need the following two propositions.

Proposition: *Suppose f is measurable and for every measurable set A we have* $\int_A f\, d\mu = 0$. *Then f = 0 almost everywhere.*

Proof: Let $A = \{x : f(x) > \varepsilon\}$. Then

$$0 = \int_A f \geq \int_A \varepsilon = \varepsilon\mu(A)$$

Since $f\chi_A \geq \varepsilon\chi_A$. Hence $\mu(A) = 0$. We use this argument for $\varepsilon = 1/n$ and $n = 1, 2, \ldots$, so $\mu\{x : f(x) > 0\} = 0$. Similarly $\mu\{x : f(x) < 0\} = 0$.

Proposition: *Suppose f is measurable and nonnegative and* $\int f\, d\mu = 0$. Then $f = 0$ almost everywhere.

Proof: If f is not almost everywhere equal to 0, there exists an n such that $\mu(A_n) > 0$ where $A_n = \{x : f(x) > 1/n\}$. But then since f is nonnegative,

$$0 = \int f \geq \int f \geq \frac{1}{n}\mu(A_n),$$

a contradiction

We give a result on qpproximating a funcation on $\mathbb{R}$ by continuous functions.

Proposition: Suppose f is a measurable function from $\mathbb{R}$ to $\mathbb{R}$ that is integrable. Let $\varepsilon > 0$.Then there exists a continuous function that is 0 outside some bounded interval such that

$$\int |f - g| < \varepsilon.$$

Proof: If we have continuous function g_1, g_2 such that $\int |f^+ - g_1| < \varepsilon/2$ and $\int |f^- g_2| < \varepsilon/2$, where $f^+ = \max(f, 0)$ and $f^- = \max(-f, 0)$, then taking $g = g_1 - g_2$ will prove our result. So without loss of generality, we may assume $f \geq 0$.

By monotone convergence $\int f\chi_{[-n,n]}$ increases to $\int f$, so by taking *n* large enough, the difference of the integrals will be less than $\varepsilon/2$ If we find g such that $\int |f\chi_{[-n,n]} - g| < \varepsilon/2$, then $\int |f - g| < \varepsilon$. Therefore we may assume that *f* is 0 outside some bounded interval.

We can find simple functions increasing to *f* whose integrals increase to $\int f$. Let s_m be a simple functions such that $s_m \leq f$ and $\int s_m \geq \int f - \varepsilon/2$. If we find *g* such that $\int s_m - g| < \varepsilon/2$, then $\int |f - g| < \varepsilon$. So it sufices to consider the case where *f* is a simple function.

If $f = \sum_{i=1}^{p} a_i \chi A_i$ and we find g_i continuos such that $\int | a_i \chi A_i - g_i | < \varepsilon / p$, then $\sum_{i=1}^{p} g_i$ will be the desired function. So we may assume *f* is a constant times a characteristic function, and by linerarity, we may assume *f* is equal to χ_A for some *A* contained in a bounded interval $[-n, n]$.

We can choose G open and F closed such that $F \subset A \subset G$ and $m(G - F) < \varepsilon$. We can replace *G* by $G \cap (-n - 1, n + 1)$.$G^c \cap [-n - 1, n + 1]$ and *F* are compact sets, so there is a minimum distance between them, say, δ.$g(x) = \max(0, 1 - \text{dist}(x, F/\delta)$.

Then *g* is continuous, $0 \leq g \leq 1$, g is 1 of F,g is 0on G^c, and *g* is 0. Let Then *g* outside of $[-n-1, n+1]$.

therefore

$$|g - \chi_A| \leq \chi_G - \chi_F,$$

So

$$\int |g - \chi_A| \leq \int (\chi_G \chi_F) = m(G - F) < \varepsilon.$$

The method of proof, where one proves a result for characteristic functions, then simple function, then non -negative functions, and then finally integrable functions is very common.

We finish this section with a comparison of the Lebesgue integral and the Riemann integral. Here we are only looking at bounded functions from $[a, b]$ into $\mathbb{R}$ .If we looking at the lebesgue integral, we write $\int f$, while, temporarily, if we are looking at the Riemann integral, we write $R(f)$. Recall that the Riemann integral on $[a, b]$ is defined as follows: if P is q partition of $[a, b]$, then

$$U(P,f)=\sum_{i=1}^{n}\overset{(\sup)}{x_{i=1\le x}} f(x))(x_i - x_{i-1})$$

$$L(P,f)=\sum_{i=1}^{n}\overset{\inf}{x_{i=1\le x\le x}} f(x))(x_{i-1}).$$

Set $\overline{R}(f)=\inf\{U(P,f\}: p$ is a partition$\}$ similarly $\underline{R}(f)$. Then the Riemann integral exists if $\overline{R}(f)=\underline{R}(f)$. and the common value is the Riemann integral, which we denote $R(f)$.

Theorem: A bounded measurable function f on $[a, b]$ is Rieamann integrable if and only if the set of points at which f is discontinuous has Lebesgue measure 0, and in that case the Riemann integral is equal in value to the Lebesgue intergral.

Proof: If P is q partition, define

$$T_P(x)=\sum_{i=1}^{n}\ (\inf_{x_{i-1}\le y\le x_i} f(y))\chi_{[x_{i-1},x_i)}(x).$$

and
$$S_P(x)=\sum_{i=1}^{n}\ (\inf_{x_{i-1}\le y\le x_i} f(y))\chi_{[x_{i-1},x_i)}(x).$$

We see that $\int T_P=U(P,f)$ and $\int S_P=L(P,f)$.

If f is Riemann integrable, there exists a sequence of partitions Q_i such that $U(Q_i f)\downarrow R(f)$ and a sequence Q_i such that $L(Q_i{}',f)\uparrow R(f)$. It is not hard toa check that adding point to a partition increases L and decreases U, so if we let $P_i=\bigcup_{j\le i}(Q_j\cup Q_j')$, then P_i is an increasing sequence of partition, $U(P_i,f)\downarrow R(f), L(P_i,f)\uparrow R(f)$.

We see also that $T_{pi}(x)$ decreases at each point, say to $T(x)$ increases at each point say, to $S(x)$. Also $T(x)\ge f(x)\ge S(x)$. Then by dominted convergence (recall that f is bounded)

$$\int(T-S)=\lim_{i\to\infty}\int(T_{P_i}S_{P_i})=\lim_{i\to\infty}(U(P_i,f)L(P_i,f))=0.$$

We conclude $T=S=f$ a.e. If x is not in the null set where $T(x)\ne S(x)$ nor in $\bigcup_i P_i$, which is countable and hence of Lebesgue measure 0, then $T_{P_i}(x) \downarrow f(x)$ and $S_{P_i}(x) \uparrow f(x)$. This implies that f is continuous at such f. Since

$$R(f) = \lim_{i\to\infty} U(p_i, f) = \lim_{i\to\infty} \int T_{p_i} \int f,$$

We see the Riemann integral and Lebesgue integral agree.

Now suppose that f is continuous a.e Let $\varepsilon > 0$ Let P_i be partition where we divide $[a, b]$ into 2^i equal parts. If x is not in the null set where f is discontinuous, nor in

$$U(P_i, f) = \int T_{P_i} \to \int f$$

and

$$L(P_i, f) = \int S_{P_i} \to \int f.$$

This does it.

Chapter 9

σ-Algebra of Borel Subsets

SIGMA-ALGEBRA

In mathematics, a σ-algebra (or sigma-algebra) (sigma is a Greek letter, upper case Σ, lower case σ) over a set X is a nonempty collection Σ of subsets of X that is closed under complementation and countable unions of its members. It is a Boolean algebra, completed to include countably infinite operations. The pair <X, Σ> is also a field of sets, sometimes called a σ-field or a measurable space.

Thus, if $X=\{a, b, c, d\}$, one sigma algebra could be $\Sigma = \{\varnothing, \{a, b\}, \{c, d\}, \{a, b, c, d\}\}$.

The main use of σ-algebras is in the definition of measures on X. The concept is important in mathematical analysis and probability theory.

DEFINITION AND PROPERTIES OF SIGMA ALGEBRA

Formally, a subset Σ of the power set of a set X is a σ-algebra if and only if it has the following properties:

1. Σ is nonempty
2. If E is in Σ then so is the complement ($X \setminus E$) of E.
3. The union of countably many sets in Σis also in Σ.

In other words, a family Σ of subsets of X is a σ-algebra if:

1. Σ contains X (or, Σ contains the empty set)
2. Σ is closed under complements
3. Σ is closed under countable unions.

From these axioms, it follows that X and the empty set are in Σ, and that the σ-algebra is also closed under countable intersections (via De Morgan's laws).

A measure on X is a function which assigns a real number to subsets of X; this can be thought of as making precise a notion of 'size' or 'volume' for sets. One might like to assign such a size to *every* subset of X, but the axiom of choice implies that when the size under consideration is standard length for subsets of the real line, then there exist sets known as Vitali sets for which no size exists. For this reason, one considers instead a smaller collection of privileged subsets of X whose measure is defined; these sets constitute the σ-algebra.

Elements of the σ-algebra are called measurable sets. An ordered pair (X, Σ), where X is a set and Σ is a σ-algebra over X, is called a measurable space. A function between two measurable spaces is called measurable if the preimage of every measurable set is measurable. The collection of measurable spaces forms a category with the measurable functions as morphisms. Measures are defined as certain types of functions from a ó-algebra to $[0,\infty]$.

σ-algebras are sometimes denoted using capital letters from a Fraktur typeface. Thus, (X,Σ)may be used to denote (X,Σ). Another common convention is to use calligraphic capital letters in place of Σ, thus (X,A)is often used in place of (X,Σ). This is handy to avoid situations where Σ might be confused for the summation operator.

GENERATED σ-ALGEBRA

If U is an arbitrary family of subsets of X then we can form a special σ-algebra from U, called the *σ-algebra generated by U*. We denote it by $\sigma(U)$ and define it as follows. First note that there is a σ-algebra over X that contains U, namely the power set of X. Let Σ be the family of all σ-algebras over X that contain U (that is, a σ-algebra Σ over X is in Σ if and only if U is a subset of Σ.) Then we define $\sigma(U)$ to be the intersection of all σ-algebras in Σ. σ (U) is then the smallest σ-algebra over X that contains U. For a simple example, consider the set $X=\{1,2,3\}$. Then the σ-algebra generated by the subset $\{1\}$ is $\sigma(\{1\}) = \{\varnothing, \{1\}, \{2,3\}, X\}$. By an abuse of notation, when the collection of subsets contains only one member, call it A, one may write $\sigma(A)$ instead of $\sigma(\{A\})$.

In general, there is no explicit description of the σ-algebra generated by a given collection. This is in contrast with a similar construction in topology: the smallest topology generated by a family of subsets can be explicitly described. For example, there can be sets in the generated algebra constructed from the generator system only in transfinite number of steps. However, it is a theorem that a generated σ-algebra is the (possibly transfinite) union of all σ-algebras generated by the countable sub-systems of the generator system.

Examples: Let X be any set, then the following are σ-algebras over X:

1. The family consisting only of the empty set and X (the minimal or trivial σ-algebra over X).

2. The full power set of X.
3. The collection of subsets of X which are countable or whose complements are countable (which is distinct from the power set of X if and only if X is uncountable.). This is the σ-algebra generated by the singletons of X.
4. If $\{\Sigma_a\}$ is a family of ó-algebras over X, then the intersection of all Σ_a is a σ-algebra over X.

Examples for Generated Algebras

An important example is the Borel algebra over any topological space: the σ-algebra generated by the open sets (or, equivalently, by the closed sets). Note that this σ-algebra is not, in general, the whole power set.

On the Euclidean space R^n, another σ-algebra is of importance: that of all Lebesgue measurable sets.

This σ-algebra contains more sets than the Borel σ-algebra on R^n and is preferred in integration theory, as it gives a complete measure space.

Definition: For S a set we let $P(S)$ be the set of all subsets of X. $\subset P(S)$ is a σ-algebra if it is closed under complements, countable unions, and countable intersections. $B \subset \mathbb{R}^N$ is Borel if it appears in the smallest σ-algebra containing the open sets.

Of course one issue is why this is even well defined. Why should there be a unique smallest such algebra? The answer is that we can take the intersection of all σ-algebras containing the open sets and this is easily seen to itself be a σ-algebra.

Definition: If S is a set and $\sum \subset P(S)$ a σ-algebra, then

$$\mu : \sum \to \mathbb{R} \cup \{\infty\}$$

is a measure if

(a) $\mu(\phi) = 0$;

(b) $\mu(A) \geq 0$ all $A \in \Sigma$;

(c) $\mu\left(\bigcup_i A_i\right)$ equals

$$\sum_{i=1}^{\infty} \mu(A_i)$$

whenever $(A_i)_{i \in N}$ is a sequence of disjoint elements of Σ.

We will take without proof the following key fact.

Theorem: Let $N \in \mathbb{N}$ and $\Sigma \subset P(\mathbb{R}^N)$ the σ-algebra of Borel subsets of N-dimensional Euclidean space. Then there is a unique measure

$$m : \Sigma \to \mathbb{R}$$

such that whenever $A = I_1 \times I_2 \times ... \times I_N$ is a rectangle, each I_n an interval of the form (a_n, b_n), $[a_n, b_n)$, $(a_n, b_n]$, or $[a_n, b_n]$ we have

$$m(A) = (b_1 - a_1) \times (b_2 - a_2) \times ...(b_N - a_N).$$

Definition: The unique measure described by the above theorem is called the Lebesgue measure.

It would take us far too afield to prove the theorem here. However it is worth mentioning a key underlying fact which can be used to give its proof.

Lemma: If B is a Borel subset of $\mathbb{R}^N$ and $\varepsilon > 0$ then there are rectangles $(A_i)_{i \in \mathbb{N}}$, $(B_i)_{i \in \mathbb{N}}$ such that

(a) $B \subset \bigcup A_i$;

(b) $\bigcup A_i \setminus B \subset \bigcup B_i$;

(c) $\sum_{i \in \mathbb{N}} m(B_i) < \in$.

We let $\Sigma *$ be the collection of all sets with properties (a)-(c). The method is then to show that $\Sigma *$ includes the open sets (obvious), is closed under complements (needs a bit of thought), and closed under countable unions (easy). Thus $\Sigma *$ is a σ-algebra containing the open sets – since the Borel sets form the smallest σ-algebra containing the open sets, they must be included in $\Sigma *$.

Definition: We say that $M \subset \mathbb{R}^N$ is measurable if there are Borel sets A, B with

$$A \subset M \subset B,$$

$$m(B \setminus A) = 0.$$

We then let $m(M) = m(A)$. This extension of m to this class of measurable sets is said to be the completion of m. It is easily checked that the measurable sets form a σ-algebra and that this completion of m, which I am simply again denoting by m, is again a measure.

Well and good. However there are other kinds of spaces and objects to which we might wish to assign something like a measure. Before embarking on the formal development, here are some examples.

Examples:

1. For a finite set X and $P(X)$ the set of all subsets of X, we could take the counting measure: $m(A) = |A|$, the size of A.

2. For $\{H, T\}^N$, which could be thought of as tossing a coin with outcome either "H" (heads) or "T" (tails), we could take the normalized counting measure:

$$\mu(A) = \frac{|A|}{2^N}$$

3. An example which will be important later on is the infinite run, $\{H, T\}^{\mathbb{N}}$. Note that this is a compact metric space in the product topology. For a cylinder set of the form $A = \{f: f(1) = S_1, f(2) = S_2, \ldots f(N) = S_N\}$ we let $\mu(A) = 2^{-N}$; more generally given $S \subset 2^N$ we let $\mu(\{f: f|_N \in S\}) = \frac{|S|}{2^{-N}}$. Since the function μ is σ-additive on the algebra generated by the cylinder sets, there is a unique extension of μ to a measure on the σ-algebra it generates – which is in fact the full collection of Borel subsets of $\{H, T\}^{\mathbb{N}}$.

4. We might instead be working with a slightly biased coin, that comes down heads 7 times out of ten. Then for each f: $\{1, 2, \ldots, N\} \to \{H, T\}$ we set

$$\mu(\{f\}) = \left(\frac{7}{10}\right)^{|\{i: f(i)=H\}|} \cdot \left(\frac{3}{10}\right)^{|\{i: f(i)=T\}|}$$

Similarly we could define this lopsided measure on the space of infinite runs.

5. Somewhat more loosely, imagine we are dealing with some experiment, such as tickling a dangerous African land animal with a long feather or shooting gamma rays into a metal alloy, and X is the space of all possible outcomes to the experiment. Let $f: X \to \mathbb{R}$ be some function which arises from measuring some property of the outcome (e.g. the decibels of the dangerous animal's laugh or the heat of the metal alloy). We could then define a measure on $\mathbb{R}$ by letting $\mu(A)$ be the probability that f assumes its value in A.

Definition: A set X equipped with a σ-algebra $\Sigma \subset P(X)$ is said to be a measure space. If $\mu : \Sigma \to \mathbb{R}$ is a measure, then we say that $M \subset X$ is measurable (with respect to μ) if there are $A, B \in \Sigma$ with

$$A \subset M \subset B$$

with $B \setminus A$ null (i.e. $\mu(B \setminus A) = 0$). A function $f: X \to \mathbb{R}$ is said to be measurable with respect to if for any open set $U \subset \mathbb{R}$ we have $f^{-1}[U] \in \Sigma$. A function $f: X \to \mathbb{R}$ is said to be measurable with respect to μ if for any open $U \subset \mathbb{R}$ we have $f^{-1}[U]$ measurable with respect to μ.

Beware: Often authors simply write "measurable" instead of "measurable with respect to μ" when context makes the intended measure clear.

Lemma: Let μ be a measure on the measure space (X, Σ). Then the measurable sets form a σ-algebra

Proof First suppose M is measurable as witnessed by $A, B \in \Sigma$, $A \subset M \subset B$. Then $A^c = X \setminus A$, $B^c = X \setminus B$ are in and have $A^c \supset M^c \supset B^c$. Since $A^c \setminus B^c = B \setminus A$, this witnesses M^c measurable.

If $(M_n)_{n\in\mathbb{N}}$ is a sequence of measurable sets, as witnessed by $(A_n)_{n\in\mathbb{N}}, (B_n)_{n\in\mathbb{N}}$, with $A_n \subset M_n \subset B_n$, then

$$\bigcup_{n\in\mathbb{N}} A_n \subset \bigcup_{n\in\mathbb{N}} M_n \subset \bigcup_{n\in\mathbb{N}} B_n$$

On the other hand

$$\bigcup_{n\in\mathbb{N}} B_n \setminus \bigcup_{n\in\mathbb{N}} A_n \subset \bigcup_{n\in\mathbb{N}} (B_n \setminus A_n),$$

and so $\bigcup_{n\in\mathbb{N}} B_n \setminus \bigcup_{n\in\mathbb{N}} A_n$ is null, as required to witness $\bigcup_{n\in\mathbb{N}} M_n$ null.

Definition: Let μ be a measure on the measure space (X, Σ). For M measurable, as witnessed by $A \subset M \subset B$ with $A, B \in \Sigma(B \setminus A) = 0$, we let $\mu^*(M) = \mu(A)(= \mu(B))$. We call μ" the completion of μ.

As with Lebesgue measure, we will commit the minor logical sin of using the same symbol for μ as its extension μ to the measurable sets. A more serious issue is to check the measure is well defined.

Lemma: The completion of μ to the measurable sets is well defined.

Proof: If $A_1 \subset M \subset B_1$ and $A_2 \subset M \subset B_2$ both witness M measurable, then

$$A_1 \Delta A_2 \subset M \setminus A_1 \cup M \setminus A_2 \subset B_1 \setminus A_1 \cup B_2 \setminus A_2.$$

and hence is null.

Lemma: Let μ be a measure on a measure space (X, Σ). Let Σ^* be the σ-algebra of measurable sets. Then the completion defined above,

$$\mu: \Sigma^* \to \mathbb{R},$$

is a measure.

Definition: A measure μ on a measure space (X, Σ) is σ-finite if X can be written as a countable union of sets in Σ on which μ is finite.

Definition: Let (X, Σ) be a measure space equipped with a measure

$$\mu: \Sigma \to \mathbb{R}.$$

A function

$$h: X \to \mathbb{R}$$

is simple if we can partition X into finitely many measurable sets $A_1, A_2, \ldots, A_n$ with h assuming a constant value a_i on each Ai. If $\mu(A_i) < \infty$ whenever $a_i \neq 0$ we say that h is integrable and let

$$\int_X h d\mu$$

be the sum of all $\mu(A_i)$. a_i for $a_i \neq 0$. For $B \subset X$ a measurable set, we define

$$\int_B h d\mu$$

to be the sum of $\mu(A_i \cap B)$. a_i for $a_i \neq 0$.

If, as in [4], we adopt the convention that ∞. 0 = 0 then we can simply write this out as

$$\int_X h d\mu = \sum_{i \leq n} \mu(A_i).a_i.$$

Definition: For (X, Σ, μ) as above, $f: X \to \mathbb{R}$ measurable and non-negative ($f(x) \geq 0$ all x " X), we let

$$\int_X f d\mu = \sup\left\{\int_S hdu : h \text{ is simple and } 0 \leq h \leq f\right\}$$

In general given $f: X \to \mathbb{R}$ measurable we can uniquely write $f = f^+ - f^-$ where f^+, f^- are both non-negative and have disjoint supports. Assuming

$$\int_X f^+ d\mu, \int_X f^- d\mu$$

are both finite we say that f is integrable and let

$$\int_X f d\mu = \int_X f^+ d\mu - \int_X f^- d\mu.$$

We have implicitly used above that f^+, f^- will be measurable. This is easy to check. You might also want to check that we could have instead used the definition $f^+ = \frac{1}{2}(|f| + f)$, and $f^- = \frac{1}{2}(|f| - f)$.

This definition of integration extends to other settings.

Definition: Let (X, Σ, μ) be a measure space equipped with a measure μ; we say that a function from X to $\mathbb{C}$ is measurable if the pullback of any open set in $\mathbb{C}$ is measurable.

For $f: X \to \mathbb{C}$ measurable, we write $f = \text{Re} f + i \text{ Im } f$, where $\text{Re } f: X \to \mathbb{R}$ and $\text{Im } f: X \to \mathbb{R}$ are the real and imaginary parts.

We say that f is integrable if both these functions are integrable in our earlier sense and let

$$\int_X f d\mu = \int_X \operatorname{Re} f d\mu + i \int_X \operatorname{Im} f d\mu .$$

Real integration. Complex integration. But there is no reason to stop here. We can integrate functions into almost any kind of vector space for which there is an appropriate notion of convergence. Thinking about the process above, it is immediately clear how to integrate a step function, as long as we can take linear combinations on the other side; after that we try to approximate a measurable f by taking suitable simple functions.

Definition: Let B be a Banach space and (X, Σ, μ) a measure space equipped with measure $\mu: \Sigma \to \mathbb{R}$.

$$h: X \to B$$

is simple if we can partition X into finitely many measurable sets $A_1, A_2, \ldots, A_n$ with h assuming a constant value a_i on each A_i. If $\mu(A_i) < \infty$ whenever $a_i \neq 0$ we say that h is integrable and let

$$\int_X f d\mu$$

be the sum in B of all $\mu(A_i)$. a_i for $a_i \neq 0$. We define the L^1-norm of an integrable simple function h by

$$\| h \| L^1 = \int_S \| h(.) \| d\mu.$$

Note then that $\| h_1 - h_2 \| < \varepsilon \Rightarrow \left| \int_S h_1 d\mu - \int_S h_2 d\mu \right| < \varepsilon$. A function $f: S \to X$ is Bochner integrable if there is a sequence (h_n) of integrable simple functions such that

$$\| h_n(x) - f(x) \| \to 0 \text{ for almost all } x \in S,$$

and such that (h_n) is Cauchy with respect to the L^1-norm. The integral of f is then defined as

$$\int_S f d\mu = \lim_{n \to \infty} \int_S h_n d\mu;$$

The theory of measurable spaces and measurable can be developed in various levels of generality. I generally take the view that most of the natural spaces in this context are either Polish spaces or standard Borel spaces.

Definition: A topological space is Polish if it is separable and admits a compatible complete metric.

We then define the Borel sets in the space to be those appearing in the smallest σ-algebra containing the open sets.

Examples:

1. Any compact metric space forms a Polish space. For instance if we take

$$2^{\mathbb{N}} = \prod_{\mathbb{N}} \{0,1\},$$

the countable product of the two element discrete space $\{0, 1\}$, then we have a Polish space. (For the metric, given $\vec{x} = (x_0, x_1,...)$, $\vec{y} = (y_0, y_1,...)$, take $d(\vec{x}, \vec{y})$ to be 2^{-n}, where n is least with $x_n \neq y_n$.)

2. $\mathbb{R}$ and $\mathbb{C}$ are Polish spaces, as are all the $\mathbb{R}^N$'s and $\mathbb{C}^N$'s.

3. Any closed subset of a Polish space is Polish.

4. Let $\mathbb{C}$ ([0, 1]) be the collection of continuous functions from the unit interval to $\mathbb{R}$. Given $f, g \in \mathbb{C}\,\mathbb{C}([0, 1])$ let $d(f, g)$ be

$$\sup_{z\in[0,1]} |f(z) - g(z)|.$$

As some of you may be aware, this metric can be shown to be complete and separable, and hence the induced topology is Polish. (More generally, any separable Banach space is Polish in the topology induced by the norm.)

Definition: A set X equipped with a σ-algebra is said to be a standard Borel space if there is some choice of a Polish topology on X which gives rise to Σ as the corresponding collection of Borel sets.

At the end of these short definitions there is a remarkable fact whose proof is too involved to present here.

Theorem: Any two uncountable standard Borel spaces are Borel isomorphic.

That is to say, if (X_1, Σ_1), (X_2, Σ_2) are uncountable Borel spaces, then there is a bijection

$$\pi: X_1 \to X_2$$

such that for all $A \subset X_2$

$$A \in \Sigma_2 \Leftrightarrow \pi^{-1}[\grave{A}] \in \Sigma_1$$

Definition: If X is a standard Borel space and $\Sigma \subset P(X)$ the σ-algebra of Borel sets, then a Borel measure on X is a measure in the earlier sense

$$\mu: \Sigma \to \mathbb{R}.$$

Again we say that $M \subset X$ is measurable if there are Borel sets A, B with

$$A \subset M \subset B,$$

$$\mu(B \setminus A) = 0.$$

We then let $\mu(M) = \mu(A)$.

Definition: A measure μ on X is said to have a point $a \in X$ as an atom if $\mu(\{a\}) > 0$. μ is said to be a probability measure if $\mu(X) = 1$. A standard Borel space equipped with a Borel probability measure is called a standard Borel probability space.

And again we have a remarkable fact:

Theorem: Any two atomless standard Borel probability spaces are isomorphic.

In other words, if (X_1, Σ_1, μ_1), (X_2, Σ_2, μ_2) are atomless standard Borel probability spaces, then there is a bijection

$$\pi: X_1 \to X_2$$

sending inducing an isomorphism of $(X_1, \Sigma_1) \cong (X_2, \Sigma_2)$ with the additional property that for $A \in \Sigma_2$

$$\mu_2(A) = \mu_1(\pi^{-1}[A]).$$

In particular, any atomless standard Borel probability space is isomorphic to the unit interval equipped with the (restriction of) Lebesgue measure.

In this course I will largely concentrate on probability spaces. In the literature, almost all work on infinite measure spaces is under the assumption of the space being σ-finite – and in this case we can partition the space into countably many pieces of measure one. In the case of finite spaces with measure other than one, we can rescale the measure by a constant to get the total mass back to one. Thus the loss of generality in working with probability spaces is very minor.

Definition: A function $f: X \to Y$ between two Polish spaces is said to be Borel if $f^{-1}[U]$ is Borel for any open $U \subset Y$.

Lemma: If $f: X \to Y$ is Borel, then for any Borel $B \subset Y$ we have $f^{-1}[B]$ Borel.

Proof: Let Σ be the collection of subsets of Y for which the pullback along f is Borel. By assumption this contains the open sets. It is a σ-algebra, since

$$f^{-1}[Y \setminus B] = X \setminus f^{-1}[B]$$

and

$$f^{-1}\left[\bigcap B_n\right] = \bigcap f^{-1}[B_n].$$

Thus it includes the Borel sets.

With this lemma in our tool kit we can go forward and define the concept of Borel function when there is no topology in sight.

Definition: Let X and Y be standard Borel spaces. $f: X \to Y$ is a Borel function if $f^{-1}[B]$ is Borel for any Borel $B \subset Y$.

Lemma: $f: X \to \mathbb{R}$ is measurable if $f^{-1}[O]$ is measurable for each open $O \subset \mathbb{R}$.

Proof: Let be the collection of all $A \subset \mathbb{R}$ with $f^{-1}[A]$ measurable. Σ includes the open sets by assumption. Since the measurable subsets of X form a σ-algebra and

$$f^{-1}[\mathbb{R} \setminus A] = X \setminus f^{-1}[A],$$

$$f^{-1}\left[\bigcup A_i\right] = \bigcup f^{-1}[A_i],$$

$$f^{-1}\left[\bigcap A_i\right] = \bigcap f^{-1}[A_i],$$

we have that Σ includes the Borel sets.

This concept of measure is apparently ethereal. It involves considering the collection of all Borel subsets and considering the behavior of the measure with respect to arbitrary sequences of Borel sets. Later in the course we will discuss a form of the Riesz representation theorem which enables us to give a concrete description of the collection of Borel probability measures on a compact metric space; this representation theorem will in particular enable us to view the collection of probability measures as forming a well behaving topological space in its own right.

RADON-NIKODYM AND CONDITIONAL EXPECTATION

One of the most important theorems in measure theory is Radon-Nikodym. It can be proved without a large amount of background and we may as well do so now. There are a few lemmas which make it possible for us to analyze the difference between two measures; we will prove these, then the Radon-Nikodym theorem, and finish with an application to conditional expectation. It all starts with the introduction of a variation on the idea of measure.

Definition: Let X be a set and $\Sigma \subset P(X)$ a σ-algebra.

$$\mu: \Sigma \to \mathbb{R}$$

is said to be a signed measure if

(a) $\mu(f) = 0$;

(b) If $(A_n)_{n \in \mathbb{N}}$ is a sequence of disjoint sets in Σ, then

$$\mu\left(\bigcup_{n \in \mathbb{N}} A_n\right) = \sum_{n=1}^{\infty} \mu(A_n)$$

Note here we are assuming finiteness of the measure and in (b) above we are demanding convergence of the series. As I have previously remarked, we will not be losing much to always think of our measures as defined on standard Borel spaces, though in the next couple of lemmas we will not be making such an assumption; it will be technically convenient to develop the results in full generality.

Lemma: If Σ is a σ-algebra on X and $\mu: \Sigma \to \mathbb{R}$ is a signed measure, then whenever $(B_n)_{n\in\mathbb{N}}$ is a sequence of sets in Σ

$$\mu\left(\bigcap_{n\in\mathbb{N}} B_n\right) = \lim_{N\to\infty} \mu\left(\bigcap_{n\le N} B_n\right)$$

Proof First some cosmetic rearrangement. Let $C_n = \bigcap_{i\in n} B_i$. So at every n we have $C_n \supset C_{n+1}$, but the sequence has the same infinite intersection. Now consider the difference sets and define $D_n = C_n \backslash C_{n+1}$; the D_n's are now disjoint and if we let $B_\infty = \bigcap_{i\in\mathbb{N}} B_i$ represent the infinite intersection we have the equalities

$$C_n = B_\infty \cup D_n \cup D_{n+1} \cup D_{n+2}\ldots$$

at every n. Thus

$$\mu(C_n) = \mu(B_\infty) + \sum_{m\ge n} \mu(D_m)$$

This in particular implies $\sum_{m=1}^{\infty} \mu(D_m)$ is convergent and

$$\sum_{m\ge n} \mu(D_m) \to 0$$

as $n \to \infty$, which is all we need to ensure $\mu(C_n) \to \mu(B_\infty)$.

Hahn Decomposition Theorem

Theorem: Let Σ be a σ-algebra on X and $\mu: \Sigma \to \mathbb{R}$ a signed measure. Then there exists $A \in \Sigma$ such that for all $B \infty \Sigma$, $B \subset A$

$$\mu(B) \ge 0$$

and for all $C \in \Sigma$, $C \subset X \backslash A$

$$\mu(C) \le 0 .$$

Proof: Let σ be the supremum of the set $\{\mu(A): A\ \Sigma\}$. Let $(B_n)_{n\in\mathbb{N}}$ be a sequence of sets in Σ with

$$\mu(B_n) \to \infty.$$

Then at each n let A_n be the algebra of sets generated by $(B_i)_{i\le n}$.

Let's pause for a moment and observe some of the properties of these A_n algbras. First, each is finite, since it is generated by finitely many sets. Moreover A_n will have "atoms" of the form

$$\bigcap_{i\in S} B_i \cap \bigcap_{i\le n, i\notin S} B_i :$$

each of these atoms contains no smaller no empty set in A_n and every element of A_n is the finite union of such atoms. Finally note that B_n is an element of A_n.

At each *n*, let C_n be element of A_n with maximum value under μ. Since $B_n \in A_n$ we have

$$\mu(B_n) \le \mu(C_n)$$

and $\mu(C_n) \to \infty$ and C_n consists of the finite union of atoms in A_n with positive measure.

Claim: At each *n*, $\mu\left(\bigcup_{i\ge n} C_i\right) \ge \mu(C_n)$.

Proof of Claim: Since at any $j \ge n, C_j \setminus \bigcup_{n\le i<j} C_i$ equals the union of finitely many atoms with positive measure. (Claim)

Now let

$$A = \bigcap_{n\in\mathbb{N}} \bigcup_{i\ge n} C_i .$$

Then by the last lemma $\mu(A) = \lim_{n\to\infty} \mu\left(\bigcup_{i\ge n} C_i\right) = \delta$.

Since *A* has attained this maximum value δ we must have every $B \subset A$ in with $\mu(B) \ge 0$ (for otherwise $\mu(A \setminus B)$ would be greater than $\mu(A)$). Similarly for any $C \in \Sigma$ disjoint to *A* we must have $\mu(C) \le 0$.

HAHN-JORDAN DECOMPOSITION

Theorem: Let Σ be a σ-algebra on a set *X*. Let $\mu: \Sigma \to \mathbb{R}$ be a signed measure. Then we can find to measures $\mu^+, \mu^- : \Sigma \to \mathbb{R}^{\ge 0}$ with

(i) $\mu = \mu^+ - \mu^-$;

(ii) μ^+, μ^- have disjoint support.

Proof: The statement of the theorem should be quickly clarified. (i) states that for any set $C \in \Sigma$ we have $\mu(C) = \mu^+(C) - \mu^-(C)$. (ii) states that we can some $A \in \Sigma$ with $\mu^+(B) = 0$ when $B \subset X \setminus A$ and $\mu^-(B) = 0$ when $B \subsetneq A$.

With this clarification in mind, the proof of the theorem is an immediate consequence of 2.2: Choose *A* for μ as there, and then let $\mu^+(B) = \mu(A) - B)$ and $\mu^-(B) = -\mu(B \setminus A)$.

Definition:. Given two measures μ, *v*: $\Sigma \to \mathbb{R} \cup \{\infty\}$, we say that μ is absolutely continuous with respect to *v*, written

$$\mu << \mu,$$

if whenever $B \in \Sigma$ has $v(B) = 0$ then $\mu(B) = 0$.

We have dealt with the concept of measurable functions in different contexts already. So there is no possibility of confusion, let us fix a convention for the entirely general context of σ-algebras.

Definition: Let be a σ-algebra on a set X. $f: X \to \mathbb{R}$ is measurable with respect to if $f^{-1}[U]$ is in Σ for any open $U \subset \mathbb{R}$.

Radon-Nikodym

Theorem: Let Σ be a σ-algebra on a set X. Let μ, $v: \Sigma \to \mathbb{R} \cup \{\infty\}$ be σ-finite measures with $\mu << v$. Then there is a measurable with respect to function

$$f: X \to \mathbb{R}$$

such that for any $C \in \Sigma$

$$\mu(C) = \int_C f(x)dv(x).$$

Proof: We can assume μ, v are finite measures, since otherwise we could partition the space X into a countable collection of elements in on which both are finite and it would suffice to prove the theorem on each of these pieces.

At each $q \in \mathbb{Q}$, $q > 0$, let $\mu_q = \mu - q.\ v$; that is to say, we define μ_q by

$$\mu_q(A) = \mu(A) - qv(A).$$

Each of these is a signed measure on (X, Σ). We can find $A_q \in \Sigma$ with $\mu_q(B) \geq 0$ all $B \subset A_q$, $B \in \Sigma$, $\mu_q(C) \leq 0$ all $C \subset X \backslash Aq$, $C \in \Sigma$.

Note that for $q_1 < q_2$ we have $A_{q_1} \setminus A_{q_2}$ null with respect to v and hence after discarding some null sets we can assume

$$q_1 < q_2 \Rightarrow A_{q_2} \subset A_{q_1}.$$

By the assumption of $\mu << v$ we get $\bigcap_{q \in \mathbb{Q}} A_q$ null with respect to both these measures – since otherwise we could let $A_\infty = \bigcap_{q \in \mathbb{Q}} A_q$ and unwinding the definitions we would have $\mu(A_\infty) > qv(A_\infty)$ all $q \in \mathbb{Q}$, which would imply $v(A) = 0$; and then again after possibly discarding a null set we can assume

$$\bigcap_{q \in \mathbb{Q}} A_q = \phi;$$

so if we let

$$f(x) = \sup\{q: x \in A_q\}$$

we obtain a measurable with respect to Σ function $f: X \to \mathbb{R}^{\geq 0}$.

For $q_1 < q_2$ let $B_{q_1,q_2} = A_{q_1} \setminus A_{q_2}$.

Claim: For $B \subset B_{q_1,q_2}$ in Σ

$$\left|\int_B f(x)d\mu(x) - \mu(B)\right| \leq (q_2 - q_1)v(B).$$

Proof of Claim: We have $B \subset A_{q_1}$

$$\therefore \mu_{q_1}(B) \geq 0$$

$$\therefore \mu(B) \geq q_1 v(B),$$

and similarly B is disjoint to A_{q_2} and

$$\therefore \mu_{q_2}(B) \leq 0,$$

$$\therefore \mu(B) \leq q_2 v(B).$$

In other words, we have the inequality

$$q_1 v(B) \leq v(B) \leq q_2 v(B).$$

Then since $f(x)$ ranges between q_1 and q_2 on B_{q_1,q_2} and hence B we obtain the parallel inequality

$$q_1 v(B) \leq \int_B f(x)dv(x) \leq q_2 v(B),$$

and hence

$$\left|\int_B f(x)d\mu(x) - \mu(B)\right| \leq (q_2 - q_1)v(B),$$

as required. (Claim)

This last observation is all we need. Given any $C \subset X$ we can first fix $\varepsilon > 0$ and let

$$C_\ell = C \cap B_{\ell-\varepsilon,(\ell+1).\varepsilon}.$$

Again we are implicitly using $\mu << v$ to see that $C = \bigcup_{\ell \in \mathbb{N}} C_\ell$.

$$\left|\int_C f(x)dv(x) - \mu(C)\right| = \left|\sum_{\ell \in \mathbb{N}} \int_{C_\ell} f(x)dv(x) - \sum_{\ell \in \mathbb{N}} \mu(C_\ell)\right|$$

$$\leq \sum_{\ell \in \mathbb{N}} \left| \int_{C_\ell} f(x) dv(x) - \mu(C_\ell) \right|$$

which by the above claim is bounded by

$$\sum_{\ell \in \mathbb{N}} \epsilon\, v(C_\ell) = \epsilon\, v(C).$$

Letting ε tend to zero we obtain $\int_C f(x) dv(x) = v(C)$.

The function f we arrived at in the theorem above is not necessarily unique, but it is almost unique. I will leave at as an exercise for you to see that if f_0 is another function with

$$\mu(C) = \int_C f_0(x) dv(x)$$

on any $C \in \Sigma$ then f_0 agrees with f off a v null set. This virtual uniqueness motivates a definition: We say that f as in the theorem is the Radon-Nikodym derivative, and is sometimes denoted by the slightly poetical notation

$$\frac{d\mu}{dv}.$$

A crucial application of Radon-Nikodym is the existence of conditional expectation. At first sight the theorem may appear abstract to the point of being ethereal. A couple of motivating examples can give a sense of its true content.

Examples:

1. Let X be a finite set – say $X = \{1, 2, 3, 4, 5, 6\}$. Let $f: X \to \mathbb{R}$. For instance, $f(n) = n^2$, just for example. Let B be a subset of X – say $B = \{2, 3, 4\}$. If someone tells you that they are thinking of a number chosen randomly from B, you would probably have an intuitive idea of the expectation of f on B:

   You would probably take the average value of f over B:
   $$\frac{1}{3}(4+9+19) = 9\frac{2}{3}.$$
2. A little bit more abstract, let us take X to be the surface of the planet earth and $f(x)$ the average temperature of the location x. Using that information you would probably be able to go ahead and calculate the average temperature at a given latitude. So in this way we could discard, as it were, some of the information carried by f and obtain another function which records averages along the latitudes alone.
3. Alternatively, you may have a formula which can precisely compute the oxygen intake of a microbe based on its size and age. In the course of an experiment perhaps only the age is known, and then the best guess as to the oxygen intake would be your expectation given the partial information available.

Intuitively then it doesn't seem outrageous to give a best guess or expectation of a function on the basis of partial information. The following slick theorem justifies this rigorously. With Radon-Nikodym already available to us, the proof is very, very short – don't blink or you will miss it.

Theorem: Let $\Sigma_0 \subset \Sigma_1$ be two σ-algebras on a set X. Let μ be a σ-finite measure on (X, Σ_1) and let

$$f: X \to \mathbb{R}$$

be measurable with respect to 1.

Then there is a function

$$g: X \to \mathbb{R}$$

which is measurable with respect to Σ_0 such that on any $B \in \Sigma_0$

$$\int_B g(x)d\mu(x) = \int_B f(x)d\mu(x).$$

Proof: First of all, we can assume $f \geq 0$, since otherwise we write $f = f^+ - f^-, f^+ = \frac{1}{2}$ $(|f| + f), f^- = \frac{1}{2}(|f| - f)$ and apply the result to these two non-negative functions in turn.

Now let v be defined on (X, Σ_0) by

$$v(B) = \int_B f(x)d\mu(x)$$

all $B \in \Sigma_0$. Let $\mu_0 = \mu \mid \Sigma_0$, the restriction of μ to the sub σ-algebra Σ_0. Thus we have v and μ_0 two measures on Σ_0. Clearly

$$v << m_0$$

since if $\mu(B) = 0$ then certainly $\int_B f(x)d\mu(x) = 0$.

Obtain $g: X \to \mathbb{R}$ which is measurable with respect to Σ_0 and has for all $B \in \Sigma_0$

$$\int_B f(x)d\mu(x) = v(B) = \int_B g(x)d\mu(x),$$

just as needed.

Definition: For $f, g, \Sigma_0, \Sigma_1, X$ as in the statement of the last theorem, we say that g is the conditional expectation of f with respect to Σ_0 and write

$$g = E(f \mid \Sigma_0).$$

Strictly speaking there is the same grammatical flaw in this terminology which we saw in the use of the term the Radon-Nikodym derivative. The conditional expectation is only defined up to null sets, but since this is good enough for our purpose we indulge the definite article.

Think of $E(f \mid \Sigma_0)$ this way: This is the function whose value at a point $x \in X$ only depends on which elements of Σ_0 the point lies inside; it is as if we are forbidden to access any information about x which uses sets in Σ_1 but not Σ_0.

BOREL AND MEASURABLE SETS AND FUNCTIONS

It is only a slight exaggeration to describe standard Borel probability spaces as the basic object of study for this course. Recall that this consists of a set X equipped with a σ-algebra and a function

$$\mu: \Sigma \to [0, 1]$$

such that there is a Polish topology on X giving rise to Σ as the Borel sets and μ is a measure with $\mu(X) = 1$.

Of course people can and do study measures on more general kinds of spaces, but for practical purposes standard Borel spaces are likely to include all the examples you will ever encounter.

The situation with the measure having mass one – $\mu(X) = 1$ – is a bit more subtle. It is natural to consider σ-finite measures – such as $\mathbb{R}$ equipped with Lebesgue measure. But even here one can partition the space into countably many pieces each having measure one. In the case that $\mu(X)$ is a finite number other than 1, the situation is sufficiently similar to a probability space for us to be unconcerned.

This section will discuss finer results on measurable functions on standard Borel spaces. We could spend a lot more time here, and a lot of the results are on the edge of my own field of descriptive set theory. Instead I will present a couple of tricks which occur over and over in certain branches of analysis. Frequently one needs to know when a function or set is measurable. Certainly Borel sets are measurable; we will also see that the projections of Borel sets are measurable as well.

Lemma: Let X be a Polish space, Σ its σ-algebra of Borel sets, and μ a Borel probability measure on X. Then for $A \in \Sigma$ we have

$$\mu(A) = \sup(\{\mu(F): F \subset A, F \text{ closed}\})$$
$$= \inf(\{\mu(O): O \supset A, O \text{ open}\}).$$

Proof: Let d be a compatible complete metric on X. Let Σ_0 be the collection consisting of all Borel sets A satisfying the conditions above. Note that for O open and $n \in \mathbb{N}$ we can let

$$F_n = \left\{ x \in O : d(x, X \setminus O) \geq \frac{1}{n} \right\},$$

where $d(x, X\setminus O) = \inf\{d(x, y): y \in X\setminus O\}$. Then

$$O = \bigcup_n F_n ,$$

and hence $\mu(O) = \lim \mu(F_n)$, and thus $O \in \Sigma_0$.

Inspecting the definition and using $\mu(X) = 1$ we have that Σ_0 is closed under complements. Finally suppose $(A_n)_{n\in\mathbb{N}}$ is a sequence of sets in Σ_0. Fix $\varepsilon > 0$. For each n we can find closed $F_n \subset A_n$ with $\mu(A_n\setminus F_n) < \frac{\varepsilon}{2^{n+1}}$. Then we can go to some large K with $\mu\left(\bigcup_{n\in\mathbb{N}} A_n\right)\setminus (A_1 \cup A_2 \cup \ldots A_K)) < \frac{\varepsilon}{2}$. It then follows that $\mu\left(\bigcup_{n\in\mathbb{N}} A_n\right)\setminus F_1 \cup F_2 .. F_k) < \varepsilon$.

Similarly if we choose open $O_n \supset A_n$ open with $\mu(O_n \setminus A_n) < \frac{\varepsilon}{2^n}$, then $\mu\left(\bigcup_{n\in\mathbb{N}} O_n\right)\setminus\left(\bigcup_{n\in\mathbb{N}} A_n\right) < \varepsilon$.

Lemma: Let X be a Polish space, Σ its σ-algebra of Borel sets, and μ a Borel probability measure on X. Then for $A \in \Sigma$ we have

$$\mu(A) = \sup(\{\mu(K): K \supset A,\ K \text{ compact}\}).$$

Proof: Fix $\varepsilon > 0$. By the last lemma we may assume A is closed. Now fix $(x_i)_{i\in\mathbb{N}}$ dense in A and at each i let B_i^1 be $\{y \in A: d(x_i, y) \leq 1\}$. Thus we have covered A by countably many balls of radius 1. We may find some N_1 such that $\mu\left(A\setminus\bigcup_{i<N_1} B_i^1\right) < \frac{\varepsilon}{2}$. Repeating we may let $B_i^2 = \{y \in A : d(x_i, y) < 1/2\}$ and find N_2 such that $\mu\left(A\setminus\bigcup_{i<N_2} B_i^2\right) < \frac{\varepsilon}{4}$, and that at each ℓ set $B_i^\ell = \{y \in A: d(x_i, y) < 2^{-\ell}\}$ and find N_ℓ such that $\mu\left(A\setminus\bigcup_{i<N_\ell} B_i^\ell\right) < \frac{\varepsilon}{2^\ell}$. If we take

$$K = \bigcap_\ell \bigcup_{i<N_\ell}$$

then K is closed and $2^{-\ell}$-bounded for each ℓ, hence compact.

Theorem: Let X be a Polish space and μ a Borel probability measure on X. If

$$f: X \to Y$$

is a Borel function into a Polish Y, then for any $\varepsilon > 0$ there is a compact $K \subset X$ with $f|_K$ continuous and $\mu(K) > 1 - \varepsilon$.

Proof: Let $\{U_\ell : \ell \in \mathbb{N}\}$ be a countable basis for Y. At each ℓ, to find open $O_\ell \supset f^{-1}[U_\ell]$ with

$$\mu(O_\ell \setminus f^{-1}[U_\ell]) < \frac{\varepsilon}{2^\ell}$$

Then it is immediate that f is continuous on

$$X \setminus \left(\bigcup O_\ell \setminus f^{-1}[U_\ell]\right).$$

The measure of this set is greater than $1 - \varepsilon$.

Definition: For (A, d) a metric space and $B \subset A$, the diameter of B, $d(B)$, is the sup $d(a, a')$ as a, a' range over B.

Notation For s a sequence of length ℓ, and a an element, $\widehat{sa}$ is the sequence of length $\ell + 1$ extending s with final term a.

Lemma: Let X, Y be Polish and

$$f: X \to Y$$

continuous. Then $f[X]$ is measurable (with respect to any Borel probability measure μ on Y).

Proof: Fix d_X and d_Y complete, compatible metrics on X and Y.

Claim: If $C \subset X$ closed and $\varepsilon > 0$, then we can find $(C_n)_{n\in\mathbb{N}}$ closed subsets of C such that

$$C \subset \bigcup C_n$$

and at each n,

$$d_X(C_n) < \varepsilon,$$
$$d_Y(f[C_n]) < \varepsilon.$$

Proof of Claim: Fix a countable basis for X. Around each $x \in C$ we can find a basic open U of diameter less than ε with $f[U]$ of diameter less than ε. We then let $(C_n)_{n\in\mathbb{N}}$ enumerate the sets of the form

$$\text{x } \overline{C \cap U}$$

as U ranges over such basic open sets. (Claim)

Iterating the above lemma we can find an array of closed sets

$$(C_s)_{s\in\mathbb{N}^{<\infty}},$$

indexed by the finite sequences of natural numbers, such that

(a) $C_\phi = X$;

(b) Each $\bigcup_{n\in\mathbb{N}} C_{s\frown n} = C_s$;

(c) If $s \in \mathbb{N}^n$ (i.e. a sequence of length n), then

$$d_X(C_s) < 2^{-n},$$
$$d_Y(f[C_s]) < 2^{-n}.$$

At each s choose Borel $B_s \supset f[C_s]$ Borel with

$$B_s \subset \overline{f[C_s]}$$

and $\mu(B_s)$ as small as possible.

Claim $B_s \setminus \bigcup_{n\in\mathbb{N}} B_{s\frown n}$ is always null.

Proof of Claim: Since

$$f[C_s] = \bigcup_{n\in\mathbb{N}} f[C_{s\frown n}] \subset \bigcup_{n\in\mathbb{N}} B_{s\frown n}$$

and we chose B_s to have minimal measure. (Claim)

We then let $A = \bigcup_{n\in\mathbb{N}^{<\infty}} (B_s \setminus \bigcup_n B_{s\frown n})$. A is the countable union of null sets, and hence null. $B_\phi \supset f[C_\phi] = f[X]$ so it suffices to show $B_\phi \subset f[C_\phi]$. If $y \in B_\phi \setminus A$, then we can choose

$$x_0 \subset s_1 \subset s_2 \ldots$$

such that each s_ℓ is of length ℓ and $y \in s_{s_\ell}$.

Then we may find $y_\ell \in f[C_{s_\ell}]$ since $B_\ell \subset \overline{f[C_{s_\ell}]}$ is non-empty. Then fix corresponding $x_\ell \in C_{s\ell}$ with $f(x_\ell) = y_\ell$. The diameters of the $(f\,[C_{s\ell}])\,\ell$ sets are approaching zero, so we have $y_\ell \to y$. Since $d_X(C_{s\ell}) < 2^{-\ell}$ we have $(x_\ell)_\ell$(Cauchy and hence there is some x with $x_\ell \to x$. By continuity, $f(x) = y$.

This lemma can in fact be proved even in the case when f is simply a Borel function. There are various ways of approaching the proof of the more general result, but I will proceed by showing any Borel function can be made continuous by an appropriate strengthening of the topology.

Lemma: Let (X, d) be a complete metric space. Then there is a compatible complete metric bounded by 1.

Theorem: Let X be a Polish space and $B \subset X$ Borel. Then there is a stronger Polish topology on X under which B becomes clopen – that is to say, both open and closed.

Proof: The usual pattern. We want to show that the collection Σ of such sets includes the open sets, is closed under complements and countable intersections.

First to see that it includes the open sets, let $O \subset X$ be open and d a compatible complete metric on X bounded by 1. For $x, y \in X \setminus O$ let $d'(x, y) = d(x, y)$. For $x \in O$, $y \in X \setminus O$, set $d'(x, y) = 2$. And finally for $x, y \in O$ let

$$d'(x,y) = d(x,y) + \left| \frac{1}{d(x, X \setminus O)} - \frac{1}{d(y, X \setminus O)} \right|.$$

It is easily seen that if $(x_n)_n$ is a d'-Cauchy sequence included in O, then it is d-Cauchy with limit inside O.

It is immediate from the structure of the definitions that Σ is closed under complements.

Finally, let $(B_n)_n$ be a sequence of sets in Σ and at each n let d_n be a complete metric which gives rise to a stronger Polish topology in which B_n is clopen. We may assume dn is bounded by 1. We can then let

$$d^*(x,y) = \sum 2^{-n} d_n(x,y).$$

It is routine to verify this is a complete metric and that the resulting topology is separable. The topology generated by d^* is at least as fine as each of d_n's, so each B_n becomes clopen. Thus $\bigcap B_n$ is closed in the new Polish topology. Going back to the second paragraph of the proof, we can find a stronger Polish topology in which $\bigcap B_n$ becomes clopen.

Corollary: Let $f: X \to Y$ be a Borel function between Polish spaces. Then there is a stronger Polish topology on X under which f becomes continuous.

Proof Let $\{U_n : n \in \mathbb{N}\}$ be a countable basis for the topology on Y. At each n let d_n be a complete metric on X given rise to a stronger Polish topology in which $f^{-1}[U_n]$ is clopen. We may assume each d_n is bounded by 1, and so the topology generated by the metric

$$d^*(x,y) = \sum 2^{-n} d_n(x,y)$$

is as required.

Corollary: If $f: X \to Y$ is Borel, $B \subset X$ Borel, then $f[B]$ is measurable in Y with respect to any Borel probability measure.

This extra piece turns out to be important in certain contexts. Roughly speaking it states that we may find a measurable selector for Borel functions – a right inverse if you will.

Theorem: **Let X and Y be standard Borel spaces, μ a standard Borel probability measure on Y, $f: X \to Y$ Borel. Then there is a measurable function $\rho: f[X] \to X$ such that**

$$f(\rho(y)) = y$$

all $y \in f[X]$.

Proof: **Fix compatible Polish topologies on X and Y. We may assume f is continuous. It suffices to define ρ on**

$$B_\phi \setminus \bigcup_s (B_s \setminus \bigcup_n B_{\widehat{sn}})$$

But now for y in this set we can successively define n_0^y to be least n with

$$y \in B_{\langle n \rangle},$$

then n_1^y least n with

$$y \in B_{\left\langle n_0^y, n \right\rangle}$$

and more generally $n_{\ell+1}^y$ to be least such

$$y \in B_{\left\langle n_0^y, n_1^y, \ldots, n_\ell^y, n \right\rangle}.$$

One verifies that for each $s \in \mathbb{N}^{<\infty}$ and $\ell \in \mathbb{N}$ the set

$$\left\{ y \in f[X] : \left\langle n_0^y, n_1^y, \ldots, n_\ell^y \right\rangle = s \right\}$$

is measurable. We then let $\rho(y)$ be the unique x in

$$\bigcap_\ell C_{\left\langle n_0^y, n_1^y, \ldots, n_\ell^y \right\rangle}$$

The function ρ is measurable since for any open $U \subset X$ we have $\rho(y) \in U$ if and only if there is some ℓ with

$$C_{\left\langle n_0^y, n_1^y, \ldots, n_\ell^y \right\rangle} \subset U.$$

Chapter 10

Geometric Measure Theory

Geometric measure theory was developed in the second half of the 20th century to treat existence and regularity questions in the calculus of variations. The lack of compactness properties of the spaces of submanifolds made it necessary to consider "generalized" submanifolds (currents, varifolds, finite perimeter sets...), with appropriate topologies, to apply the direct method of the calculus of variations. To find a minimum of some functional, take a sequence approaching the infimum.

If this sequence is compact in some topology then extract a convergent subsequence to a generalized submanifold realizing the minimum. Then prove a regularity result to recover a classical solution. In some sense, this theory is similar to Sobolev space theory, in which one enlarges the space of smooth functions to find a solution to a differential equation, and then proves regularity to recover a classical solution. These notes provide a brief overview of geometric measure theory. A few sketches of proofs will be given, with references to appropriate sources. Recommended texts include Morgan, Simon, Giusti, Massari-Miranda, Ziemer on finite perimeter sets and functions of bounded variation, and of course Federer's treatise on *Geometric Measure Theory*.

RECTIFIABLE SETS

LEBESGUE AND HAUSDORFF MEASURES

Lebesgue measure in $\mathbb{R}^n$ will be denoted by L^n. An important outer measure in $\mathbb{R}^n$, defined for any nonnegative integer m, is Hausdorff measure H^m. Let $\alpha_m = L^m(B(0, 1))$, and $A \subset \mathbb{R}^n$. Fix $\delta > 0$ and consider

$$H_\delta^m(A) = \inf\left\{\sum_j \alpha_m r_j^m : A \subset \bigcup_j S_j, \operatorname{diam} S_j \leq \delta\right\}.$$

Then define the m-dimensional Hausdorff measure of A as

$$H^m(A) = \lim_{\delta \to 0} H^m_\delta(A).$$

Note that this limit always exists, although it could be ∞. We recall that Hausdorff measure is just an outer measure, but restricted to the σ-algebra of measurable sets is a Borel-regular measure by Caratheodory's criterion.

In $\mathbb{R}^n$, we have $L^n = H^n = H^n_\delta$, for any $\delta > 0$. One may extend the definition of Hausdorff measure to any real dimension, by replacing α_m by the corresponding expression in terms of gamma functions ($\alpha_m = \pi^{m/2/} \Gamma(m/2 + 1)$). H^0 is just counting measure.

If $m < p$ then $H^m(A) < \infty$ implies $H^p(A) = 0$. If $p < m$ then $H^m(A) > 0$ implies $H^p(A) = \infty$. The Hausdorff dimension of $A \subset \mathbb{R}^n$ is defined as the real number

$$\inf \{p: H^p(A) < \infty\} = \inf \{p: H^p(A) = 0\}$$

$$= \sup \{p: H^p(A) > 0\} = \sup \{p: H^p(A) = \infty\}.$$

The m-dimensional Hausdorff measure of an m-dimensional submanifold of $\mathbb{R}^n$ coincides with the Riemannian volume of the submanifold.

DENSITIES

If $A \subset \mathbb{R}^n$, one can define the m-dimensional density of A at a by

$$\Theta(A, \alpha) = \lim_{r \to 0} \frac{H^m(A \cap B^n(a, r))}{\alpha_m r^m},$$

when this limit exists. In the formula, $B^n(a, r)$ is the closed ball of center a and radius $r > 0$ in $\mathbb{R}^n$. Similarly, one can define the density of A with respect to a measure μ in $\mathbb{R}^n$ by replacing H^m by μ in the previous formula.

The m-dimensional density of any smooth m-dimensional submanifold of $\mathbb{R}^n$ equals 1 at any of its points. If A is the boundary of a cube in $\mathbb{R}^3$ then the 2-dimensional density is 1 at every point except at the vertices, where the density is 3/4.

LIPSCHITZ FUNCTIONS

A Lipschitz function f between metric spaces is one which satisfies the inequality $d(f(x), f(y)) \leq C\, d(x, y)$ for some constant $C > 0$.

The smallest such C is called the Lipschitz constant of f. We can say that Lipschitz functions are those with controlled slope. A Lipschitz function between open sets of Euclidean spaces is differentiable almost everywhere by Rademacher's Theorem. Hence the Jacobian of a Lipschitz function can be integrated at least locally and we have the following important formulae.

Area Formula

Theorem: Consider a Lipschitz function f: $\mathbb{R}^m \to \mathbb{R}^n$ for $m \le n$.

(i) If $A \subset \mathbb{R}^m$ is an L^m-measurable set, then

$$\int_A J(f)dH^m = \int_{\mathbb{R}^n} H^0(A \cap f^{-1}(y))dH^m(y).$$

(ii) If u: $A \to \mathbb{R}$ is an L^m-integrable function, then

$$\int_A uJ(f)dH^m = \int_{\mathbb{R}^n} \sum_{x \in f^{-1}(y)} u(x)dH^m(y).$$

We have denoted the Jacobian of f by $J(f)$. This formula shows that Hausdorff measure is compatible with classical mapping area.

Coarea Formula for Lipschitz Maps

Theorem: Consider a Lipschitz function f: $\mathbb{R}^m \to \mathbb{R}^n$ with $m > n$. If $A \subset \mathbb{R}^m$ is an L^m-measurable set, then

$$\int_A J(f)dH^m = \int_{\mathbb{R}^n} H^{m-n}(A \cap f^{-1}(y))dH^n(y).$$

The coarea formula is very useful in geometric analysis since it allows slicing to recover important geometric quantities. As we shall show, it can be generalized to maps between rectifiable sets.

Tangent Cones

For a set $A \subset \mathbb{R}^n$, one may consider the tangent cone of A at a, Tan(A, a), as the cone over the subset S of the unit sphere given by

$$S = \bigcap_{\varepsilon > 0} Cl\left\{\frac{x-a}{|x-a|} : x \in A, 0 < |x-a| < \varepsilon\right\}.$$

For a C^1 m-dimensional submanifold the tangent cone is precisely the tangent m plane. We define a smaller approximate tangent cone by

$$\text{Tan}^m(A, a) = \bigcap\{Tan(S,a) : \Theta^m(A - S, a) = 0\},$$

which omits parts of Tan(A, a) with dimension less than m.

Rectifiable Sets

A set $A \subset \mathbb{R}^n$ is called (H^m, m)-rectifiable set if $H^m(A)$ is finite and H^m almost all of A is the union of the images of countably many Lipschitz maps from $\mathbb{R}^m$ to $\mathbb{R}^n$. We shall often refer to such sets as m-rectifiable sets.

In some sense, a rectifiable set behaves like a C^1 submanifold of $\mathbb{R}^n$ since it has an approximate tangent plane almost everywhere.

Proposition: If A is an m-rectifiable set in $\mathbb{R}^n$ then for H^m-almost all points $a \in A$, we have $\Theta(A, a) = 1$ and $\text{Tan}^m(A, a)$ is an m-dimensional plane.

On rectifiable sets, one can define the notion of the gradient of a function, the divergence of a vector field, and the jacobian of a map. Hence there is a generalization of the coarea formula to such sets.

General Coarea Formula

Theorem: Consider an m-dimensional rectifiable set $A \subset \mathbb{R}^n$, an μ-dimensional rectifiable set $B \subset \mathbb{R}^\nu$, $m \geq n \geq 1$, and $f: A \to B$ a Lipschitz function. Then

$$\int_A J(f)dH^m = \int_B H^{m-\mu}(f^{-1}\{z\})dH^\mu(z).$$

More generally, for any H^m-integrable function g on A

$$\int_A gJ(f)dH^m = \int_z \int_{f^{-1}\{z\}} gdH^{m-\mu}dH^\mu(z).$$

In the above result, $J(f)$ is the Jacobian of the Lipschitz map f, which is defined almost everywhere. An important consequence of coarea formula is Crofton formula.

Crofton's Formula

Corollary: If $A \subset \mathbb{R}^n$ is an m-rectifiable set, then

$$H^m(A) = \frac{1}{\beta(n,m)} \int_{P \in p} \#\{P \cap A\},$$

where $\beta(n, m)$ is a constant depending only on n and m, and P is the set of $(n - m)$ planes.

Finally we state the celebrated Federer Structure Theorem, which allows the decomposition of any set in $\mathbb{R}^n$ as the union of a rectifiable set and a set with negligable projection on almost all m-dimensional subspaces.

Federer's Structure Theorem

Theorem: Let $A \subset \mathbb{R}^n$ be an arbitrary subset with $H^m(A) < \infty$. Then A can be decomposed as the union of two disjoint sets $A = B \cup C$, where B is (H^m, m)-rectifiable and $J^m(C) = 0$.

In this theorem J^m is the m-dimensional integralgeometric measure.

Equality $J^m(C) = 0$ implies that the m-dimensional measure of the orthogonal projection of C onto a m-dimensional linear subspace is zero for almost all subspaces.

One can define an orientation on a rectifiable set, just by choosing a measurable orientation in the set of tangent planes. Recall that there is a tangent plane for almost all the points of the set. Of course one can define an uncountably number of such orientations.

CURRENTS

DEFINITION AND PROPERTIES

The concept of current was introduced by Federer and Fleming in 1960. Consider the space D^m of smooth differential m-forms with compact support in $\mathbb{R}^n$, endowed with the topology of C^∞ convergence on compact subsets. The space of continuous linear maps T: $D^m \to \mathbb{R}$ will be denoted by D_m. A *current* is an element of D_m.

An oriented m-rectifiable set $A \subset \mathbb{R}^n$ defines a current $T_A \in D_m$ in the following way. For any $a \in A$ for which there is a tangent plane, let $\vec{S}\,(a)$ denote the m-vector associated to the oriented tangent plane, and let μ be an integer-valued function in $L^1(A)$. Then

$$T_A(\varphi) = \int_A \mu \left\langle \vec{S}, \varphi \right\rangle dH^m.$$

The function $\mu(a)$ measures the multiplicity of the rectifiable set in the point a. A current induced by a *compact* oriented m-rectifible set will be called a *rectifiable current*.

The *support* of a current T is the smallest closed set C such that, for any $\varphi \in D^m$ such that $\operatorname{spt}(\varphi) \cap C = \phi$ one has $T(\varphi) = 0$.

One can define the *boundary* of a current in the following way: if $T \in D_m$ and $\varphi \in D^{m-1}$, then

$$\partial T(\varphi) = T(d\varphi).$$

Property $d^2 = 0$ of the exterior derivative of forms implies that ∂T has no boundary. The boundary of a rectifiable current T need not be rectifiable. If this is the case, then we say that T is an *integral current*.

The *mass* of a current T is

$$M(T) = \sup\{T(\varphi) \colon |\,\varphi\,| \leq 1\},$$

where $|\,\varphi\,|$ is the supremum norm of the differential form φ. If $T \in D_m$ is a rectifiable current associated to a rectifiable set A with multiplicity $\mu \equiv 1$ then $M(T) = H^m(A)$.

A natural topology in the space of currents is the weak topology. We say that $T_i \to T$ weakly in D_m if and only if $T_i(\varphi) \to T(\varphi)$ for any differential form $\varphi \in D^m$. If T_i converges weakly to T then one obtains easily

$$M(T) = \leq \liminf_i M(T_i)$$

Another seminorm in the space of curremts is the *flat norm*. It is defined as

$$F(T) = \inf\{M(A) + M(B): T = A + \partial B, A\ m\text{-rect.}, B\ (m+1)\text{-rect.}\},$$

with $T \in D_m$. If one has two m-currents T_1 and T_2, then saying that they are close in the flat norm is equivalent to say that there is an $(m+1)$-current of small mass such that its boundary is T_1 plus T_2 plus another m-current of small mass.

The flat norm is weaker in general than the mass norm, and stronger than the weak topology.

Another classical definition is that of a *normal current*, which is a current T with compact support and $M(T) + M(\partial T) < \infty$.

THE DEFORMATION THEOREM AND CONSEQUENCES

One of the most interesting results concerning integral currents is the deformation theorem, which deforms a given integral current $T \in D_m$ to a current contained in a given m-dimensional grid of prescribed mesh.

Deformation Theorem

Theorem: Let $T \in D_m$ be an integral current in $\mathbb{R}^n$. Then there are integral currents $Q \in D_m$, $S \in D_{m+1}$, and a current $P \in D_m$ such that

$$T = P + Q + \partial S,$$

so that $M(P) \leq \gamma\, M(T)$, $M(Q) \leq \varepsilon\, M(\partial T)$ and $M(S) \leq \varepsilon\gamma\, M(T)$, where $\gamma = 2n^{2m+2}$, and hence $F(T - P) \leq \varepsilon\gamma\, (M(T) + M(\partial T))$.

Moreover spt(T) is contained in a given m-dimensional 2ε grid, and spt(∂T) is contained in the $(m-1)$-dimensional 2ε grid.

Isoperimetric Inequality

Theorem: An m-dimensional cycle T ($\partial T = 0$) in $\mathbb{R}^n$ bounds an $(m+1)$-dimensional integral current S with

$$M(S)^{m/(m+1)} \leq \gamma\, M(T),$$

with $\gamma = 2n^{2m+2}$.

Almgren improved this theorem by showing that the best constant γ one can place in the above inequality corresponds to the classical isoperimetric inequality in $\mathbb{R}^{m+1}$, assuming that S is area-minimizing.

Another consequence of the deformation theorem is

Theorem: The set of integral currents T in $B(0, R)$ such that $M(T) \leq C$, $M(\partial T) \leq C$ is totally bounded in the flat norm topology.

A consequence of this result is the coincidence of the mass norm and the flat norm in the subset of integral currents T in $B(0, R)$ such that $M(T) \leq C, M(\partial T) \leq C$, for any constant $C > 0$.

Compactness Theorem

Theorem: Given $\mathbb{R} > 0, C > 0$, the set of integral currents with support in $B(0, R)$ such that $M(T) \leq C$, $M(\partial T) \leq C$ is compact in the flat norm topology.

Now we are in position of applying these results to prove existence of areaminimizing surfaces for the Plateau problem.

Corollary: Let B an $(m-1)$-rectifiable current in $\mathbb{R}^n$ with $\partial B = 0$. Then there is an m-dimensional area-minimizing rectifiable current S with $\partial S = B$.

Proof: The proof is given in several steps

1. First show that B bounds some surface by taking a cone over B.
2. Take a sequence of rectifiable currents S_i with $\partial S_i = B$ approaching the infimum of the area.
3. Project the S_i into a ball containing B. Since the projection to the boundary of the ball is distance nonincreasing, the mass is reduced.
4. By the Compactness Theorem, get a limit S.
5. By lower semicontinuity $M(S) \leq \liminf_i M(S_i)$, which shows that S is minimizing.

The limit we have obtained in this process is an integral current.

The existence of this limit is not obvious since it is not clear that there is an area-minimizing surface among the infinitely many candidates.

An example of an area-minimizing surface is a minimal graph over a convex region. This follows from projection over the right cylinder over the convex region and a calibration argument. To get the proof we consider slices of an integral current by hypersurfaces, and the Gauss-Green Theorem.

Regularity Theorems

We are interested now in seeing how regular is the solution to the Plateau problem we have obtained in the previous subsection. In this direction, the first regularity result was proved by Fleming, who showed that an area-minimizing rectifiable current in $\mathbb{R}^3$ is an embedded surface in the interior. This regularity theorem was generalized to three-dimensional surfaces in $\mathbb{R}^4$ by Almgren in 1966, and to m-dimensional surfaces in $\mathbb{R}^{m+1}$, $m \leq 6$, by Simons in 1968. In 1969, Bombieri, De Giorgi and Giusti gave an example of a seven-dimensional, area-minimizing rectifiable current in $\mathbb{R}^8$ which is not smooth at the origin (the cone over $S^3(1/\sqrt{2}) \times S^3(1/\sqrt{2})$ in $\mathbb{R}^8$). The complete regularity result was given by Federer.

Theorem: An $(n-1)$-dimensional, area-minimizing rectifiable current in $\mathbb{R}^n$ is a smooth, embedded manifold on the interior except for a singular set of Hausdorff dimension at most $n-8$.

Similarly in a smooth Riemannian manifold M^n, for $n \leq 7$, an area-minimizing rectifiable current is a smooth embedded manifold and, for $n \geq 8$, the dimension of the singular set is less than or equal to $n-8$. The same result is true with a volume constraint (almost minimizing is enough). This was proved by Gonzalez, Massari and Tamanini for finite perimeter sets. We state the equivalent result for currents.

Theorem: Let T be an $(n-1)$-dimensional rectifiable current of least area in the unit ball $B \subset \mathbb{R}^n$, with prescribed boundary in ∂B, bounding an oriented region of prescribed volume.

Then T is a smooth hypersurface of constant mean curvature in its interior,except for a singular set of dimension at most $n-8$.

Regularity in higher codimension is much weaker. It was already proved by Federer that the set of regular points is dense in the interior. On the other hand, it was known that complex analytic varieties are area-minimizing, with singular sets of codimension 2. The most conclusive result was proved by Almgren.

Theorem: An m-dimensional, area-minimizing rectifiable current in $\mathbb{R}^n$ is a smooth, embedded manifold on the interior except for a singular set of Hausdorff dimension at most $m-2$.

In 1988, Chang showed that there are only isolated singularities for 2-dimensional currents in $\mathbb{R}^n$.

The only known examples in $\mathbb{R}^4$ are complex curves. It is known that the tangent cone must consist on complex planes (with multiplicities). The only known examples of 3-dimensional surfaces in Euclidean 5-space are the 2-dimensional examples in 4-space cross a line.

Why are regularity results stronger in codimension 1? First of all one can reduce to multiplicity 1, since a nesting lemma yields decomposition into nested, multiplicity 1 area-minimizing surfaces. One then proves that each one is regular. If they touch, they must coincide by the maximum principle. All the ingredients also hold in Riemannian manifolds.

There are several open questions concerning singularities. A first one is about the structure of singularities. A second one is if area-minimizing surfaces are stratified manifolds, or could they have fractional dimensional singular sets? Also, what are the possible tangent cones?

Concerning boundary regularity, the definitive result is the one by Hardt and Simon.

Theorem: Let T be an $(n-1)$-dimensional, area-minimizing rectifiable current in $\mathbb{R}^n$, bounded by a C^2 oriented submanifold with multiplicity 1. Then at every boundary point, spt T is a C^1 embedded manifold with boundary.

In a smooth Riemannian manifold M^3 with nonempty boundary ∂M, areaminimizing surfaces S in homology classes of M (mod. ∂M) are also regular along the boundary.

The main tool of regularity theory is the monotonicity in r of (mass inside r ball)/ r^m [27, Ch. 9]. More precisely, for an m-dimensional rectifiable current T, we define the mass ratio $M(r)$ as $M(TB\,(a,\,r))/\alpha_m\, r^m$, which is an increasing function. It is inmediate that the density $\Theta(T,\,a)$ equals $\lim_{r\to 0} M(r)$. Then we have

Theorem: Let T be an area-minimizing locally rectifiable current, and let $a \in$ spt (T). Then, for $0 < r <$ dist(a, spt(T)), the mass ratio $M(r)$ is an incresing function of r.

Proof: As $f(r) = M(TB(a,\,r))$ is increasing, it is differentiable almost everywhere. By slicing we have $M(\partial(TB(a,\,r))) \leq f\,'(r)$. If C_r is the cone over $\partial(TB(a,\,r))$ with vertex a, we get $M(C_r) = (r/\,m)M(\partial(TB(a,\,r)))$ by the coarea formula. Hence

$$f(r) \leq M(C_r) \leq \frac{r}{m} f'(r),$$

and

$$\alpha_m M'(r) = \left(\alpha_m \frac{f(r)}{r^m}\right)' = \frac{m\alpha_m}{r^{m+1}}\left[\frac{r}{m} f'(r) - f(r)\right] \geq 0,$$

one easily concludes that $M(r)$ is increasing.

It is not difficult to show that equality holds if and only if T is a cone. A first corollary of the above result is that

$$M(TB(a,\,r)) \geq \Theta(T,\,a)\alpha_m\, r^m.$$

As $\Theta(T,\,a) = 1$ for almost all points of a rectifiable current T, we easily conclude that $\Theta(T,\,a) \geq 1$ everywhere on spt $T -$ spt ΘT. In particular, if T is a 2-dimensional, area-minimizing surface in $\mathbb{R}^3$, then $M(T,\,a) \geq \pi r2$.

Another interesting consequence of monotonicity is the existence of a tangent cone (limit of homothetic expansions) at any point of an area-minimimizing rectifiable current.

When proving the regularity of an area-minimizing surface there are two steps: the first one is to show that the tangent cone is a plane. The second one is to conclude that the existence of the tangent plane implies that the surface is regular.

Allard generalized monotonicity to stationary surfaces (mean curvature 0) and proved that for bounded mean curvature surfaces in a Riemannian manifold

$$M(r)\, e^{Cr}\, r^{-m}$$

is monotone, where $M(r)$ is the mass inside a ball of radius r. This result was proved in the very general context of varifolds. It is not valid for surfaces stationary for general integrands Φ , although for Φ -minimizers we still have

$$M(r) \geq Cr^m.$$

OTHER TYPES OF SURFACES

The Belgian physicist Plateau observed around 1870 two kind of singularities in soap films: either three sheets meet along a seam at 120°, or four such seams meet at a point at about 109°. The latter phenomenon corresponds to the triangles obtained from any given edge of a regular tetrahedron and from the segments which join the barycenter with the vertexes of the edge. This behaviour of soap films cannot be modelled by currents since the singular set is very large and contradicts the regularity results of the previous section. Hence other definitions of generalized surfaces are necessary to treat this problem.

For instance, to treat nonorientable surfaces one can study rectifiable currents modulo. Two rectifiable currents T_1 and T_2 are congruent modulo 2 if there is some rectifiable current Q such that $T_1 - T_2 = 2Q$. In general one can consider rectifiable currents modulo any integer number v. However these new objects are not appropiate to model the tetrahedral soap film.

We discuss now several different types of generalized surfaces.

VARIFOLDS

A varifold is a measure concentrated on a set in space and certain "tangent planes". The notion of varifold allows multiplicity, but neither cancellation nor obvious definition of orientation. Also tangent planes need not be associated with the set. More precisely, an m-dimensional *varifold* is a Radon measure on $\mathbb{R}^n \times G_m\, \mathbb{R}^n$, where $G_m \mathbb{R}^n$ is the Grassmannian of unoriented unit m-planes through the origin in $\mathbb{R}^n$.

An *integral varifold* is one associated to a rectifiable set. Let $S \subset \mathbb{R}^n$ be a rectifiable set, say with multiplicity 1, and m-tangent plane $\vec{S}$. Define

$$v(S) = H^m\{(x, \vec{S}(x)) : x \in S\}.$$

Instead of the notion of boundary we get the first variation of a varifold. $\delta V(\vec{v})$ is the initial rate of change of area in $\mathbb{R}^n$ under a smooth variation with initial velocity $\vec{v}$. A *stationary varifold* is one for which $\delta V = 0$ for all $\vec{v}$. Stationary integral varifolds include area-minimizing rectifiable currents (modulo v also), and soap films. For integral varifolds there are a compactness theorem and isoperimetric inequalities. Also the following regularity result holds.

Allard's Regularity

Theorem: An integral k-dimensional varifold with mean curvature bounded or in L^p, $p > k$, is a $C^{1,\alpha}$ submanifold at points of density 1.

An important open question for varifolds is whether a two-dimensional stationary integral varifold in an open subset of $\mathbb{R}^3$ is a smooth embedded manifold almost everywhere.

Finite Perimeter Sets and Functions of Bounded Variation

The notion of a set of finite perimeter was introduced by Cacciopoli and De Giorgi and the theory was developed independently for some time. Later it became clear that a finite perimeter set in $\mathbb{R}^n$ is (H^n, n)-rectifiable.

We shall say that $A \subset \mathbb{R}^n$ is a *finite perimeter set* if, for any vector field X with compact support in $\mathbb{R}^n$, we have

$$\int_A \operatorname{div} X dH^n \leq c(A) \,|\, X \,|,$$

where $|\, X \,|$ is the sup norm of the vector field X, and $c(A) > 0$ is a constant depending on the set A. It can be shown that this property is equivalent to saying that the characteristic function χ_A of the set A is of *bounded variation*. The perimeter $P(A)$ of the set A is defined as

$$P(A) = \sup_{|X| \leq 1} \int_A \operatorname{div} X dH^n .$$

This definition coincides with $H^{n-1}(\partial A)$ if ∂A is a smooth hypersurface.

From the definition and Riesz representation theorem, one can get the existence of an H^{n-1} -measurable vector valued function $v: A \to \mathbb{R}^n$ and a Borel measure μ on $\mathbb{R}^n$ such that

$$\int_A \operatorname{div} X dH^n = \int_{\mathbb{R}^n} \langle v, H \rangle \, d\mu.$$

Both the measure μ and the function v has support contained in the topological boundary of A.

For finite perimeter sets the natural convergence is in measure. We shall say that $A_i \to A$ if $\chi_{A_i} \to \chi_A$ in L^1. With this notion of convergence the perimeter is lower semicontinuous, i.e., if $A_i \to A$ then

$$P(A) \leq \liminf_i P(A_i).$$

Finite perimeter sets also admit slicing and isoperimetric inequalities. The interested reader should consult Giusti, Massari-Miranda, Simon, and Ziemer.

(M, e, d)-Minimal Sets

Theorem: Almgren modeled soap films as $(M, 0, \delta)$-minimal sets, which satisfy two conditions: they touch the b oundary, and tiny portions (of diameter $\leq \delta$) cannot be pinched or deformed to make the area go down. To be more precise, $S \subset \mathbb{R}^n - B$ bounded, with $H^m(S) < \infty$, is $(M, 0, \delta)$-minimal with respect to a closed set B (the boundary) if, for every Lipschitz deformation φ of $\mathbb{R}^n$ which differs from the identity map only in a δ-ball disjoint from B,

$$H^m(S) \leq H^m(\delta(S)).$$

The Lipschitz function φ need not be a diffeomorphism. For more general functions $\varepsilon(r) = C r^{\alpha}$, $\alpha > 0$, the weaker inequality,

$$H^m(S) \leq (1 + \varepsilon(r)) H^m(\varphi(S)),$$

defines (M, ε, δ)-minimal sets, which include soap bubbles with volume constraints.

These sets were treated by Almgren in his monograph. Three basic properties of (M, ε, δ)-minimal sets are (H^m, m)-rectifiability, monotonicity formula, and existence of (M, 0, δ)-minimal cone at every point. The regularity of (M, ε, δ)-minimal sets in $\mathbb{R}^3$ was studied by Taylor, who proved

Theorem: (M, ε, δ)-minimal sets in $\mathbb{R}^3$ can have two kind of singularities: the ones predicted by Plateau.

Proof: The proof consists in showing that there are only two possible singularities. To prove this, take any singularity and consider its linear approximation, which is a cone determined over a geodesic net in the sphere. This idea goes back to Lamarle. There are precisely 10 such geodesic nets (Lamarle missed one). One then shows that all but three of them are unstable. One corresponds to a great circle in the sphere, which yields regular portions, and the other two correspond to the singular sets.

Hence, for approximating cones, there are only two possible singularities. The really hard part is to show that the cone is a good enough approximation of the original soap film. Reifenberg's epiperimetric inequality is used to show that the minimizers converge rapidly to an asymptotic cone.

For singularities of $(M, 0, \delta)$-minimal sets in $\mathbb{R}^4$ one has the following conjecture.

Conjecture: A cluster in $\mathbb{R}^4$ has only the following singularities:

1. 3 hypersurfaces meeting along surfaces at 120°.
2. 4 such surfaces meeting along a curve at almost 109°.
3. 5 or 16 of such curves meeting at a point as in the cone over a regular simplex or the cone over a hypercube.

Some progress on this conjecture has been made by Brakke, Sullivan and White. Many questions remain open.

1. Boundary regularity.
2. The existence of a least-area $(M, 0, \delta)$-minimal set in $\mathbb{R}^3$ with a given smooth boundary curve.
3. The cone over the regular tetrahedron is the least-area separator of the four faces. Is there a smaller equilibrium surface not separating the four regions?
4. It is known that a smooth minimal surface is locally area-minimizing in arbitrary codimension. Is this property still true whenever the tangent cone is area-minimizing?

DOUBLE BUBBLES

AREA-MINIMIZING CLUSTERS

Bubble clusters seek the least-area way to enclose and separate several regions of prescribed volume. But they do *not* always find the absolute least area shape (they could be stable, but not minimizers).

A *cluster* is a collection of disjoint regions $R_1, \ldots, R_m$ (n-dimensional locally integral currents of multiplicity 1 in $\mathbb{R}^{n+1}$), with surface area

$$\frac{1}{2}\left(\sum_i M(\partial R_i) + M\left(\partial \sum_i R_i\right)\right).$$

A region is not assumed to be connected. The existence of an area-minimizing cluster enclosing prescribed volumes is guaranteed by Almgren's work in a very general context. A simplified proof was given by Morgan.

Theorem: In $\mathbb{R}^n$, given volumes $V_1, \ldots, V_m > 0$, there exists an area-minimizing cluster for those volumes.

The proof of this result needs some lemmas, the first of which is the following observation.

Lemma: Given any cluster, there exists $C > 0$, such that arbitrary small changes in volume may be accomplished inside small balls at a cost

$$|\Delta A| \leq C\,|\Delta V|.$$

Another interesting lemma which follows from the monotonicity formula and the isoperimetric inequality is

Lemma: An area-minimizing cluster is bounded in $\mathbb{R}^n$.

With these ingredients we are ready to prove Theorem. The proof of this result is complicated because when taking a minimizing sequence the volume could disappear at infinity. Let us take a minimizing sequence C_α with the prescribed volumes.

Step 1. Get a convergent subsequence to a non zero limit C. One can choose p_α such that

$$\text{vol}(R_{1,\alpha}) \cap B(p_\alpha, c_1)) \geq c_2,$$

where $R_{1,\alpha}$ is the first region of C_α. This is a technical result known as the Concentration

Lemma. The proof follows easily since we partition $\mathbb{R}^n$ into cubes K_j. Choosing appropiately the length of the edges we have inside K_j,

$$\text{area } (\partial R_{1,\alpha} \cap K_j) \geq \gamma \frac{vol(R_{1,\alpha} \cap K_j)}{\max_i vol(R_{1,\alpha} \cap K_i)^{1/n}}.$$

for some isoperimetric constant γ, and so

$$\text{area } (\gamma R_{1,\alpha} \cap K_j) \geq \gamma \frac{vol(R_{1,\alpha} \cap K_j)}{\max_i vol(R_{1,\alpha} \cap K_i)^{1/n}}$$

Summing over j we get which proves the desired result.

Translate to assume $p_\alpha = 0$.

Step 2. Show that the limit C is area-minimizing for its volumes. If not then one can get another (minimizing) sequence with less area than the original one. If no volume is lost then we are done.

Step 3. If some volume disappears at infinity then repeat the process with some p'_α.

Then countably many repetitions capture the total volume and yield a solution. This solution is bounded by Lemma, and so only a finite number of repetitions is needed.

Concerning regularity the first observation is that an area-minimizing cluster is (M, ε, δ)-minimal. The proof of this result relies on Lemma. Then one can apply Taylor's regularity result in $\mathbb{R}^3$, which implies that an (M, ε, δ)-minimal set consists on smooth, constant mean curvature surfaces meeting in threes at 120° along curves, in turn meeting in fours about 109°.

In higher dimensions Almgren showed that (M, ε, d)-minimal sets in $\mathbb{R}^n$, $n \geq 3$, are $C^{1,\alpha}$ almost everywhere. A simple treatment of (M, ε, δ)-minimal sets in $\mathbb{R}^2$ can be found in Morgan. White proved that they consists on constant mean curvature hypersurfaces meeting in threes at 120° along $(n-2)$-dimensional submanifolds, which meet in fours along smooth $(n-3)$-dimensional surfaces, which meet in an $(n-4)$-dimensional set.

CHARACTERIZATION OF AREA-MINIMIZING CLUSTERS

For an account of results on single bubbles the interested reader can consult. For double bubbles the only known optimal shapes are those of the double bubble in $\mathbb{R}^2$, $\mathbb{R}^3$, and $\mathbb{R}^4$. For the remanining volumes and dimensions the problem is still open.

The standard double bubble in $\mathbb{R}^n$ is composed of three $(n-1)$-spherical caps meeting along an $(n-2)$-sphere at 120°. In case we consider two equal volumes one of the spherical caps degenerates to a flat disc. The whole configuration is obtained by rotating three circles (or two circles and a segment) on a plane meeting at 120° around an $(n-2)$-dimensional subspace.

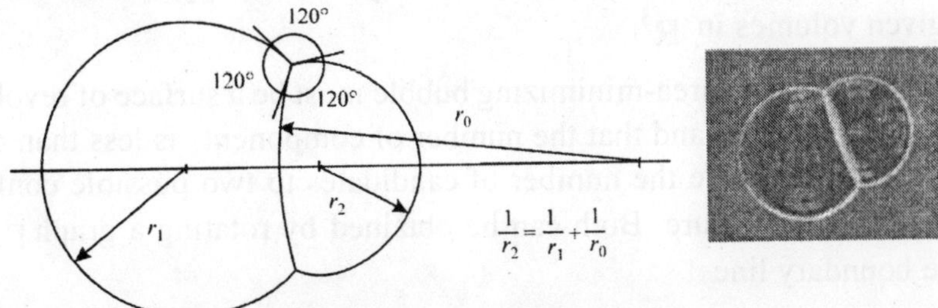

Fig. The Standard Double Bubble

The double bubble conjecture states that the standard double bubble provides the least perimeter way to enclose and separate two regions of given volumes in $\mathbb{R}^n$.

This conjecture appeared in Foisy's undergraduate thesis. The *planar* double bubble conjecture was proved in 1990 by Foisy, Alfaro, Brock, Hodges and Zimba. The double bubble conjecture in S^2 was solved by Masters in 1994.The study of double bubbles is complicated since we do not know anything about their structure. The regions can be disconnected with any number of components, and also the exterior of the bubble can be disconnected, with precisely one noncompact component, and possibly more compact ones which will be called *empty spaces* or *empty chambers*.

The first results towards proving the double bubble conjecture in Euclidean 3-space were

1. The bubble is a surface of revolution around some line in $\mathbb{R}^3$, White, Foisy, Hutchings.
2. The isoperimetric profile $A(v_1, v_2)$, which assigns to every pair of volumes (v_1, v_2) the perimeter of the area-minimizing double bubble of these volumes is a *concave* function. This implies that A is increasing in each variable, which shows that there are no empty spaces inside, Hutchings.

It was also proved in Foisy and Hutchings that the bubble must touch the axis of revolution. In order to prove this one consider the transformations $r \mapsto (r^2 - \varepsilon)^{1/2}$, where r is the distance to the axis of revolution. This deformation decreases area and keeps constant the volume enclosed. For two equal volumes. Hutchings also proved that each region is connected. By using this result Hass and Schlaffly proved, by making use of computer techniques, that

Theorem: The standard double bubble is the least perimeter way to enclose and separate two equal volumes in $\mathbb{R}^3$.

In the general case it was proved by Hutchings, Morgan, Ritore and Ros that there are at most three components (one of the regions can be disconnected and has at most two connected components). Finally they proved

Theorem: The standard double bubble is the least perimeter way to enclose and separate two arbitrary given volumes in $\mathbb{R}^3$.

We already know that an area-minimizing bubble must be a surface of revolution, that there are no empty chambers, and that the number of components is less than or equal to three. This is enough to reduce the number of candidates to two possible configurations which are depicted in the Figure. Both can be obtained by rotating a graph Γ in a half-plane about the boundary line

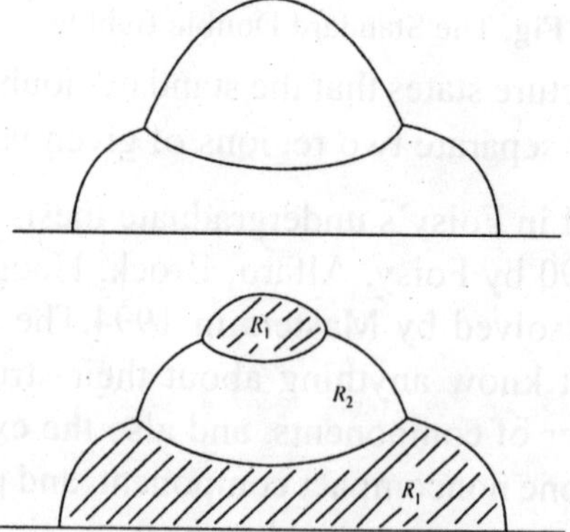

Fig. Candidate Graphs to Produce Area-minimizing Bubbles when Rotated

The main ingredient in the proof is an unstability argument, based on Courant's Nodal Domain Theorem. It essentially says that a Killing field cannot induce too many nodal regions in the bubble. If X is a Killing field, the one induced by a one-parameter group of isometries, a nodal region as a connected component of the function $\langle X, N \rangle$, where N is a unit vector field normal to the sheets of the bubble. This argument has been previously used to study stability and isoperimetric questions. From this principle one can prove the following Proposition.

Proposition: Consider a double bubble of revolution $\Sigma \subset \mathbb{R}^n, n \geq 3$, with axis L, as in Figure. Assume there is a finite number of nonsingular points $p_1, \ldots, p_k$, with $x = f(p_1) = \ldots = f(p_k)$, which separate the generating graph Γ. Then Σ is unstable.

In the above Proposition, $f(p)$ is the intersection of the normal line to p, with the axis of revolution L. Of course $f(p)$ can be ∞. This Proposition reduces the proof of the double bubble conjecture to elementary planar geometry modulo some properties of Delaunay hypersurfaces (hypersurfaces of revolution with constant mean curvature). For instance, we can easily discard the two components case. Consider Figure. Let L' be the line equidistant from a and b. Assume that this line meets the axis of revolution L at some point x. In each one of the curves joining a and b there is at least a point at maximum or minimum distance from x. Call them p_1 and p_2. Then p_1 and p_2 separate the generating graph Γ, which implies that the bubble is unstable.

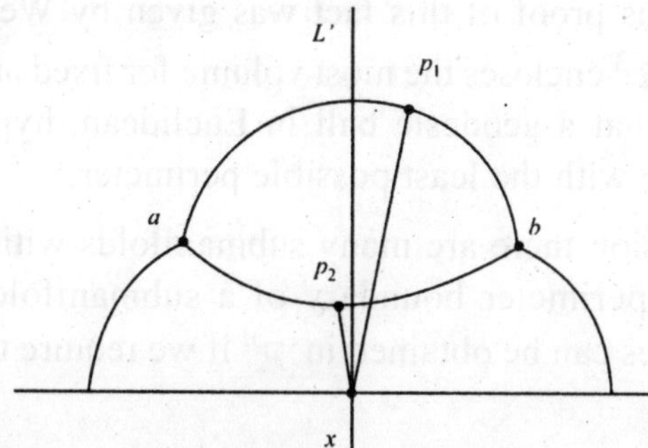

Fig. The Partition Method

The three component case is more involved, and some special properties of hypersurfaces of revolution with constant mean curvature are used. By using the arguments, Reichardt, Heilmann, Lai and Spielman were able to prove the double bubble conjecture in $\mathbb{R}^4$ and in some particular cases in $\mathbb{R}^n$.

INTEGRANDS AND ISOPERIMETRIC INEQUALITIES

INTEGRANDS

An *integrand* is a map $\Phi : \mathbb{R}^n \times \Lambda_m \mathbb{R}^n \to \mathbb{R}$ which is homogeneous and continuous. Sometimes it is required that Φ be smooth except at the origin, positive, even ($\Phi(x, \xi) = \Phi(x, -\xi)$) or convex ($\Phi(x, \xi_1 + \xi_2) \leq \Phi(x, \xi_1) + \Phi(x, \xi_2)$). An integrand defines a functional on the space of m-dimensional rectifiable sets by

$$\Phi(S) = \int_S \Phi\left(x, \vec{S}(x)\right) dH^m(x)$$

For instance the integrand $\Phi(x, \xi) = |\xi|$ defines the area functional.

Existence of Φ-minimizers follows from the compactness theorem. Interior regularity of a hypersurface minimizing a smooth elliptic integrand Φ in $\mathbb{R}^n$ or in a Riemannian manifold M^n except for a singular set of H^{n-3}-measure 0 was proved by Schoen, Simon, and Almgren. In higher codimension, regularity on an open dense set was proved by

Almgren and Federer. Boundary regularity is known just if the surface has density 1/2, Hardt.

The Φ-isoperimetric problem for curves in $\mathbb{R}^n$ seeks for curves of least Φ-energy bounding a fixed least-area. In $\mathbb{R}^2$ the optimal shape is the Wulff shape (the unit ball in the dual norm).

ISOPERIMETRIC INEQUALITIES

It is known since ancient times that the circle in $\mathbb{R}^2$ encloses the most area for fixed length. The first rigorous proof of this fact was given by Weierstrass in the late 1800s. That a round sphere in $\mathbb{R}^3$ encloses the most volume for fixed area was proved by Schwarz. Later Schmidt proved that a geodesic ball in Euclidean, hyperbolic or spherical space encloses a fixed volume with the least possible perimeter.

In higher codimension there are many submanifolds with a given boundary, so that one cannot bound the perimeter boundary of a submanifold by its volume. However isoperimetric inequalities can be obtained in $\mathbb{R}^n$ if we require that the spanning surface be area-minimizing.

Theorem: An m-dimensional area-minimizing integral current in $\mathbb{R}^n$ $(2 \leq m \leq n)$ has no more area than a round disc of the same boundary area, with equality only for the round disc itself.

Let us give the main ideas of the proof for surfaces in $\mathbb{R}^3$ (although this case was known earlier). Amongst area-minimizing surfaces in $\mathbb{R}^3$ with area π there is one Q with boundary C of least length L. We assume regularity to simplify the arguments. Of course, by direct comparison with the planar disc of area π, we get

$$L \leq 2\pi.$$

Let us see first that the curvature κ of the curve satisfies inequality $\kappa \leq 1$. Otherwise we could find a local variation of C (around some point where $\kappa > 1$) so that

$$\frac{dL}{dA} > 1$$

We use this variation to reduce the length of C reducing at the same time the area of Q. On the other hand, scaling up to restore the area we get

$$\frac{dL}{dA} = \frac{L}{2A} \leq 1,$$

so that the combination of both transformations keeps the area constant but reduce boundary length, a contradiction to the minimality of the length of C.

Finally we look at the Gauss map of the boundary B of the convex hull of C (Q lies inside by the convex hull property for minimal surfaces). The total curvature of B is 4π and the only contribution to this total curvature comes from $B \cap C$ since $B - C$ is a flat surface. Geometrical considerations imply that this contribution to the total curvature equals $(2/\pi)\kappa\alpha$, where $\alpha \in (0, \pi)$ is an angle. Hence

$$4\pi = \frac{2}{\pi}\int_C \kappa\alpha ds \le \frac{2}{\pi} 1\pi L = 2L.$$

This shows that $L = 2\pi$. If equality holds then $\kappa \equiv 1$ and $\alpha \equiv \pi$ and this is enough to show that Q is a flat disc bounded by the circle C.

To make this argument completely rigorous, Almgren considered the equidistants (sets at fixed distance) from the boundary B of the convex hull of B.

Finally we state some open questions concerning isoperimetric inequalities

1. We know that amongst area-minimizing surfaces the disc of area π is the largest minimal surface bounded by a system of curves of total length 2π. Is this still true for all minimal surfaces? Is it true for stable minimal surfaces?
2. Does any isoperimetric inequality hold for Φ -minimal surfaces?

RECTIFIABILITY

LIPSCHITZ MAPS

Let X, Y be metric spaces, and let $\lambda \in [0, \infty)$. A map $f: X \to Y$ is λ-Lipschitz if

$$d(f(x), f(x')) \le \lambda d(x, x') \text{ for all } x, x' \in X;$$

f is Lipschitz if

$$\text{Lip}(f) := \inf\{\lambda \in [0, \infty): f \text{ is } \lambda\text{-Lipschitz}\} < \infty.$$

Lemma: Suppose X is a metric space and $A \subset X$.

1. For $n \in \mathbb{N}$, every λ-Lipschitz map $f: A \to \mathbb{R}^n$ admits a $\sqrt{n}\,\lambda$-Lipschitz extension $\overline{f}: X \to \mathbb{R}^n$.
2. For any set J, every λ-Lipschitz map $f: A \to l^\infty(J)$ has a λ-Lipschitz extension $\overline{f}: X \to l^\infty(J)$.

Proof: 1. For $n = 1$, put

$$\overline{f}(x) := \inf\{f(a) + \lambda d(a, x) : a \in A\}\cdot$$

For $n \ge 2$, $f = (f_1, \ldots, f_n)$, extend each f_i separately.

2. For $f = (f_j)_{j \in J}$, extend each f_j separately.

In (1), the factor $\sqrt{n}$ cannot be replaced by a constant $< n^{1/4}$. In particular, Lipschitz maps into a Hilbert space Y cannot be extended in general. However, if X is itself a Hilbert space, one has again an optimal result:

Kirszbraun, Valentine

Theorem: If X, Y are Hilbert spaces, $A \subset X$, and $f: A \to Y$ is λ-Lipschitz, then f has a λ-Lipschitz extension $\overline{f}: X \to Y$.

The next result characterizes the extendability of partially defined Lipschitz maps from $\mathbb{R}^m$ into a complete metric space Y; it is useful in connection with the definition of rectifiable sets. We call a metric space Y Lipschitz m-connected if there is a constant $c \geq 1$ such that for $k \in \{0,..., m\}$, every λ-Lipschitz map $f: S^k \to Y$ admits a $c\lambda$-Lipschitz extension $\overline{f}: B^{k+1} \to Y$; here S^k and B^{k+1} denote the unit sphere and closed ball in $\mathbb{R}^{k+1}$, endowed with the induced metric. Every Banach space is Lipschitz m-connected for all $m \geq 0$. The sphere S^n is Lipschitz $(n-1)$-connected.

Lipschitz Maps on $\mathbb{R}^m$

Theorem: Let Y be a complete metric space, and let $m \in \mathbb{N}$. Then the following statements are equivalent:

1. Y is Lipschitz $(m-1)$-connected.
2. There is a constant c such that every λ-Lipschitz map $f: A \to Y$, $A \subset \mathbb{R}^m$, has a $c\lambda$-Lipschitz extension $\overline{f}: \mathbb{R}^m \to Y$.

Proof: It is clear that (2) implies (1). Now suppose that (1) holds, and let $f: A \to Y$ be a λ-Lipschitz map, $A \subset \mathbb{R}^m$. As Y is complete, assume w.l.o.g. that A is closed. A dyadic cube in $\mathbb{R}^m$ is of the form $x + [0, 2^k]^m$ for some $k \in \mathbb{Z}$ and $x \in (2^k\mathbb{Z})^m$. Denote by C the family of all dyadic cubes $C \in \mathbb{R}^m \setminus A$ that are maximal (with respect to inclusion) subject to the condition

$$\operatorname{diam} C \leq 2\, d(A, C).$$

They have pairwise disjoint interiors, cover $\mathbb{R}^m \setminus A$, and satisfy

$$d(A, C) < 2 \operatorname{diam} C,$$

for otherwise the next bigger dyadic cube C' containing C would still fulfill

$$\operatorname{diam} C' = 2 \operatorname{diam} C \leq 2(d(A, C) - \operatorname{diam} C) \leq 2\, d(A, C').$$

Denote by $\sum_k \subset \mathbb{R}^m$ the k-skeleton of this cubical decomposition. Extend f to a

Lipschitz map $f_0: A \cup \sum_0 \to Y$ by precomposing f with a nearest point retraction $A \cup \sum_0 \to A$. Then, for $k = 0,\ldots, m-1$, successively extend f_k to $f_{k+1}: A \cup \sum_{k+1} \to Y$ by means of the Lipschitz $(m-1)$-connectedness of Y. As $A \cup \sum_m = \mathbb{R}^m$, $\bar{f} := f_m$ is the desired extension of f.

DIFFERENTIABILITY

GATEAUX AND FRECHET DIFFERENTIAL

Suppose X, Y are Banach spaces, f maps an open set $U \subset X$ into Y, and $x \in U$.

1. The map f is Gateaux differentiable at x if the directional derivative $D_v f(x)$ exists for every $v \in X$ and if there is a continuous linear map $L: X \to Y$ such that

$$L(v) = D_v f(x) \text{ for all } v \in X.$$

Then L is the Gateaux differential of f at x.

2. The map f is (Frechet) differentiable at x if there is a continuous linear map $L: X \to Y$ such that

$$\lim_{v \to 0} \frac{f(x+v) - f(x) - L(v)}{\|v\|} = 0.$$

Then $L =: Df_x$ is the (Frechet) differential of f at x.

The map f is Frechet differentiable at x iff f is Gateaux differentiable at x and the limit in

$$L(u) = \lim_{t \to 0} (f(x+tu) - f(x))/t$$

exists uniformly for u in the unit sphere of X, i.e. for all $\varepsilon > 0$ there is a $\delta > 0$ such that

$$\| f(x+tu) - f(x) - tL(u) \| \le \varepsilon |t|$$

whenever $|t| \le \delta$ and $u \in S(0, 1) \subset X$.

Differentiable Lipschitz Maps

Lemma: Suppose Y is a Banach space, $f: \mathbb{R}^m \to Y$ is Lipschitz, $x \in \mathbb{R}^m$, D is a dense subset of S^{m-1}, $D_u f(x)$ exists for every $u \in D$, $L: \mathbb{R}^m \to Y$ is linear, and $L(u) = D_u f(x)$ for all $u \in D$. Then f is Frechet differentiable at x with differential $Df_x = L$.

In particular, if $f: \mathbb{R}^m \to Y$ is Lipschitz and Gateaux differentiable at x, then f is Frechet differentiable at x.

Proof: Let $\varepsilon > 0$. Choose a finite set $D' \subset D$ such that for every $u \in S^{m-1}$ there is a $u' \in D'$ with $| u - u' | \leq \varepsilon$. Then there is a $\delta > 0$ such that

$$\|f(x + tu') - f(x) - tL(u')\| \leq \varepsilon \, | t |$$

whenever $| t | \leq \delta$ and $u' \in D'$. Given $u \in S^{m-1}$, pick $u' \in D'$ with $| u - u' | \leq \varepsilon$; then

$$\begin{aligned} &\|f(x + tu) - f(x) - tL(u)\| \\ &\leq \varepsilon \, | t | + \|f(x + tu) - f(x + tu')\| + | t | \, \| L(u - u') \| \\ &\leq (1 + \mathrm{Lip}(f) + \| L \|) \, \varepsilon \, | t | \end{aligned}$$

for all $| t | \leq \delta$.

Rademacher

Theorem: Every Lipschitz map $f\colon \mathbb{R}^m \to \mathbb{R}^n$ is differentiable at L^m-almost all points in $\mathbb{R}^m$.

Proof: It suffices to prove the theorem for $n = 1$; in the general case, $f = (f_1, \ldots, f_n)$ is differentiable at x iff each f_i is differentiable at x.

In the case $m = 1$ the function $f\colon \mathbb{R} \to \mathbb{R}$ is absolutely continuous and hence L^1-almost everywhere differentiable.

Now let $m \geq 2$. For $u \in S^{m-1}$, denote by B_u the set of all $x \in \mathbb{R}^m$ where $D_u f(x)$ exists and by H_u the linear hyperplane orthogonal to u. For $x_0 \in H_u$, the function $t \mapsto f(x_0 + tu)$ is L^1-almost everywhere differentiable by the result for $m = 1$, hence

$$H^1((x_0 + \mathbb{R}\, u) \setminus Bu) = 0.$$

Since B_u is a Borel set, Fubini's theorem implies

$$L^m(\mathbb{R}^m \setminus B_u) = 0.$$

Now choose a dense countable subset D of S^{m-1}. Then it follows that for L^m-almost every $x \in \mathbb{R}^m$, $D_u f(x)$ and $D_{e_i} f(x)$ exist for all $u \in D$ and $i = 1, \ldots, m$; in particular, the formal gradient

$$\nabla f(x) := D_{e_i}(f(x), \ldots, D_{e_m} f(x))$$

exists. We show that for L^m-almost all $x \in \mathbb{R}^m$ we have, in addition, the usual relation

$$D_u f(x) = \langle \nabla f(x), u \rangle \text{ for all } u \in D.$$

Let $\varphi \in C_c^\infty(\mathbb{R}^m)$. By Lebesgue's bounded convergence theorem,

$$\lim_{t\to 0+}\int \frac{f(x+tu)-f(x)}{t}\varphi(x)dx = \int D_u f(x)\varphi(x)dx,$$

$$\lim_{t\to 0+}\int f(x)\frac{\varphi(x-tu)-\varphi(x)}{t}dx = -\int f(x)D_u\varphi(x)dx.$$

Substituting $x + tu$ by x in the term $f(x + tu)\,\varphi(x)$ we see that the two left-hand sides coincide. Hence,

$$\lim_{t\to 0+}\int D_u f(x)\varphi(x)dx = -\int f(x)D_u\varphi(x)dx,$$

and similarly

$$\int \langle \nabla f(x), u\rangle \varphi(x)dx = -\int f(x)\langle \nabla\varphi(x), u\rangle dx.$$

Now the right-hand sides of these two identities coincide. As $\varphi \in C_c^\infty(\mathbb{R}^m)$ is arbitrary, we conclude that $D_u f(x) = \langle \nabla f(x), u\rangle$ for L^m-almost every $x \in \mathbb{R}^m$.

Stepanov

Theorem: Every function f: $\mathbb{R}^m \to \mathbb{R}^n$ is differentiable at L^m-almost all points in the set

$$L(f) := \{x : \limsup_{y\to x} |f(y)-f(x)|/|y-x| < \infty\}$$

This generalization of Rademacher's theorem was proved in [Ste]. The following elegant argument is due to Mal′y.

Proof: It suffices to consider the case $n = 1$. Let $(U_i)_{i\in\mathbb{N}}$ be the family of all open balls in $\mathbb{R}^m$ with rational center and radius such that $f|\,U_i$ is bounded.

This family covers $L(f)$. Let a_i: $U_i \to \mathbb{R}$ be the supremum of all i-Lipschitz functions $\leq f|\,U_i$, and let b_i: $U_i \to \mathbb{R}$ be the infimum of all i-Lipschitz functions $\geq f|\,U_i$. Note that a_i, b_i are i-Lipschitz and $a_i \leq f|\,U_i - b_i$. Let

$$A_i := \{x \in U_i\text{: both } a_i \text{ and } b_i \text{ are differentiable at } x\}.$$

By Rademacher's theorem, $Z := \bigcup_{i=1}^{\infty} U_i \setminus A_i$ has measure zero. Let $x \in L(f) \setminus Z$. We show that for some i, $x \in A_i$ and $a_i(x) = b_i(x)$; then f is differentiable at x. Since $x \in L(f)$, there is a radius $r > 0$ such that $|f(y) - f(x)| \leq \lambda\,|y - x|$ for all $y \in B(x, r)$ and for some λ independent of y. Choose i such that $i \geq \lambda$ and $x \in U_i \subset B(x, r)$. Since $x \notin Z$, $x \in A_i$. By the definition of a_i and b_i,

$$f(x) - i\,|y-x| \le a_i(y) \le f(y) \le b_i(y) \le f(x) + i\,|y-x|$$

for all $y \in U_i$. Hence, $a_i(x) = b_i(x)$.

Generalizations of these results to maps between Banach spaces or even more general classes of metric spaces are a topic of current research.

Whitney

Theorem: Suppose $f: A \to \mathbb{R}$ is a function on a closed set $A \subset \mathbb{R}^m$, $g: A \to \mathbb{R}^m$ is continuous, and for all compact sets $C \subset A$ and all $\varepsilon > 0$ there is a $\delta > 0$ such that

$$|f(y) - f(x) - \langle g(x), y-x\rangle| \le \varepsilon\,|y-x|$$

whenever $x, y \in C$ and $|y - x| \le \delta$. Then there exists a C^1 function $\overline{f}: \mathbb{R}^m \to \mathbb{R}$ with $\overline{f} \mid A = f$ and $\nabla \overline{f} \mid A = g$.

C^1 approximation of Lipschitz functions

Theorem: If $f: \mathbb{R}^m \to \mathbb{R}$ is Lipschitz and $\varepsilon > 0$, then there is a C^1 function $\overline{f}$: $\mathbb{R}^m \to \mathbb{R}$ such that

$$L^m(\{x \in \mathbb{R}^m : f(x) \ne \overline{f}(x)\}) < \varepsilon$$

Proof: By Rademacher's theorem, f is almost everywhere differentiable, and $g: = \nabla f$ is a measurable function. According to Lusin's theorem, there is a closed set $B \subset \mathbb{R}^m$ with $L^m(\mathbb{R}^m \setminus B) < \varepsilon/2$ such that $g \mid B$ is continuous. For $x \in B$ and $i \in \mathbb{N}$, let

$$r_i(x): = \sup |f(y) - f(x) - \langle g(x), y-x\rangle| / |y-x|,$$

the supremum taken over all $y \in B$ with $0 < |y - x| \le 1/i$. We know that $r_i \to 0$ pointwise on B as $i \to \infty$. By Egorov's theorem, there is a closed set $A \subset B$ with $L^m(B \setminus A) < \varepsilon/2$ such that $r_i \to 0$ uniformly on compact subsets of A. Now extend $f \mid A$ to $\mathbb{R}^m$.

AREA FORMULA

Borel Partition

Lemma: Suppose $f: \mathbb{R}^m \to \mathbb{R}^n$ is Lipschitz, and B is the set of all x where Df_x exists and has rank m. Then for every $\lambda > 1$ there exist a Borel partition $(B_i)_{i\in\mathbb{N}}$ of B and a sequence of euclidean norms $\|\cdot\|_i$ on $\mathbb{R}^m$ (i.e. $\|\cdot\|_i$ is induced by an inner product), such that

$$\lambda^{-1}\|v\|_i \le |Df_x(v)|\,\lambda\,\|v\|_i,$$

$$\lambda^{-1}\|y-x\|_i \le |f(y)-f(x)| \le \lambda\|y-x\|_i$$

for all $x, y \in B_i$ and $v \in \mathbb{R}^m$.

Proof: Choose a sequence of euclidean norms $\|\cdot\|_j$ on $\mathbb{R}^m$ such that for every euclidean norm $\|\cdot\|$ on $\mathbb{R}^m$ and for every $\varepsilon > 0$ there is a $j \in \mathbb{N}$ with

$$(1-\in)\|v\|_j \le (1+\varepsilon)\|v\|_j \text{ for all } v \in \mathbb{R}^m.$$

Given $\lambda > 1$, pick $\delta > 0$ such that $\lambda^{-1} + \delta < 1 < -\delta$. For $j, k \in \mathbb{N}$, denote by B_{jk} the Borel set of all $x \in B$ with

(i) $(\lambda^{-1}+\delta)\|v\|_j \le (Df_x(v)| \le (\lambda-\delta)\|v\|_j$ for $v \in \mathbb{R}^m$,

(ii) $|f(x+v) - f(x) - Df_x(v)| \le \delta \|v\|_j$ for $|v| \le 1/k$.

To see that the B_{jk} cover B, let $x \in B$, choose $j \in \mathbb{N}$ such that (i) holds, let $c_j > 0$ be such that $|v| \le c_j\|v\|_j$ for all $v \in \mathbb{R}^m$, and pick $k \in \mathbb{N}$ such that

$$|f(x+v) - f(x) - Df_x(v)| \le (\delta/c_j)|v|$$

whenever $|v| \le 1/k$; then $x \in B_{jk}$. Now if $C \subset B_{jk}$ is a set with diam $C \le 1/k$, then

$$|f(x+v) - f(x)| \le |Df_x(v)| + \delta\|v\|_j \le \lambda\|v\|_j,$$

$$|f(x+v) - f(x)| \ge |Df_x(v)| - \delta\|v\|_j \ge \lambda^{-1}\|v\|_j$$

whenever $x, x+v \in C$. By subdividing and relabeling the sets B_{jk} appropriately we obtain the result.

Jacobian

Definition: Let $L: X \to Y$ be a linear map between two inner product spaces, where dim $X = m$. The m-dimensional jacobian $J_m(L)$ of L is the number satisfying

$$J_m(L) = H^m(L(A))/H^m(A) = \sqrt{\det(L^*\circ L)}$$

for all $A \subset X$ with $H^m(A) > 0$, where $L^*: Y \to X$ is the adjoint map.

If $\|\cdot\|$ is a euclidean norm on $\mathbb{R}^m$, we write $J_m(\|\cdot\|)$ for $J_m(L)$ where $L: \mathbb{R}^m \to (\mathbb{R}^m, \|\cdot\|)$ is the identity map.

Area Formula

Theorem: Suppose $f: \mathbb{R}^m \to \mathbb{R}^n$ is Lipschitz with $m \le n$.

1. If $A \subset \mathbb{R}^m$ is L^m-measurable, then

$$\int_A J_m(Df_x)dx = \int_{\mathbb{R}^n} \#(f^{-1}\{y\} \cap A)dH^m(y).$$

2. If u is an L^m-integrable function, then

$$\int_{\mathbb{R}^m} u(x) J_m(Df_x)dx = \int_{\mathbb{R}^n} \sum_{x \in f^{-1}\{y\}} u(x) dH^m(y).$$

The formula says, in particular, that the differential geometric volume of an injective C^1 map $f: U \to \mathbb{R}^n$, U an open subset of $\mathbb{R}^m$, equals $H^m(f(U))$.

Proof: (1) We assume w.l.o.g. that $L^m(A) < \infty$.

Case 1: $A \subset \{x: Df_x$ exists and has rank $m\}$. Let $\lambda > 1$. We find a measurable partition $(A_i)_{i \in \mathbb{N}}$ of A and a sequence of euclidean norms $\|\cdot\|_i$ on $\mathbb{R}^m$ such that $f \mid A_i$ is injective,

$$\lambda^{-1} H^m_{\|\cdot\|}(A_i) \le H^m(f(A_i)) \le \lambda^m H^m_{\|\cdot\|_i}(A_i),$$

and $\lambda^{-1}\|.\|_i \le |Df_x(.)| \le \lambda\|.\|$ for all $x \in A_i$. This last assertion yields $\lambda^{-m} J_m(\|\cdot\|_i) \le J_m(Df_x) \le \lambda^m J_m(\|\cdot\|_i)$. We conclude that

$$H^m(f(A_i)) \le \lambda^m H^m_{\|\cdot\|m}(A_i) = \lambda^m J_m(\|\cdot\|_i) L^m(A_i)$$
$$\le \lambda^{2m} \int_{A_i} J_m(Df_x)dx.$$

Since each $f \mid A_i$ is injective, summation over i gives

$$\int_{\mathbb{R}^n} \#(f^{-1}\{y\} \cap A) dH^m(y) \le \lambda^{2m} \int_A J_m(Df_x)dx.$$

Similarly,

$$\lambda^{-2m} \int_A J_m(Df_x)dx \le \int_{\mathbb{R}^n} \#(f^{-1}\{y\} \cap A) dH^m(y).$$

As this holds for all $\lambda > 1$, the two integrals are equal.

Case 2: $A \subset \{x: Df_x$ exists and has rank $< m\}$. Then $J_m(Df_x) = 0$ for all $x \in A$. For $\varepsilon > 0$, consider the map $F: \mathbb{R}^m \to \mathbb{R}^n \times \mathbb{R}^m$, $F(x) = (f(x), x)$. For $x \in A$, it follows that $\| DF_x \| \le \text{Lip}(f) + \varepsilon$ and

$$J_m(DF_x) \le \varepsilon(\text{Lip}(f) + \varepsilon)^{m-1}.$$

Applying the result of the first case to F, we get

$$H^m(F(A)) = \int_A J_m(DF_x)dx \le \varepsilon(Lip(f) + \varepsilon)^{m-1} L^m(A).$$

Since $H^m(f(A)) \le H^m(F(A))$ for all $\varepsilon > 0$, it follows that $H^m(f(A)) = 0$. Thus both integrals equal 0.

Case 3: $A \subset \{x: Df_x$ does not exist$\}$. Then

$$H^m(f(A)) \leq \text{Lip}(f)^m H^m(A) = \text{Lip}(f)^m L^m(A) = 0$$

by Rademacher's theorem. Thus both integrals equal 0.

(2) follows from (1) by approximation. 2

RECTIFIABLE SETS

The following notion is fundamental in geometric measure theory.

COUNTABLY RECTIFIABLE SET

Definition: Let Y be a metric space. A set $E \subset Y$ is countably H^m-rectifiable if there is a sequence of Lipschitz maps $f_i: A_i \to Y$, $A_i \subset \mathbb{R}_m$, such that

$$H^m\left(E \setminus \bigcup_i f_i(A_i)\right) = 0$$

It is often possible to take w.l.o.g. $A_i = \mathbb{R}^m$, e.g. if Y is a Banach space.

Countably Rectifiable sets in $\mathbb{R}^n$

Theorem: A set $E \subset \mathbb{R}^n$ is countably H^m-rectifiable if and only if there exists a sequence of m-dimensional C^1 submanifolds M_k of $\mathbb{R}^n$ such that

$$H^m\left(E \setminus \bigcup_k M_k\right) = 0$$

Proof: Suppose that $H^m\left(E \setminus \bigcup_i f_i(\mathbb{R}^m)\right) = 0$ for a sequence of Lipschitz maps $f_i: \mathbb{R}^m \to \mathbb{R}^n$.

We assume w.l.o.g. that the f_i are C^1. Let $U_i \subset \mathbb{R}^m$ be the set of all $x \in \mathbb{R}^m$ where Df_x has rank m. By the area formula, $H^m(f_i(\mathbb{R}^m \setminus U_i)) = 0$. Hence, $H^m\left(E \setminus \bigcup_i f_i(U_i)\right) = 0$. Finally, it follows from the inverse function theorem that each $f_i(U_i)$ is a countable union of C^1 submanifolds.

The other implication is clear.

Bi-Lipschitz parametrization

Theorem: Suppose $E \subset \mathbb{R}^n$ is countably H^m-rectifiable and H^m-measurable. Then for every $\lambda > 1$ there exists a sequence of λ-bi-Lipschitz maps $f_i: C_i \to f_i(C_i) \subset E$, with $C_i \subset \mathbb{R}^m$ compact, such that the $f_i(C_i)$ are pairwise disjoint and

$$H^m\left(E \setminus \bigcup_i f_i(C_i)\right) = 0$$

Proof: First we assume that E is a Borel set contained in the image of a single Lipschitz map h: $\mathbb{R}^m \to \mathbb{R}^n$.

Using Lemma Borel partition, the area formula, and the inner regularity of L^m, we find a sequence of λ-bi-Lipschitz maps g_k: $D_k \to g_k(D_k) \subset E$, with $D_k \subset \mathbb{R}^m$ compact, such that $H^m\left(E \setminus \bigcup_k g_k(C_k)\right) = 0$. Consider the Borel sets

$$D_k^{'} := D_k \setminus g_k^{-1}\left(\bigcup_{j=1}^{k-1} g_j(D_j)\right)$$

Then $\bigcup_k g_k(D_k^{'}) = \bigcup_k g_k(D_k)$, and the $g_k(D_k^{'})$ are pairwise disjoint.

For every k, choose a sequence of pairwise disjoint compact sets $C_{k,l} \subset D_k^{'}$ such that $L^m\left(D_k^{'} \bigcup_l C_{k,l}\right) = 0$. It follows that $H^m\left(E \setminus \bigcup_{k,l} g_k(C_{k,l})\right) = 0$, and the $g_k(C_{k,l})$ are pairwise disjoint.

To prove the general result, partition E into a sequence of H^m-measurable.

Sets E_j with $H^m(E_j) < 1$ and $E_j \subset h_j(\mathbb{R}^m)$ for some Lipschitz map h_j: $\mathbb{R}^m \to \mathbb{R}^n$. Then E_j contains an F set F_j with $H^m(E_j \setminus F_j) = 0$. Now apply the above result to each F_j.

Suppose X is a metric space, $A \subset X$, and $x \in X$. Recall that the m-dimensional upper density and lower density of A at x are defined by

$$\Theta^{*m}(A,x) = \limsup_{r \downarrow 0} \frac{H^m(A \cap B(x,r))}{\alpha_m r^m},$$

$$\Theta^{*m}(A,x) = \liminf_{r \downarrow 0} \frac{H^m(A \cap B(x,r))}{\alpha_m r^m},$$

where $\alpha_m := L^m(B^m)$. If the two coincide, then the common value $\Theta^m(A, x)$ is the density of A at x.

If A, $B \subset X$ are two H^m-measurable sets with $A \subset B$ and $H^m(B) < 1$, then

$$2^{-m} \leq \Theta^m(B, x) \leq 1$$

for H^m-almost all $x \in B$,

$$\Theta^m(B, x) = 0$$

for H^m-almost all $x \in X \setminus B$, and

$$\Theta^{*m}(A, x) = \Theta^{*m}(B, x),\ \Theta_*^m(A, x) = \Theta_*^m(B, x)$$

for H^m-almost all $x \in A$.

Also Recall Lebesgue's Theorem

If $u \in L^1(\mathbb{R}^m)$, then L^m-almost every point x is a Lebesgue point of u, i.e.

$$\lim_{r \downarrow 0} \frac{1}{\alpha_m r^m} \int_{B(x,r)} |u(y) - u(x)|\, dy = 0.$$

For $x \in \mathbb{R}^n$ and $r > 0$, define $T_{x,r}: \mathbb{R}^n \to \mathbb{R}^n$, $T_{x,r}(y) = (y - x)/r$. Note that $T_{x,r}$ takes $B(x, r)$ to $B(0, 1) = B^n$.

Approximate Tangent Space

Definition: Suppose $E \subset \mathbb{R}^n$ is a H^m-measurable set with $H^m(E) < \infty$. Let $x \in \mathbb{R}^n$. An m-dimensional linear subspace $L \subset \mathbb{R}^n$ is called the (H^m-)approximate tangent space of E at x if

$$\lim_{r \downarrow 0} \int_{Tx,r} \varphi dH^m = \int_L \varphi dH^m$$

for all $\varphi \in C_c(\mathbb{R}^n)$. Then we write $L =: \mathrm{Tan}^m(E, x)$.

Clearly $\mathrm{Tan}^m(E, x)$ is uniquely determined if it exists. There are various definitions of approximate tangent spaces in the literature, compare.

Existence of Tangent Spaces

Theorem: Suppose $E \subset \mathbb{R}^n$ is a countably H^m-rectifiable and Hm-measurable set with $H^m(E) < \infty$. Then for H^m-almost every $x \in E$, $\mathrm{Tan}^m(E, x)$ exists and $\Theta^m(E, x) = 1$.

Proof: Choose a sequence of m-dimensional C^1 submanifolds M_k of $\mathbb{R}^n$ such that $H^m\left(E \setminus \bigcup_k M_k\right) = 0$. Put $E_k := E \cap M_k$; then $H^m\left(E \setminus \bigcup_k E_k\right) = 0$. Since M_k is C^1, it follows that for H^m-almost every $x \in E_k$, we have $\Theta^m(Ek, x) = 1$ and $\mathrm{Tan}^m(Ek, x) = T_x M_k$. Moreover, for H^m-almost every $x \in E_k$, $\Theta^m(E \setminus E_k, x) = 0$. Combining these two properties we conclude that for H^m-almost every $x \in E_k$, $\Theta^m(E, x) = 1$ and $\mathrm{Tan}^m(E, x) = T_x M_k$.

Theorem: Suppose $E \subset \mathbb{R}^n$ is a H^m-measurable set with $H^m(E) < \infty$. If $\mathrm{Tan}^m(E, x)$ exists for H^m-almost every $x \in E$, then E is countably H^m-rectifiable.

Preiss

Theorem: Suppose $E \subset \mathbb{R}^n$ is a H^m-measurable set with $H^m(E) < \infty$. If the density $\Theta^m(E, x)$ exists for H^m-almost every $x \in E$, then E is countably H^m-rectifiable.

Finally, we state the Besicovitch–Federer projection theorem which played a very important role in the development of the theory of currents. This deep result was proved in [Bes] for $m = 1$ and $n = \in$ and for general dimensions. A set $F \subset \mathbb{R}^n$ is purely

H^m-unrectifiable if $H^m(F \setminus f(R_m)) = 0$ for every Lipschitz map $f: \mathbb{R}^m \to \mathbb{R}^n$. Every set $A \subset \mathbb{R}^n$ with $H^m(A) < 1$ can be written as the disjoint union of a countably H^m-rectifiable set E and a purely H^m-unrectifiable set F.

Besicovitch, Federer

Theorem: Suppose $F \subset \mathbb{R}^n$ is a purely H^m-unrectifiable set with $H^m(F) < 1$. Then for $\gamma_{n,m}$-almost every $L \in G(n, m)$, $H^m(\pi L(F)) = 0$. Here $\gamma_{n,m}$ denotes the Haar measure on $G(n, m)$, and $\pi_L: \mathbb{R}^n \to L$ is orthogonal projection.

NORMAL CURRENTS

VECTORS AND COVECTORS

Denote by $e_1,..., e_n$ the standard basis for $\mathbb{R}^n$ and by $e_1^*,..., e_n^*$ the dual basis for the dual space $(\mathbb{R}^n) = \{f: \mathbb{R}^n \to \mathbb{R}$ linear$\}$, such that $e_i(e_j) = i_j$ for all i, j.

For $m \in \mathbb{N}$, $\Lambda_m \mathbb{R}^n$ and $\Lambda^m \mathbb{R}^n$ denote the vector spaces of m-vectors and m-covectors of $\mathbb{R}^n$, respectively. In case $1 \leq m \leq n$, a basis of $\Lambda_m \mathbb{R}^n$ is given by

$$\{e_\lambda := e_\lambda(1) \wedge ... \wedge e_\lambda(m) : \lambda \in \Lambda(n, m)\},$$

where $\Lambda(n, m)$ denotes the set of all strictly increasing maps from $\{1,..., m\}$ into $\{1,..., n\}$. Similarly,

$$\{e_\lambda^* = e_{\lambda(1)}^* \wedge .... \wedge e_{\lambda(1)}^* : \lambda \in \Lambda(n,m)\}$$

is a basis of $\Lambda^m \mathbb{R}^n$. For $m > n$, $\Lambda m \mathbb{R} n = \Lambda m \mathbb{R} n = \{0\}$. By convention, $\Lambda_0 \mathbb{R}^n = \Lambda^0 \mathbb{R}_n = \mathbb{R}$.

An m-vector τ is simple if it can be written as a product of m vectors, $\tau = v_1 \wedge ... \wedge v_m$. Simple covectors are define analogously.

We write $\langle \tau, \omega \rangle$ for the duality product of $\tau \in \Lambda_m \mathbb{R}^n$ and $\omega \in \Lambda^m \mathbb{R}^n$, thus $\langle e_\lambda, e_\mu^* \rangle = \delta_{\lambda\mu}$ for $\lambda, \mu \in \Lambda(n,m)$.

The standard inner product $\langle .,. \rangle$ and euclidean norm $|\cdot|$ on $\mathbb{R}^n$ induce corresponding inner products and norms on $\Lambda_m \mathbb{R}^n$ and $\Lambda^m \mathbb{R}^n$ such that the above bases are orthonormal. They will be denoted by the same symbols $\langle .,. \rangle$ and $|\cdot|$. For $\tau = v_1 \wedge ... \wedge v_m \in \Lambda_m \mathbb{R}^n$,

$$|\lambda| = \sqrt{\det(\langle v_i, v_j \rangle)}.$$

The comass norm of an m-covector ω is defined by

$$||\omega|| = \sup\{\langle \tau,\omega\rangle : \tau \in \Lambda_m \mathbb{R}^n \text{ is simple and } |\tau| \leq 1\}.$$

Always $||\tau|| \leq |\omega|$, with equality iff ω is simple. The mass norm of an m-vector τ is defined by

$$|\tau| = \sup \langle \tau,\omega\rangle \{\omega \in \Lambda_m \mathbb{R}^n \text{ and } |\tau| \leq 1\}.$$

Always $||\omega|| \leq |\omega|$, with equality iff ω is simple.

By a C^∞ differential m-form ω on $\mathbb{R}^n$ we mean an m-covectorfield $\omega \in C^\infty(\mathbb{R}^n, \Lambda^m \mathbb{R}^n)$. We denote by

$$D^m(\mathbb{R}^n) := C_c^\infty\left(\mathbb{R}^n, \Lambda^m \mathbb{R}^n\right)$$

the vector space of all m-forms on $\mathbb{R}^n$ with compact support. As usual, we write $dx^\lambda = dx^{\lambda(1)} \wedge \ldots \wedge dx^{\lambda(m)}$ for the constant covectorfield mapping x to $e_\lambda^* = e_{\lambda(1)}^* \wedge \ldots \wedge e_{\lambda(m)}^*$; then $\omega \in D^m(\mathbb{R}^n)$ is of the form

$$\omega = \sum_{\lambda \in \Lambda(n,m)} \omega_\lambda dx^\lambda, \omega_\lambda \in C_c^\infty(\mathbb{R}^n).$$

We equip $D^m(\mathbb{R}^n)$ with the topology in which

$$\omega^i = \sum_\lambda \omega_\lambda^i dx^\lambda \to 0 \quad \text{for } i \to \infty$$

if and only if there exists a compact set $C \subset \mathbb{R}^n$ with spt $\omega^i \subset C$ for all i and

$$\sup_x |D^\alpha \omega_\lambda^i(x)| \to 0 \quad \text{for } i \to \infty$$

whenever $\lambda \in \Lambda(n, m)$, $\alpha = (\alpha_1, \ldots, \alpha_n)$, $\alpha_k \in \mathbb{N} \cup \{0\}$,

$$D^\alpha \omega_\lambda^i = \frac{\partial^{\alpha_1} \ldots \partial^{\alpha_n}}{(\partial x^2)^{\alpha_1} \ldots (\partial x^n)^{\alpha_n}} \omega_\lambda^i.$$

CURRENTS

General currents were introduced by deRham. The 0-dimensional currents are exactly the distributions in the sense of Schwartz.

Definition: An m-dimensional current or m-current in $\mathbb{R}^n$ is a continuous linear functional on $D^m(\mathbb{R}^n)$. $D_m(\mathbb{R}^n)$ denotes the vector space of all m-currents in $\mathbb{R}^n$.

A sequence $(T_i)_{i\in\mathbb{N}}$ in $D_m(\mathbb{R}^n)$ converges weakly to a current $T \in D_m(\mathbb{R}^n)$ if $\lim_{i\to\infty} T_i(\omega) = T(\omega)$ for all $\omega \in D^m(\mathbb{R}^n)$; we then write

$$T_i \to T.$$

The support spt T of a current $T \in D_m(\mathbb{R}^n)$ is the smallest closed set $C \subset \mathbb{R}^n$ with the property that $T(\omega) = 0$ for all $\omega \in D^m(\mathbb{R}^n)$ with spt $\omega \cap C = \phi$.

Boundary of a Current

Definition: Let $T \in D_m(\mathbb{R}^n)$, $m \geq 1$. The boundary of T is the current $\partial T \in D_{m-1}(\mathbb{R}^n)$ defined by

$$\partial T(\pi) := T(d\pi) \text{ for all } \pi \in D_{m-1}(\mathbb{R}^n).$$

Clearly $\partial \circ \partial = 0$ since $d \circ d = 0$, spt $\partial T \subset$ spt T, and $Ti \to T$ implies $\partial T_i \to \partial T$.

Total Variation Measure and Mass

Definition: Let $T \in D_m(\mathbb{R}^n)$. For $U \subset \mathbb{R}^n$ open and $A \subset \mathbb{R}^n$ arbitrary, put

$$\| T \| (U) := \sup\{T(\omega) : spt\omega \subset U, \sup_x \| \omega(x) \| \leq 1\}$$

$$\| T \| (A) := \inf\{\| T \| (U) : U \text{ is open}, A \subset U\}$$

This defines a Borel regular outer measure $\| T \|$ on $\mathbb{R}^n$.

$$M(T) := \| T \|(\mathbb{R}^n) \in [0, \infty]$$

is the mass of T. We denote by $M_m(\mathbb{R}^n)$ the vector space of all $T \in D_m(\mathbb{R}^n)$ with $M(T) < \infty$. A current $T \in D_m(\mathbb{R}^n)$ has locally finite mass if $\| T \|$ is a Radon measure, i.e. if it is finite on compact sets, and $M_{m,\text{loc}}(\mathbb{R}^n)$ denotes the vector space of all such currents.

For a fixed open set $U \subset \mathbb{R}^n$, the map $T \mapsto \| T \| (U)$ is lower semicontinuous on $D_m(\mathbb{R}^n)$ with respect to weak convergence, i.e.,

$$\| T \| (U) \leq \liminf_{i\to\infty} \| T_i \| (U) \quad T_i \to T.$$

The space $M_m(\mathbb{R}^n)$ endowed with the norm M is a Banach space. Note that

$$| T | (\omega) \leq \sup_x \| \omega(x) \| M(T)$$

for all $\omega \in D^m(\mathbb{R}^n)$.

Example: Suppose $M \subset \mathbb{R}^n$ is an oriented m-dimensional C^1-submanifold with boundary (possibly $\partial M = \phi$;), and M is a closed subset of $\mathbb{R}^n$. We view the orientation of M as a continuous function $\tau: M \to \Lambda_m \mathbb{R}^n$ such that for every $x \in M$, $\tau(x)$ is simple and represents the tangent space $T_x M$, and $| \tau(x) | = 1$. Then

$$[M](\omega) := \int_M \langle \tau(x), \omega(x) \rangle \, dH^m(x)$$

defines an *m*-current $[M] = [M, \tau] \in D_m(\mathbb{R}^n)$. Note that this integral corresponds to the usual $\int_M \omega$ in the notation of differential geometry.

Suppose ∂M is equipped with the induced orientation $\tau': \partial M \to \Lambda_{m-1}\mathbb{R}^n$, i.e. $\tau = \eta \wedge \tau'$ for the exterior unit normal η. Then we have

$$\begin{aligned}\partial[M](\pi) = [M](d\pi) \int_M \langle \tau, d\pi \rangle \, dH^m \\ = \int_{\partial M} \langle \tau', \pi \rangle \, dH^{m-1} = [\partial M](\pi)\end{aligned}$$

for all $\pi \in D^{m-1}(\mathbb{R}^n)$ by the Theorem of Stokes.

The measure $\|[M]\|$ is simply the restriction of H^m to M, $\|[M]\|(A) = (H^m \llcorner M)(A) = H^m(A \cap M)$. Whenever μ is a Radon measure on $\mathbb{R}^n$ and $\tau: \mathbb{R}^n \to \Lambda_m \mathbb{R}^n$ is locally μ-integrable, then we obtain a current $T = [\mu, \tau] \in D_m(\mathbb{R}^n)$ by defining

$$T(\omega) := \int_{\mathbb{R}^n} \langle \tau(x), \omega(x) \rangle \, d\mu(x).$$

We say that a current $T \in D_m(\mathbb{R}^n)$ is representable by integration if it admits such a representation; then clearly $T \in M_{m,\,loc}(\mathbb{R}^n)$. In fact, $M_{m,loc}(\mathbb{R}^n)$ is precisely the set of *m*-currents representable by integration:

Integral Representation

Theorem: Let $T \in M_{m,loc}(\mathbb{R}^n)$. There is a $\|T\|$-measurable function $\vec{T}: \mathbb{R}^n \to \Lambda_m \mathbb{R}^n$ such that $\|\vec{T}(x)\| = 1$ for $\|T\|$-almost every $x \in \mathbb{R}^n$ and

$$T(\omega) = \int_{\mathbb{R}^n} \left\langle \vec{T}(x), \omega(x) \right\rangle d\|T\|(x) \text{ for all } \omega \in D^m(\mathbb{R}^n).$$

In brief, $T = [\|T\|, \vec{T}]$. This follows from an appropriate version of the Riesz respresentation theorem.

The restriction of $T = [\mu, \tau] \in M_{m,\text{loc}}(\mathbb{R}^n)$ to a function $u \in L^1_{loc}(\mu)$ is the current $T u \in M_{m,\text{loc}}(\mathbb{R}^n)$ defined by

$$(T \perp u)(\omega) := \int_{\mathbb{R}^n} \langle \tau(x), \omega(x) \rangle \mu x d\mu(x).$$

If $B \subset \mathbb{R}^n$ is a Borel set and χ_B is the characteristic function, then we write $T B$ for $T\chi_B$. (A general current $T \in D_m(\mathbb{R}^n)$ can be restricted to a function $f \in C^\infty(\mathbb{R}^n)$: $(Tf)(\omega) := T(f\omega)$.)

***Weak Compactness in* $\mathbf{M}_{m,\,loc}$**

Theorem: Suppose $(T_i)_{i\in\mathbb{N}}$ is a sequence in $M_{m,\,\mathrm{loc}}(\mathbb{R}^n)$ with $\sup_i \|T_i\|(U) < \infty$ for all open sets $U \subset \mathbb{R}^n$ with compact closure. Then there is a subsequence (T_{i_j}) and a $T \in M_{m,\mathrm{loc}}(\mathbb{R}^n)$ such that $T_{i_j} \to T$.

This is an application of the Banach–Alaoglu theorem.

Example: Choose $\eta \in C_c^\infty(\mathbb{R})$ with $\eta \geq 0$ and $\int \eta dx = 1$, and let $\eta_i(x) := i\eta\,(ix)$ for $i = 1, 2,\ldots.$ Define $T_i \in D_1(\mathbb{R})$ by $T_i := [\mathbb{R}]\,\eta_i$, i.e.

$$T_i(\omega) = \int \langle 1, \omega(x)\rangle\, \eta_i(x)dx.$$

Then $M(T_i) = \int \eta_i dx = 1$ for all i, and $T_i \to T$ for the current T satisfying

$$T(\omega) = \langle 1, \omega(0)\rangle.$$

Note that

$$\partial T_i(f) = T_i(df) = \int \langle 1, df\rangle\, \eta_i dx = \int f'\eta_i dx = -\int f\eta_i' dx.$$

$$M(\partial T_i) = \int |\eta_i'|\, dx, \text{ and } \partial T(f) = f'(0), M(\partial T) = \infty.$$

Push-Forward

Definition: The push-forward of a current $T \in D_m(\mathbb{R}^n)$ under a C^∞ map f from $\mathbb{R}^n$ into $\mathbb{R}^p$ is defined as follows.

Suppose $f|\,\mathrm{spt}\,T$ is proper, i.e. $\mathrm{spt}\,T \cap f^{-1}(C)$ is compact whenever $C \subset \mathbb{R}^p$ is compact. Given a form $\omega \in D^m(\mathbb{R}^p)$, consider its pull-back $f^\#\omega$, pick a function $\zeta_\omega \in C_c^\infty(\mathbb{R}^n)$ such that $\zeta_\omega \equiv 1$ in a neighborhood of the compact set $\mathrm{spt}\,T \cap \mathrm{spt}(f^\#\omega)\ \mathrm{spt}\,T \cap f^{-1}(\mathrm{spt}\,\omega)$, and put

$$(f_\# T)(\omega) := T(\zeta_\omega f_\# \omega).$$

This defines a current $f_\# T \in D_m(\mathbb{R}^p)$; the definition is independent of the choice of the functions ζ_ω.

The following properties hold:

$$\partial(f_\# T) = f_\#(\partial T),\ \mathrm{spt}\,(f_\# T) \subset f\,(\mathrm{spt}\,T),$$

and $(g \circ f)_\# T = g_\#(f_\# T)$ whenever g is a C^∞ map from $\mathbb{R}^p$ into $\mathbb{R}^q$ such that $g \circ f|\,\mathrm{spt}\,T$ is proper. If $T \in M_{m,loc}(\mathbb{R}^n)$, $T = [\|T\|, \vec{T}]$, then

$$f_{\#}T(\omega) = \int_{\mathbb{R}^n} \left\langle Df_{x\#}\vec{T}(x), \omega(f(x)) \right\rangle d\,\|T\|(x)$$

for all $\omega \in D_m(\mathbb{R}^p)$, hence

$$\|f_{\#}T\|(V) \leq \int_{f^{-1}(V)} \|Df_{x\#}\vec{T}(x)\|\, d\,\|T\|(x)$$

for all open sets $V \subset \mathbb{R}^p$. Thus $f_{\#}T \in M_{m,loc}(\mathbb{R}^p)$.

NORMAL CURRENTS

Definition: Let $T \in D_m(\mathbb{R}^n)$, $m \geq 1$. Put

$$N(T) := M(T) + M(\partial T).$$

T is called normal if $N(T) < \infty$ and locally normal if $\|T\| + \|\partial T\|$ is a Radon measure. The respective vector spaces are denoted $N_m(\mathbb{R}^n)$ and $N_{m,loc}(\mathbb{R}^n)$. For $m = 0$, $N(T) := M(T)$, $N_0(\mathbb{R}^n) := M_0(\mathbb{R}^n)$ and $N_{0,\,\mathrm{loc}}(\mathbb{R}^n) := M_{0,\,loc}(\mathbb{R}^n)$.

Note that the space $N_m(\mathbb{R}^n)$ endowed with the norm N is a Banach space, and the compactness theorem holds with $M_{m,\mathrm{loc}}(\mathbb{R}^n)$ and $\|T_i\|(U)$ replaced by $N_{m,\,\mathrm{loc}}(\mathbb{R}^n)$ and $(\|T_i\| + \|\partial T_i\|)(U)$. (In [Fed], $T \in N_m(\mathbb{R}^n)$ means in addition that spt T is compact.)

For a compactly supported current $T \in N_m(\mathbb{R}^n)$, the definition of $f_{\#}T$ can be extended to the case that $f\colon \mathbb{R}^n \to \mathbb{R}^p$ is a locally Lipschitz map.

The idea is as follows. Pick a symmetric mollifier $\eta \in C_c^{\infty}(\mathbb{R}^n)$ (so that $\eta \geq 0$, spt $\eta \subset B(0, 1)$, $\eta(-x) = \eta(x)$, and $\int \eta dx = 1$. For $\varepsilon > 0$, put $\eta_\varepsilon(x) = \eta(x/\varepsilon)/\varepsilon^n$ and consider the mollified functions $f^{(\varepsilon)} := \eta_\varepsilon * f$,

$$(\eta_\varepsilon * f)(x) := \int \eta_\varepsilon(x-y) f(y) dy.$$

Now show that, as $\varepsilon \to 0$, $f_{\#}^{(\varepsilon)}T$ converges weakly to a current $S \in N_m(\mathbb{R}^p)$ that does not depend on the choice of η. Then define $f_{\#}T := S$. The properties

$$\partial(f_{\#}T) = f_{\#}(\partial T),\ \mathrm{spt}(f_{\#}T) \subset f(\mathrm{spt}\ T),$$

and $(g \circ f)_{\#}T = g_{\#}(f_{\#}T)$ then hold for all locally Lipschitz maps $f\colon \mathbb{R}^n \to \mathbb{R}^p$ and $g\colon \mathbb{R}^p \to \mathbb{R}^q$, and

$$M(f_{\#}T) \leq \mathrm{Lip}(f \mid \mathrm{spt}\ T)\, M(T).$$

The proof of the existence of $f_{\#}T$ uses homotopies of currents, which we describe next.

We need some notation for 0- and 1-dimensional currents. For a $\in \mathbb{R}^n$ we define $[a] \in D_0(\mathbb{R}^n)$ by

$$[a](f) := f(a) \text{ for } f \in D^0(\mathbb{R}^n) = C_c^{\infty}(\mathbb{R}^n);$$

$[a]$ corresponds to δ_a in the notation of distribution theory. For $[a, b] \subset \mathbb{R}$ we denote by $[a, b] \in D_1(\mathbb{R})$ the current satisfying

$$[a,b](\varphi dt) = \int_a^b \varphi(t)dt \text{ for all } \varphi \in C_c^{\infty}(\mathbb{R})$$

We have $\partial[a, b](f) = [a, b](df) = [a, b](f' dt) = \int_a^b f' dt = f(b) - f(a) = [b](f) - [a](f)$ for all $f \in D^0(\mathbb{R})$, i.e.

$$\partial[a, b] = [b] - [a].$$

Next, for $[a, b] \subset \mathbb{R}$, we define the cartesian product of $[a] \in D_0(\mathbb{R})$ or $[a, b] \in D_1(\mathbb{R})$ with a current $T \in D_m(\mathbb{R}^n)$. We put

$$[a] \times T := i_{a\#} T \in D_m\mathbb{R}(\times \mathbb{R}^n),$$

where i_a: $\mathbb{R}^n \to \mathbb{R} \times \mathbb{R}^n$, $i_a(x) = (a, x)$. In case $m \geq 1$,

$$\partial([a] \times T) = \partial(i_{a\#} T) = i_{a\#}(T) = [a] \times \partial T.$$

To define $[a, b] \times T \in D_{m+1}(\mathbb{R} \times \mathbb{R}^n)$, denoting the canonical coordinates on $\mathbb{R} \times \mathbb{R}^n$ by (t, x), we write $\omega \in D^{m+1}\mathbb{R}(\times \mathbb{R}^n)$ in the form $dt \wedge \omega' + \overline{\omega}$ such that $dt \wedge \omega'$ collects all terms containing dt; then

$$([a, b] \times T)(\omega) := \int_a^b ([t] \times T)(\omega')dt.$$

Interval × Current

Lemma: Let $T \in D_m(\mathbb{R}^n)$. Then $\text{spt}([a, b] \times T) = [a, b] \times \text{spt } T$,

$$\partial([a, b] \times T) = [b] \times T - [a] \times T - [a, b] \times \partial T$$

for $m \geq 1$, and $\partial([a, b] \times T) = [b] \times T - [a] \times T$ for $m = 0$. If $T \in M_{m,\text{loc}}(\mathbb{R}^n)$, then $\|[a, b] \times T\| = \|[a, b]\| \times \|T\|$.

Now suppose f, g: $\mathbb{R}^n \to \mathbb{R}^p$ are two C^{∞} maps. Let h be a smooth homotopy from f to g, i.e. a C^{∞} map from an open neighborhood of $[0, 1] \times \mathbb{R}^n$ in $\mathbb{R} \times \mathbb{R}^n$ into $\mathbb{R}^p$ with $h(0, x) = f(x)$ and $h(1, x) = g(x)$ for all $x \in \mathbb{R}^n$.

Homotopy Formula

Theorem: Suppose $T \in D_m(\mathbb{R}^n)$, and $h \mid ([0, 1] \times \text{spt } T)$ is proper. Then

$$g_\# T - f_\# T = \partial h_\# ([0, 1] \times T) + h_\# ([0, 1] \times \partial T)$$

for $m \geq 1$, and $g_\# T - f_\# T = \partial h_\#([0, 1] \times T)$ for $m = 0$. If moreover $T \in M_m(\mathbb{R}^n)$, $||Df_x||$, $||Dg_x|| \leq \lambda$ for all $x \in$ spt T, and h is the affine homotopy from f to g, i.e. $h(t, x) = (1 - t) f(x) + tg(x)$, then

$$M(h_\# ([0, 1] \times \mathrm{T})) \sup_{x \in sptT} | g(x) - f(x) | \lambda^m M(T).$$

Finally, we define the cone $[a] \times T \in D_{m+1}(\mathbb{R}^n)$ from a point $a \in \mathbb{R}^n$ over a current $T \in D_m(\mathbb{R}^n)$ with compact support. Let $h(t, x) := (1 - t)a + tx$.

Then

$$[a] \times T := h_\#([0, 1] \times T) \in D_{m+1}(\mathbb{R}^n).$$

we get

$$([a] \times T) = T - h_\#([0, 1] \times \partial T) = T - [z] \times \partial T$$

for $m \geq 1$, in particular, $\partial([a] \times T) = T$ if $\partial T = 0$.

Suppose $T \in N_m(\mathbb{R}^n)$, $m \geq 1$. Recall that then the restrictions $T\ B$ and $(\partial T)\ B$ are defined for every Borel set $B \partial \mathbb{R}^n$. If $f: \mathbb{R}^n \to \mathbb{R}$ is Lipschitz and $s \in \mathbb{R}$, define the slices

$$\langle T, f, s+\rangle := (\partial T)\ \{f > s\} - \partial(T\ \{f > s\})$$

$$= \partial(T\ \{f \leq s\}) - (\partial T)\ \{f \leq s\},$$

$$\langle T, f, s-\rangle := \partial(T\ \{f < s\}) - (\partial T)\ \{f < s\}$$

$$= (\partial T)\ \{f \geq s\} - \partial(T\ \{f \geq s\}).$$

For all but countably many s it is true that

$(\| T \| + \| \partial T \|)\ (f^{-1}\{s\}) = 0$, so $\langle T, f, s+\rangle = \langle T, f, s-\rangle$.

Note that $\partial\langle T, f, s+\rangle = -\langle \partial T, f, s+\rangle$ if $m \geq 2$.

Slices of Normal Currents

Theorem:

1. For all $s \in \mathbb{R}$, spt $\langle T, f, s+\rangle \subset f^{-1}\{s\} \cap spt\ T$.
2. For all $s \in \mathbb{R}$, $M\left(\langle T, f, s+\rangle\right) \leq Lip(f) \liminf_{h \to 0+} \| T \| (\{s < f < s+h\})/h$

3. Whenever $-\infty \leq a < b \leq \infty$, then

$$\int_a^{*b} M(\langle T, f, s+\rangle)\, dx \leq Lip(f)\, \| T \| (\{s < f < s+b\})$$

4. For almost all $s \in \mathbb{R}, \langle T, f, s+\rangle \in N_{m-1}(\mathbb{R}^n)$.

Results for *n*-currents in $\mathbb{R}^n$

We mention two special results for n-dimensional currents in $\mathbb{R}^n$.

Constancy Theorem

Theorem: Suppose $U \subset \mathbb{R}^n$ is open and connected, $T \in D_n(\mathbb{R}^n)$, and spt $\partial T \subset \mathbb{R}^n \setminus U$. Then there exists a constant $c \in \mathbb{R}$ such that spt $(T - c[U]) \cap U = \phi$.

Here $[U] := [U, e_1 \wedge ... \wedge e_n]$, i.e.

$$[U](\omega) = \int \langle e_1 \wedge ... \wedge e_n, \omega\rangle\, dx = \int f dx$$

for $\omega = f\, dx^1 \wedge ... \wedge dx^n \in D_n(\mathbb{R}^n)$.

Characterizing $N_{n,loc}(\mathbb{R}^n)$

Theorem: Suppose $T \in D_n(\mathbb{R}^n)$. Then $T \in N_{n,\text{loc}}(\mathbb{R}^n)$ if and only if $T = [\mathbb{R}^n]\, u$ for some $u \in BV_{\text{loc}}(\mathbb{R}^n)$, in which case $\| \partial T \| = | D_u |$.

INTEGRAL CURRENTS

Integer Rectifiable Currents

Definition: A current $T \in D_m(\mathbb{R}^n)$ is called locally integer rectifiable if it admits a representation of the form

$$T(\omega) = \int_E \langle \tau(x), \omega(x)\rangle\, \theta(x) dH^m(x) \text{ where}$$

1. $E \subset \mathbb{R}^n$ is a countably H^m-rectifiable and H^m-measurable set,
2. θ is a locally H^m-integrable positive integer-valued function on E,
3. τ is an H^m-measurable $\Lambda_m \mathbb{R}^n$-valued function on E such that for H^m-almost every $x \in E$, $\tau(x)$ is simple, $| \tau(x) | = 1$, and $\tau(x)$ represents the approximate tangent space $\text{Tan}^m(E, x) \in G(n, m)$.

We then write $T = [E, \tau, \theta]$, or just $T = [E, \tau]$ in case $\tau \equiv 1$ H^m-almost everywhere on E. The set of locally integer rectifiable currents in $\mathbb{R}^n$ is denoted by $I_{m,\text{loc}}(\mathbb{R}^n)$. An integer rectifiable current in $\mathbb{R}^n$ is an element of $I_m(\mathbb{R}^n) := I_{m,\text{loc}}(\mathbb{R}^n) \cap M_m(\mathbb{R}^n)$.

Here we follow the presentation in the space $I_{m,\text{loc}}(\mathbb{R}^n)$ is denoted by $R_m^{\text{loc}}(\mathbb{R}^n)$.) If $T, T' \in I_{m,\text{loc}}(\mathbb{R}^n)$, then $T + T' \in I_{m,\text{loc}}(\mathbb{R}^n)$. The mass of $T = [E, \tau, \theta] \in I_{m,\text{loc}}(\mathbb{R}^n)$ in an open set U is given by

$$\|T\|(U) = \int_{U \cap E} \theta dH^m .$$

Note that this is finite if the closure of U is compact, thus $I_{m,\text{loc}}(\mathbb{R}^n)$ $M_{m,\text{loc}}(\mathbb{R}^n)$.

If $T \in I_{m,\text{loc}}(\mathbb{R}^n)$ and f: $\mathbb{R}^n \to \mathbb{R}^p$ is a locally Lipschitz map such that $f|$ spt T is proper, then $f_\# T$ can be defined, and $f_\# T \in I_{m,\text{loc}}(\mathbb{R}^p)$.

The boundary of an integer rectifiable current need not be integer rectifiable.

Integral Current

Definition: The space of locally integral currents in $\mathbb{R}^n$ is defined by

$$I_{m,\text{loc}}(\mathbb{R}^n) := \{T \in I_{m,\text{loc}}(\mathbb{R}^n) : \partial T \in I_{m-1,\text{loc}}(\mathbb{R}^n)\}$$

if $m \geq 1$, and $I_{0,\text{loc}}(\mathbb{R}^n) := I_{0,\text{loc}}(\mathbb{R}^n)$. An integral current in $\mathbb{R}^n$ is an element of $I_m(\mathbb{R}^n) := I_{m,\text{loc}}(\mathbb{R}^n) \in N_m(\mathbb{R}^n)$.

Slices of Integral Currents

Theorem: If $T \in I_m(\mathbb{R}^n)$, $m \geq 1$, and f: $\mathbb{R}^n \to \mathbb{R}$ is Lipschitz, then $\langle T, f, s+\rangle \in I_{m-1}(\mathbb{R}^n)$ for almost all $s \in \mathbb{R}$.

THE COMPACTNESS THEOREM

Boundary Rectifiability

Theorem: If $T \in I_m(\mathbb{R}^n)$ and $M(\partial T) < \infty$, then $\partial T \in I_{m-1}(\mathbb{R}^n)$, i.e. $T \in I_m(\mathbb{R}^n)$.

Rectifiable Slices Implies Rectifiable

Theorem: Suppose $T \in M_m(\mathbb{R}^n)$ and $\partial T = 0$. Suppose further that for every $z \in \mathbb{R}^n$, $\partial(T\, B(z, r)) \in I_{m-1}(\mathbb{R}^n)$ for almost all r. Then $T \in I_m(\mathbb{R}^n)$.

Closure Theorem

Theorem: Suppose $T_1, T_2, \ldots \in I_m(\mathbb{R}^n)$, $\sup_i N(T_i) < \infty$, and $\bigcup_i \text{spt}\, T_i \subset C$ for some compact set $C \subset \mathbb{R}^n$. If $T_i \to T$ for some $T \in D_m(\mathbb{R}^n)$, then $T \in I_m(\mathbb{R}^n)$.

The original proof on the difficult structure theorem. White gave a simultaneous inductive proof of the above three theorems that does not use structure theory.

By combining with the compactness theorem for normal currents one obtains the celebrated result for integral currents:

Compactness Theorem

Theorem: Suppose $T_1, T_2,... \in I_m(\mathbb{R}^n)$, $\sup_i N(T_i) < \infty$, and $\bigcup_i \text{spt } T_i \subset C$ for some compact set $C \subset \mathbb{R}^n$. Then there is a subsequence $T_{i_1}, T_{i_2},...$ and a $T \in I_m(\mathbb{R}^n)$ such that $T_{i_j} \to T$.

Minimizing Currents

From Theorem one easily obtains various existence results for area minimizing currents. In general, a current $T \in I_{m,\text{loc}}(\mathbb{R}^n)$ is called (absolutely area) minimizing if

$$M(S) \leq M(S')$$

whenever $S, S' \in I_m(\mathbb{R}^n)$, $\| T \| = \| S \| + \| T - S \|$, and $\partial S' = \partial S$. The assumption $\| T \| = \| S \| + \| T - S \|$ means that S is a piece of T.

Plateau Problem

Theorem: Suppose $\mathbb{R} \in I_{m-1}(\mathbb{R}^n)$, spt S is compact, and $\partial S = 0$. Then there is current $T \in I_m(\mathbb{R}^n)$ with $\partial T = R$ such that $M(T) \leq M(T')$ for all $T' \in I_m(\mathbb{R}^n)$ with $\partial T' = R$.

Of course T is minimizing in the sense defined above.

Proof: Let $T := \{T' \in I_m(\mathbb{R}^n): \partial T' = R\}$; T is non-empty since it contains e.g. the cone $[0] \times T$. Choose a sequence $T_1, T_2,...$ in T such that

$$\lim_{i\to\infty} M(T_i) = b := \inf\{M(T') : T' \in T\}.$$

Let $C \subset \mathbb{R}^n$ be a compact ball that contains spt R, and let $\pi: \mathbb{R}^n \to C$ denote the nearest point retraction; π is 1-Lipschitz. Since $M(\pi_\# T_i) \leq M(T_i)$ and $\text{spt}(\pi_\# T_i) \subset C$ for all i, we assume w.l.o.g. that spt $T_i \subset C$ for all i. Since $\sup_i N(T_i) < \infty$, by Theorem there exist a subsequence $T_{i_1}, T_{i_2},...$ and a $T \in I_m(\mathbb{R}^n)$ such that $T_{i_j} \to T$. Then also $R = \partial T_{i_j} \to \partial T$, thus $\partial T = R$, and

$$M(T) \leq \liminf_{j\to\infty} M(T_{i_j}) = b.$$

This proves the result.

Let $M \subset \mathbb{R}_{n+k}$ be a compact C^1 submanifold of dimension n. For $m \geq 1$, define the abelian groups

$$Z_m(M) := \{T \in I_m(\mathbb{R}^{n+k}) : \text{spt } T \subset M, \partial T = 0\},$$

$$B_m(M) := \{\partial S : S \in I_{m+1}(\mathbb{R}^{n+k}), \text{ spt } S \subset M\} \subset Z_m(M).$$

Two cycles $T, T' \in Z_m(M)$ are homologous if $T - T' \in B_m(M)$.

Hhomologically Minimizing Cycles

Theorem: Suppose $M \subset \mathbb{R}^{n+k}$ is as above and $T' \in Z_m(M)$. Among all cycles in $Z_m(M)$ homologous to T' there is one with minimal mass. Next, we mention the most important regularity results for minimizing currents in $\mathbb{R}^n$.

Hhypersurface Interior Regularity

Theorem: Suppose $T \in I_{n-1,\text{loc}}(\mathbb{R}^n)$ is minimizing. There is a set Σ of Hausdorff dimension at most $n - 8$ such that $(\text{spt } T \setminus \text{spt}(\partial T)) \setminus \Sigma$ is a C^∞ submanifold of $\mathbb{R}^n$.

This is due to Federer. It was known previously that the bound $n - 8$ is optimal, by the following example of Bombieri–De Giorgi–Giusti.

Minimizing Cone

Example: Let $V := \{(x, y) \in \mathbb{R}^4 \times \mathbb{R}^4 : |x|^2 < |y|^2\}$. Then $\partial[V] \in I_{7,\text{loc}}(\mathbb{R}^8)$ is a minimizing hypersurface with an isolated singular point at 0.

For currents of arbitrary codimension, the following deep result of Almgren holds.

Interior Regularity

Theorem: Suppose $T \in I_{m,\text{loc}}(\mathbb{R}^n)$ is minimizing. There is a set Σ of Hausdorff dimension at most $m - 2$ such that $(\text{spt } T \setminus \text{spt}(\partial T)) \setminus \Sigma$ is an m-dimensional C^∞ submanifold of $\mathbb{R}^n$.

The bound $m - 2$ is optimal;

$$T = [E_{12}, e_1 \wedge e_2] + [E_{34}, e_3 \wedge e_4] \in I_{2,\text{loc}}(\mathbb{R}^4)$$

is a minimizing 2-current in $\mathbb{R}^4$ with an isolated singular point at 0. Here E_{i_j} denotes the coordinate plane spanned by e_i and e_j. To show that T is minimizing, let $S \in I_2(\mathbb{R}^4)$ be a piece of T, and let $S' \in I_2(\mathbb{R}^4)$ be an arbitrary current with $\partial S' = \partial S$. Then $S = [A, e_1 \wedge e_2] + [A', e_3 \wedge e_4]$ for some H^2-measurable sets $A \subset E_{12}$ and $A' \subset E_{34}$. By (a special case of) Wirtinger's inequality, the form $\omega = dx^1 \wedge dx^2 + dx^3 \wedge dx^4$ has comass norm $\|\omega(x)\| = 1$. Since ω is exact and $\partial S = \partial S'$, one concludes that

$$M(S) = S(\omega) = S'(\omega) \leq M(S'),$$

proving that T is minimizing.

A most useful tool in regularity theory is the following property of minimizing currents.

Monotonicity Formula

Theorem: Suppose $T \in I_{m,\,\mathrm{loc}}(\mathbb{R}^n)$ is minimizing.

1. For all $x \in \mathbb{R}^n$, the function $r \mapsto M(T\, B(x, r))/ r^m$ is (non-strictly) increasing on $(0, d(x, \operatorname{spt} \partial T))$.
2. For all $x \in \operatorname{spt} T$ and $r \in (0, d(x, \operatorname{spt} \partial T))$,
$$M(T\, B(x, r)) \geq \alpha_m r^m = L^m(B^m(r)).$$

Proof: 1. For $r \in (0, d(x, \operatorname{spt} \partial T))$, put $S_r := T\, B(x, r)$ and $f(r) := M(S_r)$. For almost every r, f is differentiable at r, and
$$M(\partial S_r) = M(\langle T, d(x,.), r+\rangle) \leq f'(r)$$
by Theorem. The cone $S_r' := [x] \times \partial S_r$ from x over ∂S_r has mass $M\left(S_r'\right) = \dfrac{r}{m} M\left(\partial S_r\right)$. Thus, by minimality,
$$f(r) = M(S_r) \leq M(S_r') \leq \frac{r}{m} f'(r)$$
for almost every $r \in (0, d(x, \operatorname{spt} \partial T))$. Integrating the inequality
$$\frac{f'(r)}{f(r)} \geq \frac{m}{r}$$
from r_1 to r_2 one obtains the result.

2. Is a consequence of (1) and the fact that if $T = [E, \theta, \theta]$, where $\theta > 0$, then spt T is the closure of the set $\{y \in \mathbb{R}^n : m(E, y) = 1\}$.

SOME METRIC SPACE TECHNIQUES

Earlier we discuss some metric space techniques that are useful in connection with the new metric approach to currents developed. Indispensable tools in modern geometry, they are also of independent interest.

EMBEDDINGS

We start with some basic and well-known isometric embedding theorems for metric spaces.

Kuratowski, Frechet

Lemma:

1. Every metric space X admits an isometric embedding into $l^{¥}(X) = (\{s : X \to \mathbb{R} : s$ bounded$\}, \|\cdot\|_{¥})$.

2. Every separable metric space admits an isometric embedding into $l^\infty = l^\infty(N)$.

Proof: 1. Fix $z \in X$ and define $X \to l^\infty(X)$,

$$x \mapsto s^x,\ s^x(y) = d(x, y) - d(y, z).$$

Note that $\| s^x \|_\infty = \sup y \mid s^x(y) \mid \leq d(x, z)$. Moreover,

$$\| s^x - sx' \|_\infty = \sup_y \mid d(x, y) - d(x', y)\mid \leq d(x, x'),$$

and equality occurs for $y = x'$.

2. Embed a dense countable subset of X into l^∞ by means of (1). This extends to an isometric embedding of X.

Note that if X is bounded, we do not need to subtract the term $d(y, z)$ in the definition of s^x. In this case, the embedding is canonical.

Recall that a metric space X is said to be precompact or totally bounded if for every $\varepsilon > 0$, X can be covered by a finite number of closed balls of radius ε. We call a set $Y \subset X$ ε-separated if $d(y, y')$ whenever $y, y' \in Y, y \neq y'$. Note that X is precompact if and only if for every $\varepsilon > 0$, all ε-separated subsets of X are finite. A metric space is compact if and only if it is precompact and complete.

Uniformly Precompact Family

Definition: A family $(X_j)_{j \in J}$ of metric spaces is called uniformly precompact if for all $\varepsilon > 0$ there exists a number $n = n(\varepsilon) \in \mathbb{N}$ such that each X_j can be covered by n closed balls of radius ε. The family $(X_j)_{j \in J}$ is uniformly bounded if $\sup_{j \in J} \operatorname{diam} X_j < \infty$.

Gromov Embedding

Theorem: Suppose that $(X_j)_{j \in J}$ is a uniformly precompact and uniformly bounded family of metric spaces. Then there is a compact metric space Z such that each X_j admits an isometric embedding into Z.

Proof: For $i \in \mathbb{N}$, let $\varepsilon_i := 2^{-i}$ and pick $ni \in \mathbb{N}$ such that each X_j can be covered by n_i closed balls of radius ε_i. Choose a partition of $\mathbb{N}$ into sets N_i, $i \in \mathbb{N}$, with cardinality $\# N_i = n_1 n_2 \ldots n_i$, and define a map $\pi : \mathbb{N} \backslash W_1 \to \mathbb{N}$ such that for each $i \in \mathbb{N}$,

$$\pi^{-1}(N_i) = N_{i+1} \text{ and } \# \pi^{-1}\{k\} = n_{i+1} \text{ for all } k \,\pi\, N_i.$$

In each X_j, we construct a sequence $(x_k^j)_{k \in \mathbb{N}}$ according to the following inductive scheme. For $i = 1$, the points x_k^j with $k \in N_i = N_1$ are chosen such that the n_1 balls $B(x_k^j, \varepsilon_1)$ cover X_j. For $i \geq 1$, if the $n_1 \ldots n_i$ centers x_k^j with $k \in N_i$ are selected, the $n_1 \ldots n_i n_{i+1}$ points x_l^j with $l \in N_{i+1}$ are chosen such that for each $k \in N_i$, the ball $B(x_k^j, \varepsilon_1)$ is covered by the n_{i+1} balls

$$B(x_l^j, \varepsilon_{i+1}) \subset B(x_k^j, 2\varepsilon_i)$$

with $l \in \pi^{-1}\{k\}$. This way we obtain for every $j \in J$ a dense sequence $(x_k^j)_{k\in\mathbb{N}}$ in X_j which gives rise to an isometric embedding $f_j: X_j \to l^\infty$, mapping x to $(d(x, x_k^j))_{k\in\mathbb{N}}$. Whenever $i \in \mathbb{N}$, $k \in N_i$, and $l \in \pi^{-1}\{k\}$, then

$$|d(x, x_k^j) - d(x, x_l^j)| \le d(x_k^j, x_l^j) \le 2\varepsilon_i .$$

Hence, each $f_j(X_j)$ lies in the set Z of all sequences $(s_k)_{k\in\mathbb{N}}$ with $0 \le s_k \le \sup_j \operatorname{diam} Z_j$ for all $k \in \mathbb{N}$ and

$$|s_k - s_l| \le 2\varepsilon_i \text{ whenever } i \in \mathbb{N},\ k \in N_i, \text{ and } l \in \pi^{-1}\{k\}.$$

Since the sequence $(\varepsilon_i)_{i\in\mathbb{N}}$ is summable, Z is a compact subset of l^∞.

GROMOV–HAUSDORFF CONVERGENCE

For subsets A, B of a metric space X we denote by $N_\delta(A)$ the closed δ-neighborhood of A and by

$$d_H(A, B) = \inf\{\delta \ge 0: A \subset N_\delta(B), B \subset N_\delta(A)\}$$

the Hausdorff distance of A and B; d_H defines a metric on the set C of non-empty, closed and bounded subsets of X.

Blaschke

Theorem: Suppose that $X = (X, d)$ is a metric space and C is the set of non-empty, closed and bounded subsets of X, endowed with the Hausdorff metric d_H.

1. If X is complete, then C is complete.
2. If X is compact, then C is compact.

This was first proved by Blaschke for compact convex bodies in $\mathbb{R}^3$ to settle the existence question in the isoperimetric problem.

Proof: Let $(C_i)_{i\in\mathbb{N}}$ be a Cauchy sequence in C. Then the set

$$C := \bigcap_{i=1}^{\infty} \overline{\bigcup_{j\ge i} C_j}$$

is closed and bounded. We show that

$$\lim_{i\to\infty} d_H(C_i, C) = 0 .$$

Let $\varepsilon > 0$. Choose i_0 such that $d_H(C_i, C_j) \le \varepsilon/2$ whenever $i, j \ge i_0$. Suppose $x \in C$.

Since $C \subset \overline{\bigcup_{j \geq i_0} C_j}$ there exists an index $j \geq i_0$ with $d(x, C_j) \leq \varepsilon/2$. Hence $d(x, C_i) \leq d(x, C_j) + d_H(C_i, C_j) \leq \varepsilon$ for all $i \geq i_0$. This shows that $C \subset N_\in (Ci)$ for $i \geq i_0$.

Now suppose $x \in C_i$ for some $i \geq i_0$. Pick a sequence $i = i_1 < i_2 <...$ such that $d_H(C_m, C_n) \leq \varepsilon/2^k$ whenever $m, n \geq i_k$, $k \in \mathbb{N}$. Then choose a sequence $(x_k)_{k \in \mathbb{N}}$ such that $x_1 = x$, $x_k \in C_{i_k}$ and $d(x_k, x_{k+1}) \leq \varepsilon/2k$. As X is complete, the Cauchy sequence (x_k) converges to some point y. We have

$$d(x, y) = \lim_{k \to \infty} d(x, x_k) \leq \sum_{k=1}^{\infty} d(x_k, x_{k+1}) \leq \varepsilon,$$

and y belongs to the closure of $C_{i_k} \cup C_{i_{k+1}} \cup ...$ for all k. Thus $y \in C$ and $d(x, C) \leq \varepsilon$. This shows that $C_i \subset N_\varepsilon(C)$ whenever $i \geq i_0$.

We know that C is complete since X is, so it suffices to show that C is precompact. Let $\varepsilon > 0$. Since X is precompact we find a finite set $Z \subset X$ with $N_\varepsilon(Z) = X$. We show that every $C \in C$ is at Hausdorff distance at most ε of some subset of Z, namely $Z_C := Z \cap N_\varepsilon(C)$. For every $x \in C$ there exists a point $z \in Z$ with $d(x, z) \leq \varepsilon$, so $z \in Z_C$. This shows that $C \subset N_\varepsilon(Z_C)$. Since also $Z_C \subset N_\varepsilon(C)$, we have $d_H(C, ZC) \leq \varepsilon$. As there are only finitely many distinct subsets of Z, we conclude that C is precompact.

Gromov–Hausdorff distance

Definition: The Gromov–Hausdorff distance of two metric spaces X, Y is the number

$$d_{GH}(X, Y) = \inf d_H^Z(X', Y'),$$

where the infimum is taken over all triples (Z, X', Y') such that $Z = (Z, d^Z)$ is a metric space, $X' \subset Z$ is an isometric copy of X, and $Y' \subset Z$ is an isometric copy of Y.

Alternatively, call a metric $\bar{d}$ on the disjoint union $X \cup Y$ admissible for the given metrics $d = d^X$ and $d = d^Y$ on X and Y if $\bar{d} \mid X \times X = d^X$ and $d \mid Y \times Y = d^Y$; then

$$d_{GH}(X, Y) = \inf \bar{d}_H(X, Y)$$

where the infimum is taken over all admissible metrics $\bar{d}$ on $X \cup Y$.

For instance, suppose that diam (X), diam $(Y) \leq D < \infty$. Setting $\bar{d}(x, y) = D/2$ for $x \in X$ and $y \in Y$ we obtain an admissible metric on $X \cup Y$, in particular $d_{GH}(X, Y) \leq D/2$.

We call a subset X' of a metric space X ε-dense in X if $d_H(X, X') \leq \varepsilon$, i.e. $N_\varepsilon(X') = X$. By a correspondence ~ between two sets X, Y we mean a symmetric relation between points in X and points in Y such that every element of X is related to at least one element of Y and vice-versa.

Lemma: Suppose X, Y are two metric spaces and $\varepsilon > 0$.

1. If $d_{GH}(X, Y) < \varepsilon$, then there is a correspondence $\sim$ between X and Y such that
$$|d(x, x') - d(y, y')| < 2\varepsilon$$
whenever $x, x' \in X, y, y' \in Y$, and $x \sim y, x' \sim y'$.
2. Suppose $X' \subset X$ is ε-dense in X, $Y' \subset Y$ is ε-dense in Y, and there is a correspondence $\sim$ between X' and Y' such that
$$|d(x, x') - d(y, y')| \leq \varepsilon$$
whenever $x, x' \in X', y, y' \in Y'$, and $x \sim y, x' \sim y'$. Then $d_{GH}(X, Y) < 2\varepsilon$.

Proof: Let $\bar{d}$ be an admissible metric on $X \cup Y$ such that $\bar{d}_H(X, Y) < \varepsilon$. Then for every $x \in X$ there is a $y \in Y$ with $\bar{d}(x, y) < \varepsilon$, and vice-versa. Define a correspondence in this manner.

Define an admissible metric $\bar{d}$ on $X \cap Y$ such that $\bar{d}(x, y) = \inf d(x, x') + d(y, y') + \varepsilon/2$ for all $x \in X$ and $y \in Y$, where the infimum is taken over all pairs of points $x' \in X'$ and $y' \in Y'$ with $x' \sim y'$.

Theorem:

1. d_{GH} satisfies the triangle inequality, i.e. $dGH(X, Z) \leq d_{GH}(X, Y) + d_{GH}(Y, Z)$ for all metric spaces X, Y, Z.
2. d_{GH} defines a metric on the set of isometry classes of compact metric spaces.

Assertion (2) is no longer true if 'compact' is replaced by 'complete and bounded'.

Gromov Compactness Criterio

Theorem: Suppose that $(X_i)_{i\in\mathbb{N}}$ is a uniformly precompact and uniformly bounded sequence of metric spaces. Then there exist a subsequence $(X_{i_j})_{j\in\mathbb{N}}$ and a compact metric space Z such that (X_{i_j}) Gromov–Hausdorff converges to Z, i.e. $\lim_{j\to\infty} d_{GH}(X_{i_j}, Z) = 0$.

ULTRALIMITS

By a filter ϕ on a set I we mean a monotonous set function $\phi: 2^I \to \{0, 1\}$ with $\phi(\phi) = 0$, $\phi(I) = 1$, and $\phi(A \cap B) = 1$ whenever $\phi(A) = \phi(B) = 1$. An ultrafilter is a filter that is maximal with respect to the partial order $\leq$. Equivalently, a function $\phi: 2^I \to \{0, 1\}$ is an ultrafilter iff $\phi(I) = 1$ and ϕ is finitely additive, i.e.,

$$\phi(A) + \phi(B) = \phi(A \cup B) \text{ whenever } A \cap B = \phi.$$

An ultrafilter ϕ is called free or non-principal if it is zero on finite sets; in the opposite case, ϕ has a unique atom and is said to be fixed or principal. To prove the existence of a free ultrafilter ϕ on an infinite set I, start with the Fr´echet filter which is one exactly on complements of finite sets and apply Zorn's lemma within the class of filters that vanish on finite sets.

In the following ϕ denotes a free ultrafilter on $\mathbb{N}$. Given a sequence $(X_i)_{j\in\mathbb{N}}$ in a topological space X, we call a point $x \in X$ a ϕ-limit of (x_i), and we write $\lim_\phi x_i = x$, if

$$\phi(\{i \in \mathbb{N} : x_i \in U\}) = 1$$

for every neighborhood U of x.

In case X is a metric space, $\lim \phi x_i = x$ is equivalent to

$$\phi(\{i \in \mathbb{N} : d(x, x_i) < \varepsilon\}) = 1 \text{ for all } \varepsilon > 0$$

and hence to $\lim_\phi d(x, x_i) = 0$.

If X is a compact topological Hausdorff space and $(X_i)_{i\in\mathbb{N}}$ is a sequence in X, then there exists a uniquely determined point $x \in X$ such that $\lim_\phi x_i = x$. Since ϕ is free, every neighborhood of x contains infinitely many elements of (x_i). In particular, for every bounded sequence $(a_i)_{i\in\mathbb{N}}$ of real numbers there is a unique point a such that $\lim_\phi a_i = a$, and a is the limit of some subsequence. One has

$$\lim_\phi a_i + \lim_\phi b_i = \lim(a_i + b_i),\ \lambda \lim_\phi a_i = \lim_\phi (\lambda a_i)$$

for all bounded sequences (a_i), (b_i) and all $\lambda \in \mathbb{R}$. If $a_i \leq b_i$ for all i, then $\lim_\phi a_i \leq \lim_\phi b_i$. It follows that if (x_i) and (y_i) are two bounded sequences in a metric space X, with $\lim_\phi x_i = x$ and $\lim_\phi y_i = y$, then

$$\lim_\phi d(x_i, y_i) = d(x, y)$$

since $| d(x, y) - \lim_\phi d(x_i, y_i) | \leq \lim_\phi d(x, x_i) + \lim_\phi d(y, y_i) = 0$.

Ultralimit

Definition: Suppose $(X_i)_{i\in\mathbb{N}}$ is a sequence of pointed metric spaces $X_i = (X_i, d, *_i)$ and ϕ is a free ultrafilter on $\mathbb{N}$.

Denote by $(X_i)_\infty$ the set of all sequences (x_i) with $x_i \in X_i$ and $\sup_i d(xi, *_i) < \infty$. For (x_i), $(y_i) \in (X_i)_\infty$ the sequence $(d(x_i, y_i))$ is bounded, and $d((x_i), (y_i)) := \lim_\phi d_i (x_i, y_i)$ defines a pseudometric on $(X_i)_\infty$. The ultralimit (or ultraproduct) $(X_i)_\infty$ of the sequence (X_i) is the set of equivalence classes (x_i) of elements $(x_i) \in (X_i)_\infty$, where

$$(x_i) \sim (y_i) \text{ if and only if } d((x_i), (y_i)) = 0,$$

endowed with the metric defined by $d((x_i)_\phi, (y_i)_\phi) := d((x_i), (y_i))$.

Lemma: The ultralimit $\in = (X_i)_\phi$ of a sequence (X_i) of pointed metric spaces is complete.

Proof: Pick a Cauchy sequence $(\overline{x}^\gamma)_{\gamma\in\mathbb{N}}$ in $\overline{X}$ and represent each $\overline{x}^\gamma$ by an element $(x_i^\gamma)_{i\in\mathbb{N}} \in (X_i)_\infty$. By induction, choose sets $\mathbb{N} = N_1 \supset N_2 \ldots$ with $\phi(N_\gamma) = 1$ for all γ such that

$$| d\left(\overline{x}^\alpha, \overline{x}^\beta\right) - d\left(x_i^\alpha, x_i^\beta\right) | < 2^{-\gamma}$$

whenever $i \in N_\gamma$ and $\alpha, \beta \in \{1, 2, \ldots, \gamma\}$. Then define an element $(y_i)_{i\in\mathbb{N}} \in (X_i)_\infty$ such that $y_i = x_i^\gamma$ for $i \in N_\gamma \setminus N_{\gamma+1}$. Now $\overline{x}^\gamma \to \overline{y} := (y_i)_\phi$ as $\gamma \to \infty$.

If each X_i is a Banach or Hilbert space, then $(X_i)_\phi$ has a canonical Banach or Hilbert space structure, respectively. The ultralimit $(X)_\phi$ of a constant sequence (X), $X = (X, d, *)$, is also called the ultracompletion (or ultrapower) of X. It is independent of the choice of the basepoint $*$, and the map that assigns to each $x \in X$ the equivalence class $(x)_\phi$ of the constant sequence (x) is a canonical isometric embedding of X into $(X)_\phi$. Ultrapowers have applications e.g. in Banach space theory.

For a fixed metric space (X, d), a sequence of basepoints $*_i \in X$, and a sequence of scale factors $\lambda_i > 0$ with $\lim_{i\to\infty} \lambda_i = \infty$, the ultralimit $(X_i)_\phi$ of the sequence of rescaled metric spaces $X_i = \left(X, \frac{1}{\lambda_i} d, *_i\right)$ is referred to as an asymptotic cone of X. This construction has played a significant role in recent proofs of rigidity theorems in the theory of nonpositively curved spaces.

GH-limit Versus Ultralimit

Theorem: Suppose $(X_i)_{i\in\mathbb{N}}$ is a uniformly precompact and uniformly bounded sequence of metric spaces.

1. The ultralimit $(X_i)_\phi$ is isometric to the Gromov-Hausdorff limit of some subsequence (X_{i_k}) of (X_i).
2. If the sequence $(X_i)_\phi$ Gromov–Hausdorff converges to some metric space Y, and if each X_i is contained in a pointed metric space $Z_i = (Z_i, *_i)$ with $i \in X_i$, then Y isometrically embeds into the ultralimit $(Z_i)_\phi$.

Proof: (1) We use Theorem (Gromov embedding) and assume that each X_i belongs to the set C of all non-empty compact subsets of some fixed compact metric space Z. Then the map $f: (X_i) \to Z$ that assigns to each class $(x_i)_\phi$ the limit $\lim_\phi x_i$ is a well-defined isometric embedding since

$$d(\lim_\phi x_i, \lim_\phi y_i) = \lim_\phi d(x_i, y_i) = d((x_i)_\phi, (y_i)_\phi).$$

By Theorem (C, d_H) is compact, hence the sequence (X_i) has a unique ϕ-limit $\lim_\phi X_i = Y \in C$ with respect to d_H. We show that $f((X_i)_\phi)$ coincides with Y; then $(X_i)_\phi$ is isometric to the Gromov-Hausdorff limit of some subsequence of (X_i) as claimed.

Let $y \in Y$. For every i, choose $x_i \in X_i$ such that $d(x_i, y) \leq d_H(X_i, Y)$. Since $\lim_\phi X_i = Y$, $\lim_\phi d(x_i, y) \leq \lim_\phi d_H(X_i, Y) = 0$ and hence $\lim_\phi x_i = y$. This shows that $Y \subset f((X_i)_\phi)$.

To prove the reverse inclusion, let (x_i) be a sequence with $x_i \in X_i$ for all i. Choose $y_i \in Y$ such that $d(x_i, y_i) \leq d_H(X_i, Y)$. Then $d(\lim_\phi x_i, \lim_\phi y_i) = \lim_\phi d(x_i, y_i) \leq \lim_\phi d_H(X_i, Y) = 0$, hence $\lim x_i \in Y$.

(2) Using (1) we see that the completion $\overline{Y}$ of Y is isometric to $(X_i)_\phi$, and $(X_i)_\phi$ canonically embeds into $(Z_i)_\phi$.

MEASURE THEORY AND PARTITIONS

Natural language plays an immense role in the discovery, discussion, and preservation of scientific knowledge, but it is very poorly adapted to the exact transmission of the content of this knowledge, and especially to its processing, which forms an important part of scientific thought. Natural language performs different functions, and ascribes to other values. The language of modern set-theoretic mathematics can affect the role of such a language-intermediary thanks to its unique capacity to form geometric, spatial and kinematic forms, and at the same time to provide a precise formalism for transcribing their mathematical content.

The set of all $\zeta(\Sigma)$-sets coincides with the set of all sets which can be obtained from the sets of the system Σ with the help of all possible set-theoretic operations, repeated as many times as desired. First we review some fundamental notions from measure theory. We shall follow the terminology of Mackey. Let M be a set. A *Borel structure* (or measure structure) on M is a family A of subsets of M which is closed under the operations of complementation and countable unions, *i.e.*,

$$A \in A \Rightarrow M \setminus B \in A,\ A_n \in A \Rightarrow \bigcup_{n \in Z} A_n \in A .$$

Such a family of subsets is often called a σ-field, as it can be given the algebraic structure of a field. If M has a topology, the σ-field generated by the open sets is usually called the Borel σ-field. The following properties follow easily from the definition:

$$A, B \in A \Rightarrow A \setminus B \in A,\ A_n \in A \Rightarrow \bigcap_{n \in Z} A_n \in A .$$

We shall call the pair $M = (M, A)$ a Borel or a measure space, and the subsets in A Borel sets, or measurable sets. Given a family C of subsets of M, the unique smallest σ-field containing C is called the Borel structure *generated* by F, and denoted by $F(C)$. The

Borel space $M = (M, A)$ is called *countably generated* if there exists a countable family which generates the Borel structure.

A family C of Borel subsets *separates* M if for any two distinct points $x, y \in M$ there exists a $C \in C$ such that either $x \in C$ and $y \notin C$, or $x \in C$ and $y \in C$. The Borel space M is called *countably separated* if there exist a countable family which separates M. It is clear that a countably generated space is countably separated. Let (M, A) and (N, B) be Borel spaces. A map $f: M \to N$ is called Borel, or measurable, if $f^{-1}(B) \in A$ for each $B \in B$. A Borel map $f: M \to N$ is a Borel isomorphism if f is an isomorphism of sets, and the inverse f^{-1} is a Borel map.

We now discuss standard Borel spaces. Let X be a complete separable metric space, with the Borel structure generated by the open (equivalently closed) sets.

Definition: A space (M, A) is *standard* if it is Borel isomorphic to a Borel subset of X.

Proposition: Two standard spaces M_1 and M_2 are Borel isomorphic if and only if they have the same cardinality.

In particular, any standard space is Borel isomorphic to the unit interval of the real line $I = [0, 1]$,$\subset R$. We collect here a few related results which will be used in following sections.

Proposition: Let M be standard, and A a subset of M. A is standard if and only if A is a Borel subset.

Proposition: Let M be standard, and C a countable separating family of Borel sets of M. Then C is a generating family for the Borel structure of M.

The *product* of two measure spaces (M, B) and (M', B') is the cartesian product $M \times M'$ with the Borel structure generated by sets of the form $B \times B'$, with $B \in B$ and $B' \in B'$. We recall also that the product of two topological spaces X and X' is defined as the cartesian product $X \times X'$, with the topology generated by products of open sets.

Proposition: The product of two standard spaces (M, B) and (M, B') is a standard space.

Proof: We can take M and M' to be Borel subsets of complete separable metric spaces X and X' respectively. The cartesian product of topological spaces $X \times X'$ is a complete separable metric space, and $M \times M'$ is a Borel subset.

Proposition: Let X and Y be standard spaces, and let $f: X \to Y$ be a Borel map. Then the graph of f is a Borel subset of $X \times Y$.

We note also that a countably generated Borel space is not necessarily standard.

We briefly review the notion of a measure. A measure μ is a map from the Borel sets to the extended positive real numbers $\overline{R}^{+}$, which is countably additive, *i.e.*,

$$\mu\left(\bigcup_{n\in Z^+} B_n\right)=\sum_{n\in Z^+}\mu(B_n),$$

for any countable family $\{B_n\}$ of disjoint Borel sets. A measure μ on M is σ-finite if M is the union of a countable number of pieces with finite measure. A measure μ on S is called standard if S is the union of a measurable subset which is standard, and subsets of sets of measure zero. A measure space (M, A, μ) with standard measure μ satisfying $\mu(M) = 1$ is called a standard probability space, or a *Lebesgue space*. A Lebesgue space can have at most a countable number of points of positive measure. One usually restricts the term Lebesgue to spaces which are Lebesgue as above, and in addition have no points of positive measure. An alternate term for this is smooth Lebesgue. The notion of isomorphism of Borel spaces with a measure requires also that the map preserve the measure of all Borel sets.

Proposition: Any two (smooth) Lebesgue spaces M_1 and M_2 are isomorphic.

We now study partitions of Borel spaces, or equivalently, quotient measure spaces. Let (M, A) be a Borel space. A *partition* of M is a family $\xi = \{C\}$ of nonempty disjoint subsets of M, whose union is M, $\bigcup_{C\in\xi} C = M$. The notion an equivalence relation on M is the same. This can equivalently be described as a surjection $p: M \to M/\xi$, where the points of the space M/ξ correspond to the sets C of the partition. An element of the partition is the inverse image $p^{-1}(c)$ of a point $c \in M/\xi$. The quotient Borel structure on M/ξ is defined as the pushforward under p of the Borel structure on M, *i.e.*, a subset $A \subset M/\xi$ is measurable if $p^{-1}(A) \subset M$ is measurable.

We now suppose we have a finite standard measure μ on M, *i.e.*, M is Lebesgue, and a measurable partition μ. The quotient measure μ on M/ξ is defined as the pushforward of μ under the natural projection, *i.e.*, for a Borel subset $A \subset M/\xi$, define $\mu'(A) = \mu'(p^{-1}(A))$. When we have a measure, we generally work modulo the sets of zero measure. A *partition (mod 0)* is a partition of a measurable subset of M of full measure. We shall generally not distinguish between partitions and partitions (mod 0).

Subsets of M which are unions of elements of ξ are called ξ-sets. We now discuss the notion of a basis for a partition. A *basis* for the partition ξ is a countable separating family of measurable subsets of M/ξ. A partition is called *measurable* if a basis exists, *i.e.*, if M/ξ is countably separated. The significance of the measurability of a partition lies in the following result.

Proposition: The quotient M/ξ of a Lebesgue space M by a measurable partition ξ is a Lebesgue space.

Definition: The *graph* of an equivalence relation on a space X is the subset of $X \times X$ consisting of pairs (x, y), where x is equivalent to y.

Proposition: Let ξ be a measurable partition of a standard measure space M. The graph R of the partition ξ is a Borel subset of $M \times M$.

Proof: We see that M/ξ is standard, and the map $p: M \to M/\xi$ is Borel. The graph T of the Borel function p is a Borel subset of $X \times X/\xi$ by proposition. The map

$$1 \times p: X \times X \to X \times X/\xi$$

is clearly Borel. The graph $R \subset X \times X$ of the partition ξ is the inverse image of T under $1 \times p$, so R is a Borel subset of $X \times X$.

Definition: A canonical system of conditional measures associated to the partition ξ is a family of measures $\{\mu_C, C \in \xi\}$, with the properties

(*i*)μ_C is a Lebesgue measure on C

(*ii*)for any $A \in A$, the set $A \cap C$ is measurable in C for almost all $C \in M/\xi$ and the function $\mu_C(A \cap C)$ is measurable on M/ξ, with

$$\mu(A) = \int_{M/\xi} \mu_C(A \cap C) d\mu'.$$

The measure μ' is the quotient measure on M/ξ.

Proposition: Each measurable partition has a canonical system of conditional measures which is unique (mod 0), i.e. any other system is the same for almost all $C \in M/\xi$. Conversely, the existence of a canonical system for a partition implies that the partition is measurable.

Example: We present a simple example of a canonical system of conditional measures. Consider the two-dimensional torus T^2, defined as the Cartesian product of two circles, $T_x \times T_y$. Let x (mod 1) and y (mod 1) be coordinates on T_x and T_y respectively, then (x, y) are coordinates on T^2. Consider the partition ξ of T^2 with elements $T_x \times y$, the circles on the torus with constant y-coordinate.

The space T^2/ξ is the circle T_y. ξ is a measurable partition, since as a measure space, $T^2 = T_x \times T_y$, the cartesian product of two circles, with the product measurable structure. Any basis of T_y is thus a basis for the partition ξ. It is easy to verify that the conditional measure μ_C on the element of the partition $C = T_x \times y$ is the normalised Lebesgue measure on T_x. An analogous construction is easily made for the partition of the torus into lines of rational slope, since such lines eventually close, and are thus circles.

Example: Suppose we consider the partition ξ of the torus by the lines of irrational slope $\alpha \notin Q$. It is a familiar fact that any such line passes arbitrarily close to any point in T^2, in other words, each line is dense in the usual topology of T^2. Each element of ξ is a measurable subset with measure zero of T^2. The problem of finding a basis for the partition ξ is equivalent to the problem of finding a basis for the partition η of T by taking T transversal to the lines of constant η. Two points $x, y \in T$ are in the same element of η iff there exists an integer n such that $y = x + n\alpha \pmod 1$. If we define the map

$$f_\alpha : T \to T$$
$$x \to x + \alpha \pmod 1,$$

then it follows that the elements of the partition η are precisely the orbits of f_α. The η-sets are just the f_α-invariant subsets of T. It is well known that for irrational α the map f_α is ergodic. Thus all η-sets have either measure zero or measure unity. This shows that the only measurable partition into η-sets is the trivial partition, with the one set T^2. In other words, the quotient space T^2/ξ of T^2 by the (non-measurable) partition ξ defined by lines of constant irrational slope is not a Lebesgue space, indeed it is not even countably separated.

As a topological space, it is not Hausdorff. Using the conventional techniques of integration theory, this quotient appears to be a trivial measure space. For example, all the L^p spaces are one-dimensional, and conventional measure theory is useless. Given two foliations of T^2 by lines of irrational slope α and α' with $\alpha \neq \alpha'$, conventional measure theory is incapable of distinguishingthe two quotient spaces, both appear trivial. Connes calls such spaces which consist of more than one point but are trivial measure theoretically, *e.g.*, quotients of standard spaces by an ergodic equivalence relation, *singular* measure spaces.

Connes noncommutative integration theory is developed precisely for the study of these singular spaces. Connes has recently developed techniques of non-commutative geometry for studying singular topological spaces, and singular differentiable manifolds but we shall in the sequel be interested only in the integration theory. We note in concluding the discussion of the toral example, that one can easily construct measurable partitions of T^2 for which the elements are pieces of the lines of slope $\alpha \notin Q$, but not the entire lines.

We now introduce ideas associated to increasing partitions, which will be important later. A partition η is said to *refine* the partition ξ if for almost all $x \in M \eta(x) \subset \xi(x)$. This is also written as $\eta > \xi$. Now let $f: M \to M$ be a Borel automorphism.

Definition: We say a partition η is increasing for the automorphism f if $\eta > f_\eta$.

The existence of certain types of increasing (measurable) partitions is an important tool for the ergodic theory of hyperbolic differentiable dynamical systems. In particular, increasing partitions subordinate to the unstable manifolds are an essential tool. The prototype was an increasing partition constructed by Sinai from a Markov partition. The elements of this increasing partition are pieces of unstable manifolds.

This was generalized by Bowen to Axiom A. Pesin's work used a partition of this type. The partition ξ refered to in the discussion of proposition is of this type. We shall need in a later chapter some detailed results of Ledrappier and Strelcyn concerning partitions of this type.

Chapter 11

Special Results of Measure Theory

The purpose of the present report is to record in writing some results of measure theory that are known to many people but do not seem to be in print, or at least seem to be dicult to nd in the printed literature. Let X be compact with a Radon measure μ. Let f be a map from X to a metric space Y such that for every open set $S \subset Y$ the inverse image, $f^{-1}(S)$ is Radon measurable. Then, if the cardinality of $f(X)$ is not outlandishly large, there is a subset $X_0 \subset X$ such that $\mu(X \setminus X_0) = 0$ and $f(X_0)$ is separable. Precise definition of what outlandishly large means will be given below.

The theorem may not appear very useful. However, after seeing it, one usually looks at empirical measures and processes in a dierent light. The theorem could be stated briefly as follows: A measurable image of a Radon measure in a complete metric space is Radon. The second part of the paper is an elaboration on certain classes of measures that are limits in a suitable sense of\molecular" ones, that is measures carried by nite sets. It ties together several possible formulations of relations of measures and integrals of uniformly continuous functions. It also puts Prohorov's theorem: relative compactness is equivalent to tightness, in a very dierent light. Typically, the conditions it gives for relative compactness are enormously weaker (in appearance at least) than tightness. It would be good to see whether they can be used instead of tightness to prove relative compactness.

Using some of the results of that second part, it is easy to see that several properties of measures on paracompact spaces can be checked on their continuous images in metric spaces. This allows us to replace the term "metric" by "paracompact" in the rst part and gives many other results.

The results of the second and third part were established. There were also related results of Berezanskii, Granirer and several other authors. The results were further rened by Z. Frolik. In the intervening years they may have became obsolete. We have not checked. They are reproduced here because the proofs are reasonably simple and because they might prod other investigations.

MEASURABLE IMAGES OF RADON MEASURES

Let X be a compact set. A positive Radon measure on X is a curious object, constructed from positive linear functionals on the space $C(X)$ of continuous functions on X. One considers a positive linear functional, denoted $\langle \mu, f \rangle = \int d\mu$ for $f \in C(X)$. Then one extends the denition to lower semicontinuous functions by $\int f d\mu =$ $\sup\left\{\int \gamma d\mu; \gamma \leq f, \gamma \in C(X)\right\}$ and finally to all functions f that for, each $\varepsilon > 0$, can be squeezed between an upper semicontinuous g_ε and a lower semicontinuous h_ε so that $g \leq f \leq h$ and $\left|\int h_\varepsilon d\mu - \int g_\varepsilon d\mu\right| \leq \varepsilon$, with $\int g_\varepsilon = -\int(-g_\varepsilon)$. This gives a functional dened on a large space of functions on X. The sets called "measurable" for μ are those whose indicator has an integral dened by that process.

Note that part of the denition of μ is its domain. One might be able to extend μ even further, but such extensions would not be called "Radon". An illustration is given by a famous theorem of Kakutani and Oxtoby: Let λ be the (usual) Lebesgue measure on [0, 1]. It is a Radon measure. There is an extension v of it that has the following properties: a) It is invariant by all one to one pointwise transformations of [0, 1] that left λ invariant and b) its domain is such that the Hilbert space of v-square integrable functions it generates has a basis of cardinality 2^c.

Here c is the cardinality of the continuum. The Lebesgue measure itself give a Hilbert space with a countable basis. The measure v extends λ to a much larger class of sets, though not to all subsets of [0, 1]. However it is not a Radon measure. A characteristic of positive Radon measures is that the measure of a set is the supremum of the measures of the compact sets contained in it.

That is, if μ is Radon and A is μ-measurable then for each $\varepsilon > 0$ there is a compact $K_\varepsilon \subset A$ with $\mu(A \setminus K_\varepsilon) < \varepsilon$. Besides Radon measures, we shall need some information on "measurable cardinals". A measurable cardinal is a set S that admits a nontrivial probability measure μ defined for all subsets of S and taking only two values, so that $\mu(A) = [\mu(A)]^2$. By "nontrivial" is meant that each point has measure zero.

To see how large a measurable cardinal must be, let us rst look at the first innite cardinal $\aleph_0$, cardinal of an innite countable set. It has two remarkable properties as follows:

1. If $n < \aleph_0$ then $2^n < \aleph_0$
2. A set S of cardinality $\aleph_0$ cannot be written as a union $\bigcup_{j \in J} A_j$ where card $A_j < \aleph_0$ and card $J < \aleph_0$.

A cardinal $\aleph$ is said to be (strongly) inaccessible if it is uncountable and has property (*A*). It is weakly inaccessible if it is uncountable and has property (B).

Now property (*A*) is very strong. A (strongly) inaccessible cardinal $\aleph > \aleph_0$ must also be larger than $c = 2^{\aleph_0}$ and 2^c and so forth for any innite cardinal$n < \aleph$. That means that if one restricts oneself to accessible cardinals one can carry out all the usual set operations such as unions, products, powers (because $\in \leq m \leq n$ implies $m^n = 2^n$ for innite n). One will never get any inaccessible cardinal by such operations. In other words, one can ignore inaccessible cardinals for all usual and statistical purposes.

As to measurability; it has been proved by Ulam and Tarski that a measurable cardinal must be inaccessible. Thus if one lives in the land or universe of accessible cardinals one can say that measurable cardinals are "outlandishly large", giving a precise meaning to the sentence used in the Introduction.

Property (B) is not as strong as (A). Its implications may be weaker, depending on what axioms one adds to the usual set theory ones. Here the usual axioms will mean the axioms of Zermelo-Fraenkel with the axiom of choice, refered to as ZFC.

In a system such as ZFC, or the Bernays-Godel system, one can list the infinite cardinals in increasing order using the ordinals as indices for the list. It would look like $\aleph_0 < \aleph_1 <... < \aleph_\omega < \aleph_{\omega+1} <...$

In such a list it is always true that card $\alpha \leq \aleph_\alpha$. However equality is not ruled out. One can readily obtain ordinals such that card $\alpha = \aleph_\alpha$. For instance dene a sequence of pairs $(\omega(n), Z_n)$ as follows. Let Z_1 = @0 and let $\omega(1)$ be the rst ordinal that has cardinality $\aleph_0$. If $(\omega(n), Z_n)$ has been constructed, let $\omega(n + 1)$ be the rst ordinal that has cardinality Z_n and let $Z_{n+1} = \aleph_{\omega(n+1)}$. The limit = $\sup_n \omega(n)$ is such that $\aleph_\alpha = \sup_n Z_n$ is card α. However that $\aleph_\alpha$ is clearly the sum of a countable set of cardinals $Z_n < \aleph\alpha$. (One could continue further, say up to the rst uncountable ordinal Ω if each time one gets a limit such as α and $\aleph_\alpha$ with card $\alpha = \aleph_\alpha$ one replaces $(\alpha, \aleph_\alpha)$ by $(\alpha, \aleph_{\alpha+1})$ in the list). Now if α is a limit ordinal and if $\aleph_\alpha$ is weakly inaccessible then $\aleph_\alpha$ must equal the conality of α. The conality of α is the smallest cardinality of a set of ordinals $\omega_\xi < \alpha$ such that $\sup_\xi \omega_\xi = \alpha$.

Such a set cannot be obtained by the recursive process described above. It is a fact of life that the existence of such alephs cannot be proved in ZFC.

Now what has that to do with measurability? Dene a weakly measurable cardinal as one of a set *S* that admits a probability measure 0 defined on all its subsets giving mass zero to all points. This is the same as before except that (*A*), $A \subset S$ is allowed to be any value in [0, 1] instead of being just zero or one as in our previous denition. In a world where there are no strongly inaccessible cardinals a weakly measurable cardinal would be one that admits an atomless probability measure dened on all its subsets.

Now Ulam (1930) has proved that if $\aleph_\alpha$ is not weakly measurable $\aleph_{\alpha+1}$ is not weakly measurable either. So $\aleph_1, \aleph_2, \ldots \aleph_\omega, \ldots \aleph_\Omega \ldots$ are not weakly measurable. In fact Ulam's result imply that either a weakly measurable set is already (two-valued) measurable, or there is some weakly measurable cardinal $n \leq c = 2^{\aleph_0}$. That second option is ruled out by the ordinary continuum hypothesis $c = \aleph_1$. Thus the rst option is consistent with ZFC. However one could also assume that on the contrary c admits an atomless probability measure dened for all its subsets. This will be responsible for some of the precautions in the statements given below.

Consortium

Theorem: Let X be a compact space. Let μ be a positive Radon measure on X. Let f be a map from X to a metric space Y. Assume that for every open set $G \subset Y$ the inverse image $f^{-1}(G)$ is μ-measurable (that is, in the domain of the Radon measure μ).

Assume also that the cardinal of $f(X)$ does not admit any nontrivial two valued probability measure defined for all subsets. Then there is a subset X_0 of X such that $(X_{n\backslash} X_0) = 0$ and such that $f(X_0)$ is separable.

Remark: As we shall see in the proof, the result becomes much easier if one assumes that the cardinal of $f(X)$ is not even weakly measurable. Then the Radon structure of is not essential. We shall elaborate later on what this means for possible denitions of "distributions" for such items as ordinary empirical cumulatives. Because of this the preceding Theorem, although peculiar, is not entirely uninformative.

Proof: One can without loss of generality assume $f(X) = Y$. Let $\{G_j, j \in J\}$ be a covering of Y by open sets G_j. Assume that the index set J is well ordered. For any $j \in J$ let $U_j = \bigcup_i \{G_i; i < j\}$ and $A_j = G_j \backslash U_j$. This is an intersection of an open with a closed set. Deleting the empty A_j, one can assume that each A_j is nonempty. Let B_j, $j \in J$ be other subsets of Y. Form the set $\bigcup_j (A_j \cap B_j)$. This is called the result of operation (M) on the B_j. It is a theorem of D. Montgomery that if the B_j are Borel subsets of Y of one of the transnite classes called $F\alpha$ or G_α so is $\bigcup_j (A_j \cap B_j)$. In particular any union $\bigcup (A_j; j \in S)$ where S is an arbitrary subset of J is a Borel set. Now the measure μ has an image $f(\mu)$ on Y, but by writing $v(S) = \{f^{-1}[\cup(A_j, j \in S)]\}$ one obtains a measure dened on all the subsets of the index set J. This is the image of μ by the map $g \circ f$ where g maps Y into J by $g(y) = j$ if $y \in A_j$.

Let us look at the measure v on J. It may have atoms. That is there may be sets $S \subset J$ such that $v(S) > 0$ but such that for any subset S' of S one has either $v(S') = 0$ or $v(S') = v(S)$. Take such an atom S_0 and delete the rest of J. This gives a measure v_0 carried by S_0 taking only two values and defined for all subsets of S_0. Since we have assumed that the cardinality of $f(X) = Y$ is not measurable, the same applies to J (since we have deleted the empty A_j's), hence also to S_0. Thus the measure v_0 carried by S_0 must be carried by a particular point $s_0 \in S_0$. Doing this for each atom of v obtains a countable set $C \subset J$ that carries the atomic part of v. Map back C into X taking $(g \circ f)^{-1}(C) = D$, say. This is a μ-measurable set.

Now replace μ by μ_1 dened by $\mu_1(V) = (V \cap D^c)$. This is still a Radon measure on X and its image by $g \circ f$ has no atoms. Let us first deal with this particular μ_1. If it is not zero one can assume for simplicity that it is a probability measure so that its image v_1 by $g \circ f$ is also a probability measure. Since v_1 is atomless, one can divide J into two sets, say $D_{1,0}$ and $D_{1,i}$ so that $v_1(D_{1,i}) = \dfrac{1}{2}$. Dividing each set in equal parts one gets sets $D_{2,i}$, $i = 0, 1, 2, 3$ such that $v_1(D_{2,i}) = \dfrac{1}{2^2}$. Proceed on dividing each set into equal parts each time. It is clear that this gives a map, say φ, of J into the interval [0, 1]. For the Borel subsets of [0, 1] the image λ_1 of v_1 coincides with the Lebesgue measure λ. However λ_1 is dened on all the subsets of [0, 1].

Now consider the composition of map $\omega = \varphi \circ g \circ f$. It is a map from X to [0, 1). By construction it is "measurable" at least in the sense that the inverse image of any Borel subset of [0, 1] is a μ_1-measurable set in X.

This implies in particular that ω is also "measurable" in the sense of Bourbaki's denition of that term. Specically, for every $\varepsilon > 0$ there exists a compact subset K_ε of X such that $\mu_1(X \setminus K_\varepsilon) < \varepsilon$ and such that when restricted to K_ε the map ω is *continuous.* Now consider an arbitrary subset $T \subset [0, 1]$. Its inverse image $\omega^{-1}(T)$ is μ_1-measurable. Thus, as above, for every $\varepsilon > 0$ there is a compact K_ε $\omega^{-1}(T)$, such that $\mu_1[\omega^{-1}(T) \setminus K_\varepsilon] < \varepsilon$ and such that when restricted to K_ε the map ω is continuous. If so the image $\omega(K_\varepsilon)$ is also a compact. It is such that $\omega(K_\varepsilon) \subset T$ and that $\lambda_1[T \setminus \omega(K_\varepsilon)] < \varepsilon$. This is true for every $\varepsilon > 0$. Thus T must be Lebesgue measurable. Since there do exist non Lebesgue measurable sets, we have reached a contradiction.

This leads to the conclusion that the measure 1 must in fact be equal to zero. Equivalently one can say that for every arbitrary covering $\{G_j ; j \in J\}$ and partition $A_j = G_j \setminus [\cup(G_i; i < j)]$ the image of μ by f is carried by a countable subfamily of the A_j, or equivalently again, by a countable subfamily of the G_j. In particular one can take a cover by open balls $B_{j,m}$ of center $y_{j,m}$ and radius $\dfrac{1}{m}$. Here j is in a certain set J_m which may be highly innite. However the set $J_{m,1}$ of these j's such that $\mu[f^{-1}(B_{j,m})] > 0$ is countable. Consider the subset of Y formed by the centers $y_{j,m}$ for which $j \in J_{m,1}$. Taking all these sets for all m one obtains a countable subset of Y. Let W be the closure of that set. It has the property that for each integer m the union W_m of the open balls of radius $1/m$ centered at elements of W contains all the measure for the image $f(\mu)$. Thus W, intersection of the W_m must also have full measure for $f(\mu)$. The inverse image $X_0 = f^{-1}(W)$ satises the conditions of the theorem. Hence the result.

Remark:

1. If instead of assuming that the cardinal of $f(X)$ is non measurable (for two-valued measures) one had instead assumed that it cannot carry arbitrary atomless probability measures, the proof could have been simplified. One could remove the part of it that disposed of the purely non atomic m_1 and avoid the reference to Lebesgue measure and μ-measurability in the Bourbaki sense.
2. One could also state the conclusion of the theorem by saying that if Y is complete the image $f\mu$ extends to a Radon measure.
3. The theorem could have been stated in a dierent form: If f satises the conditions of Theorem, it is already μ-measurable in the sense of Bourbaki. That is, for each $\varepsilon > 0$ there is a compact K with $\mu(X \setminus K_\varepsilon) < \varepsilon$ and with f continuous when restricted to K_ε.
4. Theorem says something applicable to empirical measures or cumulatives. To see this, let λ be the Lebesgue measure on $X = [0, 1]$. Define a map f from $[0, 1]$ to the space B of bounded functions on $[0, 1]$ as follows. If $x \in [0, 1]$ then $f(x)$ is the function $t \to F_x(t)$ such that $F_x(t) = 0$ for $t < x$ and $F_x(t) = 1$ for $t \geq x$. Metrize the space B by it sup norm

$$\| u \| \sup_t | u(t) |.$$

In this case the image $f(X)$ is a discrete subset of B. If $x_1 \neq x_2$ then $\| f(x_1) - f(x_2) \| = 1$. Any Lebesgue set of positive measure must have the cardinality c of the continuum. Since $c, 2^c, \ldots$ are all non measurable, Theorem implies that there must be closed subsets S of $f(X)$ such that $f^{-1}(S)$ is not λ-measurable. This is not too surprising. Every subset of $f(X)$ is closed. The map f from X to $f(X) = Y \subset B$ is one to one and the sets that map back to Lebesgue measurable sets are the images of these Lebesgue sets. If instead of taking just one observation on $[0, 1]$ for λ one would take many, say n, from some joint distribution on $[0, 1]^n$ and form the corresponding cumulative distribution in B the conclusion would be the same. Assuming that μ is not purely atomic, there will be closed subsets of B whose inverse image is not μ-measurable. This statement assumes that on $[0, 1]^n$ is a Radon measure so that its measurable sets dier from Borel sets by negligible sets.

Now take $[0, 1]^n$ for our set X and take for $f(x)$ the corresponding cumulative as element of the space B of bounded functions on $[0, 1]^n$. In this way the "distribution" of the cumulative may be thought of as the image $f(\mu)$ of μ on B. It is dened on some σ-field F of subsets of B (or of $Y = f(X)$). By Theorem this σ-field cannot contain the Borel eld of $f(X)$. However Theorem does not say anything about the possibility of *extending* the definition of $f(\mu)$ to the Borel field. This is a dierent question. Here the image $Y = f(X)$ is again a discrete space with the cardinality c of the continuum. All its subsets are closed.

The question of possible extension of $f(\mu)$ then boils down to the following: Does the continuum c admit an atomless probability measure dened for all subsets.

If there was such a measure say P it would not be hard to match it with $f(\mu)$ where this is dened. This is easier to see if $n = 1$ and μ is Lebesgue on $[0, 1]$. Using the partitioning by half already used in the proof of Theorem to construct our map φ, one can readily match the σ-field F on which $f(\mu)$ is defined with a corresponding σ-field of subsets of c, matching at the same time the respective measures. This is clear for Lebesgue on $[0, 1]$ and works in the same manner for any non atomic μ on $[0, 1]^n$ and any non atomic P on c.

Any extension of the non atomic $f(\mu)$ to all subsets of $Y = f(X)$ would necessarily be non atomic. Indeed c is nonmeasurable. Thus atoms would be points and these are already in the domain of $f(\mu)$.

Thus the question becomes: Does there exist a purely non atomic probability measure P dened on all the subsets of c. In other words is c weakly measurable?

We have already pointed out that this depends on where c is located in the string of alephs. The answer is negative if c is strictly inferior to the first aleph, $\aleph_{\alpha_0}$, such that α_0 has for conality $\aleph_{\alpha_0}$ itself. In such a case "weak measurability" and "measurability" would be equivalent.

Thus, for instance, if $c \leq \aleph_\Omega$ where Ω is the rst uncountable ordinal, then our $f(\mu)$ on $f(X)$ does not admit extensions to all the Borel subsets of $f(X)$.

However, the location of c in the alephs depends on what axioms of set theory one is willing to assume, while Theorem does not depend on such assumptions. (It depends, however on the existence of sets that are not Lebesgue measurable. These sets do not exist in some systems where the axiom of choice is suitably restricted). We presume that most statisticians would cheerfully accept the hypothesis that c is not weakly measurable, or even the hypothesis $c < \aleph_{\alpha_0}$, even though this puts a crimp in what one can do with the distribution of empirical cumulatives.

Another implication of Theorem can be described as follows. Take a stochastic process $Z: t \to Z(t)$ dened on some non degenerate interval T of the line. Assume that the trajectories of the process are bounded, by, say, 0 and 1. The "distribution" of the process Z is then some measure m on some σ-field of subsets of $[0, 1]^T$. According to Kakutani [1943] there exist a Radon measure μ on the compact set $[0, 1]^T$ (for the topology of pointwise convergence) such that μ and m have the same values on sets dened by finitely or countably many Borel restrictions on the coordinates. Here the cardinality of $[0, 1]^T$ is clearly not a measurable one.

Thus, according to Theorem, if we map $X = [0, 1]^T$ into a metric space Y by a function f there will necessarily exist closed subsets of Y whose inverse images are not μ-measurable unless f has almost separable range. (This means that there is a subset X_0 of X with $(X \backslash X_0) = 0$ and $f(X_0)$ separable). Thus, for instance, take for Y the set $[0, 1]^T$ itself but with the supremum norm. Take for f the identity map. These will be closed subsets of Y that are not μ-measurable unless the measure is concentrated on a separable subset of Y. Thus, although the Kakutani extension is much richer than the usual product σ-field of Kolmogorov, it leaves out many sets.

Molecular Measures and their Limits

Often probability spaces come with more structure than the standard triplet (Ω, F, P). For instance they may be metric or topological. It is then pleasant if the properties of the measure and the metric or topology are somewhat related. Here we study a class of measures that are nicely related to the uniform structure of the space on which they live. We have taken uniform spaces instead of topological spaces because they aord more flexibility. A uniform space is given by a set X and a lter V of "vicinities" of the diagonal Δ of $X \times X$. This filter is assumed to be such that the diagonal Δ is included in every $V \in V$. Also, if a $V \in V$, its symmetric $\{(x, y): (y, x) \in V\}$ is also in V. Finally for each $V \in V$ there is a $W \in V$ such that $W^2 \subset V$. Here W^2 is defined as the set of pairs (x, z) for which there is a y such that $(x, y) \in W$ and $(y, z) \in W$.

A pseudo metric ρ on X denes a uniform structure, by taking for basis of the vicinities the sets of the form $\{(x, y); \rho(x, y) < \varepsilon\}$ for $\varepsilon > 0$. Conversely it can be shown that every uniform structure can be generated in this manner by some family $\{\sigma_\alpha; \in A\}$ of pseudo metrics.

A function f from a uniform space (X_1, V_1) to another (X_2, V_2) is called uniformly continuous if for every $V_2 \in V_2$ there is a $V_1 \in V_1$ such that $(f(x), f(y)) \in V_2$ for all $(x, y) \in V_1$.

A family $\{f_{\alpha;}\ \alpha \in A\}$ of such functions is called uniformly equicontinuous if for $V_2 \in V_2$ there is a $V_1 \in V_1$ such that $(f_\alpha(x), f_\alpha(y)) \in V_2$ for all pairs $(x, y) \in V_1$ and all $\alpha \in A$. The real line IR is a uniform space if one takes for base of the vicinities the sets $\{(x, y): |x - y| < \varepsilon\}$ for $\varepsilon > 0$.

A separated uniform space is one in which if $x \neq y$ then there is a $V \in V$ such that

$(x, y) \notin V$. Separated uniform spaces admit uniformly continuous real valued functions in large quantities. Let us denote by D the space of all bounded real valued uniformly continuous functions on (X, V). It is a Banach space with dual D^* if one gives it the sup norm $\| f \| = \sup_x | f(x) |$. We shall be interested in various subspaces of D^* that are naturally linked to the uniform structure of (X, V).

To describe them, let us introduce some particular subsets of D. A set $B \subset D$ will be called a UEB set if it is bounded and uniformly equicontinuous. "Bounded" means that there is some number $\alpha \in (0, \infty)$ such that $| f(x) | \leq \alpha$ for all $x \in X$ and all $f \in B$.

A set $S \subset X$ is called *precompact* if for every $V \in V$ the set S can be covered by a finite family of sets $A_j, j = 1, \ldots, n$ that are small of order V, that is such that $(x, y) \in V$ for every pair (x, y) of elements of A_j. The set D can be given the uniform structure u of uniform convergence on the precompact subsets of X. This is usually weaker than the structure generated by the sup norm.

The subspaces of D^* that will be of interest below are as follows

1. The space M_s of linear functionals with nite support on X. Specifically a linear functional φ on D belongs to M_s if there is a nite set $\{x_j; j = 1, \ldots, n\}$ and coecients $c_j \in IR$ such that $\langle \varphi, \gamma \rangle = \sum_j c_j \gamma(x_j)$ for all $\gamma \in D$. Because a mass c_j at x_j looks like an atom and because φ is a nite sum of such atoms, the elements of M_s have also been called *molecular measures*.
2. The space M_p of linear functionals whose restrictions to the unit ball $B_1 = \{\gamma: \gamma \in D, \| \gamma \| \leq 1\}$ are continuous for the precompact convergence u.
3. The space M_u closure of M_s in D^* for the structure of uniform convergence on the UEB subsets of D.

   This uniform structure will be called $[V]$ to recall that it came from V.
4. The space M_u of elements of D^* whose restrictions to UEB subsets of D are continuous for u.

Note that we stay in D^*. Thus each μ in any of these spaces has a finite norm $\| \mu \| = \sup \{ | \langle \mu, \gamma \rangle |; \gamma \in D, | \gamma | \leq 1 \}$.

The spaces dened above clearly satisfy the relations

$$Ms \subset M_p \subset M_u \subset M_u \subset D^*.$$

Another important property is given by the following result

Proposition: Each one of the spaces listed above is *a* band in D^*. The spaces M_u and $\bar{M}_u$ always coincide. Balls of the type $\{\mu : \| \mu - \varphi \| \leq a\}$ and the positive cone M_u^+ of M_u are complete for the structure $[V]$. If V is metrizable, then M_p is also equal to M_u.

This will be proved below in several steps.

One of the reasons for the importance of Proposition is the completeness statement. It will allow us to use the compactness criteria of Grothendieck [1952]. The identity $M_p = M_u$ when (X, V) is metrizable also allows another characterization of M_u: Let (Y, V_1) be another uniform space whose uniform structure is metrizable. Let f be a uniformly continuous map from (X, V) to (Y, V_1). If $\mu \in M_u$ for (X, V) then its image $f(\mu)$ by f is in M_p for (Y, V_1).

It turns out that if $\mu \in D^*$ maps this way into M_p for every metrizable (Y, V_1) and every uniformly continuous map f then $\mu \in M_u$. Now an element of M_p on (Y, V_1) is just something that extends to a Radon measure on the completion of (Y, V_1). This will also allow us to extend the result of Theorem to images of Radon measures into uniform spaces: Suppose that K is compact and that μ is a Radon measure on K.

Let f be a map from K to a uniform space (X, V). Assume that for every $\gamma \in D$ the function $\gamma \circ f$ is μ-integrable (in the Radon sense). Then, in a world that does not possess any measurable cardinals, the image of μ is in M_u.

Note that we have not mentioned countable additivity of the measures. A simple example will show why. Let X_1 be the interval [0, 1] of the line with its standard uniformity. Let $X \subset X_1$ be the set of rational numbers, with its standard uniformity V inherited from X_1. The set D of uniformly continuous functions on (X, V) is the set of restrictions to X of continuous functions on X_1. Every positive linear functional on the set $C(X_1)$ of continuous functions on X_1 can also be identied as a positive linear functional on D. Since in this example X is precompact, our space M_u^+ coincides with the set of linear functionals that can be obtained that way. However if our arbitrary uniform space (X, V) is *complete* the elements of M_u are -smooth on D in the sense that if $\mu \in M_u$ and if $u_n \in D$ decreases pointwise to zero on X, then $\langle \mu, u_n \rangle \to 0$. This will be a consequence of the results proved below.

Note however that M_u does depend on the uniform structure V on X, not only of the topology it generates. There are usually many uniform structures yielding the same topology. Two of them are particularly interesting. Assuming (X, V) separated, for simplicity, one can dene on X the smallest structure $\underline{V}$ that make the elements of D uniformly continuous. Then the completion $\overline{X}$ of X for $\underline{V}$ is a compact set. The set D becomes the set of restrictions to X of the continuous functions $C(\overline{X})$ on $\overline{X}$. The corresponding $M_u(X, \underline{V})$ can be canonically identied to the Radon measures on $\overline{X}$.

Another interesting structure is $\overline{V}$, the "universal structure" attached to (X, V). It is the one dened by all the pseudo-metrics ρ defined on X and such that $\rho(x, y)$ is (jointly) continuous. The corresponding set $D(X, \bar{V})$ is the set $C^b(X)$ of bounded continuous functions on (X, V). The corresponding $M_u(X, \bar{V})$ is typically much smaller than $M_u(X, V)$ and therefore smaller than $M_u(X, \underline{V})$.

If X is separated and complete for V, its V precompact sets have compact closure. Hence the structure u becomes the structure of uniform convergence on compacts. Since compactness is a topological property a set that is compact for V is also compact for $\bar{V}$. By the Stone Weierstrass theorem $D(X, V)$ is dense for u in the set $C_b(X)$ of bounded continuous functions on X. The elements of $M_p(X, V)$ are also in $M_p(X, \bar{V})$. They extend to Radon measures on X. However the conclusion does not extend to $M_u(X, V)$.

Before leaving these generalities and studying more closely the structure of M_u let us note the following. Start with an (X, V) and dene the corresponding M_u with its uniformly $[V]$ of uniform convergence on the UEB sets of D.

The set X itself can be identied to a subset X' of M_u by associating to each x the Dirac measure x, probability measure concentrated at x.

The structure induced by $[V]$ on X' is the initial structure V since V can be defined by the family of pseudo-metric $\rho_B(x, y) = \sup_\gamma \{| \gamma(x) - \gamma(y) |: \gamma \in B\}$ as B ranges through UEB sets. Indeed if $V \in V$ is such that $\gamma \in B$ implies $| \gamma(x) - \gamma(y) | \leq \varepsilon$ for $(x, y) \in V$ then $\rho_B(x, y) \leq \varepsilon$ on V. Conversely if ρ is one of the pseudo-metrics used to dene V and $\rho \leq 1$, take the set B of functions $\gamma \in D$ such that $| \gamma(x) - \gamma(y) | \leq \rho(x, y)$ and $| \gamma | \leq 1$. Then B is UEB and $\rho(x, y) = \sup \{| \gamma(x) - \gamma(y) |; \gamma \in B\}$. (Take $\gamma_z(x) = (x, z), z \in X$).

This shows that the notation V and $[V]$ will not lead to confusion.

The Structure of $\bar{M}_u$

In this section we consider a xed separated uniform space (X, V) with its space D of bounded uniformly continuous functions and the attached spaces of linear functionals $M_s \subset M_p \subset M_u \subset \bar{M}_u$.

Consider first the UEB subsets of D. If B_1 and B_2 are two such sets, so is their union B_1 and B_2. The convex hull of a UEB set B is also UEB, so is the convex symmetrized hull

$$\{\gamma : \gamma = \sum c_j \gamma_j, \sum | c_j | \leq 1, \gamma_i \in B\}$$

In addition the pointwise closure of a UEB set is UEB. A pointwise closed UEB is compact for pointwise convergence and on it the topology of pointwise convergence coincides with the topology induced by the structure u of uniform convergence on precompact sets. We shall also need the following simple observation.

Lemma: Let S be a UEB set that is compact for pointwise convergence on X. Then S is also compact for the weak topology $W(D, \bar{M}_u)$.

Proof: Take a compact (Hausdor) space B and let $C(B)$ be the space of continuous real functions on B. Then B is also compact for the weakest topology that makes all $\gamma \in C(B)$ continuous because that topology is Hausdorff and certainly weaker than the initial one on B and therefore identical to it. On our compact UEB set the elements of $\bar{M}_u$ are continuous for the pointwise topology since it coincides with to u-topology.

A consequence of this state of aairs is as follows.

Corollary: The dual of $\bar{M}_u$ for $[V]$ is D. That is every linear functional defined on $\bar{M}_u$ u and $[V]$ continuous is given by the evaluations $\langle \varphi, \gamma \rangle$ for $\varphi \in \bar{M}_u$ and some $\gamma \in D$.

Proof: This is well-known fact in the duality theory for locally convex spaces. Briefly if B is a convex symmetric compact UEB then its second polar B^{00} in the dual of ($\bar{M}_u$, $[V]$) is the closure of B. However B being compact is already closed.

It is clear from the denition that both M_u and $\bar{M}_u$ are closed subsets of D^* for the structure V. Thus they are also closed for the stronger topology defined by the norm of D^*. Thespace M_p is also closed for the norm topology. Note also the following property.

Lemma: Let M be any one of the spaces M_s, M_p, M_u or $\bar{M}_u$. If μ is an element of M then its positive part μ^+ also belongs to M.

Proof: One can dene μ^+ by the relation $\langle \mu^+, \gamma \rangle = \sup_u \{\langle \mu, u \rangle ; 0 \leq u \leq \gamma, u \in D\}$ for any $\gamma \in D$ that is in the positive cone D^+ of D. Let $v \in D$ be such that $0 \leq v \leq 1$ and such that $\langle \mu^+, 1 \rangle \leq \langle \mu, v \rangle + \varepsilon$. Now $\langle \mu, v \rangle + \varepsilon$ may be written $\langle \mu^+, v \rangle - \langle \mu^-, v \rangle + \varepsilon$, giving $\langle \mu^+, 1-v \rangle + \langle \mu^-, v \rangle \leq \varepsilon$. For any element γ of D one can write

$$\langle \mu^+, \gamma \rangle = \langle \mu^+, v\gamma \rangle - \langle \mu^-, v\gamma \rangle + \langle \mu^-, v\gamma \rangle + \langle \mu^+, (1-v\gamma) \rangle$$

$$= \langle \mu, v\gamma \rangle + \langle \mu^-, v\gamma \rangle + \langle \mu^+, (1-v)\gamma \rangle$$

$$\leq \langle \mu, v\gamma \rangle + \varepsilon \| \gamma \|$$

Define a new functional $v \times \mu$ by $\langle v \times \mu, \gamma \rangle = \langle \mu, v\gamma \rangle$. The foregoing inequality says that $\| v \times \mu - \mu^+ \| \leq 2\varepsilon$. Nowif B is an UEB set, so is the set $\{v\gamma; \gamma \in B\}$. Thus $v \times \mu$ belongs to M_u (resp $\bar{M}_u$, M_p) whenever μ does. It follows that μ^+, limit of the $v \times \mu$ for the norm topology, is also in the same space. The case of M_s is clear, hence the result.

Lemma: The spaces M_p, M_u and $\bar{M}_u$ are bands in D^*.

Proof: Since all these spaces are closed in D^* for its norm topology and since Lemma applies to them, it is sucient to show that if $0 \leq v \leq \mu \in M_p$ then $v \in M_p$ and similarly for M_u and $\bar{M}_u$.

In all cases, if $0 \leq v \leq \mu$ and if $\varepsilon > 0$, there is some $\gamma \in D$ such that $\| v - \gamma \times \mu \| < \varepsilon$. Another procedure would be to complete X for the smallest uniform structure $\underline{V}$ that makes the elements of D uniformly continuous. On the completion v and μ become Radon measures such that $v \leq \mu$. Therefore there is a measurable f such that $0 \leq f \leq 1$ and $v = f \times \mu$. Then, for every $\varepsilon > 0$ thee is a $\gamma \in D$ such that its extension to the completion satisfies $\int |\gamma - f| \, d\mu < \varepsilon$.

Thus since v can be approximated as closely as one wishes in the norm by $\gamma \times \mu$ one concludes as in Lemma that if $\mu \in \bar{M}_u$ and $0 \leq v \leq \mu$ then $v \in \bar{M}_u$. Similarly for M_p or M_u. Hence the result.

Theorem: The positive cone $\bar{M}_u^+$ of $\bar{M}_u$ is the $[V]$ closure in D^* of the positive cone M_s^+ of M_s. The space M_u and $\bar{M}_u$ are the same. The balls $\{\mu; \|\mu\| \leq b\}$ of M_u are complete for $[V]$ and so is M_u^+.

Proof: Let $C = M_s^+$ be the positive cone of M_s and let $\bar{C}$ be its closure for $[V]$ in $\bar{M}_u$. Let μ be a positive element of $\bar{M}_u^+$. Suppose that $\mu \neq \bar{C}$. Then there is a $[V]$ continuous linear functional γ and numbers $a < a + \varepsilon$ such that

$$\langle \mu, \gamma \rangle = a < a + \varepsilon \leq \langle \varphi, \gamma \rangle$$

for all $\varphi \in C$.

According to the corollary of Lemma, this linear functional γ is in fact an element of D. The inequality $a + \varepsilon \leq \langle \varphi, \gamma \rangle$ for all $\varphi \in C$ implies that $\gamma \geq 0$, fairly obviously. Thus we have $a + \varepsilon \leq \inf_\varphi \{ \langle \varphi, \gamma \rangle ; \varphi \in C\} = 0$. This implies $\langle \mu, \gamma \rangle = a < 0$. This contradict the positivity of μ and γ. Thus $M_u \supset \bar{M}_u^+$ hence also $\bar{M}_u = \bar{M}_u^+ - M_u$. Since $M_u \supset \bar{M}_u$, it follows that $M_u = \bar{M}_u$ as claimed.

For the completeness statement let S be the ball $S=\{\mu :\|\mu\| \leq b, \mu \in M_u\}$. If $\mu \in S$ then μ^+ and μ^- are also in S. By the rst part of the argument μ^+ is a $[V]$ limit of elements $\varphi \in M_s^+$ such that $\|\varphi\| \leq \mu+$. A similar statement applies to μ^-. Hence S is the $[V]$ closure of the ball $\{\varphi :\|\varphi\| \leq b; \varphi \in M_s\}$ in D^*. Now the corresponding ball of D is $w(D^*, D)$ compact, hence $w(D^*, D)$ complete, hence also $[V]$ complete. Thus S being closed in that ball is also $[V]$ complete. The argument for the completeness of M_u^+ is the same.

Actually one can prove a better completeness result.

Theorem: The space $\bar{M}_u$ is complete for the structure of uniform convergence on the UEB subsets of D.

Proof: Note that we have worked from the start with subsets of D^*, the space of bounded linear functionals on D. However according to an observation of E. Caby a linear functional whose restriction to UEB's is continuous for precompact convergence (or uniform convergence) is already bounded. Indeed any sequence $\{\gamma_n\}$, $\gamma_n \in D$ such that $\|\gamma_n\| \to 0$ is UEB.

Thus the set $\bar{M}_u$ is exactly the space of (arbitrary) linear functionals whose restriction to UEB sets are continuous for u. Such a space is complete for $[V]$ according to a theorem of Grothendieck..

Note: The completeness statement in Theorem may seem inconsequential. Nevertheless it will allow us to use the compactness criteria given by Grothendieck.

Another property of M_u can be stated as follows.

Proposition: Let μ be a positive element of D^*. The condition $\mu \in M_u$ is equivalent to the statement that every lter on $(D^*)^+$ that converges for $w(D^*, D)$ already converges for $[V]$. On M_u^+ the weak topology $w(M_u, D)$ and the topology induced by $[V]$ are the same.

Proof: Let $\mu \in D^*$ be positive. It is the $w(D^*, D)$ limit of some filter F on M_s^+. If F converges also for $[V]$ then $\mu \in M_u$, by denition of M_u. To prove the converse let S be a UEB subset of D. Let $\rho(x, y) = \sup \{|\gamma(x) - \gamma(y)| : \gamma \in S\}$ and let S_m be the set of functions γ such that $|\gamma| \leq m$ and $|\gamma(x) - \gamma(y)| \leq m\rho(x, y)$. Note that $S \subset S_1$ and that each S_m is also UEB.

Let μ be an element of M_u^+ such that $\|\mu\| = 1$. Then for a given $\varepsilon > 0$ and a given integer m there is some $v \in M_s^+$ such that $\|v\| = 1$ and such that $|\langle \mu, \gamma \rangle - \langle v, \gamma \rangle| < \frac{\varepsilon}{8}$ for every $\gamma \in S_m$. By denition of M_s, this v has a finite support, say F. Let G be the set

$G=\left\{x:\rho(x,F)>\frac{1}{m}\right\}$. Define a function f by $f(x) = \rho(x, G)\,[\rho(x, F) + \rho(x, G)]^{-1}$. This f is such that $0 \le f \le 1$. It is unity on F and zero on G. Also $\rho(x, F) + \rho(x, G) \ge \frac{1}{m}$ and a simple computation shows that $f \in S_m$. Since by construction $f(x) = 1$ for $x \in F$ one has $\langle v, f\rangle = 1$, hence $\langle \mu, f\rangle \ge 1 - \varepsilon/8$.

Now take a lter F on $[D^*]^+$ and assume that F converges to μ for $w(D^*, D)$. There is then a set $A \in F$ such that $|\langle\varphi, f\rangle - \langle\mu, f\rangle| < \varepsilon/8$ for all $\varphi \in A$. Thus $\langle\varphi, f\rangle \ge 1 - \varepsilon/4$ for all $\varphi \in A$.

Returning to the nite set F and the set S_1 one can nd a nite family $\{\gamma_j : j = 1,\ldots,n\}$ of functions $\gamma_j \in S_1$ such that $\inf_j \sup_x \{|\gamma(x) - \gamma_j(x)| : x \in F\} < \varepsilon/4$ for every $\gamma \in S_1$. Now by denition of S_1 and ρ, the inequality $\sup_x \{|\gamma(x) - \gamma_j(x)|; x \in F\} < \gamma/4$ implies $\sup_x \{|\gamma(x) - \gamma_j(x)| : x \in Gc\} < \varepsilon/4 + \frac{2}{m}$. From this it follows that for each $\gamma \in S_1$ one has $\inf_j \|f\gamma - f\gamma_j\| \le \frac{\varepsilon}{4} + \frac{2}{m}$. By assumption $\lim_F |\langle\varphi, f\gamma\rangle - \langle\mu, f\gamma\rangle| \le \frac{\varepsilon}{4} + \frac{2}{m}$ for each j. This implies

$$\limsup_F \sup_{\gamma\in S_1} |\langle\varphi, f\gamma\rangle - \langle\mu, f\gamma\rangle| \le \frac{\varepsilon}{4} + \frac{2}{m}$$

and finally

$$\limsup_F \sup_{\gamma\in S_1} |\langle\varphi, \gamma\rangle - \langle\mu, \gamma\rangle| \le \varepsilon + \frac{2}{m}$$

Since ε and m are arbitrary this implies

$$\lim F \sup_{\gamma\in S_1} |\langle\varphi, \gamma\rangle - \langle\mu, \gamma\rangle| = 0$$

hence the result.

FUNCTIONALS DENED ON A SUBLATTICE

The main result of the present section is Theorem stated below after a few preparatory lemmas. What it says in effect is that, if the structure V of the set (X, V) is metrizable, a bounded linear functional continuous for the structure u of precompact convergence on UEB sets of D is already u-continuous on the ball of D.

To prove it we start by examining properties of certain lattices of numerical functions on X. Here the algebraic and lattice operations refer to operations carried out *pointwise* on the functions. The basic object will be a set S of numerical functions dened on a set X and subject to the following assumptions

1. The constant functions $\gamma \equiv a$ belong to S for every $a \in [0, 1]$ and $\gamma \in S$ implies $0 \leq \gamma \leq 1$.
2. The set S is a lattice for the operations $\gamma_1 \wedge \gamma_2$ and $\gamma_1 \vee \gamma_2$ carried out pointwise.
3. The set S is convex.
4. If a is a number $a \in [-1, +1]$ and $\gamma \in S$ then $a + \gamma \in S$ if $0 \leq a + \gamma \leq 1$

   Similarly $a - \gamma \in S$ if $0 \leq a - \gamma \leq 1$.

To such a lattice S we shall attach on X a pseudo metric S or simply by the prescription.

$$\rho(x, y) = \sup\{|\gamma(x) - \gamma(y)|; \gamma \in S\}.$$

The norm $\|\gamma\|$ of a function will be the sup norm $\|\gamma\| = \sup_x |\gamma(x)|$.

Lemma: Let S satisfy conditions (1) to (4) and let ρ be the pseudo metric S defined by S. Let A and B be two subsets of X such that for some $m \geq 1$.

$$\inf\{\rho(x, y), x \in A, y \in B\} > \frac{1}{m}.$$

Let $\overline{S}$ be the closure of S in the space $F(X, R)$ of numerical functions on X for the topology of uniform convergence on the ρ-precompact subsets of X.

Then there is an element $f \in m\overline{S}$ such that $0 \leq f \leq 1$, $f(x) = 0$ for $x \in A$ and $f(y) = 1$ for $y \in B$.

Proof: Let $\{x_1, x_2, ..., x_n\}$ be a finite subset of A and let y be a specied element of B. Consider a pair (x_j, y). By denition of ρ there is some $\gamma \in S$ such that $|\gamma(x_j) - \gamma(y)| > \frac{1}{m}$. One can assume $0 \leq (x_j) < g(y) \leq 1$. If the reverse inequality held on would replace γ by $1 - \gamma$. Let $c_j = [\gamma(y) - \gamma(x_j)]^{-1}$. Define v_j by $c_j \{[(\gamma \vee a) \wedge b] - a\}$ where $a = \gamma(x_j)$ and $b = \gamma(y)$. This v_j is an element of mS. Now let $u = \min_j v_j$. Then $u \in mS$, $0 \leq u \leq 1$. Also $u(x_j) = 0$ for each x_j and $u(y) = 1$.

Fix the value of y and take the pointwise inmum of all such functions for all nite subsets $\{x_1, ..., x_n\}$ of A. Let it be g. Then g vanishes on A and $g(y) = 1$. Thus g also belongs to $m\overline{S}$ since in mS pointwise convergence implies precompact convergence. This procedure can be repeated for all $y \in B$. The pointwise supremum of all functions obtained in this way still belongs to $m\overline{S}$. It gives the desired function f.

The reader may have noticed the similarity of this proof with the usual proof of the Stone - Weierstrass theorem. The similarity will become even more apparent in our next result (which however will not be needed for our main arguments!)

Lemma: Let S satisfy (1) to (4) and let ρ be the pseudo-metric it defines. Then $\overline{S}$ consists exactly of those functions γ defined on X that satisfy $0 \leq \gamma \leq 1$ and $| \gamma(x) - (y) | \leq \rho(x, y)$ for all pairs (x, y) of elements of X.

Proof: Taking equivalence classes, if necessary, one may assume that ρ is a metric. One can also complete X for this metric getting a completion $\overline{X}$ where each precompact subset of X has a compact closure in $\overline{X}$. The function $\gamma \in S$ extend by continuity to all of $\overline{X}$. They still satisfy the Lipschitz condition $| \gamma(x) - \gamma(y) | \leq \rho(x, y)$ for all pairs (x, y) of elements of $\overline{X}$.

Now the standard argument in the proof of the Stone -Weirstrass theorem says that if K is a compact subset of $\overline{X}$ in order that a function f defined on K be approximable *uniformly on* K by elements of S it is necessary and sucient that for every pair (x, y) of elements of K and every $\varepsilon > 0$ there exist some element $\gamma_{x,y,\varepsilon}$ of S such that $| f(x) - \gamma_{x,y,\varepsilon}(x) | < \varepsilon$ and $| f(y) - \gamma_{x,y,\varepsilon}(y) | < \varepsilon$.

The argument showing this is the same as the argument carried above for Lemma.

Now suppose that f satises $| f(x) - f(y) | \leq \rho(x, y)$ for all pairs (x, y) of elements of X and that $0 \leq \rho \leq 1$. In order to prove that for each triplet (x, y, ε) with x and y in $K \subset \overline{X}$ one can obtain the two-point approximation described above, it is sucient to show that it can be obtained for pairs (x, y) of elements of X. This is true of course only if one extends f to $\overline{X}$ by continuity.

Now if x and y belong to X there is some $\gamma_1 \in S$ such that $| \gamma_1(x) - \gamma_1(y) | \geq \gamma(x, y) - \frac{\varepsilon}{4}$. Since f satisfies the desired Lipschitz condition and since S is convex, there is some $\gamma_2 \in S$ such that $|| \gamma_2(x) - \gamma_2(y) | - | f(x) - f(y) || < \frac{\varepsilon}{2}$. Suppose for instance $f(x) = a < b = f(y)$, and replacing γ_2 by $1 - \gamma_2$ if necessary suppose $\gamma_2(x) \leq \gamma_2(y)$. One can also suppose that $0 \leq [f(y) - f(x)] - [\gamma(y) - \gamma(x)] < \frac{\varepsilon}{4}$. Now let $\alpha = \gamma_2(x) \leq \gamma_2(y) = \beta$ and let $\gamma_3 = (\gamma_2 \vee \alpha) \wedge \beta - \alpha$. This is still in S. It is zero at x and equal to $\beta - \alpha$ at y. Also $0 \leq \gamma_3 \leq \beta - \alpha \leq f(y) - f(x)$. Let $\gamma_4 = a + \gamma_3$. Then $\gamma_4(x) = f(x)$ and $\gamma_4(y) = a + (\beta - \alpha) \leq f(y) \leq a + (\beta - \alpha) + \varepsilon$. Also $0 \leq \gamma_4 \leq 1$ hence $\gamma_4 \in S$ completing the proof of the result.

As already said above this lemma is not essential to the rest of our proof. It was included to give a clearer picture of what is happening. By contrast the following result will be very useful.

Lemma: Let f be a real valued function defined on X and such that $0 \le f \le 1$. Assume also that for a particular pair (k, m) of integers the inequality $\rho(x, y) \le \frac{1}{m}$ implies $|f(x) - f(y)| \le \frac{1}{k}$. Then there is a $\gamma \in m\bar{S}$ such that $0 \le g \le f \le g + \frac{2}{k}$.

Proof: Let $B_j = \{x: f(x) \le j/k\}$ for $j = 1, 2, \ldots, k$. For each $j < k$ consider the pair $(B_j, B^c_{j=1})$. If $x \in B_j$ and $y \in B^c_{j+1}$ then $\rho(x, y) > \frac{1}{m}$ since $f(y) - f(x) > \frac{1}{k}$. Thus, by Lemma there is a $u_j \in \bar{S}$ such that $0 \le u_j \le 1$ and such that $u_j(x) = 0$ for $x \in B_j$ and $u_j(y) = 1$ for $y \in B^c_{j+1}$. Let $u_0 = 0$ and let $g = \frac{1}{k}\sum_{j=0}^{k-1} u_j$. We claim that $0 \le g \le f \le g + \frac{2}{k}$. To see this consider a particular point x element of a set $B^c_i \cap B_{i+1}$. For every $j \ge i + 1$ one has $u_j(x) = 0$. For $0 < j \le i - 1$ one has $u_j(x) = 1$. Therefore if $i \le k - 1$

$$\sum_{j=0}^{k-1} u_j(x) \sum_{j=0}^{i} u_j(x) = (i-1) + u_i(x),$$

so that $\frac{(i-1)}{k} \le \frac{i}{k} \le f(x) \le \frac{i+1}{k}$. This gives the desired result.

Consider now a real valued function φ defined on the set S. We shall call such a function *linear* if $\varphi[\alpha u + \beta v) = \alpha\varphi(u) + \beta\varphi(v)$ for all systems (α, β, u, v) such that u and v and $\alpha u + \beta v$ are in S. It will be called *positive* if $u \in S$, $v \in S$ and $u \le v$ implies $\varphi(u) \le \varphi(v)$. Let L be the linear space spanned by $\bar{S}$. One can write L as $H - H$ where $H = \bigcup\{m\bar{S}; m \ge 1\}$ is the convex cone generated by $\bar{S}$.

Lemma: Let φ be a linear functional that is defined and positive on $\bar{S}$. Then it possesses a unique positive linear extension to the space D_ρ of all bounded uniformly continuous functions for the pseudo-metric ρ attached to S.

Proof: The functional φ extends to $H = \bigcup\{m\bar{S}; m \ge 1\}$ by writing $\varphi(f) = m\varphi\left(\frac{1}{m} f\right)$ if $f \in \bar{S}$. Because of the linearity assumption if $f \in n\bar{S}$ for $n > m$ then $m\varphi\left(\frac{1}{m} f\right) = n\varphi\left(\frac{1}{n} f\right)$ since $\frac{1}{n} f = \frac{m}{n}\left(\frac{1}{m} f\right)$ are all in $\bar{S}$. Thus the definition is consistent. Now suppose that $f \in D_\rho$ is such that $0 \le f \le 1$. According to Lemma for any $\varepsilon > 0$ there is a $g \in H$ such that $0 \le g \le f \le g + \varepsilon$. This implies that $\bar{\varphi}(f) = \inf\{\varphi(h), h \in H, h \ge f\}$ and $\underline{\varphi}(f) = \sup\{\varphi(g);$

$g \in H, g \leq f\}$ are equal. One can easily check that the extension of φ to H satises our "linearity" requirement. So does the extension $\overline{\varphi}$ to positive elements of D_ρ. The extension to $L = H - H$ is immediate. Hence the result.

Lemma does not say anything about the continuity of the exténsion of φ to D_ρ. This will be the subject of Theorem below. Before we state the theorem, let us note the following.

Lemma: A positive linear functional φ is u-uniformly continuous on S if and only if it is u-continuous at zero. The same applies to H, D_ρ^+ or D_ρ.

Proof: The result would be an immediate consequence of known results in functional analysis if we had assumed that S was symmetric. We have not assumed that, but it is strong enough to imply a sort of "symmetry". Take a positive φ and two elements f and g of S. Then $\frac{1}{2}[f + (1-g)] \in S$ and so does $\left[\frac{1}{2}(f-g)+\frac{1}{2}\right] \vee \left(\frac{1}{2}\right)$ and $\left[\frac{1}{2}(f-g)+\frac{1}{2}\right] \wedge \frac{1}{2}$.

Subtracting $1/\in$ one sees that $\frac{1}{2}(f-g)^+$ and $\frac{1}{2}(f-g)^-$ both belong to S. Therefore if $0 \leq \gamma \leq \alpha$ on a precompact set K implies $\langle\varphi,\gamma\rangle < \varepsilon/2$ for $\gamma \in S$, the inequality $|f-g| < \alpha/\in$ on K will imply $\langle\varphi, f-\gamma\rangle < \varepsilon$ for all pairs (f, g) of elements of S. The same applies to the other spaces listed: H, D_ρ^+ etc. Hence the statement.

Lemma: Let φ be positive on D_ρ in order that φ be u-continuous it is sucient that for each $\varepsilon > 0$ there be a precompact set K such that $\gamma \in D_\rho$, $0 \leq \gamma \leq 1$, $\gamma(x) = 0$ for $x \in K$ implies $\langle\varphi,\gamma\rangle < \varepsilon$.

Proof: Suppose $f \in D_\rho$ is such that $|f| < \varepsilon$ on K. Then $(f \vee \varepsilon) - \varepsilon$ vanishes on K. However $\langle\varphi, f\rangle \leq \langle\varphi,(f \vee \varepsilon)\rangle = \langle\varphi,(f \vee \varepsilon\rangle - \varepsilon\langle\varphi,1\rangle$. Hence the conclusion.

Theorem: Let ρ be the pseudo-metric attached to S and let D_ρ be the corresponding set of bounded uniformly continuous functions on X. On D_ρ, or subsets of it, let u be the structure of uniform convergence on the ρ-precompact subsets of X.

Every positive linear functional dened on S and u-continuous there admits a (unique) positive linear extension that is u-continuous on the balls of D_ρ.

Furthermore, let Φ be a bounded set of positive linear functionals defined on D_ρ. The following conditions are all equivalent:

(a) The restriction of Φ to S is u-equicontinuous at zero.

(b) The restriction of Φ to each UEB subset of D_ρ is u-equicontinuous

(c) The restriction of Φ to a ball $\{\gamma: \gamma \in D_\rho; \|\gamma\| \leq 1\}$ of D_ρ is u-equicontinuous.

Remark: For statements (b) and (c) one should really say u-uniformly equicontinuity. However, this is equivalent to equicontinuity at zero.

Proof: A φ that is u-continuous at zero is already uniformly continuous on S. Thus it has an extension by continuity to $\overline{S}$. Lemma says that this extension has a unique positive linear extension to D_ρ. Thus, below, we shall make no notational dierence between φ dened on S or on the whole of D_ρ.

It is obvious that $(c) \Rightarrow (b) \Rightarrow (a)$ and that this implies the continuity statement for an individual φ. Thus, it will sucient to show that $(a) \Rightarrow (c)$. To do this we shall use the following notation. For $u \in D_\rho$ the symbol $\varphi.\, u$ will denote the functional dened by $\langle \varphi.u, f\rangle = \langle \varphi, uf\rangle$ for all uf in the φ.

If (a) holds on S then it also hold on $\overline{S}$ for the continuous extensions of the elements of Φ. Thus we can assume Φ defined on $\overline{S}$ and u-equicontinuous.

Now choose an $\varepsilon > 0$ such that $\varepsilon \in (0, 1/2)$ and an integer m_1 such that $m_1\varepsilon > 1$. Then there is a precompact set F_1 such that $f \in m_1\overline{S}$, $0 \le f \le 1$ and $f = 0$ on F_1 implies $\langle \varphi, f\rangle < \frac{\varepsilon}{2}$ for all $\varphi \in \Phi$. Continuing sequentially, if $(m_1, F_1), (m_2, F_2) \ldots (m_{n-1}, F_{n-1})$ have been selected, take $m_n > m_{n-1}$ so that $m_n\varepsilon^n > 1$ and select a precompact set F_n so that $F_{n-1} \subset F_n$ and so that $f \in \overline{S}$, $0 \le f \le 1$, $f = 0$ on F_n implies $\langle \varphi, f\rangle < \varepsilon^n/2$ for all $\varphi \in \Phi$.

For any set F let $F(\varepsilon) = \{x: \rho(x, F) < \varepsilon\}$. Let us consider the set, $B_1 = F_1$, $B_2 = \{F_2 \cap [B_1(\varepsilon)]\} \cap B_1, \ldots$ and so forth so that

$$B_{j+1} = B_j \cup [B_j(\varepsilon^j) \cap F_{j+1}].$$

These are precompact sets and their union $K = \bigcup_j B_j$ is also precompact since every element of K is within $(1-\varepsilon)^{-1}\varepsilon^j$ of the precompact set B_j. For each integer j let u_j be a function $u_j \in m_j\overline{S}$ such that $0 \le u_j \le 1$ and such that $u_j(x) = 1$ for $x \in B_j$ but $u_j(x) = 0$ for $x \in [B_j(\varepsilon^j)]^c$. Let $v_j = u_1 \wedge u_2 \wedge \ldots \wedge u_j$.

Then $v_j \in m_j\overline{S}$. Consider the dierence $v_j - v_{j+1} = v_j - (v_j \wedge u_{j+1})$. Since $u_{j+1}(x) = 1$ for $x \in B_{j+1}$ the dierence $v_j - v_{j+1}$ vanishes on B_{j+1}. Also $v_j \le u_j$ vanishes on $[B_j(\varepsilon^j)]^c$. Thus $v_j - v_{j+1}$ vanishes on $B_{j+1} \cup [B_j(\varepsilon^j)]^c$. Since B_{j+1} contains $B_j(\varepsilon^j) \cap F_{j+1}$, this implies that $v_j - v_{j+1}$ vanishes on F_{j+1}. Both v_j and v_{j+1} belong to $m_{j+1}\overline{S}$. Thus arguing as in Lemma, one sees that

$$\langle \varphi, v_j - v_{j+1}\rangle < \varepsilon^{j+1}$$

Now consider the decreasing sequence $\varphi \geq \varphi.\, v_1 \geq \varphi.\, v_2 \geq \ldots \geq \varphi.\, v_j \geq \varphi.\, v_{j+1} \geq \ldots$ It has a limit in norm, say ψ, and this ψ is such that $\|\psi - \psi\| \leq \sum \varepsilon_j \leq (1-\varepsilon)^{-1}\varepsilon$.

Let f, $0 \leq f \leq 1$ be an element of $H = \bigcup_m (m\bar{S})$ such that $f(x) = 0$ for $x \in K$. However f vanishes on $B_n \subset K$. Thus, for $\varphi \in \Phi$ one has $\langle \psi, f \rangle < \varepsilon^n$. Since n is arbitrary this implies $\langle \psi, f \rangle = 0$. This entails that $\langle \varphi, f \rangle = \langle \psi, f \rangle + \langle \varphi, \psi \rangle f \leq 0 + \|\varphi - \psi\| < (1-\varepsilon)^{-1}\varepsilon$. The desired result follows then from Lemma at least for the balls of D_ρ^+. However that is enough to imply u-equicontinuity on the unit ball of D_ρ. This completes the proof.

Corollary: If the structure V of (X, V) is metrisable then the spaces M_p and M_u are the same.

Indeed, let ρ, $\rho \leq 1$ be a metric compatible with V. The set S of positive functions f such that $0 \leq f \leq 1$ and $|f(x) - f(y)| \leq \rho(x, y)$ satises all the properties (A1) to (A4).

Corollary: Let (X, V) be a uniform space. The space M_u is identical with the space of bounded functionals μ such that for every metric space Y and every uniformly continuous map f of X into Y the image $f\mu$ of μ belongs to M_p of Y.

Proof: One can decompose into its positive and negative parts. Thus it is enough to prove the result assuming that $\mu \geq 0$. By Theorem, $f\mu$ that belongs to M_u of Y must also be in M_p of Y. Conversely if S is a UEB of X, it denes a pseudo metric ρ. By passage to a quotient one obtains a metric space Y whose Lipschitz functions of coecient unity reproduce on X a set that contains S. Thus the condition is certainly sucient.

RELATIONS WITH RADON MEASURES

There are various extensions of the denition. For completely regular spaces and bounded measures. For more general topological spaces and unbounded measures. The natural definition in the context of the present paper is that of a *tight* linear functional. A linear functional φ dened on a convex symmetric set of bounded numerical functions Γ on a topological space X is called tight on Γ if for every $\varepsilon > 0$ there is a compact $K \subset X$ and a $\delta > 0$ such that $f \in \Gamma$, $|f| \leq 1$ and $|f(x)| < \delta$ for $x \in K$ implies $|\langle \varphi, f \rangle| < \varepsilon$.

Note that this denition refers only to the set Γ and the compact subsets of X. It does not say anything about the domain on which φ might be dened or extended.

Now take a set X with a separated uniform structure V and with set of bounded uniformly continuous functions D. A positive linear functional φ is tight on D for V if it is continuous at zero on $\{\gamma\colon \gamma \in D, |\gamma| \leq 1\}$ for the structure of uniform convergence on *compact* subsets of X. If (X, V) is complete this is the same as continuity for the structure U of uniform convergence on the *precompact* subsets of (X, V).

If φ is positive tight on D for V it admits an extension by the Bourbaki-MacShane procedure. One takes functions u that are pointwise supremum of subsets $S \subset D$ and write

$$\langle \varphi, u \rangle = \{\langle \varphi, \gamma \rangle ; \gamma \in D, \gamma \leq u\}$$

This gives an extension to lower semicontinuous functions that are bounded from below. Similarly if v is upper semicontinuous bounded above, one writes $\langle \varphi, u \rangle = \inf\{\langle \varphi, \gamma \rangle ; \gamma \in D, \gamma \geq v\}$. Finally if f is such that for every $\varepsilon > 0$ there is a lower semicontinuous u (bounded from below) and an upper semicontinuous v (bounded from above) such that $v \leq f \leq u$ and $\varphi(u) - \varphi(v) < \varepsilon$ one lets $\langle \varphi, f \rangle$ be the intersection of all the brackets $[\varphi(v), \varphi(u)]$ obtained in the procedure just described. This extends φ to a positive linear functional on a large space of numerical functions on X. We shall call it the Radon extension (or the Bourbaki extension) of the original φ. It is easily seen that the bounded functions in the Bourbaki-Radon extension of a φ tight on (D, V) can also be obtained as follows. One completes X for the smallest uniform structure that makes the elements of D continuous and extend D to the compact completion $\bar{X}$ where it becomes $C(\bar{X})$. Then φ is tight on (D, V) if and only if X is the Bourbaki-Radon domain of φ extended from $C(\bar{X})$ and if $\varphi(\bar{X} \setminus X) = 0$. Thus making a Radon extension from (X, D) or from $(\bar{X}, C(\bar{X} X))$ gives the same domain of extension as far as subsets of X are concerned. This can be summarized by saying that φ on (X, D) admits a Bourbaki-Radon extension if and only if it is tight on (X, D) for V.

There are other functionals that admit extension by the Bourbaki-MacShane procedure. All the positive linear functionals that are φ-smooth on (X, D) admit such extensions, but they are not necessarily such that $\varphi(\bar{X} \setminus X) = 0$, just such that each compact subset of $\bar{X} \setminus X$ has measure zero.

The word "tight" as originally used in Le Cam [1957] was meant to apply to a *set* of linear functionals. A set Φ of bounded linear functionals on D was called "tight" if it was uniformly bounded and uniformly continuous at zero on the unit ball of D for the structure of uniform convergence on the *compacts* of (X, V). Thus if X is complete for V a bounded set of linear functionals is tight on D if and only if it is u-equicontinuous on the unit ball of D. The word tight was applied to single linear functionals by abuse of language, regarding a single linear functional a the set that consists of that one functional.

Theorem: Let (X, V) be an arbitrary uniform space. Let Φ be a bounded set of positive linear functionals on D. Then Φ is equicontinuous for u on the unit ball of D if and only if for every uniformly continuous map f of (X, V) into a complete metric space (Y, ρ) the image of Φ by f is tight on (Y, ρ, D_ρ).

Theorem: Let (X_1, V_1) be a separated uniform space with a positive finite Radon measure μ. Let (X_2, V_2) be another uniform space with space of bounded uniformly continuous functions D_2. Let f be a map from X_1 to X_2. Assume that

1. For every $\gamma \in D_2$ the composed map $\gamma \circ f$ is in the Bourbaki-Radon domain of μ.
2. The cardinal of the image $f(X_1)$ is not measurable (no two valued nontrivial probability measure).

Then the image $f\mu$ of μ by f belongs to M_u on (X_2, V_2).

Indeed all the uniformly continuous images of $f\mu$ in metric spaces must belong to $M_u = M_p$ there.

It would be nice to have an extension of such a theorem to maps between linear functionals that are not necessarily obtained from a function f.

Suppose for instance that (X_i, V_i) are two uniform spaces with respective sets of bounded uniformly continuous functions D_i. Consider a map A from D_2 to bounded integrable functions for a Radon measure μ on (X_1, V_1). Suppose A is positive, such that $A_1 = 1$ and such that if a sequence $\{\gamma_n\}, \gamma_n \in D_2$ decreases pointwise to zero on X_2 the images $A\gamma_n$ do the same on X_1.

One might expect that such a map would be given by a Markov kernel mapping X_1 into elements of M_u of X_2. However, even if one lives in a universe where there are no measurable cardinals (or strongly inaccessible ones) this might not be the case.

Suppose for instance that there is a non atomic probability measure π defined on all the subsets of a discrete set X_2 with the cardinality of the continuum. Consider the map from X_1 to linear functional on D_2 that assigns to each $x \in X_1$ the same $\pi_x = \pi$ on D_2. A $\gamma \in D_2$ will be transformed to the constant $\int \gamma d\pi_x$, element of D_1. However $\pi = \pi_x$ is not in M_u of X_2, neither is $\int \pi_x \mu(dx)$ for any non zero positive Radon measure μ on X_1.

However it is possible to obtain an extension of Theorem, to certain Markov kernels. For instance one can prove the following.

Theorem: Let (X, V) be a separated uniform space with a positive finite Radon measure μ. Let (Y, V_2) be another uniform space and let $x \to \pi_x$ be a map from X to probability measures on Y that are elements of M_u on (Y, V_2).

Assume that for every $\gamma \in D(Y, V_2)$ the image $x \to \pi_x = \int \gamma(y)\pi_x(dy)$ is in the Radon-Bourbaki domain of λ. Assume also that the cardinal of Y does not admit nontrivial two valued probability measures.

Then the image of λ defined by $v = \int \pi_x \lambda(dx)$ (that is $(v, \gamma) = \int \left[\int \gamma(y)\pi_x(dy) \right] \lambda(dx)$ for $\gamma \in D(Y, V_2)$) is an element of M_u on (Y, V_2).

$$\Delta\Xi^-(s) = \Xi\frac{\rho^-(s+1-\varepsilon)}{\Xi^+(s+1-\varepsilon)},$$

$$\Delta\Xi^+(s) = -\Xi\frac{\rho^+(s+\varepsilon)}{\Xi^-(s+\varepsilon)},$$

with boundary conditions

$$\Xi^\pm(\mp\infty) = \Xi,$$

$$\Xi^\pm(\pm\infty) = 1,$$

wherewe have introduced the difference operator Δ, definedby $\Delta f(s) \equiv f(s+1) - f(s)$. This system is exactly solvable and is equivalent to the original recursion relations. In order to solve we have to combine the two equations of the system, taking into account the discrete Leibnitz rule $\Delta(fg)(s) = f(s+1)\Delta g(s) + g(s)\,\Delta f(s) = g(s+1)\,\Delta f(s) + f(s)\,\Delta g(s)$.

This yields

$$\Delta(\Xi^-\Xi^+)(s) = \Xi[\rho^-(s+1-\varepsilon) - \rho^+(s+\varepsilon)],$$

whose solution, taking into account the boundary conditions (2.17), is

$$\Xi^-(s)\Xi^+(s) = \Xi\left\{1 + \sum_{r=-\infty}^{s-1}[\rho^-(r+1-\varepsilon) - \rho^+(r+\varepsilon)]\right\},$$

or, using the definitions,

$$\Xi^-(s)\Xi^+(s) = \Xi\left(1 - \sum_\alpha \sum_{r=s-a_\alpha+1-\varepsilon}^{s+a_\alpha-1+\varepsilon} \rho_\alpha(r)\right).$$

This equation permits us to decouple the system and solve it for $\Xi^\pm(s)$. After a little algebra and index manipulation, the solution has the form

$$\frac{\Xi^-(s)}{\Xi} = [1 - n^{(0)}(s-\varepsilon)] \prod_{r=s-\varepsilon}^{\infty} \frac{1-n^{(1)}(r)}{1-n^{(0)}(r)},$$

$$\Xi^+(s) = \prod_{r=s}^{\infty} \frac{1-n^{(0)}(r)}{1-n^{(1)}(r)},$$

where we have introduced the weighted densities, $n^{(k)}(s)$ $(k = 0, 1)$, defined as

$$n^{(k)}(s) = \sum_\alpha \omega_\alpha^{(k)} * \rho_\alpha(s),$$

$$\omega_\alpha^{(k)}(s) = \begin{cases} 1 & \text{if } -a_\alpha - k - \varepsilon < s < a_\alpha, \\ 0 & \text{otherwise.} \end{cases}$$

We have basically solved our problem, we can write the grand potential, $\beta\,\Omega \equiv -\ln \Xi$, as

$$\beta\Omega = \sum_{s\varepsilon\mathbb{Z}} \ln\left(\frac{1-n^{(1)}(s)}{1-n^{(0)}(s)}\right).$$

Finally, the free-energy density functional, βF, defined as

$$\beta F[\rho_\alpha] = \beta\Omega[\rho_\varepsilon] + \sum_\alpha \sum_{s\varepsilon\mathbb{Z}} \rho_\alpha \ln \omega_\alpha(s),$$

takes, the form

$$\beta F[\rho_\alpha] = \sum_\alpha \sum_{s\varepsilon\mathbb{Z}} \rho_\alpha(s) \ln \rho_\alpha(s) + \beta\Omega[\rho_\alpha] - \sum_\alpha \sum_{s\varepsilon\mathbb{Z}} \rho_\alpha(s)$$

$$\times\left[\sum_{r=s-a_\alpha-\varepsilon}^{s+a_\alpha-1} \ln(1-n^{(1)}(r)) - \sum_{r=s-a_\alpha-\varepsilon+1}^{s+a_\alpha-1} \ln(1-n^{(0)}(r))\right]$$

For reasons that will be made clear later, we prefer to write this functional in the equivalent, but more convenient, form

$$\beta F[\rho_\alpha] = \beta F^{id}[\rho_\alpha] + \sum_{s\varepsilon\mathbb{Z}} [\Phi_0(n^{(1)}(s)) - \Phi_0(n^{(0)}(s))],$$

where $\beta F^{id} = \sum_\alpha \sum_{s\varepsilon\mathbb{Z}} \rho_\alpha(s)[\ln \rho_\alpha(s) - 1]$ refers to the ideal part of the free energy and $\Phi_0(\eta) \equiv \eta + (1-\eta)\ln(1-\eta)$ is the zero-dimensional excess free energy for a zero-dimensional cavity with mean occupancy.

NONADDITIVE MIXTURE

As we have mentioned above, the exact density functional for this special system can be derived as a particular case of the additive functional. Let us suppose that we want to study a system of hard rodswhose diameters are $\sigma_\alpha = 2a_\alpha + \varepsilon_\alpha$, where $a_\alpha \in \mathbb{N}$ and $\varepsilon_\alpha = 0, 1$. Because of the α-dependence in ε_α, the factorization cannot be applied. This difficulty can be overcome if we notice that this nonadditive mixture can be mapped to an additive mixture of hard rods whose diameters are $\tilde{\sigma}_\alpha = 2\sigma_\alpha$, while their positions are restricted to lie on the even sites of the lattice (i.e. $\tilde{\rho}_\alpha(s) = 0$ if $s = 2r + 1$ with $r \in \mathbb{Z}$; this amounts to setting an infinite external potential on the odd sites, while keeping it arbitrary on the even sites). Therefore, the density functional for the nonadditive system will arise by introducing in the density profile

$$\tilde{\rho}_\alpha(s) = \sum_{r \in \mathbb{Z}} \rho_\alpha(r)\delta_{s,2r},$$

where $\rho_\alpha(s)$ is the density profile corresponding to the original nonadditive system. This yields for the weighted densities

$$n_0^{(k)}(s) \equiv \tilde{n}^{(k)}(2s) = n_c^{(k)} + n_o^{(1)}(s-1)$$

$$n_1^{(k)}(s) \equiv \tilde{n}^{(k)}(2s+1) = n_c^{(k)} + n_o^{(k)}(s),$$

where $s \in \mathbb{Z}$, $k = 0, 1$ and the subindex $\{e, o\}$ states that the sum over species in the definition is restricted to those with even or odd diameter, respectively.

There are thus four instead of two weighted densities for this special case. An alternative way of expressing theseweighted densities is $n_j^{(k)}(s) = \sum_\alpha \omega_{j,\alpha}^{(k)} * \rho_\alpha(s)(k, j = 0,1)$, if the following nonadditive weights are introduced:

$$\omega_{j,\alpha}^{(k)}(s) \equiv \begin{cases} \omega_\alpha^{(j+k(1-j))}(s) & \text{if } \sigma_\alpha \text{ is even,} \\ \omega_\alpha^{(1-j+kj)}(s-1+j) & \text{if } \sigma_\alpha \text{ is odd.} \end{cases}$$

With these definitions, the excess free-energy density functional for the nonadditive mixture (the ideal part remains the same) adopts the form

$$\beta F^{ex}(\rho_\alpha) = \sum_{s \in \mathbb{Z}} [\Phi_0(n_1^{(1)}(s)) + \Phi_0(n_0^{(1)}(s) - \Phi_0(n_1^{(0)}(s)) - \Phi_0(n_0^{(0)}(s))].$$

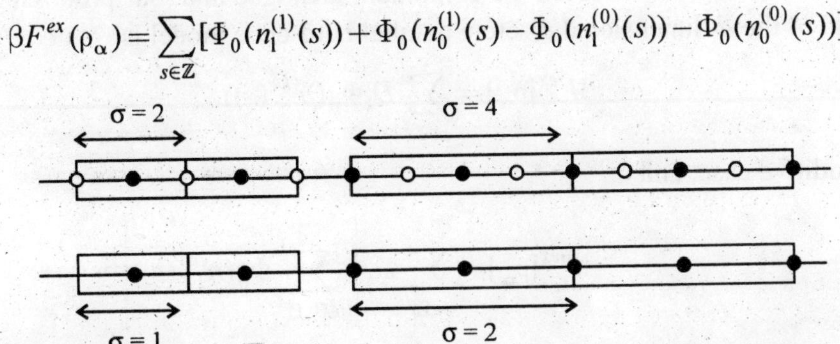

Fig. An Additive Mixture of hard rods of Diameters 2 and 4 can be made to have the same Configurations as a Nonadditive Mixture of rods with Diameters 1 and 2. The Positions of the upper System are Restricted to be on the black sites, while the white sites are Forbidden.

As far as we know, functionals have never been reported in the literature. The additive case is just an exercise of rewriting the original derivation by Vanderlick *et al* for the continuum, but this nonadditive case cannot be directly obtained with their method.

Besides, even the functional (which for the monocomponent case reduces to that obtained several times in the literature) has been written in a form that will make of the obtention of an approximate functional in higher dimensions an easy task.

FROM ZERO-DIMENSIONAL CAVITIES TO HIGHER DIMENSIONS

Cuesta and Mart´1nez-Rat´on proved that the one-dimensional excess free energy functional in the continuum model can be generated from the zero-dimensional one by just applying a differential operator with respect to the particle sizes. The form in which the functionals and have been written proves that this is also true for the discrete case, but now, due to the discrete nature of the system, the differential operator becomes a difference operator. This suggests rewriting as

$$\beta F^{ex}[\rho_\alpha]=\sum_{s\in\mathbb{Z}} D_k\Phi_0(n^{(k)}(s)),$$

$$\beta F^{ex}[\rho_\alpha]=\sum_{s\in\mathbb{Z}} D_k\left[\sum_{j=0,1}\Phi_0(n_j^{(k)}(s))\right],$$

respectively. Here D_k is the difference operator $D_k f(k)\equiv f(1)-f(0)$. Cuestaand Mart´1nez-Rat´on also proved that the FM functional for the d-dimensional system of parallel hard cubes followed the same simple recipe, i.e. it can be generated from the zero-dimensional functional by successive applications of a differential operator with respect to the particle size along every coordinate axis. The functional so obtained satisfies the zero-dimensional reduction requirement. For the equivalent lattice model it seems natural to propose a similar recipe. Then, taking into account the continuum analogue and, our proposal for the FM functional of the d-dimensional discrete parallel hard-cube model is

$$\beta F^{ex}[\rho_\alpha]=\sum_{s\in\mathbb{Z}^d} D_k\Phi_0(n^{(k)}(s))$$

for the additive case, and

$$\beta F^{ex}[\rho_\alpha]=\sum_{s\in\mathbb{Z}^d} D_k\left[\sum_{j\in(0,1)^d}\Phi_0(n_j^{(k)}(s))\right]$$

for the nonadditive one, where $k\equiv(k_1,\ldots,k_d)$, $j\equiv(j_1,\ldots,j_d)$ are vector indices, $D_k\equiv\prod_{i=1}^d D_{k_i}$ is the difference operator and the weighted densities are defined as $n^{(k)}(s)\equiv\sum_\alpha \omega_\alpha^{(k)}*\rho_\alpha(s)$ with $\omega_\alpha^{(k)}(s)\equiv\prod_{i=1}^d\omega_\alpha^{(k_i)}(s_i)$ for the additive system, and as $n_j^{(k)}(s)\equiv\sum_\alpha\omega_{j,\alpha}^{(k)}*\rho_\alpha(s)$ with $\omega_{j,\alpha}^{(k)}(s)\equiv\prod_{i=2}^d\omega_{j_i,\alpha}^{(k_i)}(s_i)$ for the nonadditive one.

Functionals are the main results of this work since they constitute, to the best of our knowledge, the first FM functionals for a lattice model. In this sense, it can be checked that they depend on the density profile through a set of weighted densities defined from one-particlemeasures which are compatible with the decomposition of the Mayer function into a sumof their convolutions, the virial expansion of the direct correlation function is exact up to first order in the density and they consistently reduce to any lower dimension down to $d = 0$.

Also, it must be noticed that their continuum counterpart is recovered in the continuum limit of vanishing lattice spacing $\delta \to \infty$ and infinite size $\sigma_\alpha \to \infty$ with $\sigma_\alpha \delta =$ const. (notice that the additive–nonadditive differentiation becomes immaterial in this limit). Points (i) and (ii) are easy to check, so we shall focus in what follows in the crucial point (iii) and its connection with an alternative way of obtaining functionals.

As mentioned in the introduction, there is an alternative method to Rosenfeld's original one to construct a FM functional. This method will be referred to as the *zero-dimensional cavity method* because it is based on the exact dimensional reduction of the functional to $d = 0$. Pursuing the same idea but having in mind the form of functionals, we have devised a similar method suitable for lattice models. The method amounts to :

(i) Directly extending the exact zero-dimensional functional to higher dimensions (note here a difference from the continuumcase , where it is the one-dimensional functional that is extended to higher dimensions)

(ii) Adding the appropriate extra terms in order for the functional to recover the exact form when applied to zero-dimensional cavities.

To illustrate this procedure we must first recall that we are dealing with a mixture, sometimes nonadditive, and, in contrast to themonocomponent case, it is not very clear what a zero-dimensional cavity means for a mixture. Let us therefore start by defining what we shall understand by this.

In the monocomponent case, a zero-dimensional cavity is simply a set of connected points in the d-dimensional lattice such that if a particle (its centre ofmass) is placed at one of them no other particle can be placed at any other point of this set. For one-dimensional hard rods of length $2a$, the largest zero-dimensional cavity ('maximal' cavity) is the interval $[-a + 1, a]$, whereas for hard rods of length $2a + 1$ it is $[-a + 1, a + 1]$. In d dimensions the maximal cavity is the set $[-a + 1, a]^d$ or $[-a + 1, a + 1]^d$, depending on the length parity.

For a mixture, a zero-dimensional cavity can be defined as a collection of sets, one for each species, such that if a particle of any species occupies one of the points of its corresponding set, no other particle of the same or different species can be placed at any point of its corresponding set. In $d = 1$ and in the additive case, this means the collection of sets $\{[-a_\alpha + 1, a_\alpha]\}_\alpha$, if $2a_\alpha$ is the rod length of species α, or $\{[-a_\alpha + 1, a_\alpha + 1]\}_\alpha$, if this rod length is $2a_\alpha + 1$; in the nonadditive case, however, indexing by α the species of even length and by α' those of odd length, there are two maximal cavities: $\{[-a_\alpha + 1, a_\alpha], [-a_\alpha, a_\alpha]\}_{\alpha\alpha'}$ and $\{[-a_\alpha + 1, a_\alpha], [-a_\alpha + 1, a_\alpha + 1]\}_{\alpha,\alpha'}$ S.

In d dimensions the cavities for the additive case generalize as in the monocomponent case, while in the nonadditive case there will appear $2d$ different cavities, depending on which of the two choices above we select for each dimension. Because of this proliferation of maximal cavities in the nonadditive case, it is convenient to discuss the two types of mixture separately.

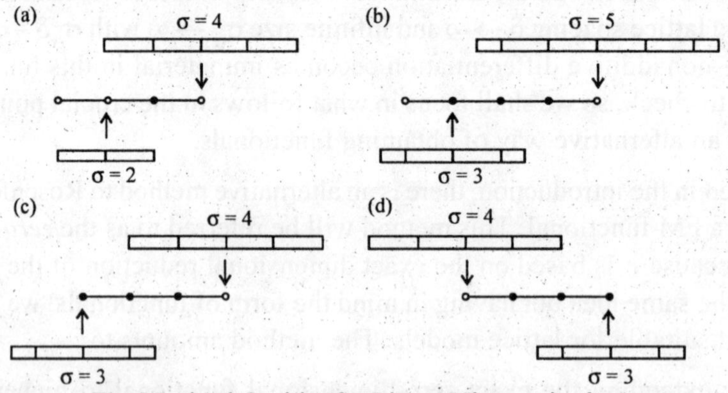

Fig. Maximal Cavities for Binary Mixtures in $d = 1$.

ADDITIVE MIXTURE

For density profiles constrained to be zero outside the points of a maximal zero-dimensional cavity it is easy to check that the weighted densities satisfy the relationships

$$n^{(1)}(s) = n^{(0)}(s+1) \quad (s < 0),$$
$$n^{(1)}(s) = n^{(0)}(s) \quad (s > 0).$$

These are the keystone of the method. As explained, the method starts off from the straightforward extension of the zero-dimensional excess functional to $d = 1$, $\Phi_0^{(1d)}$. As $n^{(1)}(0) = \eta$, the occupancy probability of the cavity,

$$\Phi_0^{(1d)}[\rho] = \sum_{s \in \mathbb{Z}} \Phi_0(n^{(1)}(s)).$$

Now we apply this functional to a zero-dimensional cavity and, by means of (3.4), rewrite it as

$$\Phi_0^{(1d)}[\rho] = \Phi_0(n^{(1)}(0) + \sum_{s<0} \Phi_0(n^{(0)}(s+1)) + \sum_{s>0} \Phi_0(n^{(0)}(s))$$
$$= \Phi_0(n^{(1)}(0)) + \sum_{s \in \mathbb{Z}} \Phi_0(n^{(0)}(s)).$$

However, $\Phi_0(n^{(1)}(0))$ on the rhs of is the exact zero-dimensional excess free energy for the cavity, so the method dictates that in order to obtain the excess functional for the onedimensional system we must subtract from the form we started with all the spurious terms.

This leads to

$$\beta F^{ex}[\rho] = \sum_{s\in\mathbb{Z}} \Phi_0(n^{(1)}(s)) - \sum_{s\in\mathbb{Z}} \Phi_0(n^{(0)}(s)).$$

Let us now illustrate how the method works for higher dimensions as well. For the sake of simplicity we shall restrict ourselves to $d = 2$.

(a) $W_1^{(1)}$ $W_0^{(1)}$ $W_1^{(0)}$ $W_0^{(0)}$

(b) $= n_1^{(1)}(0)$ $= n_1^{(1)}(0)$ $= n_1^{(1)}(0)$ (c) $= n_1^{(1)}(0)$ $= n_0^{(1)}(0)$ $= n_1^{(0)}(0)$

Fig. Diagrammatic representation of the weights corresponding to a binarymixture of rods of lengths $\sigma_1 = 3$ and $\sigma_2 = 2$.

The corresponding relationships between the weighted densities of density profiles vanishing outside a zero-dimensional cavity are

$$n^{(k,1)}(s_1, s_2) = n^{(k,0)}(s_1, s_2 + 1) \ (s_2 < 0),$$
$$n^{(k,1)}(s_1, s_2) = n^{(k,0)}(s_1, s_2) \ (s_2 > 0),$$

with $k = 0$, 1, and similar ones if the coordinate axes are interchanged. Using the same notation as for the one-dimensional case and repeating the steps for s_2 we obtain

$$\Phi_0^{(2d)}[\rho] = \sum_{s\in\mathbb{Z}^2} (n^{(1,1)}(s)) = \sum_{s_1\in\mathbb{Z}} \Phi_0(n^{(1,1)}(s_1, 0) + \sum_{s\in\mathbb{Z}^2} \Phi_0(n^{(1,0)}(s))$$

Repeating the trick with $\sum_{s\in\mathbb{Z}^2} \Phi_0(n^{(0,1)}(s))$ we obtain

$$0 = \sum_{s\in\mathbb{Z}^2} \Phi_0(n^{(0,1)}(s)) - \sum_{s_1\in\mathbb{Z}} \Phi_0(n^{(0,1)}(s_1, 0)) - \sum_{s\in\mathbb{Z}^2} \Phi_0(n^{(0,0)}(s))$$

Adding the equations we get

$$\sum_{s_1\in\mathbb{Z}} \Phi_0(n^{(1,1)}(s_1, 0)) - \sum_{s_1\in\mathbb{Z}} \Phi_0(n^{(0,1)}(s_1, 0)) = \Phi_0(n^{(1,1)}(0))$$

yields

$$\Phi_0^{(2d)}[\rho] = \Phi_0(n^{(1,1)}(0) + \sum_{s\in\mathbb{Z}^2} [\Phi_0(n^{(1,0)}(s)) - \Phi_0(n^{(0,0)}(s))].$$

Again, $\Phi_0(n^{(1,1)}(0)$ is the exact zero-dimensional excess functional, therefore

$$\beta F^{ex}[\rho] = \sum_{s \in \mathbb{Z}^2} [\Phi_0(n^{(1,1)}(s)) - \Phi_0(n^{(1,0)}(s)) - \Phi_0(n^{(0,1)}(s)) + \Phi_0(n^{(0,0)})(s))]$$

which coincides for $d = 2$. This provides the rationale for for arbitrary dimension, and proves in passing that functional has a consistent dimensional reduction to any lower dimension.

NONADDITIVE MIXTURE

In the additive case it is always theweighted density $n^{(1)}(0)$ that gives the occupancy probability, ·, of the (unique)maximal zero-dimensional cavity, so extending the zero-dimensional excess functional is straightforward as equation shows. In the nonadditive case, however, this is not as simple. Due to the degeneracy of maximal cavities, which weighted density yields the occupancy probability depends on which cavity we choose.

Before proceeding with the construction of the functional we therefore need a unique expression in terms of the onedimensional weighted densities that provides the excess free energy of *any* maximal cavity. One such expression is

$$\Phi_0^{(0d)}[\rho] = \Phi_0(n_1^{(1)}(0)) + \Phi_0(n_0^{(1)}(0)) - \Phi_0(n_1^{(0)}(0))$$

In order to make it easy to understand why this is so and the discussions to come, we have introduced a diagrammatic notation for a particular case. We shall consider a binary mixture of rods of lengths $\sigma_1 = 3$ and $\sigma_2 = 2$, which is prototypical of this case. For thismixture we represent the weights as a chain of empty and shaded circles. The effect of convoluting a weight with a density profile is to select the density at those sites overlapped by the empty circles for the large rods and by the shaded circles for the small rods, and add them up.

The result of this operation is the correspondingweighted density. The application of these weights to each maximal cavity is depicted respectively. Figure illustrates that $n_0^{(1)}(0) = \eta$ and $n_1^{(1)}(0) = n_1^{(0)}(0)$, so the equation reduces to Φ_0, as expected. On the other hand, figure illustrates that $n_1^{(1)}(0) = \eta$ and $n_0^{(1)}(0) = n_1^{(0)}(0)$, so again reduces to $\Phi_0(\eta)$.

We can now carry on with the method as in the additive case and propose

$$\Phi_0^{(1d)}[\rho] = \sum_{s \in \mathbb{Z}} [\Phi_0(n_1^{(1)}(s)) + \Phi_0(n_0^{(1)}(s)) - \Phi_0(n_1^{(0)}(s))]$$

Whichever the maximal cavity we choose, from definitions and the relationships the following analogous relationships between the nonadditive weighted densities hold:

$$n_0^{(1)}(s) = n_1^{(0)}(s), \qquad n_1^{(1)}(s) = n_0^{(0)}(s+1) \quad (s < 0)$$

$$n_0^{(1)}(s) = n_0^{(0)}(s), \qquad n_1^{(1)}(s) = n_1^{(0)}(s) \qquad (s > 0)$$

Proceeding as we did for the additive case, we use the above relationships to rewrite as

$$\Phi_0^{(1d)}[\rho] = \Phi_0^{(0d)} + \sum_{s<0} \Phi_0(n_0^{(0)}(s+1)) + \sum_{s>0} \Phi_0(n_0^{(0)}(s)) = \Phi_0^{(0d)}[\rho] + \sum_{s\in\mathbb{Z}} \Phi_0(n_0^{(0)}(s)),$$

where we have made use of, which gives the zero-dimensional excess functional; so the method gives

$$BF^{ex}[\rho_\alpha] = \sum_{s\in\mathbb{Z}} [\Phi_0(n_1^{(1)}(s)) + \Phi_0(n_0^{(1)}(s)) - \Phi_0(n_1^{(0)}(s)) - \Phi_0(n_0^{(0)}(s))],$$

In order to extend the cavity method to dimensions higher than one, we have to generalize expression. Due to the cubic symmetry of our system this can be done just by applying the same rule on each coordinate axis. For the sake of simplicity, we can define the operator T_k^j as $T_k^j\ f(k, j) \equiv f(1,1) + f(1,0)$ and its d-dimentional version as $T_k^j \equiv \prod_{i=1}^d T_{k_i}^{j_i}$.

Then the natural generalization of is simply

$$\Phi_0^{(0d)}[\rho] = T_k^j \Phi_0(n_j^{(k)}(0)),$$

and the extension of the zero-dimensional excess functional to d dimensions

$$\Phi_0^{(d)}[\rho] = \sum_{s\in\mathbb{Z}^d} T_k^j \Phi_0(n_j^{(k)}(s))$$

Again, as a representative of a higher dimension, we shall apply the cavity method for $d = 2$. The diagrammatic representation of the weights for a binary mixture of squares of edge lengths $\sigma_1 = 3$ and $\sigma_2 = 2$, as well as all maximal cavities for this system.

Expression for $d = 2$ accounts for all cavities in figure. For example, if we take the cavity at the top the application of the weights in figure gives $n_{0,0}^{(1,1)}(0) = \eta$, $n_{1,1}^{(1,1)}(0) = n_{1,1}^{(0,0)}(0), n_{1,0}^{(1,1)}(0) = n_{1,0}^{(0,1)}(0)$, $n_{0,1}^{(1,1)}(0) = n_{0,1}^{(1,0)}(0)$ and $n_{1,1}^{(1,0)}(0) = n_{1,1}^{(0,1)}(0)$, so reduces to $\Phi_0(\eta)$. By symmetry the same happens for any of the cavities.

Proceeding with the method, from we shall start with

$$\Phi_0^{(2d)}[\rho] = \sum_{s\in\mathbb{Z}^2} [\Phi_0(n_{1,1}^{(1,1)}(s)) + \Phi_0(n_{1,0}^{(1,1)}(s)) - \Phi_0(n_{1,1}^{(1,0)}(s)) + \Phi_0 n_{0,1}^{(1,1)}(s)) + \Phi_0(n_{0,0}^{(1,1)}(s))$$

$$-\Phi_0(n_{0,1}^{(1,0)}(s)) - \Phi_0(n_{1,1}^{0,1}(s)) - \Phi_0(n_{1,0}^{(0,1)}(s)) - \Phi_0(n_{1,1}^{(0,0)}(s))]$$

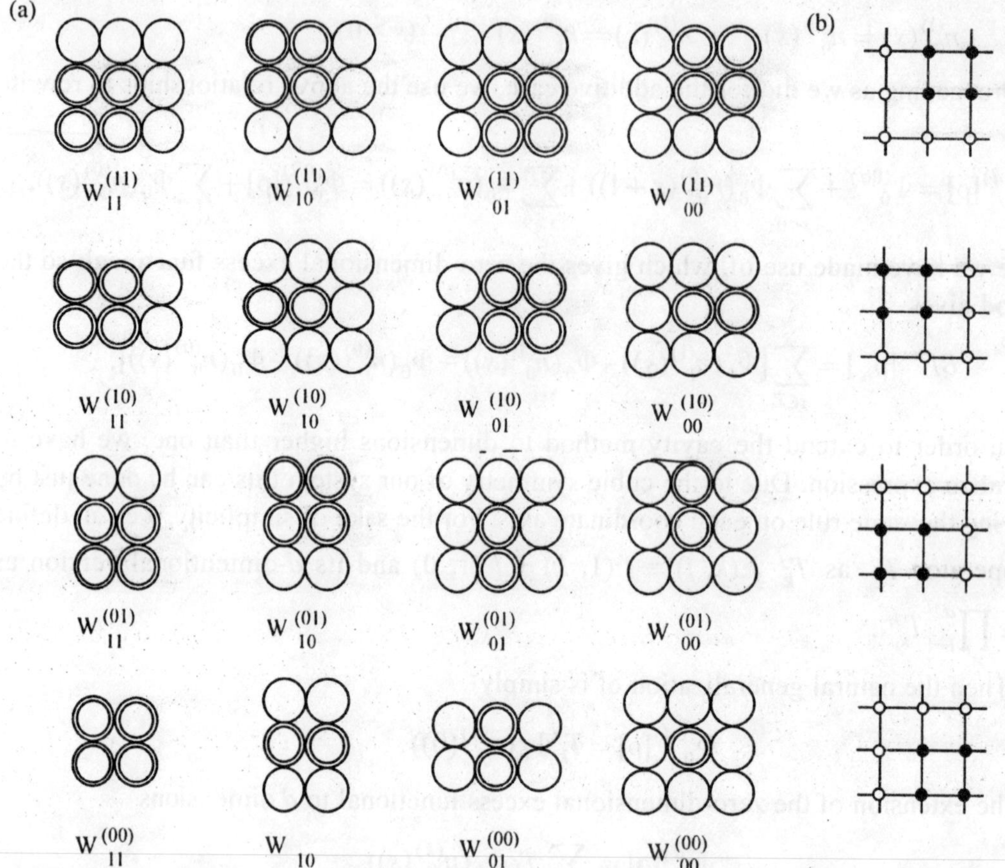

Fig. (a) Diagrammatic Representation of the Weights Corresponding to a Binary Mixture of Squares of Edge Lengths $\sigma_1 = 3$ and $\sigma_2 = 2$. (b) Maximal Cavities for this System.

Now, the relationships become

$$n_{j,0}^{(k,1)}(s_1,s_2) = n_{j,1}^{(k,0)}(s_1,s_2), \qquad n_{j,1}^{(k,1)}(s_1,s_2) = n_{j,0}^{(k,0)}(s_1,s_2+1) \qquad (s_2+0),$$

$$n_{j,0}^{(k,1)}(s_1,s_2) = n_{j,0}^{(k,0)}(s_1,s_2), \qquad n_{j,1}^{(k,1)}(s_1,s_2) = n_{j,1}^{(k,0)}(s_1,s_2) \qquad (s_2>0),$$

and those obtained by interchanging the coordinate axes. As before, by making use of these relationships in we obtain

$$\Phi_0^{(2d)}[\rho] = \sum_{s_1\in\mathbb{Z}} T_k^j \Phi_0(n_j^{(k)}(s_1,0)) + \sum_{s\in\mathbb{Z}^2} [\Phi_0(n_{1,0}^{(1,0)}(s)) + \Phi_0 n_{0,0}^{(1,0)}(s)) - \Phi_0(n_{1,0}^{(0,0)}(s))],$$

where itmust be noticed that the second termin the rhs of is just $\sum_{s\in\mathbb{Z}^2} T_k^j \Phi_0(n_{j,0}^{(k,0)}(s))$.

$$0 = \sum_{s\in\mathbb{Z}^2} T_k^j \Phi_0(n_{0,j}^{(0,k)}(s)) - \sum_{s_1\in\mathbb{Z}} T_k^j \Phi_0(n_{0,j}^{(0,k)}(s_1,0)) - \sum_{s\in\mathbb{Z}} \Phi_0(n_{0,0}^{(0,0)}(s)).$$

By the same argument, we have.

Adding the equations we get the zero-dimensional reduction of the one-dimensional case proved above

$$\sum_{s_1\in\mathbb{Z}}[T_k^j\Phi_0(n_j^{(k)}(s_1,0))-T_k^j\Phi_0(n_{0,j}^{(0,k)}(s_1,0))]=\Phi_0^{(0d)}[\rho],$$

we obtain

$$\Phi_0^{(2d)}[\rho]=\Phi_0^{(0,d)}[\rho]+\sum_{s\in\mathbb{Z}^2}[\Phi_0(n_{1,0}^{(1,0)}(s))+\Phi_0(n_{0,0}^{(0,0)}(s))]$$

$$+\sum_{s\in\mathbb{Z}^2}[\Phi_0(n_{0,1}^{(0,1)}(s))+\Phi_0(n_{0,0}^{(0,1)}(s))]$$

$$-\sum_{s\in\mathbb{Z}^2}[\Phi_0(n_{1,0}^{(0,0)}(s))+\Phi_0(n_{0,1}^{(0,0)}(s))]+\Phi_0(n_{0,0}^{(0,0)}(s))].$$

Thus, the cavity method produces

$$\beta F^{ex}[\rho]=\Phi_0^{(2d)}(\rho)-\sum_{s\in\mathbb{Z}^2}[\Phi_0(n_{1,0}^{(1,0)}(s))+\Phi_0(n_{0,0}^{(1,0)}(s))]$$

$$-\sum_{s\in\mathbb{Z}^2}[\Phi_0(n_{0,1}^{(0,1)}(s))+\Phi_0(n_{0,0}^{(0,1)}(s))]$$

$$+\sum_{s\in\mathbb{Z}^2}[\Phi_0(n_{1,0}^{(0,0)}(s))+\Phi_0(n_{0,1}^{(0,0)}(s))+\Phi_0(n_{0,0}^{(0,0)}(s))],$$

which is for $d = 2$.

The extension of this argument to arbitrary dimension proves that is the functional produced by the cavity method. In addition, its consistent dimensional reduction to any lower dimension is also proved.

APPLICATIONS

In order to show the applicability of the theory exposed in the previous sections, we present the results obtained for some particular two- and three-dimensional systems.

HARD-SQUARE-LATTICE GAS

The hard-square-lattice gas is just the lattice gas with first- and second-nearest-neighbour exclusion in the simple square lattice. This system has been widely studied but no definitive conclusion has been reached about its phase behaviour. The results reported cover thewhole spectrum, depending onwhich theory has been used to study it. Some authors have claimed it to have a second-order transition, others a weaker (third-order) transition and others even no transition at all. Nowadays, the only thing that seems clear is the structure of the ordered phase: periodic along one coordinate axis while uniform along the other (i.e. columnar). However, no simulations have ever been performed in order to clarify the disputed order of the transition.

The system is found to have a second-order transition from a fluid phase to a columnar phase. The value of the packing fraction at the transition is $\eta_c = 3-\sqrt{5} = 0.764$ and the chemical potential at this point $(\beta_\mu)_c = 2.41$. This result is compatible with that obtained by Bellemans and Nigam using the method of Rushbrooke and Scoins.

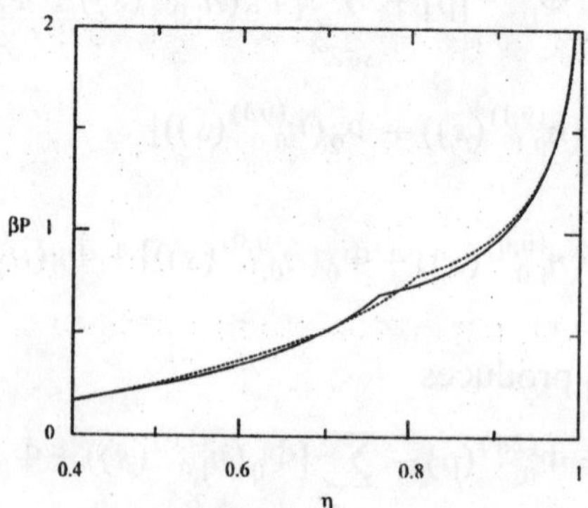

Fig. Equation of State from FM Theory (Solid Curve) and from the Method of Rushbrooke and Scoins.

They also found a second-order transition, but at $\eta_c = 0.807$ and $(\beta\mu)_c = 2.85$. The equation of state they obtained with this method is compared with the FM result in figure. There is good agreement between the two theories; only in the critical region do the two curves deviate from each other.

It is worth mentioning that within the FM approximation this system is analytically solvable. Omitting the calculations and denoting $L(x) \equiv x \ln x$, we have obtained for the Helmholtz free-energy density of the fluid branch

$$\beta\Phi(\eta) = L(\eta/4) + L(1-\eta/4) - 2L(1-\eta/2) + L(1-\eta),$$

and for the columnar branch

$$\beta\Phi(\eta) = \frac{1}{2}[L(\eta_1/2) + L(1-\eta_1/2) + L(1-\eta/2+\eta_1/2)$$

$$-2L(1-\eta/2) - L(1-\eta+\eta_1) + 2L(1-\eta)],$$

where the occupancy probability of a sublattice is given by

$$\eta_1(\eta) = \frac{\eta}{2} + \sqrt{\frac{\eta^3 - 8\eta^2 + 16\eta - 8}{4\eta}}$$

Clearly more work, either by means of simulations or by other methods, is needed in order to resolve the nature of the transition. This notwithstanding, we do not expect good agreement except at low or high pressures; critical points are difficult (sometimes impossible) to capture, even approximately, with mean-field-like theories such as this one. Nevertheless, we believe the FM approach to be valuable in the sense that it is possible to obtain an analytic approximation, the result obtained is compatible with one previously reported using a different theory and it correctly predicts the structure of the ordered phase.

MONOCOMPONENT AND BICOMPONENT HARD-CUBE-LATTICE GAS

We have also applied the theory to three-dimensional systems. In this section we present the results obtained for the monocomponent systems of parallel hard cubes of edge lengths $\sigma = 2$ and 6, as well as for their binary mixture. These being three-dimensional systems, there are three ordered phases that compete in the phase diagrams: smectic (one-dimensional order), columnar (two-dimensional order) and solid (three-dimensional order).

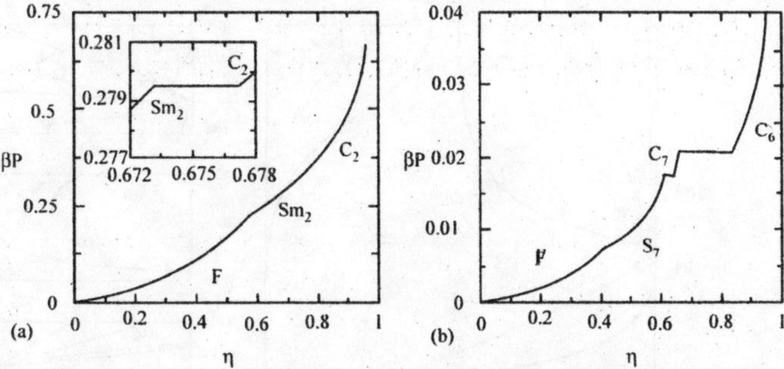

Fig. Equation of state Obtained from FM Theory for the Monocomponent Hard-cube-Lattice gas.

The period of the ordering is determined from the wavevector at which the first divergence of the structure factor occurs. In the first case this is π, hence the period is two. As this is also the period of the closest-packed phase it is the only one that is found in stable phases. In the second case the wavevector is $2\pi / 7$, implying a period of seven; however, in this case the closest-packed phase has period six, so both period-six and period-seven phases must be considered. The same holds for the binary mixture.

The results for the monocomponent systems are shown in figures. The first system undergoes, at $\eta_c = 0.568$, a second-order transition from a fluid phase (F) to a period-two smectic one (Sm_2). At a higher value of the packing fraction the system is found to have a very narrow first-order phase transition, where the smectic phase at $\eta_{Sm} = 0.673$ coexists with a period-two columnar phase (C_2) at $\eta_C = 0.677$. For the $\sigma = 6$ system we find a more complex scenario with several ordered phases, due to the proximity of their free energies.

There are six ordered phases in competition: smectic, columnar and solid with periods both six and seven. Only three of them, the two columnar and the period-seven solid, are stable in some region. The other three are just metastable. Again, the columnar phase is the most stable at high density (notice the difference of the high-density ordered phase of these lattice systems with respect to that of the continuum model. Upon increasing density, the sequence of transitions is as follows: first, a second-order one at $\eta_c = 0.402$ from a fluid phase (F) to a period-seven solid phase (S_7); then a first-order transition in which the S_7 solid coexists, at $\eta_{S_7} = 0.617$, with a period-seven columnar phase (C_7), at $\eta C_7 = 0.631$; finally, another first-order transition with the C_7 columnar at $\sigma C_7 = 0.656$ coexisting with a period-six columnar phase (C_6) at $\eta C_6 = 0.827$. This latter phase extends up to the closest packing.

The competition of different ordered phases in the latter system is remarkable, but it produces a real mess when the two kinds of cube are mixed. The phase diagram for this binary mixture (pressure versus composition). All possible phases are present there. A few particularly complex regions are shown in more detail. Remarkable features of this phase diagram are the very wide $F - C_6$ phase separation dominating the lower part of the diagram.

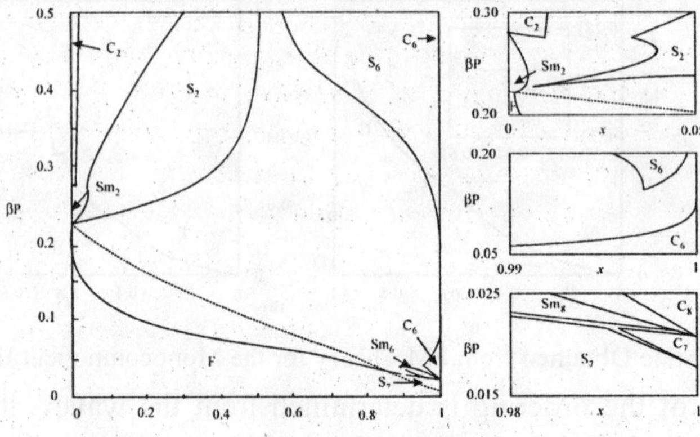

Fig. Phase Diagram (Reduced pressure, βP, Versus Composition, x) of the Lattice Binary Mixture of Hard Cubes.

The extraordinarily narrow $S_6 - C_6$ coexistence, extremely close to the monocomponent fluid of large cubes. The latter tells us about the proximity of the free energy of the solid phase to that of the columnar in this monocomponent system (a small amount of 'impurities' breaks the translational symmetry along the columns of the columnar phase, transforming it into a solid). The former has a very important consequence concerning the demixing scenario for this system.

This binary mixture has been reported, from simulations, to demix into two fluid phases. As a matter of fact, this is, as far as we know, the only existing evidence of an entropy-driven fluid–fluid demixing in an additive binary mixture.

In all other systems of the same kind, e.g. hard spheres or parallel hard cubes in the continuum, fluid–fluid demixing is preempted by freezing of (at least) the largest component. Our results contradict this simulations and fit in what seems to be the standard scenario of demixing for additive binary mixtures, namely fluid-ordered phase demixing (the fact that the ordered phase is a columnar is just a peculiarity of the model). A more detailed analysis of this result can be found in Lafuente and Cuesta.

In this article we have proposed a formulation of FM theory for lattice models. As mentioned in the introduction, it is not possible just to carry on with Rosenfeld's original formulation but in discrete space.

This is so because of the lack of a scaled-particle theory for lattice models. In fact, FM theory in the continuum can be regarded as a generalization of scaled-particle theory to inhomogeneous phases. For this reason, we believe that the knowledge of a LFMT could guide the derivation of a lattice scaled-particle theory. We are currently working in this direction.

In the recent reformulations of FM theory the scaled-particle equation of state is no longer an input but an output of the theory. Two fundamental ideas appear as the keystone of these formulations: the exact density functional must have a consistent dimensional reduction to any lower dimension and the exact zero-dimensional limit is necessary for the correct description of the freezing transition.

The latter turns out to be a very stringent constraint for the functional and seems to contain most of the information needed for the construction of a higher-dimensional functional. In this context, the zero-dimensional cavity method appears as a scheme for the construction of FM functionals.

The LFMT we have proposed is the generalization of this method suggested by the exact form of the functional of the one-dimensional mixture of hard rods on a lattice. Although this system has been previously studied in the literature, the originality of our derivation is that this is the first time that the distinction between additive and nonadditive mixtures have been noticed and resolved and the exact functional is written in a form in which the zero-dimensional functional arises naturally.

The cavity method can be extended to other particle shapes and other lattice structures. This will be the content of a future work.

We have applied the functional to two-and three-dimensional models and found very reasonable results.

In two dimensions these are comparable to those previously obtained by other means, and in three dimensions they agree with previous simulation data, in the case of the mixture, although with an utterly different interpretation, which has yet to be confirmed. One remarkable feature of all three-dimensional models is the very rich and complex phase diagrams they possess (specially the mixture), a complexity which the present theory has nevertheless allowed us to tackle without problems.

Encyclopaedia of

MEASURE THEORY

Encyclopaedia of

MEASURE THEORY

Volume 2

R.K. Pandey

ANMOL PUBLICATIONS PVT. LTD.

NEW DELHI - 110 002 (INDIA)

ANMOL PUBLICATIONS PVT. LTD.

H.O.: 4374/4B, Ansari Road, Daryaganj,
New Delhi-110 002 (India)
Ph.: 23278000, 23261597

B.O.: No. 1015, Ist Main Road, BSK IIIrd Stage
IIIrd Phase, IIIrd Block,
Bangalore - 560 085 (India)
Visit us at: www.anmolpublications.com

Encyclopaedia of Measure Theory

First Edition, 2009

ISBN 978-81-261-3921-7 (Set)

PRINTED IN INDIA

Printed at Mehra Offset Press, Delhi.

Contents

Preface

Measure Theory has become increasingly essential to *Geometry* as well as *Numerous and Varied Physical Applications*. Over the past thirty years, this theory has contributed to major advances in many classical branches of *Mathematics*. This Encyclopaedia is an introduction to Measure Theory. It is designed for graduate students in a variety of fields (*Mathematics, Statistics, Economics, Management, Finance, Computer Science, and Engineering*) who require a working knowledge of Measure Theory that is Mathematically Precise, but without excessive Technicalities. The text provides complete proofs of all the essential introductory results. Nevertheless, the treatment is focused and accessible, with the Measure Theory and Mathematical details presented in terms of intuitive Probabilistic concepts, rather than as separate, imposing subjects. The text strikes an appropriate balance, rigorously developing *Probability Theory* while avoiding unnecessary detail.

Author

List of Symbols

± ⊄ ⊂ ⊃ ⊆ ∈ ∉ ∂

≠ ≡ ≈ ∠ ∇ ∑ ∏ Ω

ϕ α β χ ε γ η λ

μ ν ⊥ π θ ρ σ τ

ℜ ω ξ ψ ζ ϒ ∃ ∫

∬ ∭ ∮ × √ ∴ ∵ Δ

≥ ≤ ⇒ ⇓ ⇔ ∞ ∝ ℝ

ℤ ℕ ℂ ℚ

Chapter 1

Category Theory

In mathematics, category theory deals in an abstract way with mathematical structures and relationships between them. Categories now appear in most branches of mathematics and in some areas of theoretical computer science and mathematical physics, and have been a unifying notion. Categories were first introduced by Samuel Eilenberg and Saunders Mac Lane in 1942-1945, in connection with algebraic topology.

Category theory has several faces known not just to specialists, but to other mathematicians. "General abstract nonsense" refers, perhaps not entirely affectionately, to its high level of abstraction, compared to more classical branches of mathematics. Homological algebra is category theory in its aspect of organising and suggesting calculations in abstract algebra. Diagram chasing is a visual method of arguing with abstract 'arrows'. Topos theory is a form of abstract sheaf theory, with geometric origins, and leads to ideas such as pointless topology.

The study of categories is an attempt to capture what is commonly found in various classes of related mathematical structures. Consider the following example. The class Grp of groups consists of all objects having a "group structure". More precisely, Grp consists of all sets *G* endowed with a binary operation satisfying a certain set of axioms. One can proceed to prove theorems about groups by making logical deductions from the set of axioms. For example, it is immediately proved from the axioms that the identity element of a group is unique.

Instead of focusing merely on the individual objects (e.g. groups) possessing a given structure, category theory emphasizes the morphisms — the structure-preserving mappings — between these objects. It turns out that by studying these morphisms, we are able to learn more about the structure of the objects. In the case of groups, the morphisms are the group homomorphisms. A group homomorphism between two groups "preserves the group structure" in a precise sense — it is a "process" taking one group to another, in a way that carries along information about the structure of the first group into the second group. The study of group homomorphisms then provides a tool for studying general properties of groups and consequences of the group axioms. A similar type of investigation occurs in many mathematical theories, such as the study of continuous maps between topological spaces in topology and the study of diffeomorphisms in manifold theory. The notion of a category is an *axiomatic* formulation of this idea of relating mathematical structures to the structure-preserving functions between them. A systematic study of categories then allows us to prove general results from the axioms of a category.

A category is itself a type of mathematical structure, so we can look for 'processes' which preserve this structure in some sense. Such a process is called a functor. It associates to every object of one category an object of another category; and to every morphism in the first category a morphism in the second.

By studying categories and functors, we are not just studying a class of mathematical structures and the morphisms between them, we are studying the *relationships between various classes of mathematical structures*. This is a fundamental idea, which first surfaced in algebraic topology. Difficult *topological* questions can be translated into *algebraic* questions which are often easier to solve.

Basic constructions, such as the fundamental group of a topological space, can be expressed as functors in this way. Constructions are often "naturally related", a vague notion at first sight. This leads to the clarifying concept of natural transformation, a way to "map" one functor to another. Many important constructions in mathematics can be studied in this context. 'Naturality' is a principle, like general covariance in physics, that cuts deeper than is initially apparent.

A reoccurring technique in pure mathematics is to take a well known mathematical structure and find an abstraction of this structure that captures its key properties. As new structures are developed, further abstractions become possible, leading to deeper insights. In this dissertation we develop an abstraction of measure theory (which is itself an abstraction of integration theory). The framework that we use to do this is category theory. More precisely, we use the apparatus of categorical logic to establish connections between the analytic ideas of measure theory and the geometric ideas of sheaf theory. We start with some of the key definitions from these three areas of mathematics. Results in these sections will be presented without proof, as they are part of the standard literature of the respective fields. After establishing these definitions, we present the structure of this dissertation.

SOME CATEGORY THEORY

We can study a class of algebraic objects by investigating the functions between members of this class that that preserve the algebraic structure. For example, we can study groups by investigating group homomorphisms, we can study sets by investigating functions, and we can study topological spaces by investigating continuous functions.

Categories are algebraic structures that capture the relationships between similar types of objects, and so allow us to formalize this notion.

The study of category theory allows for the development of techniques that can apply simultaneously in all of these settings.

Definition: A category $\mathbb{C}$ consists of a collection $O_{\mathbb{C}}$ of objects and a collection of M_C of arrows, or morphisms, such that

1. Each arrow f is assigned a pair of objects; the domain of f, written dom (f), and the codomain of f, written cod(f). If A = dom(f), and B = cod(f), then we write $f : A \to B$.
2. If $f : A \to B$ and $g : B \to C$ are two arrows in $\mathbb{C}$, then there is an arrow $g \circ f : A \to C$, called the composition of g and f.
3. Every object A is associated with an identity arrow $\mathrm{id}_A : A \to A$. This arrow is the identity with respect to composition, so that if $f : A \to B$ and $g : Z \to A$, then $f \circ \mathrm{id}_A = f$, and $\mathrm{id}_A \circ g = g$.

There are many examples of categories. The prototypical example is the category SET. The objects of SET are sets, and the arrows are functions, with domain, codomain, composition, and identity defined in the obvious ways. More generally, any model of ZFC constitutes a category in this way. Two other important examples are the category GRP, whose objects are groups and whose arrows are group homomorphisms, and the category TOP, whose objects are topological spaces, and whose arrows are continuous functions.

These are all examples of categories where the objects can be considered as "sets with structure" (although in the case of SET, the structure is trivial). Not all categories have this property. Categories are classified according to the following taxonomy :

Definition: Let $\mathbb{C}$ be a category.

1. $\mathbb{C}$ is called small if the collection of arrows M_C is a set (and not a proper class).
2. $\mathbb{C}$ is called large if $\mathbb{C}$ is not small.
3. $\mathbb{C}$ is called locally small if for any pair of objects C and D, the collection of arrows with domain C and codomain D is a set (and not a proper class).

For a locally small category $\mathbb{C}$, and objects C_1 and C_2 of $\mathbb{C}$, we refer to the set of arrows of $\mathbb{C}$ with domain C_1 and codomain C_2 as the "homomorphism set", or "hom set", denoted $\mathrm{Hom}_{\mathbb{C}}(C_1, C_2)$.

The categories S_{ET}, G_{RP}, and T_{OP} share the same taxonomic classification from Definition; they are all large, locally small, categories. They are also all examples of concrete categories (categories whose objects are "sets with structure" and whose arrows are functions from these underlying sets). However, there are categories that are small, and there are categories that are not concrete.

For example, let $G = \langle G, \oplus, -, e \}$ be a group. Then we can represent G as a category with one object *, and whose arrows are elements of G. Composition of arrows corresponds to the group operation, so that the composition $g \circ h$ is just $c \oplus h$. Note that the identity arrow is just e. This idea can obviously also be applied to represent monoids as categories.

As another example, let $(P, \leq \Rightarrow$ be any poset. Then we can view P as a category. The objects of P are just the elements of P, and the arrows are witnesses to the "$\leq$" relation. Between any two elements of P, there is at most one arrow.

For example, $\mathbb{N}$, the natural numbers, constitute a category:

$$0 \longrightarrow 1 \longrightarrow 2 \longrightarrow 3 \longrightarrow \cdots$$

Note that there are other implicit arrows here, for example from 0 to 2. This arrow is the composition of the arrows from 0 to 1 and from 1 to 2.

Given a category $\mathbb{C}$, there is category $\mathbb{C}^{op}$, called the dual, or opposite category of $\mathbb{C}$. The dual category has the same objects and arrows as $\mathbb{C}$, but the domain and codomain operations are reversed.

Definition: Let $f : C \to D$ and $g : D \to C$ be two arrows in $\mathbb{C}$ such that $g \circ f = \mathrm{id}_C$ and $f \circ g = \mathrm{id}_D$. Then we say that f and g are isomorphisms, and that $C \cong D$.

Since every element of a group has an inverse, it follows that if we represent the group G as a category, every arrow is an isomorphism.

This observation leads to the following definitions: A category $\mathbb{C}$ is called a groupoid if every arrow of $\mathbb{C}$ is an isomorphism. $\mathbb{C}$ is called a group if $\mathbb{C}$ is a groupoid with only one object.

The concept of an isomorphism is essential in category theory. The cancellation properties of isomorphisms, together with the fact that the idiosyncratic properties of objects are inaccessible to a categorical analysis except insofar as they are reflected in the arrows of the category, mean that in category theory critical objects are only defined up to isomorphism.

In the category SET, for example, isomorphisms are just bijections. Since in SET every set is isomorphic to exactly one cardinal, we can think of every set represented by its cardinality. As an illustration of this, every singleton is a terminal object. There is therefore a proper class of terminal objects in SET.

However, the terminal object is unique, up to isomorphism. From a categorical point of view it doesn't matter which singleton we are considering, only that the set is indeed a singleton.

Definition: An arrow $f : C \to D$ in $\mathbb{C}$ is called a monomorphism if for any $g, h : B \to C$ such that $f \circ g = f \circ h$, we must have $g = h$. In this case we call C (or more properly the diagram $f : C \to D$) a subobject of D.

Monomorphisms are indicated by the special arrow "→", so that we write $f : C \rightarrow D$.

In the category SET, monomorphisms correspond to injections. Thus we say that, $f : A \rightarrow B$ is a subobject of B, even though A need not be an actual subset of B.

However, it does follow that A is isomorphic to a subobject of B. In fact, in the category S_{ET}, A is a subobject of B (for some monomorphism) if and only if $|A| \leq |B|$.

Group homomorphisms are functions that preserve the group structure. A corresponding role in category theory is taken by functors.

Definition: Let $\mathbb{C} = \langle O_C, M_C \rangle$ and$D = \langle O_D, M_D \rangle$ be two categories. Afunctor $F : \mathbb{C} \rightarrow D$ consists of two functions, $F_O : O_C \rightarrow O_D$, and $F_M : M_C\, !M_D$, such that all the categorical structure (domain, codomain, composition, identity) is preserved.

There aremanyexamples of functors. For any concrete category $\mathbb{C}$, there is a "forgetful" functor $U : \mathbb{C} \rightarrow SET$, which takes every "set with structure" to the underlying set. If P_1 and P_2 are two posets, viewed as categories, then a functor from P_1 to P_2 is just an order preserving map.

One way to think of a functor $F : J \rightarrow \mathbb{C}$ is as a diagram. F describes a copy of the category J inside $\mathbb{C}$. For example, suppose that J is the category given by the following diagram:

$$J \xrightarrow{\;j\;} L \xleftarrow{\;k\;} K$$

then we have a diagram inside $\mathbb{C}$:

$$F(J) \xrightarrow{\;F(j)\;} F(L) \xleftarrow{\;F(k)\;} F(K)$$

Using this terminology, we can define limits in $\mathbb{C}$.

Definition: Let $F : J \rightarrow \mathbb{C}$ be a functor.

1. A cone for F consists of an object C of $\mathbb{C}$, together with a family of arrows

$$f = \langle f_J : C \rightarrow f(J) \mid J \in O_j \rangle$$

such that for any arrow $j : J \rightarrow K$ in J, the following diagram commutes:

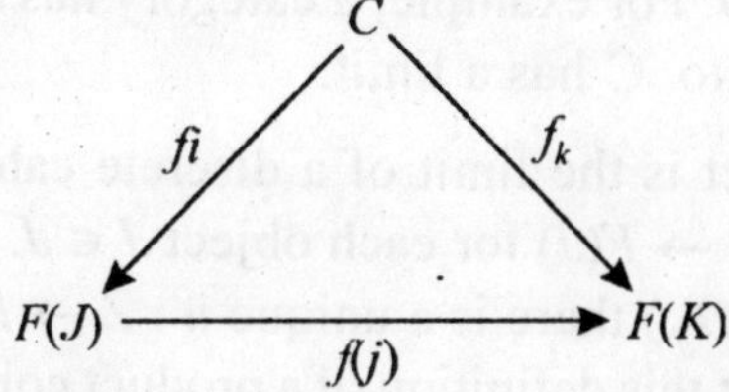

2. A limiting cone for F is a cone (C, f) is a cone such that for any other cone (D, g) ther is a unique arrow $h : D \rightarrow C$ such that for any $J \in O_J$ we have $f_J \circ h = gJ$. Such an arrangement looks something like this:

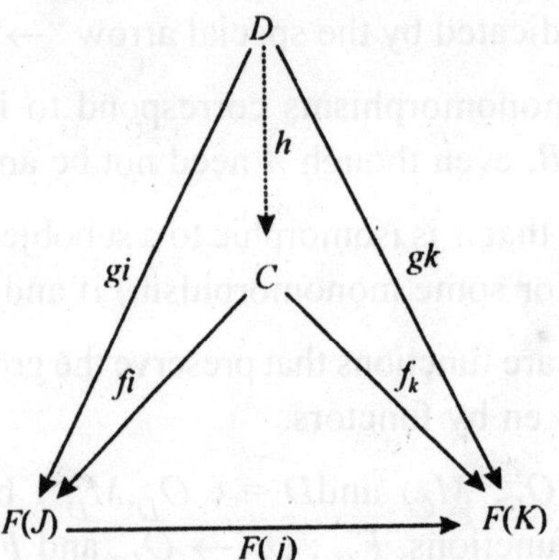

Definition: Let $F : f \to \mathbb{C}$ be a functor. Then, viewing F as a diagram in $\mathbb{C}$, the limit of the diagram, denoted

$$\lim_{\leftarrow J} F$$

is the limiting cone.

Limits are also sometimes called "inverse limits". *A* limit in $\mathbb{C}^{op}$ is called a colimit in $\mathbb{C}$, or a "direct limit". *A* colimit can be thought of as a limiting cocone:

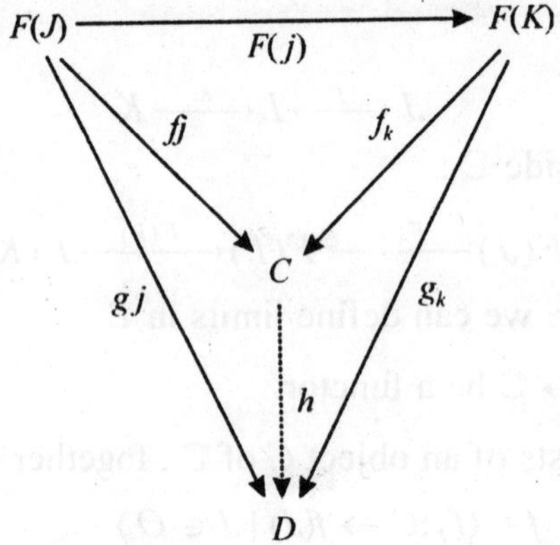

Sometimes we will refer to a category having all limits of a certain class. This usually refers to the index category *J*. For example, a category has all finite limits if every functor *F* from a finite category *J* into $\mathbb{C}$ has a limit.

Definition: 1. *A* product is the limit of a discrete category *J*. It consists of a single object *P* and an arrow $\pi_J : P \to F(J)$ for each object $J \in J$. such that for any object *Z*, and arrows $\langle f_J : Z \to F(J) \mid J \in O_J \rangle$ there is a unique $u : Z \to P$ such that for each *J*, we have $p_J \circ u = f_J$. It is easy to see that this definition of a product coincides with the usual definition of the product in S_{ET}, in G_{RP}, and in T_{OP}. In a poset category *P*, the productof elements *A* and *B* is their greatest lower boundArrows into products are generally written by pairing the arrows into the factors. For example:

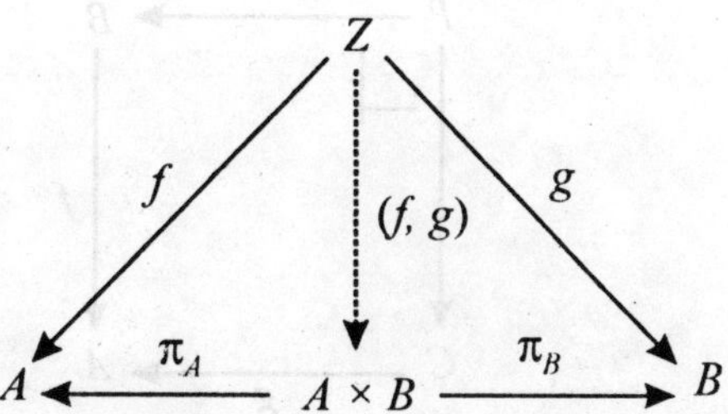

Thus we write $\langle f, g\rangle : Z \to A \times B$. Occasionally we will have arrows from oneproduct to another. In this case, we will sometimes drop the projection maps. For example, in the following diagram, we will write timesg : $A \times B \to C \times D$, rather than

$$\langle f \circ \pi_A, g \circ \pi_2 \rangle : A \times B \to C \times D\rangle.$$

$$\begin{array}{ccccc} A & \xleftarrow{\pi_A} & A \times B & \longrightarrow & B \\ \downarrow f & & \downarrow f \times g & & \downarrow g \\ C & \xleftarrow{\pi_C} & C \times D & \xrightarrow{\pi_D} & D \end{array}$$

2. *A* terminal object is a special product. It is the limit of the empty diagram. Since everyobject of $\mathbb{C}$ is a cone for the empty diagram, the terminal object is just an object **1** suchthat for any object C of $\mathbb{C}$ for which there is a unique arrow $! : C \to 1$. In a poset,the terminal object, if it exists, is the top element.

In S_{ET} any singleton is a terminalobject. In G_{RP}, the terminal objects are the trivial groups; that is, groups with only oneelement. Note that although there may be more than one terminal object, all of theterminal objects in $\mathbb{C}$ are isomorphic to one another.

3. An equalizer is a limit of a diagram of the form

$$A \overset{f}{\underset{g}{\rightrightarrows}} B$$

A cone for such a diagram consists of an object *Z* together with an arrow $z : Z \to A$,such that $f^\circ = g^\circ z$. Hence an equalizer consists of an object *E* and an arrow $e : E \to A$ such that for any such *Z*, there is a u : $Z \to E$ such that $z = e \circ u$. The arrow *e* is alwaysa monomorphism. In the category of sets, *E* is the set $\{x \in A \mid f(x) = g(x)\}$.

4. A pullback is a limit of a diagram of the form

$$B \xrightarrow{f} A \xleftarrow{g} C$$

The pullback is usually expressed as a commutative square:

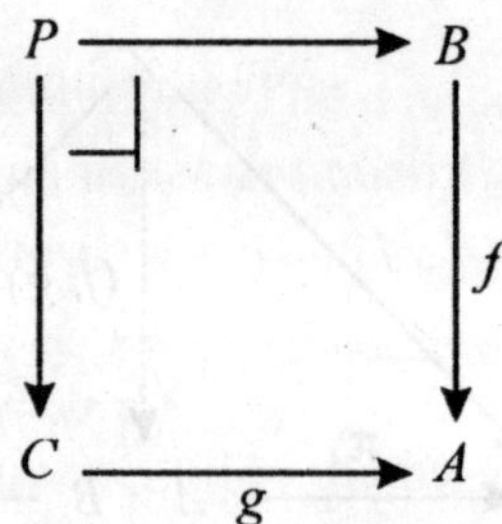

in the category of sets, the pullback is the subset of $B \times C$ given by

$$P = \{\langle b, c\rangle \in B \times C \mid f(b) = g(c)\}$$

A functor *F* from $C^{op} \to D$ is sometimes called a "contravariant functor" from *C* to *D*. This terminology is something of a misnomer, as *F* is not a functor from *C* to *D*.

Given two categories, *C* and *D*, there is a category D^C, whose objects are the functors from $\mathbb{C}$ to *D*. In order to understand this category, we need a notion of an arrow from one functor to another. Such an arrow is called a "natural transformation".

Definition: Given two functors $F, G : \mathbb{C} \to D$, a natural transformation $\eta : F \to G$ consists of a family of arrows $\langle \eta^C | C \in O_C\rangle$such that for any $f : C_1 \to C_2$ in *C*, the following square commutes in *D* :

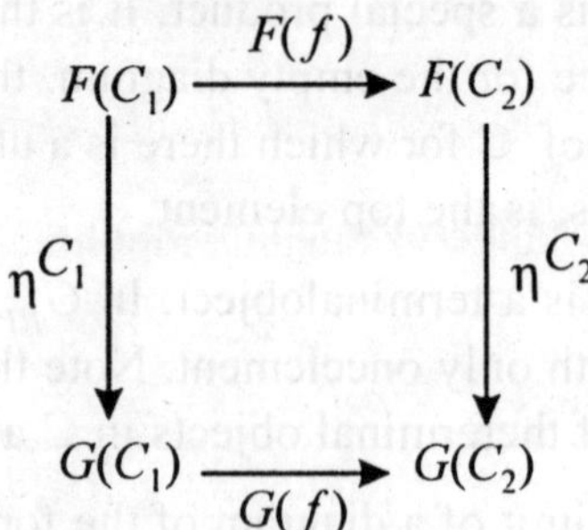

The arrow η^C is called the "component of at *C*".

Suppose that $\mathbb{C}$ and *D* are two categories, and $F : \mathbb{C} \to D$ and $G : D \to \mathbb{C}$ are two functors. Then we can compute the composites, to get to functors $GF : \mathbb{C} \to \mathbb{C}$ and $FG : D \to D$. These compositions are objects in the categories $\mathbb{C}^{\mathbb{C}}$ and D^D respectively. Each of these categories also has an identity functor, $id_{\mathbb{C}} : \mathbb{C} \to \mathbb{C}$, in $\mathbb{C}^{\mathbb{C}}$ and $id_D : D \to D$ in *DD*.

Definition: If $F : \mathbb{C} \to D$ and $G : D \to \mathbb{C}$ are functors such that *GF* is isomorphic to $id_{\mathbb{C}}$ (in $\mathbb{C}^{\mathbb{C}}$), and *FG* is isomorphic to id_D (in D^D), then we say that $\mathbb{C}$ and *D* are equivalent and write $C \simeq D$.

From above, it is clear that $S_{ET} \simeq C_{ARD}$, where C_{ARD} is the subcategory of S_{ET} who objects are cardinals. It is often said that an equivalence is "isomorphic to an isomorphism".Equivalence is a special case of a more general relation between functors.

Let $\mathbb{C}$ and D be two categories, and let $F : \mathbb{C} \to \mathrm{D}$ and $G : D \to \mathbb{C}$ be two functors. We say that F is the left adjoint of G, or the G is the right adjoint of F (written $F \perp G$), if for any objects $C \in O_{\mathrm{C}}$ and $D \in O_D$ there is an isomorphism between $\mathrm{Hom}_{\mathrm{C}}(C,GD)$ and $\mathrm{Hom}_{\mathrm{D}}(FC,D)$ (natural in both C and D). Given an adjunction $F \perp G$ there are two natural transformations $\eta : id_C \to GF$ and $\in : FG \to id_D$, called the unit and counit of the adjunction respectively.

The unit and counit are universal, in the sense that for any objects C in $\mathbb{C}$ and D in D, and every arrow $f : C \to G(D)$, there is a unique arrow $h : F(C) \to D$ in D such that the following diagram commutes:

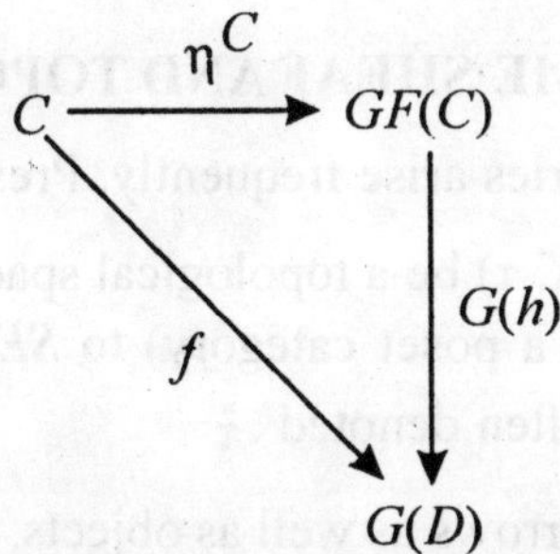

Adjunctions occur in many contexts. For example, the "forgetful" functor $U : G_{RP} \to S_{ET}$ has a left adjoint F, which takes a set X to the free group on X. The unit of this adjunction embeds a set X into the underlying set of the free group on X. The counit takesan element of the free group on the letters taken from the underlying set of a group G (which is a string of elements of G) to the product of that string in G.

One specific example of an adjunction that is important here is in the constructionof exponentials. Let $\mathbb{C}$ be a category, and fix an object C in $\mathbb{C}$. Then there is a functor $PC : \mathbb{C} \to \mathbb{C}$ with the action $\mathrm{B} \mapsto B \times C$.

If this functor has a right adjoint, that adjoint isan exponentiation functor, E_C, given by the action $B \mapsto C^B$. All of the key properties of an exponential are deduced from the properties of the adjunction.

The counit of this adjunction is particularly interesting. For a given $B, \in^{\mathrm{B}}$ is an arrow from $B^C \times C$ to B. In the S_{ET}, this arrow represents function application. An element of B^C is a function f from C to B, so an element of $B^C \times C$ can be thought of as an ordered pair $\langle f, c\rangle$. Then η^B applied to this pair is just $f(c) \in B$.

This counit also has an important role in a Lindenbaum algebra of logical formulas.In this case, the product is conjunction, and the exponential is the conditional. Hence we write $B \wedge C$, rather than $B \times C$, and $C \Rightarrow B$, in place of B^C. In this context, arrows correspondto the provability predicate, so we get the inferential law modus ponens.

$$C \wedge (C \Rightarrow B) \perp B$$

The unit also has familiar interpretations. The component of the unit at C takes B to $(B \times C)^C$. Interpreting this in S_{ET} gives us the following

$$B \to (B \times C)^C$$

$$b \mapsto \lambda x \langle b, x \rangle$$

Applying the unit in the Lindenbaum algebra gives us the following inference (a form of implication introduction):

$$B \perp C) \Rightarrow (B \wedge C)$$

SOME SHEAF AND TOPOS THEORY

Certain functor categories arise frequently. Presheaves are an example:

Definition: Let $X = (X, \tau)$ be a topological space. A presheaf on X is a contravariant functor from τ(viewed as a poset category) to *SET* The category of presheaves on τ is$S_{ET}{}^{\tau op}$. This category is often denoted . $\hat{\tau}$

Since functors act on arrows as well as objects, a presheaf P can be thought of as a τ-indexed family of sets, together with functions between them. Since the arrows in τ correspond to subsets, if follows that if $V \subseteq U$, then P includes a function $\rho_v^u : p(u) \to p(V)$. This function is called a "restriction map".

In fact, presheaves can be studied more generally. If $\mathbb{C}$ is any small category, then the category $S_{ET}{}^{\mathbb{C}op}$ is called the category of presheaves on $\mathbb{C}$, and is usually denoted $\widehat{\mathbb{C}}$.

As the name suggests, one reason for the significance of presheaves is that they are related to sheaves. Unfortunately, it is difficult to give a single definition of a sheaf, as different settings require different languages. Here we give three presentations of the definition of a sheaf, in order of increasing generalization.

The most specific of these examples is a sheaf on a topological space. To understand this concept, we start with the idea of a matching family.

Definition: Suppose that P is a presheaf on τ, and $U \in \tau$.

1. Asieve I on τ is any family of open subsets of U which is "downward closed", meaning that if $W \subseteq V \subseteq U$ and $V \in I$, then $W \in I$.
2. A sieve I covers U if $\bigcup_{V \in I} V = U$.
3. A family for P and I is a element $x \in \prod_{V \in I} P(V)$
4. A family $x = \langle x_V \mid V \in I \rangle$ is a matching family if for any $V, W \in I$ we have

$$\rho^{V}_{V\cap W}(x_V) = \rho^{W}_{V\cap W}(x_W)$$

5. $x \in P(U)$ is an amalgamation for a matching family x if for every $V \in I$ we have

$$\rho^{W}_{V}(x) = x_v$$

6. P is a sheaf if for every $U \in \tau$ and for every covering sieve I of U, and for everymatching family x for I, there is a unique amalgamation $x \in P(U)$.

The arrows of $Sh(\tau)$ are just natural transformations between sheaves, so that $Sh(\tau)$ is a full subcategory of $\hat{\tau}$. The inclusion of $Sh(\tau)$ into $\hat{\tau}$ is a functor i:

$$\text{i: sh}(\tau) \rightarrow \hat{\tau}$$

This functor has a left adjoint, *a,* called the associated sheaf, or sheafification functor. Thecomponent at P of the unit of this adjunction is a natural transformation $\eta P : P \rightarrow a^P$. Itis immediate from the definition of the unit of an adjunction that for any sheaf F and any natural transformation $\phi : P \rightarrow F$, there is a unique natural transformation : $\bar{\phi} : a^P \rightarrow F$ such that the following diagram commutes:

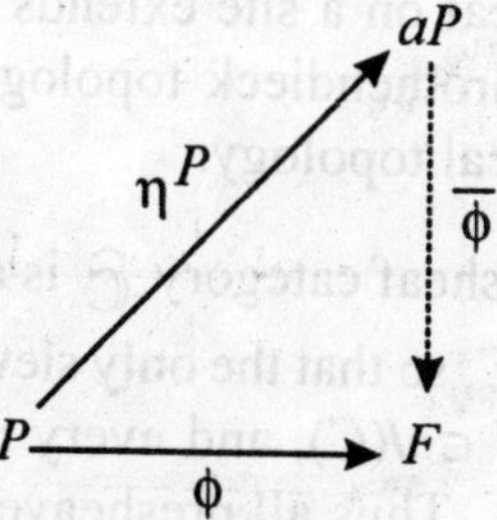

The concept of a sheaf on a topological space can be generalized. Let $\mathbb{C}$ be any small category. *A* sieve on an object C is a set I of arrows, all with codomain C, such that if $f \in I$, and g and h are any arrows such that $g = f^\circ h$, then g is also in I. In the case that $\mathbb{C}$ is a poset category, a sieve is just a lower or downward closed set.

Definition: Ω is the presheaf of sieves on $\mathbb{C}$. Hence $\Omega(C)$ is the set of sieves on C. The restriction operation is given by

$$\rho_f(I) = \{g \in M_C \mid f \circ g \in I\}$$

$\rho_f(I)$ can be thought of as the arrows of I that factor through f. Note that if $f \in I$, then $\rho_f(I)$ is the maximal sieve on dom (f).

Definition: *A* Grothendieck topology is a subfunctor $J \rightarrow \Omega$ that assigns to each object$C$ of $\mathbb{C}$ a set of sieves on C that cover C. In order to be a Grothendieck topology, J mustsatisfy the following axioms:

1. Maximality: The maximal sieve $I = \{f \in M_C | \text{cod}(f \Rightarrow = C\}$ is a cover. (Note that the maximal sieve on C is the principal sieve generated by id_C.)
2. Transitivity: If $I \in J(C)$ and for each $I = \{f \in M_C \mid cod(f) = C\}$, then

$$\bigcup_{f \in I} (f^\circ g \mid g \in f_f)$$

is a cover for C.

It is usually required that J also satisfy the stability condition:

If $I \in J(C)$ and $f: D \to C$, then $\{g \in M_C | f^\circ g \in I\} \in J(D)$

However, this follows directly from the fact that J is a subfunctor of Ω.

A small category $\mathbb{C}$, together with a Grothendieck topology J is called a site or a coverage. Given a site $(\mathbb{C}, J)$ a sheaf on the site isdefined in a way that is analogous to the way that a sheaf on a topological space; a sheafis a presheaf that has unique amalgamations for every matching family for every cover.

It is easy to verify that the usual notion of a cover of an open set is a Grothendieck topology. Hence the definition of a sheaf on a site extends the definition of a sheaf on a topological space. In fact, the usual Grothendieck topology on a topological space has aspecial name; it is called the canonical topology.

It is also worth noting that any presheaf category $\widehat{\mathbb{C}}$ is also a sheaf category. Let J be the smallest Grothendieck topology on C, so that the only sieve that covers C is the maximal sieve. Then for every covering sieve $I \in J(C)$, and every matching family x for P there must be an amalgamation, namely xid_C. Thus all presheaves are sheaves.

Definition: Let $\mathbb{C}$ be a category, *A* subobject classifier in $\mathbb{C}$ is an object Ω, together with an arrow $\perp : 1 \to \Omega$, called "true", such that for any monomorphism $S \to A$ in $\mathbb{C}$, there is a unique arrow $\chi_S : A \to \Omega$ such that for any arrow $f : Z \to A$, there is a unique arrow u making the following diagram commute:

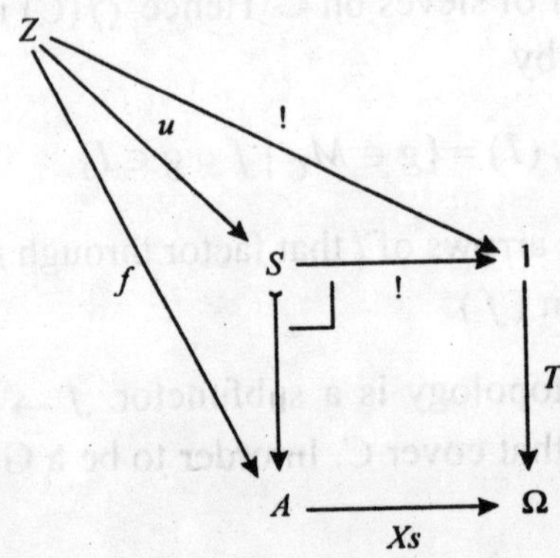

In other words, S is the pullback of "true" along χ_S.

In S_{ET}, the subobject classifier is just the two point set $\{\perp, \perp\}$, and the characteristic maps are just the usual characteristic functions. More generally, we think of Ω as the object of truth values in ε. The subobject classifier is the key to building an internal logic inside a topos. We can now give a formal definition of a topos:

Definition: A topos is a category ε such that ε has all finite limits and colimits, exponential objects, and a subobject classifier. *A* topos that is equivalent to the topos of sheaves on some site ($\mathbb{C}$, *J*) is called a "Grothendieck topos".

Since we can take exponentials in a topos, we can compute Ω^A, the "power object" of *A*. In S_{ET}, this is just the set of all characteristic functions of subsets of *A*. Note that the counit in this case is just the "element of" relation, thus justifying the use of the letter "ε" to denote the counit. Rather than writing it as an exponential, we denote the power object of *A* by *pA*.

The internal logic of a topos is higher order intuitionistic logic. The objects of $\mathbb{C}$ represent the types of the logic. There are arrows of the form $\Omega \times \Omega \rightarrow \Omega$ corresponding to and $\vee \wedge$ and $\Rightarrow$, together with a negation arrow $\neg : \Omega \rightarrow \Omega$. An arrow from $\mathbb{C} \rightarrow \Omega$ represents a logical formula with free variable of type *C*. If $\phi : C \rightarrow \Omega$ is some formula, then the arrow corresponding to the negation ofϕ is just the following composition:

$$C \xrightarrow{\phi} \Omega \xleftarrow{\neg} \Omega$$

Likewise, if $\phi : A \rightarrow \Omega$ and $\psi : B \rightarrow \Omega$ are two formulas, then we find the formula $\phi(a) \wedge \psi(b)$ by means of the following diagram:

$$A \times B \xrightarrow{\phi \times \psi} \Omega \times \Omega \xleftarrow{\wedge} \Omega$$

Since toposes have a lot of features in common with S_{ET}, it is not surprising that toposes have a similar feel to S_{ET}. In particular, viewed internally, we can think of a topos as a model for an intuitionistic set theory. A topos will not, in general, satisfy all of ZFC. However, a Grothendieck topos will satisfy most of the set theory IZF , except for the axiom of foundation. Hence we can reason about a topos by considering the objects to be sets, and using intuitionistic logic.

It is possible to study the internal logic of more general categories that do not have subobject classifiers. In a presheaf topos, the subobject classifier is just Ω, the presheaf of sieves. Let (*L*, *J*) be a site.

A sieve $I \in \Omega(C)$ is called "closed" if every for every $f : D \rightarrow C$ such that $\{g \in O_{\leq} | f \circ g \in I\} \in j(D)$, we have $f \in I$. In other words, a sieve is closed if it contains all that arrows that it covers. The object of closed sieves is denoted Ω_j, and is a subobject of Ω. In fact, Ω_j is a sheaf, and is the subobject classifier in Sh ($\leq$, *J*).

Since Ω_j is a subobject of Ω, it follows that there is a characteristic map $j = \chi\Omega_j : \Omega \to \Omega$. This map is called the "closure map", and is the key to the most general notion of a sheaf.

Definition: Let e be a topos, and let Ω be the subobject classifier of ε. Then $j : \Omega \to \Omega$ is called a Lawvere-Tierney topology if the following diagrams commute:

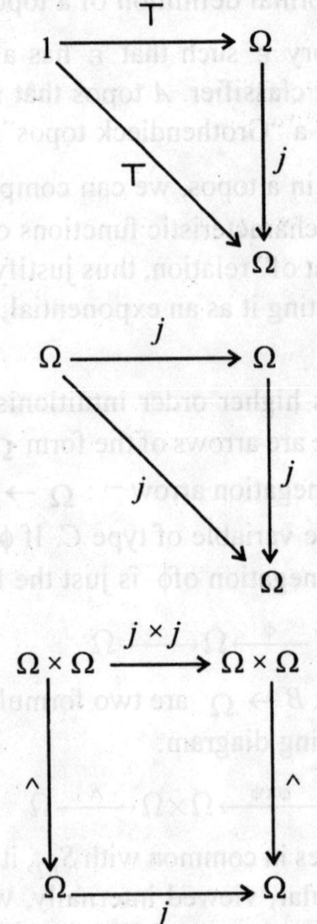

Definition: An object E of e is a j-sheaf if for any $S \alpha E$ such that $j \circ c_{\Sigma} = \mathrm{T}$ and for any arrow $f : S \to F$, there is a unique arrow $\bar{f} : E :\to F$ making the following diagram commute:

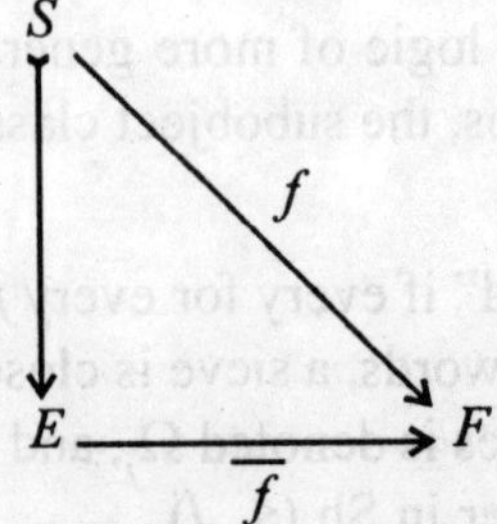

The topos of *j*-sheaves in e is denoted $Sh_j(\varepsilon)$.

As before, the inclusion functor $i : Sh_j(\varepsilon)$ a ε has a left adjoint α, called the "associated sheaf" or "sheafification" functor. The subobject classifier in $Shj(\varepsilon)$ is Ω_j, which is given by the following equalizer:

$$\Omega_j \mapsto \Omega \overset{id_\Omega}{\underset{j}{\rightrightarrows}} \Omega$$

If ε is a topos, and *j* is a Lawvere-Tierney topology in ε, then we say that $Shj(\varepsilon)$, the topos of *j*-sheaves, is a "subtopos of ε".

It turns out that if ε is a presheaf topos, then the Grothendieck topologies on e correspond to the Lawvere-Tierney topologies, and the two notions of sheaf coincede.Therefore, this new notion of a sheaf does indeed generalize the notion of a sheaf on a site.

There is one very special class of toposes that arise frequently in this dissertation. A locale is a type of lattice (specifically, a complete Heyting algebra). Locales arise often in topology, as the algebra of open sets in a topological space is a locale. Point-free topology is generally construed as the study of locales. However, locales need not be spatial. To recognize this, we use the symbols "g","l", and "$\preceq$" to refer to the lattice operations of a locale l , and "*T*" and "$\perp$" to refer to the top and bottom elements of l.

In any Grothendieck topos, the object Ω forms an internal locale object. However, the significance of locales does not stop there. Any topos that is equivalent to the topos of sheaves on a locale (with the canonical topology) is called a "localic topos". Localic toposes have many useful properties, but the most useful here is that if IP is any poset, and *J* is any Grothendieck topology on IP, then Sh(IP, *J*) is a localic topos, and moreover, is equivalent to the topos of sheaves on the locale of closed sieves in IP.

Another important class of toposes with which we need to be familiar are the Boolean toposes. A topos e is Boolean if the internal logic satisfies the law of the excluded middle:

$$\varepsilon * \phi \vee (\neg\phi)$$

This is *y* equivalent to the subobject classifier of ε being an internal Boolean algebra object. For any topos, the "double negation arrow" $\neg\neg : \Omega \to \Omega$ is a Lawvere-Tierney topology, and the resulting subtopos is Boolean. This construction is related to the double negation translation between intuitionistic and classical logic.

SOME MEASURE THEORY

Classical measure theory begins with the following definitions.

Definition: Let X be a set. Then $F \subseteq PX$ is called σ-field on *X* if

1. F is clos ed under complements.
2. F is closed under countable unions.

Note that $\phi \in F$, since and $X \in F$, since $X = \neg\varepsilon$.

Definition: Let F be σ-field. Then a function $\mu : F \to [0, \infty]$ is called a measure if for any countable antichain $A = \langle A_i \mid 0 < \alpha \leq \omega \rangle$ in F,

$$\sum_{i<\alpha} u(Ai) = \left(\bigcup_{i<\alpha} A_i \right)$$

Note that it is a consequence of this that $\mu(f) = 0$.

Definition: A measure space consists of a triple (X, F, μ), where X is a set, F is σ-field on X, and μ is a measure on F.

There are a number of special subclasses of the set of measures on F. The most important for our needs is the class of σ-finite measures.

Definition: Let (X, F, μ) be a measure space. Then μ is called σ-finite if there is a countable partition of of $\langle X_i \in F \mid i < \omega \rangle X$ such that for each i, $\mu(X_i) < \infty$.

Definition: Let $X = (X, F, \mu)$ and $Y = (Y, G, v)$ be two measure spaces. A measurable function from X to y is a function $f : X \to Y$ such that for any $G \in G$, $f^{-1}[G] \in F$.

Measure theory can also be studied in a point-free way .

The point-free approach to measure theory focuses on the algebraic properties of the σ-field. Correspondingly, the underlying sets X and Y are de-emphasized. The distinction is made explicit in the following Definition:

Definition: "σ-algebra" is a countably complete Boolean algebra.

Many authors use the terms "σ-algebra" and "σ-field" interchangeably, usually to mean what we have referred to as σ-field. Our terminology here echoes the distinction between a Boolean algebra, and a field of sets (that is, a collection of subsets of some universe X that contains f; and is closed under the operations of union, intersection, and complementation. Every field of sets is a Boolean algebra, but the converse is not true. Likewise, σ-field is necessarily σ-algebra, but σ-algebras are not necessarily σ-fields.

The well known Stone representation theorem shows that every Boolean algebra B is isomorphic to a field of sets (the underlying set being the set of ultrafilters of B). There is no direct analogue for the relationship between σ-algebras and σ-fields. The closest that we can get is the Loomis-Sikorski theorem. This theorem says that every σ-algebra is isomorphic to the quotient of some σ-field F by some countably complete ideal $I \subseteq F$.

In order to emphasize that the s-algebras that we refer to are not necessarily spatial, we use the symbols "*", "*", and "*" to denote the meet and join operations, and the partial ordering in σ-algebra F, and "$\perp$" and "T" to denote the smallest and largest elements of F. In the special case where F is s-field, we revert to the usual set theoretic symbols: "$\cup$" "$\cap$", etc.

If (X, F, μ) is a measure space, and $f \to [0, \infty)$ is a measurable function, (when IR is equipped with the σ-field of Lebesgue measurable sets, and the Lebesgue measure), then we can find the integral $\int fd\mu$. This integral is itself a measure v, given by

$$v(A) = \int_A fdu$$

The process of calculating the integral, Lebesgue integration, takes several steps. The integral of a constant function is found through multiplication:

$$\int_A cdu\ \chi.\mu(A)$$

The integral of a measurable function with a finite range (ie, a simple function) is computed by exploiting the additive property of measures: Suppose that $\langle X_i \mid = ...n \rangle$ is a partition of X, and that for all $x \in X_i$, $S(x) = s_i$ *Then*

$$\int_A sdu = \sum_{i-1}^{n} s_i.u(x_i \cap A)$$

Finally, the integral of a measurable function f is calculated by taking the limit of the integrals of an increasing sequence of simple functions converging to f.

In addition to the usual (pointwise) partial ordering on the measures, there is also an important preordering, the "absolute continuity" ordering:

$$v \ll u \Leftrightarrow ((U(A) = 0 \Rightarrow v(A) = 0)$$

This ordering allows us to state the Radon-Nikodym Theorem, one of the central results in Measure Theory.

Theorem: If "$v \ll \mu$ are two σ-finite measures, then there is a measurable function f such that

$$v(-) = \int_- fdu$$

The function f is called "the Radon-Nikodym derivative of with respect to μ", and is often denoted $\frac{dv}{du}$. It is important to note that the derivative is not necessarily unique.

Two functions f_1 and f_2 can both be derivatives of with respect to v if

$$u(\{\in X \mid f_1(x) \neq f_2(x)\}) = 0$$

Consequently, we say that the Radon-Nikodym derivative is defined only up to "almost everywhere" equivalence.

A number of connections have been observed between the measure theory of s-algebra F, and the geometry of the topos Sh(F) (where the Grothendieck topology is the countable join topology). Breitsprecher observed that the functor $M : F^{\text{op}} \to \text{SET}$ of measures is in fact an object of Sh(F). Scott showed that the Dedekind real numbers object in Sh(F) is the sheaf of measurable real valued functions. Combining these two observations, it is obvious that integration can be represented as a natural transformation $\int : D \times M \to M$, where D is the sheaf of non-negative measurable real numbers. More recently, Wendt showed that the notion of almost everywhere equivalence corresponds to a certain Grothendieck topology.

Between them, these results suggest that there are some strong connections between measure theory and the topos of sheaves on σ-algebra. In this dissertation, we ground these connections in the internal logic of the sheaf topos, and then extend them to create a measure theory for an arbitrary localic topos.This measure theory is based around the object of measures, the sheaf of measurable real numbers, and an integration arrow. The object of measures is constructed in the presheaf topos L, but is a sheaf. Thus the measure theory of λ exists in Sh(λ),

Simultaneously, we show that when l is the locale of closed sieves in the σ-algebra F (in other words, when Sh($F \Rightarrow \simeq$ Sh(λ)), this localic measure theory restricts to the usual measure theory on F. We also show that when the constructions of the sheaf of measures and the integration arrow are carried out in $\hat{\lambda}$ and Sh(F), we arrive at the same objects of Sh(λ) as we did when building a localic measure theory.

The construction ofMstarts in the presheaf topos with the construction of a "presheaf of semireals". These objects act as functionals from the underlying locale l to [0, ∞]. We construct the sheaf of measures M by taking only those semireals that are both additive and semicontinuous. The construction of ∫ mimics the usual construction of the Lebesgue integral, starting with constant functions, proceeding to locally constant functions, and then, by limits, to measurable functions.

One immediate generalization of classical measure theory that follows fromthis framework is that it is possible to consider integration theory for non-spatial σ-algebras. Since Dedekind real numbers take the role of measurable functions, there is no need to have an underlying set in order to integrate.

We investigate subtoposes of Sh(L), and Sh (F). We generalize Wendt's construction of the "almost everywhere" topology so that it has a more natural interpretation in localic toposes. Equipped with this topology, we prove a generalization of the Radon-Nikodym

Theorem: A locally finite measure μ that induces a Boolean ubtopos has all Radon-Nikodym derivatives.

The next step in the construction would be the expression of the martingale property.

MEASURE THEORY AND CHANGE OF BASIS

Measure theorists have a well known change of basis technique. If (X,F) and (Y,G) aremeasurable spaces, and $f: X \rightarrow Y$ is a measurable function, then f can "carry" a measureμ on (X,F) to (Y,G) by means of the equation

$$\nu(G)\mu\left(f^{-1}[G]\right)$$

This argument suggests that there is some connection between the sheaf of measuresand geometric morphisms between sheaf toposes on different locales, or σ-algebras. Wendt [29] has looked at the change of basis operation between sheaf toposes over-σ–algebras in various categories of measure and measurable spaces, although he has not studied the measure theories of these sheaf toposes.

EXTENSIONS TO WIDER CLASSES OF TOPOSES

Many of the results in this dissertation apply in an arbitrary topos, with a designated topology *j*. The most significant exception is Theorem, the proof of which makes explicit reference to the fact that ε is the topos of presheaves on a σ-algebra. Strengthening this result to apply to a wider class of Grothendieck toposes would provide an immediate extension of a great deal of measure theory to a wide class of toposes.

In a similar vein, we know that in an arbitrary topos, the double negation topology induces a Boolean subtopos. We can find the object of measures in such a topos. This object is not necessarily a sheaf (that is, an object of the Boolean subtopos), but might n netheless have interesting properties, especially with regards to differentiation, as all measures would be Boolean.

Chapter 2

Measures on a Locale

In computing, locale is a set of parameters that defines the user's language, country and any special variant preferences that the user wants to see in their user interface. Usually a locale identifier consists of at least a language identifier and a region identifier. The definition of a measure on a σ-algebra can be extended to a locale:

Definition: Let $(L, \leq, \bot, T)$ be a locale. Then a function $\mu: L \to [0, \infty]$ is a called a measure if it satisfies the following conditions:

1. μ is order preserving
2. $\mu(A) + \mu(B) = \mu(A \wedge B) + \mu(A \vee B)$
3. For any directed family $D \subseteq L$ we have

$$\mu\left(\bigvee_{D \in D} D\right) = \bigvee_{D \in D} \mu(D)$$

Note that the last condition implies that $\mu(\bot) = 0$, since $\bot = \bigvee \phi$;.

In order to justify calling such things measures, there needs to be some sort of connection between these localic measures and traditional σ-algebra measures.

Let $(F, \subseteq, \bot, \mathrm{T})$ be a σ-algebra. A countably complete sieve in F is a set $I \subseteq F$ which is downward closed and closed under countable joins. The collection of all countably closed sieves forms a locale L. Clearly all subsets of F of the form $\downarrow A = \{B \in F \mid B \subseteq A\}$ are countably closed, so we have an embedding $F \to L$.

Lemma: Let μ be a measure on L, and let μ' be the restriction of μ to F (so that $\mu'(A) = \mu(\downarrow A)$. Then μ' is a measure on F.

Proof. We need to show that μ' satisfies the following conditions:

1. $\mu'(\bot) = 0$

2. If $A \cap B = \bot$ then $\mu'(A) + \mu'(B) = \mu'(A \cup B)$

3. If $A = \langle A_i \mid i < \omega \rangle$ is a countable increasing sequence, then

$$\mu'\left(\bigcup_{i<\omega} A_i\right) = \bigvee_{i<\omega} \mu'(A_i)$$

For the first condition, note that $\downarrow\bot = \bot$. Therefore $\mu'(\bot) = \mu(\bot) = 0$, as required.

For the second condition, take $A, B \in F$ with $A \cap B = \bot$.

$$\mu'(A) + \mu'(B) = \mu(\downarrow A) + \mu(\downarrow B)$$
$$= \mu((\downarrow A) \wedge (\downarrow B)) + \mu((\downarrow A) \vee (\downarrow B))$$

But $(\downarrow A) \wedge (\downarrow B) = \bot$ and $(\downarrow A) \vee (\downarrow B) = \downarrow(A \cup B)$, so we get

$$\mu'(A) + \mu'(B) = \mu(\bot) + \mu(\downarrow(A \cup B))$$
$$= \mu'(A \cup B)$$

For the final condition, let $A = \langle A_i \mid i < \omega \rangle$ be an increasing sequence in F. Observe that

$$\gamma_{i<\omega} A_i = \downarrow\left(\bigcup_{i<\omega} A_i\right)$$

Using this observation, we can write:

$$\mu'\left(\bigcup_{i<\omega} A_i\right) = \mu\left(\downarrow\left(\bigcup_{i<\omega} A_i\right)\omega\right)$$

$$= \mu\left(\gamma_{i<\omega} (\downarrow A_i)\right)$$

$$= \bigvee_{i<\omega} \mu(\downarrow A_i)$$

$$= \bigvee_{i<\omega} \mu'(A_i)$$

and we are done.

Lemma: Let μ be a measure on F. Define $\bar{\mu}: L \to [0, \infty]$ by

$$\bar{\mu}(I) = \gamma_{A \in I} \mu(A)$$

Then $\bar{\mu}$ is a measure on the locale L.

Proof: It is obvious that $\bar{\mu}$ is order preserving.

To see that $\bar{\mu}$ satisfies the additivity condition, start by taking two countably complete sieves $I, J \in L$, and set $\varepsilon > 0$. Then there exist $B_I \in I$ and $B_J \in J$ such that

$$\bar{\mu}(I) < \mu(B_I) + \frac{\varepsilon}{2}$$

$$\bar{\mu}(J) < \mu(B_J) + \frac{\varepsilon}{2}$$

Furthermore, there exist $B_{I \vee J} \in I \vee J$ and $B_{I \wedge J} \in J$ such that

$$\bar{\mu}(I \vee J) < \mu(B_{I \vee J}) + \frac{\varepsilon}{2}$$

$$\bar{\mu}(I \wedge J) < \mu(B_{I \wedge J}) + \frac{\varepsilon}{2}$$

Since $I \vee J$ is the set of all elements of F that can be expressed as the join of an element of I and an element of J, we know that there exist $B^1_{I \vee J} \in I$ and $B^2_{I \vee J} \in J$ such that

$$B^1_{I \vee J} \cup B^2_{I \vee J} = B_{I \vee J}.$$

Furthermore, since $I \wedge J = I \cap J$, we know that $B_{I \wedge I} \in I \cap J$. Now, let $B_1 \in I$ and $B_2 \in J$ be defined by

$$B_1 = B_I \cup B^1_{I \vee J} \cup B_{I \wedge J} \quad B_2 = B_J \cup B^2_{I \vee J} \cup B_{I \wedge J}$$

Now

$$\begin{aligned}
&\mu(B_1) + \mu(B_2) \le \bar{\mu}(I) + \bar{\mu}(J) \\
&\le \mu(B_1) + \mu(B_2) + \varepsilon \\
&\mu(B_1) + \mu(B_2) = \mu(B_1 \cup B_2) + \mu(B_1 \cap B_2) \\
&\le \bar{\mu}(I \vee J) + \bar{\mu}(I \wedge J) \\
&\le \bar{\mu}(B_1 \cup B_2) + \mu(B_1 \cap B_2) + \varepsilon \\
&= \mu(B_1) + \mu(B_2) + \varepsilon
\end{aligned}$$

Hence

$$|\bar{\mu}(I \vee J) + \bar{\mu}(I \wedge J)) - (\bar{\mu}(I) + \bar{\mu}(J)| \le \varepsilon$$

and so $\bar{\mu}$ satisfies the additivity condition.

Now, to see that $\bar{\mu}$ satisfies the semicontinuity condition, take a directed family $S = \langle I_i \mid i \in I \rangle$ of countably complete sieves in L, and let $I = \gamma\, S$ be the join of the I_i's.

We know that $A \in I$ if and only if there is a countable sequence $\langle A_\alpha \mid i \in I \rangle$ contained in $\bigcup_{i \in I} I_i$ such that $\bigcup_{\alpha \in \omega} A_\alpha = A$.

Take $\varepsilon > 0$. Then there is an $A \in I$ such that $\bar{\mu}(I) \leq \mu(A) + \varepsilon$. Let $C = \langle A_\alpha \mid \alpha < \omega \rangle$ be the sequence described in the above paragraph. We may assume without loss of generality that C is an directed sequence. Then since C is countable, we can write

$$\begin{aligned} \bigvee_{\alpha<\infty} \mu(A_\alpha) &\leq \bigvee_{i \in I} \bar{\mu}(T_i) \\ &\leq \bar{\mu}(I) \\ &\leq \bar{\mu}(A) + \varepsilon \\ &= \bigvee_{\alpha<\infty} \mu(A_\alpha) + \varepsilon \end{aligned}$$

Hence

$$\bigvee_{i \in I} \bar{\mu}(I_i) = \bar{\mu}\left(\gamma_{i \in I} I_i \right)$$

and so $\bar{\mu}$ is a (localic) measure.

Theorem: The operations in Lemmas are inverse to one another. Hence the set of measures on L is isomorphic to the set of measures on F.

Proof: Let μ be a measure on L and let v be a measure on F. We must show that $\bar{\mu}' = \mu$ and $((\bar{v})') = v$.

For the first of these, take $I \in L$. Then

$$\begin{aligned} \bar{\mu}'(I) &= \bigvee_{A \in I} \mu'(A) \\ &= \bigvee_{A \in I} \mu(\downarrow A) \end{aligned}$$

However, it is immediate that

$$I = \gamma_{A \in T} \downarrow A$$

and that $\{\downarrow A \mid A \in I\}$ is a directed set, and so we have

$$\bar{\mu}'(I) = \mu(I)$$

Now, take $A \in F$. Then

$$(\bar{\nu})'(A) = \bar{\nu}(\downarrow A)$$
$$\bigvee_{B \sqsubseteq A} \nu \ (B)$$
$$= \nu(A)$$

As a consequence of this Theorem, we know that we can study measures on locales in a way that generalizes the study of measures on σ-algebras.

Theorem tells us that the notion of a measure on a locale generalizes the notion of a measure on a σ-algebra. It is natural to ask a related question: If L is the locale of open sets in some topological space (X,L) and μ is a measure on L, can μ be uniquely extended to the measure space $(X, \sigma(L))$, where $\sigma(L)$ is the smallest σ-field on X containing L, namely the Borel algebra?

The following Theorem gives sufficient conditions for the measures on L to correspond with the measures on $\sigma(L)$.

Theorem: Let (X,L) be a metrizable Lindel¨of space. Then every locally finite measure μ on L can be uniquely extended to $\sigma(L)$.

Proof: Take a locally finite μ on L.

Since (X,L) is Lindel¨ of, and since μ is locally finite, it follows that there is a countable cover of X, with μ finite on each part. We work in the subspace induced by one of these μ finite sets.

L is closed under finite intersections and is thus π-system generating $\sigma(L)$. We can therefore apply Dynkin's π–λ theorem to conclude that if μ has an extension, it must be unique. We work to extend μ recursively, through the Borel heirachy.

Definition:

1. Σ_0 is the set of open sets in (X,L)
2. Π_α is the set $\{X \setminus A \mid A \in \Sigma_\alpha\}$
3. $\Sigma_{\alpha+1}$ is the set of countable unions of subsets of Π_α
4. When γ is a limit ordinal, then $\Sigma_\gamma = \bigcup_{\alpha<\gamma} \Sigma_\alpha$

Note that there is a duality here between the $\sum_\alpha S$ and the $\coprod_\alpha S$. We could just have well taken Π_0 to be the set of closed sets, defined Σ_α as the set of complements of elements of Π_α and Π_{α_1} as the intersections of countable subsets of Σ_α.

The following properties of the Borel heirachy are useful:

Proposition:

1. For any $\alpha < \beta$ we have
$$(\Sigma_\alpha \cup \Pi_\alpha) \subseteq (\Sigma_\beta \cap \Pi_\beta)$$
2. Σ_α and Π_α are closed under finite unions and intersections.
3. $\sigma(L) = \Sigma_{\alpha_1}$

Proof: It is immediate that $\Pi_\alpha \subseteq \Sigma_{\alpha+1}$ and that $\Sigma_\alpha \subseteq \Pi_{\alpha+1}$. We prove that $\Sigma_\alpha \subseteq \Sigma_{\alpha+1}$ and $\Pi_\alpha \subseteq \Pi_{\alpha+1}$ by induction.

Put $\alpha = 0$. Then Σ_1 is the set of F_α sets, that is, countable unions of closed sets. It is well known that every open set in a metric space is F_α. Hence $\Sigma_0 \subseteq \Sigma_1$. Similarly, since Π_1 is the set of G_δ sets, it follows that $\Pi_0 \subseteq \Pi_1$.

Now suppose that
$$(\Sigma_\alpha \cup \Pi_\alpha) \subseteq (\Sigma_{\alpha+1} \cap \Pi_{\alpha+1})$$
An element A of $\Sigma_{\alpha+1}$ is the union of a countable family elements of Π_α, and hence the union of a countable family of elements of $\Pi_{\alpha+1}$. Thus A is an element of $\Sigma_{\alpha+2}$. Likewise, an element B of $\Pi_{\alpha+1}$ is the intersection of a countable family in Σ_α, and hence the intersection of a countable family in $\Sigma_{\alpha+1}$. Therefore B is an element of $\Pi_{\alpha+2}$.

Thus we have shown that
$$(\Sigma_{\alpha+1} \cup \Pi_{\alpha+1}) \subseteq (\Sigma_{\alpha+2} \cap \Pi_{\alpha+2})$$
It only remains to show that
$$\Sigma_\alpha \cup \Pi_\alpha \subseteq \Sigma_\beta \cap \Pi_\beta$$
for $\alpha < \beta$, where β is a limit.

It is immediate that $\Sigma_\alpha \subseteq \Sigma_\beta$. If we can show that $\Pi_\alpha \subseteq \Pi_\beta$, we will be finished. But this is also immediate, since an element of Π_α is the complement of an element of Σ_α, and thus the complement of an element of Σ_β, as required.

We start by showing that Σ_α is closed under finite intersections, and Π_α is closed under finite unions. We proceed by induction. The result is immediate for $\alpha = 0$, since Σ_0 is the set of open sets, and Π_0 is the set of closed sets. Assume that Π_α is closed under finite unions. Then it follows from DeMorgan's laws that Σ_α is closed under finite intersections. Likewise, if we assume that Σ_α is closed under finite intersections, it follows that Π_α is closed under finite unions.

To check the results at limits, suppose that γ is a limit ordinal. Since Σ_γ is the union of an expanding sequence of sets, each closed under finite intersections, it follows that Σ_γ is also closed under finite intersections. The fact that Π_γ is closed under finite unions follows directly.

Now to Σ_α verify that is closed under finite unions, and that Π_α is closed under finite intersections. We again proceed by induction. The base case is immediate. For the successor case, observe that each $\Sigma_{\alpha+1}$ is the union of countably many elements of Π_α, it is trivial that $\Sigma_{\alpha+1}$ is closed under finite unions. Likewise, it is immediate that $\Pi_{\alpha+1}$ is closed under finite intersections. The limit case is similar.

In fact, we have shown that Σ_α is closed under countable unions, and that Π_α is closed under countable intersections, except possibly at limit stages.

Since the cofinality of ω_1 is uncountable, it follows that Σ_{ω_1} is closed under countable unions and complements. Therefore Σ_{ω_1} is a σ-field containing $L = \Sigma_0$. Hence

$$\sigma(L) \subseteq \Sigma_{\omega_1}$$

It is easy to prove, by induction, that each Σ_α is a subset of $\sigma(L)$, and so we have

$$\Sigma_{\omega_1} \subseteq \sigma(L)$$

Now, let μ be a finite measure with $\mu(X) = M$. We extend μ through the heirachy.

1. $\mu_0 : \Sigma_0 \to [0,M]$ is just μ
2. $\mu^*_\alpha : \Pi_\alpha \to [0,M]$ is given by

   $\mu^*_\alpha(F) = M - \mu_\alpha(X/F)$
3. $\mu_{\alpha+1} : \Sigma_{\alpha+1} \to [0,M]$ is given by

   $$\mu_{a+1}\left(\bigcup_{i=1}^{\infty} F_i\right) = \vee\left\{\mu^*_a(F) \,\middle|\, F \in \amalg_\alpha \wedge \left(F \subseteq \bigcup_{i=1}^{\infty} F_i\right)\right\}$$
4. For a limit $\beta, \mu_\beta(A) = \mu_\alpha(A)$ for some $\alpha < \beta$ satisfying $A \in \Sigma_\alpha$

We must verify that this construction of $\mu_{\omega 1}$ is well defined, and is indeed a measure (in the σ-algebra sense). Note that $\mu_{a+1}(A)$ does not depend on the choice of countable family $\langle F_i \mid i < \omega \rangle$ in Π_α.

We start by proving that all the $\mu_\alpha s$ are additive, in the sense that

$$\mu_\alpha(A) + \mu_\alpha(B) = \mu_\alpha(A \cup B) + \mu_\alpha(A \cap B)$$

It is immmediate that μ_0 is additive, as it is a measure (in the localic sense) on $L = \Sigma_0$. Assume that μ_α is a additive. Then it is immediate from DeMorgan's laws that μ_α^* is also additive.

Suppose that μ_α is additive, and consider $A, B \in \Sigma_{\alpha+1}$. Then

$$\begin{aligned}
&\mu_{\alpha+1}(A) + \mu_{\alpha+1}(B) \\
&\bigvee\{M - \mu_\alpha(F) \mid F \in \Sigma_\alpha \wedge F \cap A = \theta\} \\
&+ \bigvee\{M - \mu_\alpha(G) \mid G \in \Sigma_\alpha \wedge G \cap B = \theta\} \\
&= 2M - \bigwedge\{\mu_\alpha(F) + \mu_{\alpha(G)} \mid F, G \in \Sigma_\alpha \wedge (F \cap A) = (G \cap B) = \theta\} \\
&= 2M - \bigwedge\{\mu_\alpha(F \cup G) + \mu_{\alpha(F \cap G)} \mid F, G \in \Sigma_\alpha \wedge (F \cap A) = (G \cap B) = \theta\} \\
&= 2M - \bigwedge\{\mu_\alpha(D) + \mu_{\alpha(E)} \mid D, E \in \Sigma_\alpha \wedge (D \cap (A \cup B) = (E \cap (A \cap B)) = \theta\} \\
&= \bigvee\{M - \mu_\alpha(D) \mid D \in \Sigma_\alpha \wedge D \cap (A \cup B) = \theta\} \\
&+ \bigvee\{M - \mu_\alpha(E) \mid E \in \Sigma_\alpha \wedge E \cap (A \cap B) = \theta\} \\
&= \mu_{\alpha+1}(A \cap B) + \mu_{\alpha+1}(A \cup B)
\end{aligned}$$

The fact that μ_α is additive at limit stages is immediate.Now that we have shown that the $\mu_\alpha S$ are additive, it is immediate that for $\alpha < \beta, \mu_\beta$ extends μ_α. In turn, this result shows that μ_β is well defined for limit ordinals β. Finally, the fact that the μ_αs have the required continuity condition is also immediate from the definition, and the fact that the $\Pi_\alpha s$ are closed under finite unions.

THE PRESHEAF S

In this section, we make the following notational conventions. ε is a topos (with natural numbers object), $\mathbb{Q}$ is the object of positive rational numbers in ε, Ω is the subobject classifier in ε, $(L \leq, \perp, \perp)$ is a locale, (possibly, although not necessarily, the locale of countably complete sieves on some σ-algebra), and $\hat{L}$ is the topos of presheaves on L.

We construct an object S of ε. S is called the semireal numbers object.

Definition: The object S of semireals in ε is the subobject of $P\mathbb{Q}$ characterized by the formula

$$\phi(s) \equiv \forall q \in \mathbb{Q}(q \in S \Leftrightarrow \forall r \in \mathbb{Q} q + r \in S)$$

These objects are called semireal numbers because they contain half the data of a Dedekind real; they have an upper cut, but no lower cut. Johnstone and Reichman call them semicontinuous numbers, but that terminology is confusing here as we are using a different notion of semicontinuity to discuss measures.

The justification for calling these numbers "semicontinuous" stems from the fact that if they are interpreted in the topos of sheaves on a topological space, then these numbers do indeed correspond to semicontimuous real valued functions, just as Dedekind real numbers correspond to continuous real valued functions.

Although the semireals can be interpreted in any topos (with natural numbers object), they have a special interpretation in the topos of presheaves over some poset P.

Definition: Let $(p, \leq)$ be a poset, and let $\widehat{p}$ be the topos of presheaves on P. Say that S is the presheaf of order preserving functionals on P if

$$S(P) = \{s : \downarrow P \to [0,\infty] \mid A \Rightarrow s(A) \leq s(B)\}$$

Theorem: Let P be a poset, and let $\widehat{p}$ be the topos of presheaves on P. Then inside $\widehat{p}$ we have

$$S \cong S'$$

In order to study the elements of $S(P)$, we use the following Lemma:

Lemma: Assume that $\varepsilon = \widehat{p}$ for some poset P. A subfunctor $S \to \mathbb{Q}$ is a semireal if and only if for every $\varepsilon \in P$, $S(A)$ is a topologically closed upper segment of the positive rationals.

Note that a "topologically closed upper segment of the positive rationals" is the same thing as "the set of all positive rationals greater than or equal to some extended real $x \in [0,\infty]$".

Proof: S is a subobject of $P\mathbb{Q}$. Therefore if $S \in S(A)$, then S is a subfunctor of $\mathbb{Q}$ satisfying $S(B) \subseteq S(C)$, whenever $C \leq B \leq A$ (and $S(D) = \theta$ for and $D \not\leq A$). We can interpret the formula $\phi(S)$ that characterizes S by using Kripke-Joyal sheaf semantics. In this framework, we can write $P \Vdash \phi(S)$ for $S \in S(P)$.

$$p \Vdash \forall q \in \dot{\mathbb{Q}}(q \in S \Leftrightarrow \forall r \in \mathbb{Q} q + r \in S)$$

$\leftrightarrow$ for every $\mathbb{Q} \leq$ P and every $q \in \mathbb{Q}$

$$\mathbb{Q} \Vdash q \in S' \Leftrightarrow \forall r \in \mathbb{Q} q + r \in S'$$

where S' is the restriction of S to Q

$\leftrightarrow$ for every q $\in \mathbb{Q}$ and every $Q \leq P$ we have

$$Q \Vdash (q \in S) \Rightarrow (\forall r \in \mathbb{Q} q + r \in S) \text{ and}$$

$$Q \Vdash \forall r \in \mathbb{Q}(q + r \in S \Rightarrow q \in S)$$

If $Q \Vdash (q \in S) \Rightarrow (\forall r \in \mathbb{Q} q + r \in S)$ then for every $R \leq \mathbb{Q}$ such that $q \in S(R)$, we must have

$$R \Vdash \forall r \in \mathbb{Q}(q + r) \in S'$$

But this is just equivalent to saying that if $\mathbb{Q}$ is an element of $S(R)$ then all rationals greaterthan $\mathbb{Q}$ are also elements of $S(R)$. Hence $S(R)$ is an upper segment of rationals.

Now, if $Q \Vdash \forall r \in \mathbb{Q}(q + r) \in S \Rightarrow q \in S$, then for every $R \leq \mathbb{Q}$ such that

$$\forall r \in \mathbb{Q}(q + r) \in S(R)$$

We must have $q \in S(R)$. This means that if all the rationals greater than $\mathbb{Q}$ are elements of $S(R)$, then $\mathbb{Q}$ must also be an element of $S(R)$. Hence $S(R)$ is (topologically) closed. Since R is an arbitrary element of $\downarrow P$, it follows thta $S(P)$ is a topologically closed upper segment of rationals. We can now prove Theorem.

Proof: Fix $P \in P$. We construct a bijection between $S(P)$ and $S(P)$. Take $S \in S(P)$. Then let the order preserving functional $s{:}\downarrow P \to [0,\infty]$ be given by

$$s(Q) = \wedge S(Q)$$

Now, given an order preserving functional $t \in S'(P)$, we define $T \in S(P)$ by

$$T(Q) = \{q \in \mathbb{Q} | t(Q) \leq q\}$$

It is immediate that the two operations are inverse to one another. The fact that s is an order preserving map is a consequence of the fact that S is a subobject $yA \times \mathbb{Q}; R \leq Q \leq p$ implies that $S(R) \supseteq S(Q)$, and so that $s(R) \rightharpoonup s(Q)$.

As with other number systems, the semireals have a number of important properties.

Proposition: There is an embedding $Q \to S$ given by

$$q \mapsto \{r \in \mathbb{Q} \mid q \leq r\}$$

Proof: First note that if we are working in the topos of presheaves on a poset, then the result is an immediate consequence of Lemma.

We work internally in ε. Fix $q \in \mathbb{Q}$. We show that $\{r \in \mathbb{Q} \mid q \leq r\}$ is a semireal.

$$q = \{r \in \mathbb{Q} | q \leq r\} = \{r \in \mathbb{Q} | q < r \vee q = r\}$$

We need to show that $(r \in q) \Leftrightarrow \forall s \in \mathbb{Q}(r+s \in q)$. Suppose that $r \in q$. Then $\mathcal{Q} < r$ or $\mathcal{Q} = r$. In either case, $\mathcal{Q} < r + s$, and so $r + s \in q$.

Conversely, suppose that $\forall s \in \mathbb{Q}(r + s) \in q$. To show that r—q, we exploit the fact that the rationals are totally ordered, and so satisfy the following formula:

$$\forall r \in \mathbb{Q}(r < q) \vee (q \leq r)$$

Suppose that $r < q$. Then let $s = \frac{q-r}{2}$.2. Then $r + s < q$. Hence $(r + s) \notin q$. This is a contradiction, and so we must have $\leq \mathbb{Q}\ r$, as required. In view of Theorem, it would obviously be convenient to have some form of evaluation operation for the functionals.

Unfortunately, there is no natural way to do this directly. Suppose we were to try for a morphism of the form $\Omega \times S \rightarrow \mathbb{R}$, where is the subobject classifier, and $\mathbb{R}$ the object of Dedekind real numbers. In order for such a morphism to be a natural transformation, we would need the following diagram to commute (for $R \leq Q$):

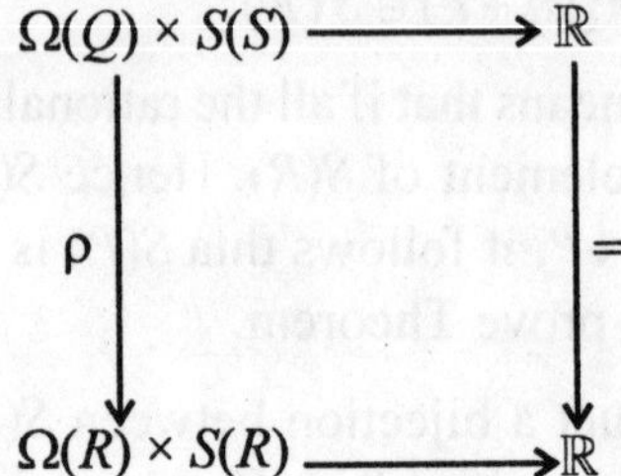

However, the object of Dedekind real numbers in a presheaf topos is just the constant functor, and so the right hand restriction here is just equality. But since $s(Q) \neq s(R)$, in general, our evaluation map wold not be compatible with thisrestriction.

Lemma: Let $\widehat{\mathbb{C}}$ be a presheaf topos. Then the object $\mathbb{R}$ of Dedekind reals in $\widehat{\mathbb{C}}$ is a constant functor whose value at every object C of $\mathbb{C}$ is just the set $\mathbb{Q}$ of real numbers.

Proof: It is well known that in a presheaf topos $\widehat{\mathbb{C}}$, the rational numbers object $\mathbb{Q}$ is just the presheaf ΔQ, whose action at every object C of $\mathbb{C}$ is just the set $\mathbb{Q}$ of rationals.

Let D be the Dedekind real numbers object of $\widehat{\mathbb{C}}$.

Then an element of $D(C)$ is a pair $\langle L, U\rangle$ of subfunctors of $yC \times \mathbb{Q}$. For any object D, $L(D)$ is a family $\langle S_f \mid f \in \hom(D,C)\rangle$ open lower sets of rationals. Likewise $U(D)$ is a family $\langle T_f \mid f \in \hom(D,C)\rangle$ of open upper sets of rationals. Following the arguments in Theorem we can construct functionals l and u from $t_{C,}$ the maximal sieve on C, to R, the set of reals. These functional are given by

$$l(f) = \vee s_f \quad s(f) = \wedge T_f$$

(Note that S_f and T_f are members of $L(\mathrm{dom}(f))$ and $U(\mathrm{dom}(f))$ respectively.) There is a preordering on t_C. Write $f \leq g$ if there is an h: $\mathrm{dom}(f) \to \mathrm{dom}(g)$ such that

$$f^{\circ} h = g:$$

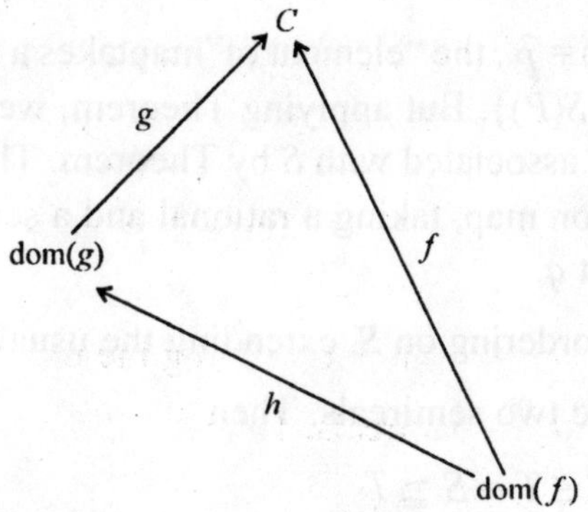

Note that id_C is the top element of this preorder. If $\mathbb{C}$ has an initial object **0**, then theinitial map!:: $0 \to C$ is the minimal element. It is easy to see that l is an order reversing functional, and that u is order preserving.

We have used most of the axioms of a Dedekind real in order to build l and u. However, we have not used the disjointness and apartness axioms:

$$\widehat{\mathbb{C}} \Vdash \forall q \in \mathbb{Q} \neg (q \in L \wedge q \in U)$$

$$\widehat{\mathbb{C}} \Vdash \forall q, r \in \mathbb{Q} (q < r) \Rightarrow (q \in L \vee r \in U)$$

Since elements of $\mathbb{Q}$ are just constant rational numbers, we can represent a rational as a constant functional $\tilde{q} : t_C \to R$. The above conditions can now be rewritten in terms ofthe functionals l and u:

$$\widehat{\mathbb{C}} \models \forall q \in \mathbb{Q} \neg (l < \tilde{q}) \wedge (\tilde{q} < u))$$

$$\widehat{\mathbb{C}} \models \forall q, r \in \mathbb{Q} \neg (q < r) \Rightarrow ((\tilde{q} < l) \vee (\tilde{r} < u)$$

Since $\widehat{\mathbb{C}}$ is a presheaf topos, it follows that for every arrow $f \in t_C$, we must have

$$\forall q \in Q \neg (l(f) < q) \wedge (q < u(f)))$$

$$\forall q, r \in Q (q < r) \Rightarrow ((q < l(f) \vee (r < u)(f)))$$

But this implies that for any f, $\langle S_f, T_f \rangle$ is a Dedekind real in S_{ET} (that is, a real number). Furthermore, S_f and T_f must be independent of f, as $\langle S_{idc}, T_{idc} \rangle$ is also a Dedekind real, and we must have

$$S_f \supseteq S_{idc} T_f \supseteq T_{idc}$$

So, we cannot have a direct evaluation map for the semireals. However, we do have an indirect evaluation map. We can use the following composition:

$$Q \times S \mapsto Q \times PQ \xleftarrow{\in} \Omega$$

In the special case where $\varepsilon = \widehat{p}$, the "element of"maptakes a rational $\mathbb{Q}$ and a semireal S to the sieve $I = \{P \in P | q \in S(P)\}$. But applying Theorem, we see that $I = \{P \in P | s(\mathrm{P}) \leq \mathrm{q}\}$, where s is the functional associated with S by Theorem. This map, the "element of" map will serve as our evaluation map, taking a rational and a semireal to the sieve where the the semireal is smaller that q.

There is a natural partial ordering on S, extending the usual ordering on $\mathbb{Q}$.

Definition: Let S and T be two semireals. Then

$$S \leq T \equiv S \supseteq T$$

Note that in the event that $\varepsilon = \widehat{P}$, then this coincides with the usual ordering of functionals on P.

This ordering is just the reverse of the inclusion inherited from $P\mathbb{Q}$. It turns out that with this ordering, S is internally a complete lattice:

Proposition: Take $S \subseteq S$ and define $\vee S$ and $\wedge S$ by

$$\vee \mathrm{s} = \{\mathrm{q} \in \mathbb{Q} \mid \forall \mathrm{S} \mathrm{q} \in \mathrm{S}\}$$

$$\wedge \mathrm{S} = \{\mathrm{q} \in \mathbb{Q} \mid \forall \mathrm{r} \in \mathbb{Q} \exists \mathrm{S} \in \mathrm{S} \mathrm{q} + \mathrm{r} \in \mathrm{S}\}$$

Then $\vee$ and $\wedge$ are the supremum and infimum operators on S respectively.

Proof: It is immediate from Definition that if $\vee \mathrm{s}$ and $\wedge \mathrm{s}$ are indeed semireals then they must be the supremum and infimum of S respectively. Hence it suffices to show that they are semireals. But this is immediate from their definitions.

There are also a number of algebraic operations defined on S: Semireals can be added together, multiplied by a rational, and restricted to a truth value (or sieve, when working externally). All of these operations are defined using the internal logic of ε, treating semireals as certain sets of rationals. We define addition first:

$$S + T + = \{q \in \mathbb{Q} \mid \forall r \in \mathbb{Q} \exists s \in S \exists t \in T(s + t = q + r)\}$$

Proposition: The addition of two semireals, as defined above, does indeed yield a semireal.

Proof: Let S and T be two semireals.

Firstly, we show that if $q \in S + T$ and $u \in \mathbb{Q}$, then $(q + u) \in (S + T)$. Take $r \in \mathbb{Q}$. Then there exist $s \in S$ and $t \in T$ such that $\mathbb{Q} + (u + r) = s + t$. But this is all that is needed to show that $(q + u) \in (S + T)$. This shows that $S + T$ is an upper segment.

Now, assume that $(q + u) \in (S + T)$ for every $u \in \mathbb{Q}$. We need to show that Take $r \in \mathbb{Q}$. We know that $\left(q+\frac{r}{2}\right) \in (S+T)$, so there must be $s \in S$ and $t \in T$ such that

$$\left(q+\frac{r}{2}+\frac{r}{2}\right)=s+t\left(q+\frac{r}{2}+\frac{r}{2}\right)=s+t$$

Consequently,

$$\forall r \in \mathbb{Q} \exists s \in S \exists t \in T (q+r)=(s+t)$$

But this implies that $q \in S + T$, as required.

Multiplication of a semireal by a rational is also defined internally:

Definition: Take $a \in \mathbb{Q}$ and $S \in S$, Then the product $a \times S$ is given by:

$$(a \times S)=\left\{q \in \mathbb{Q} \mid \frac{q}{a} \in S\right\}$$

Note that the right hand side here is just $\{a \cdot q | q \in S\}$.

Proposition: Multiplication of a semireal S by a rational a as described above does indeed yield a semireal.

Proof: Take $q \in a \times S$, and $r \in \mathbb{Q}$. Since $\frac{q}{a} \in S,$ it follows that $\frac{q+r}{a} \in S.$ But this is just what is needed to prove that $(q + r) \in a \times S$.

Now assume that for every $r \in \mathbb{Q}$ we have $(q + r) \in (a \times S)$. Then for every $r \in \mathbb{Q}$ we have $\frac{q}{a}+\frac{r}{a} \in S.$ Putting $s=\frac{r}{a}$, this is equivalent to saying that for every $s \in \mathbb{Q}$ we have $\frac{q}{a}+s \in S.$ Since S is a semireal, this in turn implies that $\frac{q}{a} \in S$, whence $q \in a \times S$, as required.

It is clear that $\mathbb{Q}$ is a commutative division semiring (a field, except without additive inverses, and without zero).

Proposition: The object S of semireals is a semimodule over $\mathbb{Q}$, with the operations of addition and scalar multiplication as defined above.

Proof: $\langle S,+\rangle$ is clearly an abelian monoid (associativity of addition is easy to check). Thus we need only show that for any a, $b \in \mathbb{Q}$ and $S, T \in S$, we have

1. $a \times (S + T) = (a \times S) + (a \times T)$
2. $(a + b) \times S = a \times S + b \times S$

3. $a \times (b \times S) = (a \cdot b) \times S$

4. $1 \times S = S$

5. Suppose that $q \in a \times (S + T)$. Then $\frac{q}{a} \in S + T$.. This means that for any $r \in \mathbb{Q}$, there exists $s \in S$ and $t \in T$ such that $s + t = \frac{q}{a} + r$. Consequently, $a \cdot s \in a \times S$ and $a \cdot t \in a \times T$.Hence

$a \cdot s + a \cdot t = \mathcal{Q} + a \cdot r$

which, since a is fixed, and r is arbitrary, shows that $q \in a \times S + a \times T$.

For the converse direction, suppose that $q \in a \times S + a \times T$. Then for any $r \in \mathbb{Q}$, there exist $s \in q \times S$ and $t \in a \times T$ such that $s + t = \mathcal{Q} + r$. Since $\frac{s}{a} \in S$, and $\frac{t}{a} \in T$, it follows that $\frac{q}{a} \in S + T$, whence $q \in a \times (S + T)$, as required.

7. Suppose that $q \in a \times S + b \times S$. Then for any $r \in \mathbb{Q}$ we know that there exist s_1, $s_2 \in$ such that $as_1 + bs_2 = \mathcal{Q} + r$. Since the rationals are totally ordered, we may assume without loss of generality that $s_1 \leq s_2$, so that $s_2 = s_1 + d$. Hence $(a + b)s_1 \leq \mathbb{Q} + r$ and so, for every $r \in \mathbb{Q}$, we have

$$S_1 \leq \frac{q+r}{a+b}$$

Therefore $\mathcal{Q} + r \in (a + b) \times S$, whence $q \in (a + b) \times S$, as required.

For the converse direction, suppose that $q \in (a + b) \times S$. The $s = \frac{q}{a+b} \in S$.a+b $\in$ S. It will suffice to find s_1 and s_2 in S such that $a \cdot s_1 + b \cdot s_2 = q$. Put $s_1 = s_2 = s$. Then

$$a \cdot s_1 + b \cdot s_2 = a \cdot s + b \cdot s$$
$$= (a + b) \cdot s$$
$$= (a + b)\frac{q}{a+b}$$ as required.

8. This is immediate.

9. This is also immediate.

With this semimodule structure established, we can now study the restriction operation.

Definition: The restriction operator $\rho : S \times \Omega \to S$ is defined internally:

$$\rho(I, S) = \{q \in \mathbb{Q} | I \Rightarrow q \in S\}$$

Lemma: Take $I \in \Omega$ and $S \in S$. Then $\rho(S,I)$, as described above, is indeed a semireal.

Proof:

$$I \Rightarrow q \in S \leftrightarrow I \Rightarrow (\forall r \in \mathbb{Q} q + r \in S)$$
$$\leftrightarrow \forall r \in \mathbb{Q}[I \Rightarrow (q + r \in S)]$$
$$\leftrightarrow \forall r \in \mathbb{Q}[q + r \in \rho(S,I)]$$

The restriction operation, $\rho : S \times \Omega \rightarrow S$ can be thought of as an Ω indexed family of linear maps from the semimodule S to itself.

Proposition:

1. For any $I \in \Omega$, the operation $\rho : (-,1) : S \rightarrow S$ is a linear map.
2. For a fixed $S \in S$, the operation $\rho : (S,) : \Omega \rightarrow S$ is an order preserving map.
3. For a fixed $S \in S$, we have $\rho : (S, \perp)$ and $\rho(S, \perp) = 0$, where 0 is the bottom element of S.

Proof: To see that $\rho(-,I)$ preserves sums, note that the following argument is intuitionistically valid:

$$q \in \rho(S+T, I)$$
$$I \Rightarrow (q \in S + T)$$
$$\leftrightarrow I \Rightarrow \forall r \in \mathbb{Q} \exists s,t \in \mathbb{Q}(s \wedge St \in T) \wedge (s + t = q + r)$$
$$\leftrightarrow \forall r \in \mathbb{Q} \exists s,t \in \mathbb{Q}[I \Rightarrow (s \in S \wedge t \in T] \wedge [s + t = r)$$
$$\equiv q \in \rho(S,I) + \rho(T,I)$$

The fact that $\rho(-,I)$ preserves scalar multiplication is immediate.

Suppose that $I \leq J$ Then $J \Rightarrow q \in S$, implies that $I \Rightarrow q \in S$, so that $\rho(S,I) \supseteq (S,J)$, as required.

This is immediate.

Our goal is to provide a logical construction (in ε) of M, the object of measures. We take as our data not just ε, but also a topology j on ε. This means that we say that M is the measure object of ε, relative to the topology j. In the special case where $\varepsilon = \hat{L}$, and j is the canonical topology on L, then M is a j-sheaf.

It turns out that we will only ever need to take the restriction to closed truth values (or closed sieves, in the external view). Hence we take ρ to have $\Omega_j \times S \rightarrow \Omega \times S$ as its domain.

In the case that $\varepsilon = \widehat{P}$ (of course, this case subsumes the case where $\varepsilon = \widehat{L}$) all of the operations that we have defined on S have the natural interpretations when applied to the associated functionals.0

Proposition: Suppose that $\varepsilon = \widehat{P}$ is the topos of presheaves on some poset P. Take $S, T \in S(P)$,

$\{S_i | i \in I\} \subseteq S(p), I \in \Omega(P)$ and $a \in \mathbb{Q}(P)$, and let s, t, $\{s_i | i \in I\}$ be the associated functionals. Then:

1. The associated functional of $S + T$ is $s + t$
2. The associated functional of $a \times S$ is $a \cdot s$
3. The associated functional of $\rho(S, I)$ is given by

$$\rho(s, I)(Q) = \bigvee_{R \in \downarrow Q \cap I} s(R)$$

4. $S \leq T$ if and only if $s \leq t$ s.
5. The associated functional $\bigvee_{i \in i} S_i$ is $\bigvee_{i \in i} s_i$

Proof: Except for part 3, this is immediate from Theorem.

For part 3, we can use sheaf semantics. Recall that $s(P) \leq q \leftrightarrow P \Vdash q \in S$. Then

$$\rho(s, I)(P) \leq q \leftrightarrow P \Vdash q \in \rho(S, I) \ \$ \ \mathrm{P\,I} \Rightarrow \mathrm{q} \in \mathrm{S}$$

$$\leftrightarrow \text{for all } R \in I \text{ such that } R \leq P, R \Vdash q \in S$$

for all $\leftrightarrow R \in I$ such that $R \leq P, S(R) \leq q$

Note that in the case where P is a meet semilattice, part 3 can be rewritten

$$\rho(s, I)(Q) = \bigvee_{R \in I} S(R \wedge Q)$$

Furthermore, if P is a locale, and I is a closed sieve, then there is an $I \in L$ such that $I = \downarrow I$ and we can write

$$\rho\,(s, I)(Q) = s(I \wedge Q)$$

This observation provides the motivation for ρ calling the "restriction" operation.

THE CONSTRUCTION OF M

In this section, we work in a topos ε (with natural numbers object). Ω is the subobject classifier in ε, $\mathbb{Q}$ is the object of positive rationals in ε, and S is the object of semireal numbers in ε. We assume that there is a designated topology $j: \Omega \to \Omega$, which induces the sheaf topos $Sh_j(\varepsilon)$. The subobject classifier in $Sh_j(\varepsilon)$ is denoted j. We refer to ε as the "presheaf topos", and $Sh_j(\varepsilon)$ as the "sheaf topos". Sometimes, we make the additional assumption that ε is the topos of presheaves on some locale ε. In this case, *j* will be the canonical topology on *L*.

In this section we construct a subobject *M* of *S*. In the special case where $\varepsilon = \hat{L}$, *M* is the presheaf of measures, and is in fact a sheaf.

To construct *M*, we find logical formulas that pick out those semireals satisfying the additivity and semicontinuity conditions.

We start with additivity.

Definition: The additive semireals are semireals satisfying the following formula:

$$\phi(S) \equiv \forall IJ \in \Omega_j[\rho(S,I) + \rho(S,J) = \rho(S, I \wedge J) + \rho(S, I \vee J)]$$

(where " $\wedge$ " and " $\vee$ " are the meet and join in Ω_j.)

The presheaf of additive semireals is denoted $\mathcal{S}_A \to \mathcal{S}$

Proposition: Suppose that $\varepsilon = \hat{L}$ is the topos of presheaves on some locale *L*. Let *j* be the cannonical topology. Then a semireal $S \in \mathcal{S}(A)$ is additive if and only if the associated functional $s : \downarrow A \to [\infty, 1]$ satisfies

$$s(B) + s(C) = s(B \wedge C) + s(B'' \vee C)$$

Proof: This direction follows immediately from Proposition, by considering the closed sieves $\downarrow B$ and $\downarrow C$.

$\Leftarrow$ For the reverse direction, we can use the fact that any closed sieves *I* and *J* are in fact principal sieves $\downarrow I$ and $\downarrow J$ respectively. Then taking an arbitrary $A \in L$, we have

$$\begin{aligned}
\rho(s, I)(A) + \rho(s, J)(A) &= \rho(s, \downarrow I)(A) + \rho(s, \downarrow J)(A) \\
&= s(A \wedge I) + s(A \wedge J) \\
&= s((A \wedge I) \wedge (A \wedge J)) + s((A \wedge I) \vee (A \wedge J)) \\
&= s(A \wedge (I \wedge J)) + s(A \wedge (I \vee J)) \\
&= \rho(s, \downarrow (I \wedge J))(A) + \rho(s, \downarrow (I \vee J))(A) \\
&= \rho(s, I \wedge J)(A) + \rho(s, I \vee J)(A)
\end{aligned}$$

Characterizing the semicontinuity condition requires some preparatory steps.

Definition: $I \in \Omega$ is called "directed" if it satisfies the condition:

$$\forall J,K \in \Omega_j J \vee K \le I \Rightarrow \overline{J \vee K} \le I$$

Using this formula, we find an object $\Omega_D \to \Omega$ of directed truth values. It is easy to see that in $\hat{L}, \Omega_D(A)$ consists of the ideals in $\downarrow$ A.

Definition: A sieve $I \in \Omega$ is called "directed closed" if it satisfies the condition:

$$\forall J \in \Omega_D J \le I \Rightarrow \overline{J} \le I$$

If $\varepsilon = \hat{L}$ then the directed closed sieves are those that are closed under directed joins.

As an example of a directed closed sieve that is not closed, fix A, $B \in L$, and let

$$I = (\downarrow A) \cup (\downarrow B)$$

The following is immediate:

Proposition:

$$\Omega_J \to \Omega_{DC} \to \Omega_D \to \Omega$$

Before we describe the semicontinuity condition, we need to introduce a notational convention. One feature of topos logic is that logical formulas are themselves arrows in the topos (arrows into the subobject classifier Ω).

Since the Ω is object of truth values, the arrow into Ω can be thought of as representing the truth value of a formula. To avoidambiguity, when we refer to the truth value of a formula ϕ,we shall use Scott brackets: $[\![\phi]\!]$. For example, if $\mathbb{Q}$ is a rational number, and S is a semireal, then $[\![q \in S]\!]$ is the truth value of the formula $q \in S$. If we are working in a presheaf topos $\hat{L}$, then $[\![q \in S]\!]$ is the sieve of those $A \in L$ such that $q \in S(A)$.

We can now describe the semicontinuity condition for semireals.

Definition: The object $\mathbb{S}_C$ of semicontinuous semireals is defined by the following formula

$$\mathbb{S}_C \{S \in \mathbb{S} \mid \forall q \in \mathbb{Q} [\![q \in S]\!] \in \Omega_{DC}\}$$

In the case where $\varepsilon = \widehat{P}$, an order preserving functional $s : \downarrow A \to [0,\infty]$ corresponds to an element of $\mathbb{S}_C(A) \subseteq \mathbb{S}(A)$ if for every rational q, the set

$$\{B \le A \mid s(B) \le q\}$$

is closed under directed joins.

It follows immediately that:

Theorem: Let $\varepsilon = \hat{L}$ be the topos of presheaves on a locale L. Then M, the presheaf of measures is defined by the following pullback:

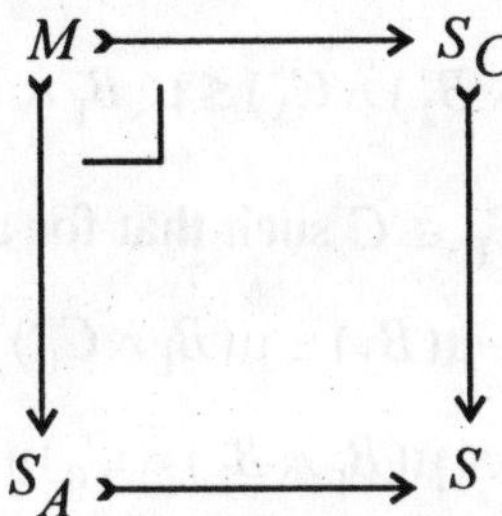

Equivalently, a measure is a semireal that is both additive and semicontinuous. Internally, this can be written:

$$M = \mathcal{S}_A \cap \mathcal{S}_C \subseteq \mathcal{S}$$

Theorem: Let M be the presheaf of measures in the topos $\varepsilon = \hat{L}$. Then M is a sheaf.

The are two approaches to proving this Theorem. The first option is to use the external interpretation of M as the presheaf of measures, and show that measures can be amalgamated in the appropriate ways. The second approach is to use the logical characterization of M.

We use a combination of the two approaches. Explicit reference is not made to the fact that the elements of $\mathcal{S}$ (A) are measures. However, we do make use of the fact that $Sh_j(\varepsilon)$ is a localic topos.

Proof: We must show that given $A \in L$, a cover C for A, and a matching family $M = M = \langle \mu_C \in M(C) | C \in C \rangle$ and C, there is a unique $\mu \in M(A)$ which is an amalgamation for M. We do this by looking at the nature of C. Recall that in a locale L, a sieve $C \subseteq L$ is a cover for A if and only if $\gamma C = A$.

First, consider the case where C is directed. In this case, since measures must satisfy the semicontinuity condition, it follows that if μ is an amalgamation, we must have

$$\mu(B) = \vee\{\mu(B \wedge C) | C \in C\}$$

Hence, if an amalgamation exists, it must be unique. We now show that the μ defined above is a measure, and that it is an amalgamation of M. The latter follows immediately from the fact that M is a matching family. To see that μ is a measure, take $B_1, B_2 \leq A$. Then

$$\mu(B_1) + \mu(B_2) = \vee\{\mu(B_1 \wedge C) | C \in C\} + \vee\{\mu(B_2 \wedge C) | C \in C\}$$

Fix an $\varepsilon > 0$. Then there exists C_1, C_2, C_3, C_4 in C such that

$$\mu(B_1 \wedge C_1) \leq \mu(B_1) \leq \mu(B_1 \wedge C_1) + \frac{\varepsilon}{2}$$

$$\mu(B_2 \wedge C_2) \leq \mu(B_2) \leq \mu(B_2 \wedge C_2) + \frac{\varepsilon}{2}$$

$$\mu((B_1 \wedge B_2) \wedge C_3) \leq \mu(B_1 \wedge B_2) \leq \mu((B_1 \wedge B_2) \wedge C_3) + \frac{\varepsilon}{2}$$

Since C is directed, there is a $C_0 \in C$ such that for all $i \in \{1,..., 4\}$, $C_i \leq C_0$. Then

$$\mu(B_1) + \mu(B_2) \leq \mu(B_1 \wedge C_0) + \mu(B_1 \wedge C_0) + \varepsilon$$

$$\mu(B_1 \vee B_2) + \mu(B_1 \wedge B_2) \leq \mu(B_1 \wedge B_2) \wedge C_0) + \mu((B_1 \vee B_2) \wedge C_0) + \varepsilon$$

Since $\mu(B_1 \vee C_0) + \mu(B_1 \wedge C_0) = \mu(B_1 \wedge B_2) \wedge C_0) + \mu((B_1 \vee B_2 \wedge C_0)$ it follows that the difference between $\mu(B_1) + \mu(B_2)$ and $\mu(B_1 \wedge B_2) + \mu(B_1 \vee B_2)$ must be less than ε. Hence the amalgamation is additive. We must show that μ satisfies the semicontinuity condition. Take an increasing chain

A in. $\downarrow A \downarrow A$. Then

$$\mu(\gamma A) = \vee\{\mu((\gamma A) \wedge C) | C \in C\}$$

$$= \vee\{\mu(\gamma A \wedge C) | C \in C\}$$

$$= \vee\{\vee\{\mu(A \wedge C) | C \in C\} | A \in A\}$$

$$= \vee\{\vee\{\mu(A \wedge C) | C \in C\} | A \in A\}$$

$$= \vee\{\{\mu(A) | A \in A\}$$

This shows that if I is a directed sieve, then the sieve of elements of L for which a matching family for Mon I can be amalgamated contains $\bar{I}$.;

We now show that given any sieve I, and matching family, that family can be uniquely extended to some directed sieve that contains I, namely the closure of I under the finite join topology. To do this it suffices to show that if B and C are elements of L satisfying $B \vee C = A$, and μ_B and μ_C are measures on $\downarrow B$ and $\downarrow C$ respectively, which match on $(\downarrow B) \wedge (\downarrow C)$, then there is a unique $\mu \in M(A)$ such that the restriction of μ to B is μ_B and the restriction of μ to C is μ_C. Define μ by

$$\mu(D) = \begin{cases} \{\mu_B(B \wedge D) + \mu_C(C \wedge D) - \mu_B(B \wedge C \wedge D); \text{ if } (B \wedge C \wedge D) < \infty \\ \infty \quad \textit{otherwise} \end{cases}$$

It is clear that this is the only possible amalgamation. We merely need to verify that this is indeed a measure. However, both the additivity and semicontinuity conditions follow immediately from the fact that μ_B and μ_C satisfy these conditions.

Hence, given a matching family for M on a sieve I, that matching family may be uniquely extended to a certain directed sieve containing I, and then amalgamated to γI. Hence M is a sheaf.

Corollary: Let F be σ-algebra, and let M be the presheaf of measures on F. Then F is a sheaf with respect to the countable join topology.

Proof: Breitsprecher's original proof of this result involved an argument explicitly using measures on σ-algebra. However, the result can now be proved using the fact that the presheaf of measures on a locale is a sheaf.

Let F be a -algebra, and let L be the locale of countably complete sieves on F. Then every presheaf on L can be restricted to a presheaf on F. When this restriction is carried out on a sheaf on L, the result is a sheaf on F. In fact, this operation is one direction of the equivalence $Sh(F \Rightarrow \simeq Sh(L)$.

But, according to Theorem, applying this restriction to M, the presheaf of measures on L yields the presheaf of measures on F. Since M is in fact a sheaf, it follows that the corresponding presheaf of measures on F is a sheaf on F. We can go further.

Theorem: Let F be a σ –algebra, and let j be the countable join topology. Then carrying out the logical construction of M in the topos $\widehat{F}$ (relative to the topology j) yields the sheaf of (σ- algebra) measures.

Proof: The argument is the same as for showing that in $\widehat{L}$, M is the (pre)sheaf of localic measures: M_A is easily seen to be the presheaf of finitely additive measures, and M_C is the sheaf of semicontinuous measures.

Corollary: Let B be a Boolean algebra, and let j be the finite join topology. Then carrying out the logical construction on M in the topos $\widehat{B}$ (relative to the topology j) yields the presheaf of finitely additive measures. Furthermore,M is a sheaf.

PROPERTIES OF *M*

Throughout this section we work in a fixed elementary topos ε(with natural numbers object). We will designate a Lawvere-Tierney topology j: $\Omega \rightarrow \Omega$, which induces a subtopos $Sh_j(\varepsilon)$. ε shall be referred to as the presheaf topos, and $Sh_j(\varepsilon)$ as the sheaf topos.Ω shall denote the subobject classifier in ε, and Ω_j shall denote the subobject classifier in $Shj(\varepsilon)$. \$ will denote the object of semireals in ε, and M the object of measures as defined in the previous Sections.

We will assume that M is a j-sheaf. As a consequence of Theorem, we know that the case where ε is the topos of sheaves on a locale L, and j is the canonical topology on ε, then M is automatically a sheaf, and so working in ε and $Sh_j(\varepsilon)$ can be thought of as generalizing this case.

Beyond the assumption that M is a sheaf, we will not assume anything about the structure of ε or $Shj(\varepsilon)$.

M is a subobject of $\mathcal{S}$ (in ε), and inherits many of its properties:

Lemma: The following arrows factor through $M \to S$:

$$M \times \Omega_j \mapsto S \times \Omega_j \xleftarrow{\rho} S$$

$$M \times M \mapsto S \times S \xleftarrow{+} S$$

$$Q \times M \mapsto Q \times S \xleftarrow{\times} S$$

Proof: Fix $I \in \Omega_j$, and $\mu \in M$. We want to show thatr $\rho(\mu, I) \in M$. To do this, we must show that (μ, I) is both additive and semicontinuous.

For additivity, first note that for arbitrary $J \in \Omega_j$, we have

$$\rho(\rho(\mu, I), J) = \{q \in \mathbb{Q} \mid J \Rightarrow (I \Rightarrow q \in \mu)\}$$
$$= \{q \in \mathbb{Q} \mid (I \wedge J \Rightarrow q \in \mu)\}$$
$$\rho(\mu, I \wedge J)$$

Now, fix $J, k \in \Omega_j$

$$\rho(\rho(\mu, I), J) + \rho(\rho(\mu, I), K)$$
$$\rho(\mu, I \wedge J) + \rho(\mu, I \wedge K)$$
$$\rho(\mu, (I \wedge J) \wedge (I \wedge K) + \rho(\mu, (I \wedge J) \vee (I \wedge K))$$
$$\rho(\mu, I \wedge (J \wedge K)) + \rho(\mu, I \wedge (J \vee K))$$
$$\rho(\rho(\mu, I), J \wedge K) + \rho(\rho(\mu, I), J \vee K)$$

To verify that $\rho(\mu, I)$ is semicontinuous, we need to show that for an arbitrary $q \in \mathbb{Q}$, the truth value

$$[\![q \in (\rho(\mu, I)]\!]$$

is directed closed.

We start by noting that

$$[\![q \in (\rho(\mu, I)]\!] = [\![I \Rightarrow q \in \mu]\!]$$

$$= [\![I]\!] \Rightarrow [q \in \mu]$$

Since $\llbracket I \rrbracket$ is closed, and $\llbracket q \in \mu \rrbracket$ is directed closed, the result will follow from the following Lemma:

Lemma: Let I be a closed truth value, and let J be a directed closed truth value. Then $I \Rightarrow J$ is also directed closed.

Proof: Since we are working explicitly with truth values, we do not need the Scott brackets to distinguish between formulas and their truth values. In this proof, the objects of discourse are truth values. The argument is similar to an argument in propositional logic. It is possible to notate the following argument in terms of the " $\leq$ " rather than the "$\Rightarrow$" symbol. Likewise, we could write "=" for "$\Leftrightarrow$ ". Such substitutions would be natural when thinking of the truth values as elements of a Heting algebra. However, for the sake of consistency, we work here with logical connectives. The topology j is a unary logical connective satisfying the axioms:

$$I \Rightarrow jI$$

$$jjI \Rightarrow jI$$

$$j(I \wedge J) \Leftrightarrow jI \wedge jJ$$

A truth value I is closed if and only if

$$jI \Rightarrow I$$

Recall that a truth value K is directed if for any closed truth values A and B we have

$$[(A \Rightarrow K) \wedge (B \Rightarrow K)] \Rightarrow [j(A \vee B) \Rightarrow K]$$

A truth value Z is directed closed, if for all directed K such that $K \Rightarrow Z$, we have $j(K) \Rightarrow Z$.

Start by taking truth values I and J such that I is closed and J is directed closed. In order to show that $I \Rightarrow I$ is directed closed, we take a directed truth value K, assume that $K \Rightarrow (I \Rightarrow J)$, and prove that $jK \Rightarrow (I \Rightarrow J)$. But

$$K \Rightarrow (I \Rightarrow J) \leftrightarrow (I \wedge K)\ J$$

and

$$j\mathrm{K} \Rightarrow (I \Rightarrow J) \leftrightarrow j(I \wedge K) \Rightarrow J$$

since I is closed.

Hence, in order to show that $I \Rightarrow J$ is directed closed, we just need to show that

$$[(I \wedge K) \Rightarrow J] \Rightarrow j(I \wedge K) \Rightarrow J$$

Since J is assumed to be directed closed, it suffices to prove that $I \wedge K$ is directed.

But this is trivial since I and J are both directed (since closed sieves are directed closed, and directed closed sieves are closed).

Take $\mu, v \in M$. To see that $\mu + v$ is a measure, we must verify that $\mu + v$ is both additive and semicontinuous.

For additivity, take $I, J \in \Omega_j$. Then

$$\rho(\rho(\mu, I), j) + \{q \in \mathbb{Q} | J \Rightarrow (I \Rightarrow \mathbb{Q} \in \mu)\}$$

$$= \{q \in \mathbb{Q} \mid (I \wedge J) \Rightarrow q \in \mu\}$$

$$= \rho(\mu, I \wedge J)$$

Now, fix $J, K \in \Omega_j$.

$$\rho(\rho(\mu, I), J + \rho(\rho(\mu, I), K)$$

$$= \rho(\mu, I \wedge J) + \rho(\mu, I \wedge K)$$

$$= \rho(\mu, (I \wedge J) \wedge (I \wedge K)) + \rho(\mu, (I \wedge J) \vee (I \wedge K))$$

$$= \rho(\mu, I \wedge (J \wedge K)) + \rho(\mu, I \wedge (J \vee K))$$

$$= \rho(\rho(\mu, I), J \wedge K) + \rho(\rho(\mu, I), J \vee K)$$

To verify the $\rho(\mu, I)$ semicontinuity condition, fix $q \in \mathbb{Q}$, the truth value

$$[\![q \in \rho(\mu, I)]\!]$$

is directed closed.

We start by noting that

$$[\![q \in \rho(\mu, I)]\!] = [\![I \Rightarrow q \in \mu]\!]$$

$$= [\![I]\!] \Rightarrow [\![q \in \mu]\!]$$

Since $[\![I]\!]$ is closed, and $[\![q \in \mu]\!]$ is directed closed, the result will follow from the following Lemma:

Lemma: Let I be a closed truth value, and let J be a directed closed truth value. Then $I \Rightarrow J$ is also directed closed.

Proof: Since we are working explicitly with truth values, we do not need the Scott brackets to distinguish between formulas and their truth values.

In this proof, the objects of discourse are truth values. The argument is similar to an argument in propositional logic. It is possible to notate the following argument in terms of the "$\leq$", rather than the "$\Rightarrow$" symbol. Likewise, we could write "=" for "$\Leftrightarrow$". Such substitutions would be natural when thinking of the truth values as elements of a Heting algebra. However, for the sake of consistency, we work here with logical connectives.

The topology j is a unary logical connective satisfying the axioms:

$$I \Rightarrow jI$$

$$jjI \Rightarrow jI$$

$$j(T \wedge J) \Leftrightarrow jI \wedge jJ$$

A truth value I is closed if and only if

$$jI \Rightarrow I$$

Recall that a truth value K is directed if for any closed truth values A and B we have

$$[(A \Rightarrow K) \wedge (B \Rightarrow K)] \Rightarrow [j(A \vee B) \Rightarrow K]$$

A truth value Z is directed closed, if for all directed K such that $K \Rightarrow Z$, we have $j(K) \Rightarrow Z$. Start by taking truth values I and J such that I is closed and J is directed closed. In order to show that $I \Rightarrow I$ is directed closed, we take a directed truth value K, assume that $K \Rightarrow (I \Rightarrow J)$, and prove that $jK \Rightarrow (I \Rightarrow J)$. But

$$K \Rightarrow (I \Rightarrow J) \leftrightarrow (I \wedge K) \Rightarrow J$$

and

$$jK \Rightarrow (I \Rightarrow J) \leftrightarrow j(I \wedge K) \Rightarrow J$$

since I is closed.

Hence, in order to show that $I \Rightarrow J$ is directed closed, we just need to show that

$$[(I \wedge K) \Rightarrow J] \Rightarrow [j(I \wedge K) \Rightarrow J]$$

Since J is assumed to be directed closed, it suffices to prove that $I \wedge K$ is directed. But this is trivial since I and J are both directed (since closed sieves are directed closed, and directed closed sieves are closed).

Take $\mu, \nu \in M$. To see that $\mu + \nu$ is a measure, we must verify that $\mu + \nu$ is both additive and semicontinuous.

For additivity, take $I, J \in \Omega_j$. Then

$$\begin{aligned}\rho(\mu+\nu,I)+\rho(\mu+\nu,J) &= \rho(\mu,I)+\rho(\nu,I)+\rho(\mu,I)+\rho(\nu,J)\\ &= \rho(\mu+I\wedge J)+\rho(\mu,I\vee J)+\rho(\mu,I\wedge J)+\rho(\nu,I\vee J)\\ &= \rho(\mu+\nu,I\wedge J)+\rho(\mu+\nu,I\vee J)\end{aligned}$$

To verify the semicontinuity condition, fix $q \in \mathbb{Q}, \mu, \nu \in M$, and a directed family of truth values $D \subseteq \Omega$, such that for each $D \in D$, we have

$$D \Rightarrow \mathcal{Q} \in (\mu + \nu)$$

Fix an $r \in \mathbb{Q}$. Then for each $D \in D$, there must exist rationals m_D and n_D such that

$$\left(m_D + n_D = q + \frac{r}{3}\right) \wedge [D \Rightarrow (m_D \in \mu) \wedge (n_D \in v)]$$

Let k and l be the smallest natural numbers such that

$$\forall D \in D \exists E \in D (D \Rightarrow E) \wedge m_E < k.\frac{r}{3}$$

$$\forall D \in D \exists E \in D (D \Rightarrow E) \wedge n_E < k.\frac{r}{3}$$

Put $m = \frac{kr}{3}$ and $n = \frac{lr}{3}$. Note that for every $\mathbb{Q} \in D$, we have

$$E \Rightarrow (m \in \mu) \wedge (u \in v)$$

Furthermore, we have $m < M_E + \frac{r}{3}$ and $n < n_E + \frac{r}{3}$.

Our goal is to show that $jD \Rightarrow \mathbb{Q} \in \mu + v$. We know that

$$m + n \leq m_E + n_E + \frac{r}{3}$$

$$m_E + n_E = q + \frac{r}{3}$$

and so $m + n \leq \mathbb{Q} + r$.

Finally, we just need to show that $jD \Rightarrow (m \in \mu) \wedge (n \in v)$. But for every $D \in D$, we have

$$D \Rightarrow (m \in \mu) \wedge (n \in v)$$

Since both μ and v are measures, and hence semicontinuous, the result holds.

Suppose that $\mu \in S$ is a measure, and $\mathbb{Q} \in \mathbb{Q}$ is a rational. To see that $\mathbb{Q} \times \mu$ is a measure, we must show that

$$\theta = \left\{ r \in \mathbb{Q} \middle| \frac{r}{q} \in \mu \right\}$$

is both additive and semicontinuous.

Take $I, J \in \Omega_j$. Then

$$s \in \rho(\theta, I) + \rho(\theta, J)$$

$$\leftrightarrow \exists t, u \in \mathbb{Q}[I \Rightarrow t \in \theta] \wedge (J \Rightarrow u \in \theta) \wedge (t + u = s)]$$

$$\leftrightarrow \exists t,u \in \mathbb{Q}\left[\left(I \Rightarrow \frac{t}{q} \in \mu\right) \wedge \left(J \Rightarrow \frac{u}{q} \in \mu\right) \wedge (t+u=s)\right]$$

$$\leftrightarrow \exists t,u \in \mathbb{Q}\left[\left(I \Rightarrow \frac{t}{q} \in \mu\right) \wedge \left(J \Rightarrow \frac{u}{q} \in \mu\right) \wedge \left(\frac{t}{q}+\frac{t}{q}=\frac{s}{q}\right)\right]$$

$$\leftrightarrow \exists v,w \in \mathbb{Q}\left[(I \Rightarrow w \in \mu) \wedge (J \Rightarrow w \in \mu) \wedge \left(v+w=\frac{s}{q}\right)\right]$$

$$\leftrightarrow \frac{s}{q} \in \rho(\mu, I)+\rho(\mu, J)$$

$$\leftrightarrow \frac{s}{q} \in \rho(\mu, I \vee J)+\rho(\mu, I \wedge J)$$

$$\leftrightarrow s \in \rho(\theta, I \vee J)+\rho(\theta, I \wedge J)$$

Hence θ is additive.

To verify that θ is semicontinuous, note that

$$[\![r \in \theta]\!] = \left[\!\!\left[\frac{r}{q} \in \mu \right]\!\!\right]$$

where the right hand side is known to be directed closed, since μ is semicontinuous.

Proposition: *M* is a semimodule over $\mathbb{Q}$, and for any $I \in \Omega_j$, $\rho(-, I)$ is a linear operator on *M*. *M* also has certain joins. Let $D \mapsto \Omega_j^M$ be the presheaf of totally ordered subsheaves of *M*. To define a "supremum" arrow $\bigvee : D \mapsto M$, we just restrict the usual infimum arrow from $\Omega^s \to S$:

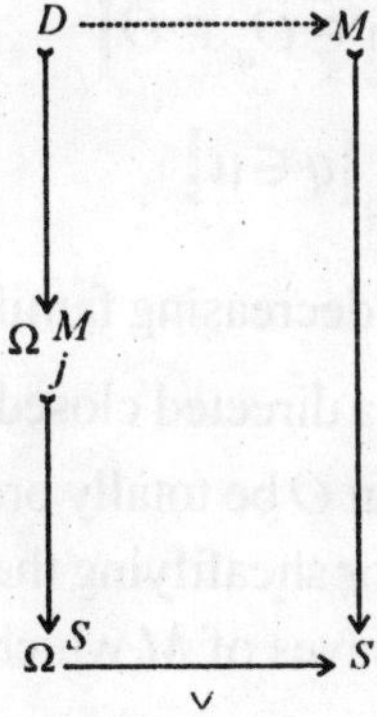

Proposition: The supremum of an ordered family of measures, computed in $\widehat{F}$, is again a measure. In other words, there is an arrow making the following diagram commute:

Proof: Let D be the set of totally ordered subsets of M. The supremum of some $O \in D$ is just the intersection:

$$\bigvee O = \{q \in \mathbb{Q} \mid \forall \mu \in O q \in \mu\} = \bigcap O$$

We just need to show that this set of rationals is a measure.

For additivity, take closed truth values I and J. We need to show that

$$q \in \rho\left(\bigvee O, I\right) + \rho\left(\bigvee O, J\right) \Leftrightarrow q \in \rho\left(\bigvee O, I \vee J\right) + \rho\left(\bigvee O, I \wedge J\right)$$

But

$$q \in \rho\left(\bigvee O, I\right) + \rho\left(\bigvee O, J\right)$$

$$\leftrightarrow \exists a, b \in \mathbb{Q}(a + b = q) \wedge \left(I \Rightarrow a \in \bigvee O\right) \wedge \left(J \Rightarrow b \in \bigvee O\right)$$

$$\leftrightarrow \exists a, b \in \mathbb{Q}(a + b = q) \wedge (\forall \mu \in OI \Rightarrow a \in \mu) \wedge (\forall \mu \in OJ \Rightarrow b \in \mu)$$

$$\leftrightarrow \exists a, b \in \mathbb{Q}(a + b = q) \wedge \forall \mu \in O(I \Rightarrow a \in \mu) \wedge (J \Rightarrow b \in \mu)$$

$$\leftrightarrow \exists c, d \in \mathbb{Q}(c + d = q) \wedge \forall \mu \in O((I \vee J) \Rightarrow c \in \mu) \wedge ((I \wedge J) \Rightarrow d \in \mu)$$

$$\leftrightarrow \exists c, d \in \mathbb{Q}(c + d = q) \wedge (\forall \mu \in O(I \vee J) \Rightarrow c \in \mu) \wedge (\forall \mu \in O(I \wedge J) \Rightarrow d \in \mu)$$

$$\leftrightarrow q \in \rho\left(\bigvee O, (I \vee J)\right) + \rho\left(\bigvee O, (I \wedge J)\right)$$

Now, to verify the semicontinuity condition, we must show that $\left[\!\left[q \in \bigvee O\right]\!\right]$ is directed closed. But

$$\left[\!\left[q \in \bigvee O\right]\!\right]$$

$$= \left[\!\left[\forall \mu \in O_q \in O\right]\!\right]$$

$$= \bigwedge_{\mu \in O} [\![q \in \mu]\!]$$

Since this is itself the meet of a decreasing family of truth values, each of which is directed closed, it follows that $\left[\!\left[q \in \bigvee O\right]\!\right]$ is a directed closed truth value, as required. In fact, we can do slightly better. We do not require that O be totally ordered in ε, but only that it be locally totally ordered (totally ordered in $\mathrm{Sh}_j(\varepsilon)$), for sheafifying the arrow $\bigvee : D \rightarrow M$ yields an arrow whose domain is the sheaf of those subsheaves of M which are totally ordered in $\mathrm{Sh}_j(\varepsilon)$.

INTEGRATION

In order to generalize measure theory, it is necessary to find a way to discuss integration.

With the framework that has been set up here, we can mimic the standard approach to defining the Lebesgue integral. We work in ε, with a designated topology $j: \Omega \to \Omega$, and we assume that M is a j-sheaf.

In this Section, we build an arrow in $\mathrm{Sh}_j(\varepsilon)$ that captures the operation of integration. We know that when $\varepsilon = \widehat{F}$ is the topos of presheaves on a σ-algebra F, M is the sheaf of measures on F. Our goal in this Section is find a logical characterization of the integration arrow in $\widehat{F}$. In the next Section we use this logical characterization to find elementary properties of this arrow.

The classical approach to integrating measurable functions is to first integrate constant functions, then locally constant ("simple") functions, and finally to integrate measurable functions.

In the topos of sheaves on a σ-algebra F, the sheaf of measurable real valued functions on the measurable space (X, F) is just the Dedekind real numbers object in $\mathrm{Sh}(F)$. We will therefore consider integration as acting on the sheaf D of nonnegative Dedekind real numbers:

$$f: D \times M \to M$$

$$\langle f, \mu \rangle \mapsto \int_{-} df\mu$$

Since we are working with positive Dedekind reals, we modifify the definition slightly. A Dedekind real consists of a pair $\langle L, U \rangle$ of subsheaves of the sheaf of positive rationals. We do not assume that L is nonempty, as the pair h$\langle \phi, \mathbb{Q} \rangle$ corresponds to the zero function. We do retain the assumption that U is non-empty, so that the corresponding measurable function is locally finite.

First, we verify that f, as described above, is indeed a natural transformation. This is an immediate consequence of the following well known result:

Lemma: Let (X, F) be a measurable space, take $B \subseteq A$ in F, let μ be a measure on (X, F), and let f be a positive real valued measurable function defined on the subspace $(A, \downarrow A)$. Then for any $C \subseteq B$ in F, we have

$$\int_C f d\mu = \int_C (f \uparrow_B) d(\mu \uparrow_B)$$

where $(f \uparrow_B)$ is the restriction of f to B and $(\mu \uparrow_B)$ is the restriction of μ to B.

Proof: Both the left and right hand sides of the above equation can be rewritten as

$$\int_X f . \chi_C d\mu$$

where χ_C is the characteristic function of C.

Corollary: If F is a σ-algebra, then working in the topos Sh(F), the operation of Lebesgue integration is a natural transformation $\int : D \times M \to M$.

In our framework, constant functions can be considered to be elements of the presheaf $\mathbb{Q}$, and locally constant functions as elements of $a\mathbb{Q}$. As we extend the definition of the integral, we need to ensure that at each stage, the interpretation of the integral coincides with the usual definition of the Lebesgue integral in the toposes $\widehat{F}$ and Sh(F). With this goal, the following Lemmas provide a logical characterization of the integration arrow in $\widehat{F}$ and Sh(F).

Lemma: The arrow describing integration of constant functions is just multiplication. This is expressed by stating that the following diagram commutes:

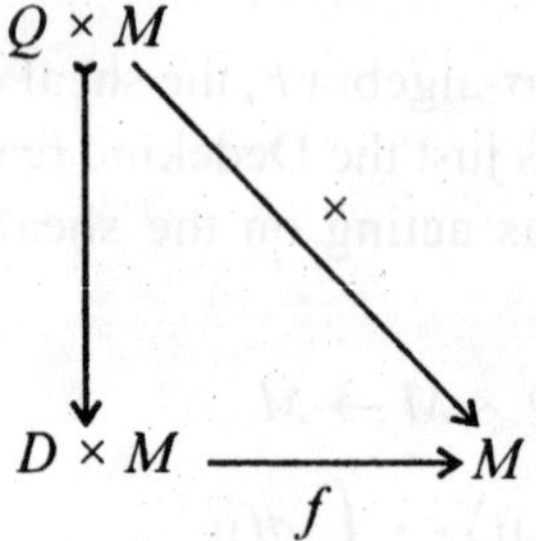

The embedding $\mathbb{Q} \mapsto D$ is given by

$$q \mapsto \langle \{r \in \mathbb{Q} \mid r < q\}, \{x \in \mathbb{Q} \mid q < s\} \rangle$$

Note that D is a sheaf, and that $\mathbb{Q}$ is a presheaf. The logical description $\langle \{r \in \mathbb{Q} \mid r < q\}, \{s \in \mathbb{Q} \mid q < s\} \rangle$ is interpreted in Sh(F).

Proof: Just as D is the sheaf of measurable real valued functions, $\mathbb{Q}$ is the presheaf of constant, rational valued functions, and the embedding $\mathbb{Q} \to D$ is the natural embedding.

By definition,

$$\int_B q d\mu = q.\mu(B)$$

But since the multiplication arrow $\mathbb{Q} \times S \to S$ is just pointwise multiplication of the corresponding functional, the right hand side of the above equation is the product of q with μ, and we are done.

Lemma: The arrow describing integration of locally constant rational valued functions is the image of the multiplication arrow under the sheafification functor. This is expressed by stating that the following diagram commutes:

The following Corollary exploits the fact that integration of rationals is just sheafified multiplication, and makes this relationship explicit:

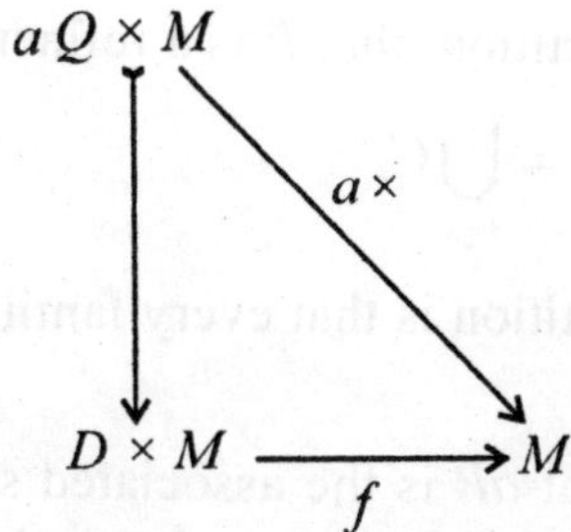

Corollary: Working in the presheaf topos $\widehat{F}$, take q, $r \in \mathbb{Q}$ and $\mu \in M$. Then

$$q \in \int(r,\mu) \Leftrightarrow \frac{q}{r} \in \mu$$

Before we can prove Lemma, we first prove the following useful fact about sheafification in σ-algebras.

Lemma: Let $(F \subseteq, \bot, \mathrm{T}\neg)$ be a σ-algebra, and let P be a presheaf on F. Then aP is given by

$$a(P)(A) = \left. \coprod_{P \in B(A)} \left(\prod_{B \in P} P(B) \right) \right/ \sim$$

where $B(A)$ is the set of all countable partitions of A, and

$$\langle x_B \mid B \in P_1 \rangle \sim \langle x_C \mid C \in P_2 \rangle$$

if for any $B \in P_1$ and $C \in P_2$ we have $\rho^B_{B \cap C}(x_B) = \rho^C_{B \cap C}(x_C)$.0 This is just the usual notion of equivalence of two matching families.

The substance of this Lemma is in two parts. Firstly, it says that to find the associated sheaf of a presheaf, you need only apply the Grothendieck "+" construction one time. Secondly, it says that we need only consider countable partitions of A, rather than all countably generated covers of A.

Proof: The fact that we need only investigate partitions follows directly from the observation that any countable cover in a σ-algebra can be refined to form a partition, since then the countable partitions will form a basis for the countable join topology.

Let $C = \langle C_i \mid i < \omega \rangle$ be a countable cover for A. Let $P = \langle P_i \mid i < \omega \rangle$ be defined recursively:

$$P_0 = C_0$$

$$P_{k+1} = C_{k+1} \cap \left(\bigcap_{i=0}^{k} \neg C_k \right)$$

It is immediate that P is a partition, that P is a refinement of C, and that for any k,

$$\bigcup_{i\leq k} P_i = \bigcup_{i\leq k} C_i$$

The advantage of using a partition is that every family on the partition is a matching family,

So now, in order to show that aP is the associated sheaf of P, all we must show is that aP is a sheaf. Since the countable partitions form a basis for the topology, and since a single application of the Grothendieck "+" construction provides a separated presheaf, it suffices to show that given any countable partition P of A, and any family $x = \langle x_B \in aP(B) \mid B \in P \rangle$, there is an amalgamation for x in $aP(A)$.

But each x_B is itself (or at least, can be represented by) a family $\langle x_C \in P(C) \mid C \in P_B \rangle$ for some countable partition P_C of C. The union of these families over all $B \in P$ provides a matching family for the partition

$$\bigcup_{B\in P} P_B$$

But this union is itself a family over a countable partition, and so corresponds to an element of $aP(A)$, as required.

Corollary: Let (X, F) be a measurable space, let $\mathbb{Q}$ be the presheaf of positive rational numbers (in $\widehat{F}$), and let **a** be the sheafification functor for the countable join topology. Then $a\mathbb{Q}$ is the sheaf of measurable rational valued functions.

Proof: We know from Lemma that $a\mathbb{Q}(B)$ is the set of matching families of rational numbers for partitions (modulo a trivial equivalence). Let P be such a partition. Then a matching family for P consists of a family $\langle q_P \in \mathbb{Q} \mid P \in P \rangle$ of rational numbers. But this is equivalent to the locally constant rational valued function

$$q(x) = q_P$$

whenever $x \in P$.

Proof: Since $a\mathbb{Q} \times M$ is the associated sheaf of $\mathbb{Q} \times M$, we know that there is a limiting map $a\times$: $a\,\mathbb{Q} \times M \rightarrow M$ making the following diagram commute:

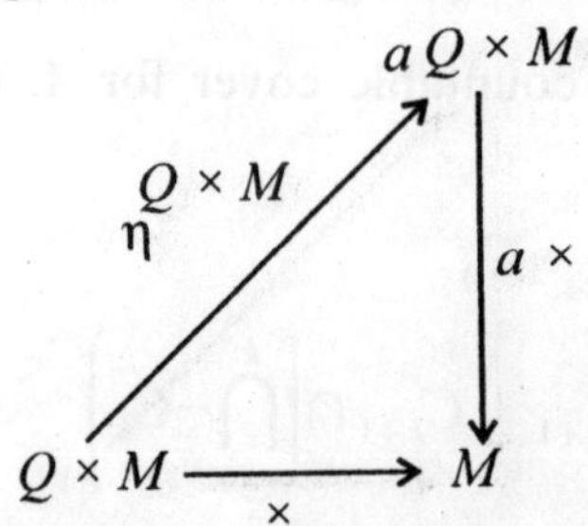

Given $q = \langle q_P \mid P \in P \rangle \in a\mathbb{Q}(A)$, the natural transformation $a \times$ agrees with $\times$ on each element $P \in P$. Hence whenever there is a $P \in P$ such that $B \subseteq P$, we have

$$\int (q,\mu)(B) = \int (q_P,\mu)(B)$$

We now extend $\int (q,\mu)$ to all measurable $B \subseteq A$, by using the countable additivity property of M:

$$\sum_{P \in P} \int (q_P,\mu)(B \cap P) = \int (q,\mu)(B)$$

But this is the usual definition of the integral of a simple function, and so $a \times$ does indeed coincide with the classical notion of the integral.

We can now define the integral of a nonnegative Dedekind real. A Dedekind real consists of a pair $f = \langle L,U \rangle$ of subsheaves of $\mathbb{Q}$ (in Sh(F)). Since $a\mathbb{Q}$ is totally ordered in Sh(F), and since integration preserves order, it follows that for fixed $f = \langle L,U \rangle$ and μ, the set of measures

$$\left\{ \int q,\mu \mid q \in L \right\}$$

is totally ordered in Sh(F). Hence this family of measures has a supremum.

Definition: Let $f = \langle L,U \rangle$ be a Dedekind real, and let μ be a measure. Then set:

$$\int (f,\mu) = \bigvee_{q \in L} \int (q,\mu) = \bigvee_{q \in L} a \times (q\mu)$$

Note that the right hand side here is makes sense in an arbitray topos ε with topology j.

INTEGRABILITY AND PROPERTIES OF INTEGRATION

One might ask why we define the integral of a Dedekind real $f = \langle L,U \rangle$ as the supremum of the integrals of the rationals in L, rather than as the infimum of the rationals in U. The difficulty is that in general, it is hard to construct the infimum of an ordered family of measures. The approach that we used to construct the supremum of such a family was to construct the supremum in S, and then show that this supremum is indeed in M. The corresponding argument is not valid for infima.

However, we can work around this problem for the special case of a Dedekind real:

Theorem: Let μ be a measure, and let $f = \langle L,U \rangle$ be a Dedekind real satisfying $\mathrm{Sh}_j(\varepsilon)$ $\models \exists q \in L$. Then working in S, we have

$$\bigwedge_{r\in U}\int(r,\mu)=\bigvee_{q\in L}\int(q,\mu)$$

Corollary: Let μ be a measure, and let $f=\langle L,U\rangle$ be a Dedekind real satisfying

$$\mathrm{Sh}_j(\varepsilon)\mid=(\exists q\in L)$$

Then the semireal

$$\bigwedge_{r\in U}\int(r,\mu)$$

is a measure.

Proof: Proposition asserts that

$$\bigvee_{q\in L}\int(q,\mu)$$

is a measure, and Theorem asserts that

$$\bigwedge_{r\in U}\int(r,\mu)=\bigvee_{q\in L}\int(q,\mu)$$

So now, to prove Theorem:

Proof: We work using the internal logic of the "presheaf topos" ε. Let $\langle L,U\rangle$ be a Dedekind real in the sheaf topos $\mathrm{Sh}_j(\varepsilon)$, and let μ be a measure. Note that L and U are subpresheaves of $\mathbb{Q}$, as is μ.

For simplicity of notation, let v_L and v_U be the semireals given by

$$v_L=\bigvee_{q\in L}\int(q,\mu)$$

$$v_U=\bigwedge_{r\in U}\int(r,\mu)$$

Then it follows from Proposition that

$$v_L=\{s\in\mathbb{Q}\mid\forall q\in L s\in\int(q,u)\}=\left\{s\in\mathbb{Q}\mid\forall q\in L\frac{s}{q}\in\mu\right\}$$

Likewise,

$$v_U=\{s\in\mathbb{Q}\mid\forall t\in\mathbb{Q}\exists r\in U(s+t)\in\int(r,u)\}=\left\{s\in\mathbb{Q}\mid\forall q\in\mathbb{Q}\exists r\in U\frac{s+t}{r}\in\mu\right\}$$

It is immediate that

$$v_L\leq v_U$$

Hence we need only show the reverse inequality, or equivalently:

$$v_L \subseteq v_U$$

where both the terms in the above expression are viewed as subobjects of $\mathbb{Q}$.

Since $\mathrm{Sh}_j(\varepsilon) \models \exists r \in U$, we (still working in ε) know that there is an $I \in \mathbb{Q}$ such that $jI = T$, and

$$I \Rightarrow (\exists q \in L \wedge \exists r \in U)$$

Let q_0 and r_0 be any witnesses to this statement, so

$$I \Rightarrow (q_0 \in L \wedge r_0 \in U)$$

It follows that $q_0 < r_0$ since if $q_0 \geq r_0$, we would have $q_0 \in L \cap U$, which cannot happen. We need to show that any $s \in v_L$, satisfies $s \in v_U$. Start by taking $t \in \mathbb{Q}$. We must find $r \in U$ such that

$$\frac{s+t}{r} \in \mu$$

We know that $\langle L, U \rangle$ is a Dedekind real in Sh(*L*). Therefore we know that

$$j(\forall_q, r \in \mathbb{Q} q < r \Rightarrow q \in L \vee r \in U)$$

In other words, for any pair of rationals $q_1 < r_1$, there is a dense $I_1 \in \Omega$ for which the following apartness property holds:

$$I_1 \Rightarrow q_1 \in L \vee r_1 \in U$$

Using q_0 as our starting point, we will construct two pair of recursively defined sequences: The first pair, is given by

$$m_n = \frac{s}{q_n}$$

$$q_{n+1} = \frac{s+t}{m_n}$$

We can rewrite q_{n+1} as

$$q_{n+1} = \frac{s}{m_n} + \frac{t}{m_n} = q_n + \frac{t}{m_n}$$

so it is immediate that $\langle q_n \rangle$ is an increasing sequence, and $\langle m_n \rangle$ is a decreasing sequence. Furthermore,

$$q_{n+1} \geq q_n + \frac{t}{m_0}$$

$$\geq q_0 + \frac{(n+t)t}{m_0}$$

The second pair of sequences $\langle q_n' \rangle$ and $\langle m_n' \rangle$ are given by the same recurrence relation. The only difference is that q_0' is chosen such that

$$q_0 < q_0' < q_1$$

and $q_0' \in L$. (We know that there is a $q' \in L$ such that $q_0 < q'$. Let q_0' be the minimum of q' and $\frac{q_0 + q_1}{2}$.)

An easy induction argument shows that for any n, we have

$$q_n < q_n' < q_{n+1} \quad m_{n+1} < m_n' < m_n$$

Now, suppose that n satisfies $n \geq \frac{m_0(r_0 - q_0)}{t}$. It follows from above that $q_n \geq r_0$, whence $q_n \in U$. Likewise, for such an n, we would have $q_n' \in U$.

For each n, since $q_n < q_{n+1}$, we know that we can find an I_{n+1} such that jI_{n+1} = T, and

$$I_{n+1} \Rightarrow (q_n \in L) \vee (q_{n+1} \in U) \wedge I_n$$

Likewise, we can always find J_{n+1} such that jJ_{n+1} = T, and

$$J_{n+1} \Rightarrow (q_n' \in L) \vee (q_{n+1}' \in U) \wedge J_n$$

(let J_0 = T).

Let N be the smallest natural number such that $J_{N+1} \Rightarrow q_{N+1}' \in U$.

We first show that

$$I_{n+1} \wedge J_{N+1} \Rightarrow q_N \in L$$

For convenience, let $K = I_{N+1} \wedge J_{N+1}$. Note that we have jK = T.

We know that either $K \Rightarrow q_N \in L$ or $K \Rightarrow q_N' \in U$. But from our choice of N, it follows that q_{N+1}' is the first term in the sequence $\langle q_n' \rangle$ satisfying $K \Rightarrow q_n' \in U$. Hence we cannot have $K \Rightarrow q_N' \in U$, and so $K \Rightarrow q_N \in L$.

Now, since $K \Rightarrow q_N \in L$, we must have $K \Rightarrow \frac{s}{q_N} \in \mu$, since $K \Rightarrow s \in v_L$. But this means that $K \Rightarrow \frac{s+t}{q_{N+1}} \in \mu$, since

$$\frac{s}{q_N} = m_N$$

$$= \frac{s+t}{q_{N+1}}$$

Hence, we have shown that for arbitrary $s \in v_L$, and $t \in \mathbb{Q}$, there is a dense $K \in \Omega$ such that $K \Rightarrow s + t \in U$. Hence $s \in v_U$, as required.

It is possible to extend the definition of the integral arrow a little further. The object $\mathbb{R}_M$ of McNeille real numbers is somewhat more general than the object of $\mathbb{R}_D$ of Dedekind real numbers.

Like a Dedekind real number, a McNeille real number consists of a pair $\langle L, U \rangle$ of subsheaves of $\mathbb{Q}$. Most of the axioms for a McNeille real number are the same as for a edekind real number. The exception is the "apartness condition". For a Dedekind real number, this is stated as

$$\forall q, r \in \mathbb{Q}(q < r) \Rightarrow (q \in L \vee r \in U)$$

The equivalent condition for McNeille real numbers is the conjunction of the following two formulas

$$\forall q, r \in \mathbb{Q}(q < r \wedge q \notin L) \Rightarrow r \in U$$

$$\forall q, r \in \mathbb{Q}(q < r \wedge q \notin U) \Rightarrow q \in L$$

It is obvious from this definition that every Dedekind real number is a McNeille real number. Hence $\mathbb{R}_D \mapsto \mathbb{R}_M$.

The converse is not intuitionistically valid. In fact, $\mathbb{R}_D \cong \mathbb{R}_M$ if and only if DeMorgan's law holds:

$$\neg(A \wedge B) \neg A \vee \neg B$$

(The other three of DeMorgan's laws are intuitionistically valid.) DeMorgan's law is strictly weaker than the law of the excluded middle, so all classical logical systems satisfy DeMorgan's law, but conversely, there are non-classical toposes where DeMorgan's law is satisfied.

The principal reason that McNeille real numbers are studied is that they are the order completion of the Dedekind real numbers. Hence DeMorgan's law holds if and only if the Dedekind reals numbers satisfy Bolzano–Weierstrass completeness.

This also allows us to construct McNeille real numbers that are not Dedekind real numbers. Let $(\mathbb{R}, L)$ be the measurable space consisting of the real numbers and the Lebesgue measurable functions. Let Z be a non-Lebesgue measurable set.

Then for each $z \in Z$, the characteristic function $\chi_{\{z\}}$ is measurable, and so is a Dedekind real number in Sh(L). However, the supremum of these characteristic functions is the characteristic function χ_Z, which is evidently not a measurable function, and so not a Dedekind real number.

As a result, we could define the integral of a McNeille real f (relative to a measure μ) as the supremum of the integrals of the rationals in the lower cut of f. However Theorem does not apply in such a case, and so a McNeille real is not integrable, in the usual sense.

Finally, we present some important properties of the integration arrow.

Proposition: Integration is an order preserving map.

Proof: We need to show two things here. Firstly, if $f \leq g$, then

$$\int (f,\mu) \leq \int (g,\mu)$$

and secondly, if $\mu \leq v$, then

$$\int (f,\mu) \leq \int (g,\mu)$$

For the first, note that if $f = \langle L_f, U_f \rangle$ and $g = \langle L_g, U_g \rangle$, then

$$f \leq g \Leftrightarrow L_f \subseteq U_f$$

Since

$$\int (f,\mu) = \bigvee_{q \in L_f} \int (q,\mu)$$

and

$$\int (g,\mu) = \bigvee_{r \in L_g} \int (r,\mu)$$

it follows that

$$\int (f,\mu) \leq \int (g,\mu)$$

as required.

Now suppose that $\mu \leq v$. Then for each $q \in L_f$, it follows immediately fromCorollary that

$$\int (q,\mu) \leq \int (q,v)$$

Hence

$$\bigvee_{q \in L_f} \int (q,\mu) \leq \bigvee_{q \in L_f} \int (q,v)$$

as required.

Monotone Convergence Theorem

Theorem: Suppose that $f_\alpha \uparrow f$ is an increasing family of Dedekind reals, converging to another Dedekind real f. Then

$$\int (f_\alpha,\mu) \rightarrow \int (f,\mu)$$

Proof: We first show that the result holds for increasing families of rationals, and then for increasing families of Dedekind reals.

Let $\mathbb{Q} = \langle q_i \mid i \in I \rangle$ be a directed family of rational numbers, and let $q = \bigvee Q$. Every rational number a can be represented by the semireal $\{s \in \mathbb{Q} | a \leq s\}$. The arrow $\int$ is just multiplication, so

$$\begin{aligned}
a \in \bigvee_{i \in I} \int (q_i,\mu) &\leftrightarrow \forall i \in I a \in \int (q_i,\mu) \\
&\leftrightarrow \forall i \in I \exists m \in \mu a = mq_i \\
&\leftrightarrow \forall i \in I \frac{a}{q_i} \in \mu \\
&\leftrightarrow \forall d \in \mathbb{Q} \frac{a}{q} + d \in \mu \\
&\leftrightarrow \frac{a}{q} \in \mu \\
&\leftrightarrow a \in \int (q,\mu)
\end{aligned}$$

Now, let $D = \langle f_i \mid i \in I \rangle$ be a directed family of Dedekind real numbers (defined in Sh(L)), and let $f = \langle L,U \rangle = \bigvee D$. Then

$$a \in \bigvee_{i \in I} \int (f_i,\mu) \leftrightarrow \forall i \in I a \int (f_i,\mu)$$

$$\leftrightarrow \forall i \in I \left[\forall q \in L_i \left(a \in \int (q,\mu) \right) \right]$$

$$\leftrightarrow \forall i \in I \left[\forall q \in L_i \left(\frac{a}{q} \in \mu \right) \right]$$

$$\leftrightarrow \left[\forall q \in \bigcup_{i \in I} L_i \left(\frac{a}{q} \in \mu \right) \right]$$

$$\leftrightarrow \forall q \in \bigcup_{i \in I} L_i \left(\frac{a}{q} \in \mu \right)$$

$$\leftrightarrow \forall q \in L \left(\frac{a}{q} \in \mu \right)$$

$$\leftrightarrow \forall q \in L \left(a \in \int (q,\mu) \right)$$

$$\leftrightarrow a \in \int (f,\mu)$$

Theorem: M is a semimodule over the semiring D, with the action of scalar multiplication given by integration.

Proof: We know from Proposition that S is a semimodule over the semiring of positive rationals. Since $(M, +)$ is a subalgebra of $(S, +)$, it follows that M is an abelian monoid. Hence we just need to show the following, for arbitrary μ, $v \in M$ and $f=\langle L_1,U_1\rangle$, $g=\langle L_2,U_2\rangle \in D$:

1. $\int (f,(\mu+v)) = \int (f,\mu) + \int (f,v)$
2. $\int ((f+g),\mu) = \int (f,\mu) + \int (g,\mu)$
3. $\int (f, \int (g,\mu) = \int (f.g,\mu)$
4. $\int (1,\mu) = \mu$

To prove these:

$$\int (f,\mu+v) = \bigvee_{q \in L_1} \int (q,\mu+v)$$

$$= \bigvee_{q \in L_1} \int (q,\mu) + \int (q,v)$$

Since L_1 is totally ordered, we can apply distributivity here, to get

$$\bigvee_{q \in L_1} \int (q,\mu) + \int (g,v) = \bigvee_{q \in L_1} \int (q,\mu) + \bigvee_{q \in L_1} \int (q,v)$$

$$= \int (q,\mu) + \int (g,v)$$

2. First, note that $f + g = \langle L_1 \oplus L_2, U_1 \oplus U_2 \rangle$, where

$$A \oplus B = \{a + b | \langle a,b \rangle \hat{\text{I}}\ A \times B\}$$

$$\int (f+g,\mu) = \bigvee_{s \in L_1 \oplus L_2} \int (s,\mu)$$

$$= \bigvee_{q \in L_1} \bigvee_{r \in L_2} \int (q+r,\mu)$$

$$= \bigvee_{q \in L_1} \bigvee_{r \in L_2} \int (q,\mu) + \int (r,\mu)$$

$$= \left(\bigvee_{q \in L_1} \bigvee_{r \in L_2} \int (q,\mu) \right) + \left(\bigvee_{q \in L_1} \bigvee_{r \in L_2} \int (r,\mu) \right)$$

$$= \left(\bigvee_{q \in L_1} \int (q,\mu) \right) + \left(\bigvee_{r \in L_2} \int (r,\mu) \right)$$

$$= \int (f,\mu) + \int (g,\mu)$$

3. This time, note that $f \cdot g = \langle L_1 \otimes L_2, U_1 \otimes U_2 \rangle$, where

$$A \otimes B = \{a \cdot b | \langle a,b \rangle \in A \times B\}$$

$$\int \left(f \int (g,\mu) \right) = \bigvee_{q \in L_1} \int \left(q, \int (g,\mu) \right)$$

$$= \bigvee_{q \in L_1} \int (g.g,\mu)$$

$$= \bigvee_{q \in L_1} \bigvee_{r \in L_2} \int (q.r,\mu)$$

$$= \bigvee_{s \in L_1 \otimes L_2} \int (s,\mu)$$

$$= \int (f.g,\mu)$$

4. Observe that the Dedekind real number 1 is a rational number, and so we can apply Proposition here.

Chapter 3

Localic Topos

In mathematics, a topos is a type of category that behaves like the category of sheaves of sets on a topological space. A traditional axiomatic foundation of mathematics is set theory, in which all mathematical objects are ultimately represented by sets (even functions which map between sets). More recent work in category theory allows this foundation to be generalized using toposes; each topos completely defines its own mathematical framework. The category of sets forms a familiar topos, and working within this topos is equivalent to using traditional set theoretic mathematics. But one could instead choose to work with many alternative toposes.

A standard formulation of the axiom of choice makes sense in any topos, and there are toposes in which it is invalid. Constructivists will be interested to work in a topos without the law of excluded middle.

If symmetry under a particular group G is of importance, one can use the topos consisting of all G-sets. It is also possible to encode an algebraic theory, such as the theory of groups, as a topos. The individual models of the theory, i.e. the groups in our example, then correspond to functors from the encoding topos to the category of sets that respect the topos structure.

ELEMENTARY TOPOSES

Let $\mathbb{C}$ be an arbitrary category, let $\widehat{\mathbb{C}}$ be the topos of presheaves on $\mathbb{C}$, and let denote the subobject classifier in $\widehat{\mathbb{C}}$. Then a Grothendieck topology on $\widehat{\mathbb{C}}$ is a presheaf $J \mapsto \Omega$ of "covering sieves"; a sieve $I \in \Omega(C)$ is a cover for C if and only if $I \in J(C)$. Grothendieck topologies correspond to the Lawvere-Tierney topologies $j: \Omega \to \Omega$ which characterize them. A (Lawvere-Tierney or Grothendieck) topology induces a subtopos of $\widehat{\mathbb{C}}$, the topos of sheaves on the site $(\mathbb{C}, J)$, denoted $\mathrm{Sh}(\mathbb{C}, J)$. Such sheaf toposes have been extensively studied.

This result connects Grothendieck topologies on a presheaf topos with Lawvere-Tierney topologies on the same presheaf topos. A more general situation is given by the case where

$\varepsilon = \mathrm{Sh}(\mathbb{C}, J)$ is a Grothendieck topos, and j is a Lawvere-Tierney topology in ε. In this case, there is a relationship between Lawvere-Tierney topologies in ε and certain Grothendieck topologies on $\mathbb{C}$.

Definition: Let j and k be two Lawvere-Tierney topologies in an elementary topos ε. Then k is said to be finer than j if $k \circ j = k$.

Lemma: Suppose that k is finer than j. Then $j \circ k = k$.

Proof: It is obvious that the composition of Lawvere-Tierney topologies is also a Lawvere- Tierney topology. Start by assuming that $k \circ j = k$, and let l be the topology $j \circ k$.

$$\begin{aligned} l \circ k &= j \circ k \circ k \\ &= j \circ k \\ &= l \end{aligned}$$

Hence l is a finer topology than k.

Conversely, consider $k \circ l$:

$$\begin{aligned} k \circ l &= k \circ j \circ k \\ &= k \circ k \\ &= k \end{aligned}$$

Hence k is finer than l. Since the "finer" relationship on topologies is inherited from the usual ordering on Ω, it follows that the "finer" relation is in fact a partial ordering. Hence $k = l$, whence

$$k = j \circ k$$

The condition $j \circ k = k$ is exactly the condition needed to infer that k factors through i: $\Omega_j \to \Omega$. The reason for this is given by the following diagram:

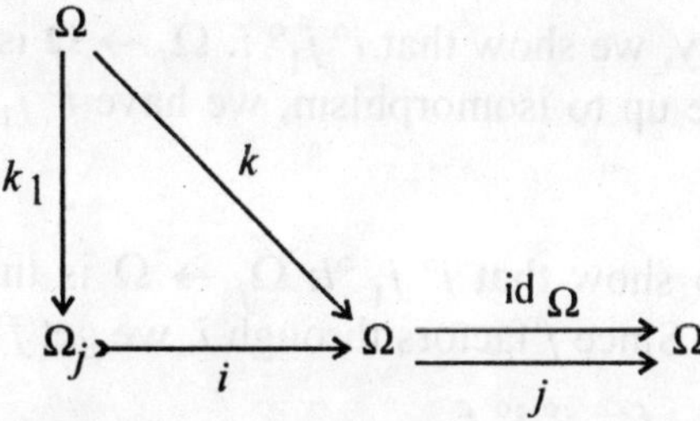

Since i is the equalizer of the arrows j, id: $\Omega \to \Omega$, it follows that $j \circ k = k$ if and only if k factors through i.

Proposition: Let ε be an elementary topos, and let j, k: $\Omega \to \Omega$ be two Lawvere-Tierney topologies on ε, such that k is finer than j. Let Ω_j be the subobject classifier in $\mathrm{Sh}_j(\varepsilon)$, and define k' as the composition $k_1 i$, where k_1 and i are given by the following diagram:

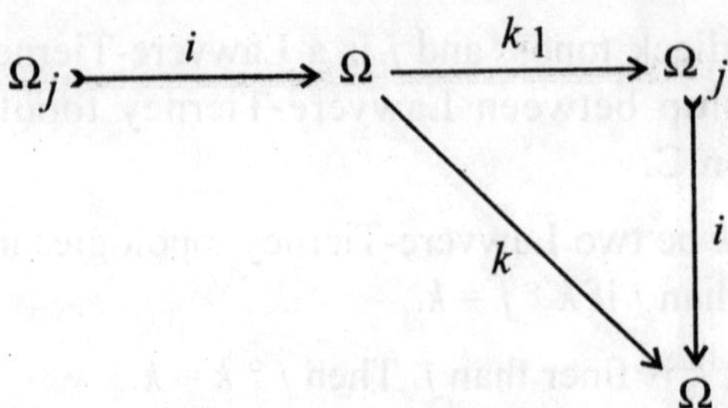

Then k' is a Lawvere-Tierney topology in the topos $\text{Sh}_j(\varepsilon)$.

Proof: To see that k' is a Lawvere-Tierney topology, we need to show that k' satisfies the usual three commutative properties in $\text{Sh}_j(\varepsilon)$.

To do this, we introduce an arrow $j_1 : \Omega \to \Omega_j$. We know that $i: \Omega_j \rightarrowtail \Omega$ is the equalizer of the arrows $j: \Omega \to \Omega$ and $\text{id}_\Omega : \Omega \to \Omega$. Since $j \circ j = j$, there must be an arrow j_1 making the following diagram commute:

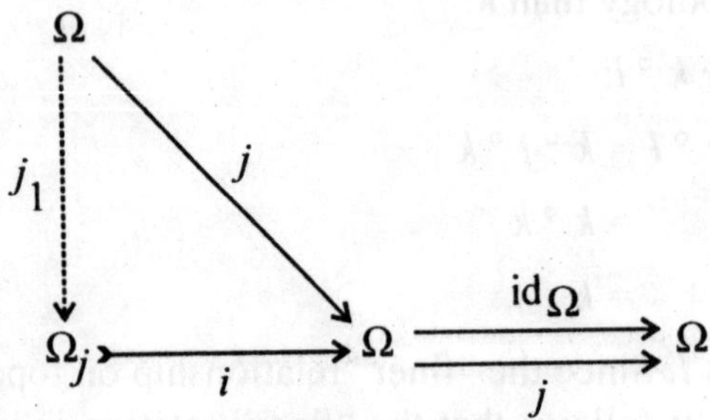

First we observe that j_1 has the following properties:

Lemma:

$$i \circ j_1 = j \; j_1 \circ i = id_{\Omega_j}$$

Proof: The first property is immediate from the above diagram.

For the second property, we show that $i \circ j_1 \circ i : \Omega_j \to \Omega$ is an equalizer for j and id_Ω. Since equalizers are unique up to isomorphism, we have $i \circ j_1 \circ i = i$. Finally, since i is a monic, we get $j_1 \circ i = id_{\Omega_j}$.

So, it only remains to show that $i \circ j_1 \circ i : \Omega_j \to \Omega$ is the required equalizer. Take $f: Z \to \Omega$ such that $j \circ f = f$. Since f factors through i, we get $f = i \circ f_1$, so

$$\begin{aligned} f &= j \circ i \circ f_1 \\ &= i \circ j_1 \circ i \circ f_1 \end{aligned}$$

But this tells us that f factors through $i \circ j_1 \circ i$, as required.

Note also that i and k_1 satisfy the following properties:

$$k' = k_1 \circ i \; k = i \circ k_1$$

The first condition that we need to show is inflationarity:

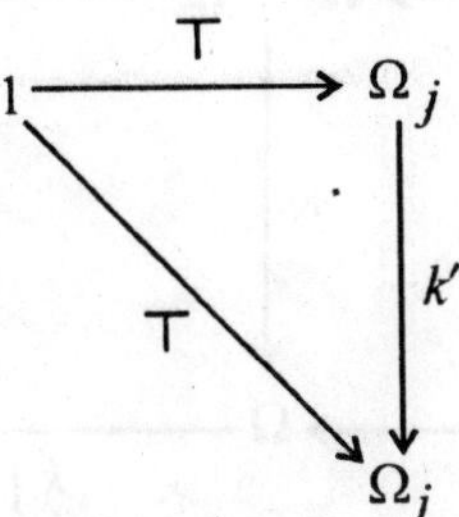

Note that 1 is also the terminal object in ε, and $i \circ \mathrm{T}: 1 \to \Omega$ is the "top" map in ε.

The result follows from the following diagram chase:

$$\begin{aligned} k' \circ \mathrm{T} &= k_1 \circ i \circ \mathrm{T} \\ &= j_1 \circ i \circ k_1 \circ i \circ T \\ &= j_1 \circ k \circ i \circ \mathrm{T} \\ &= j_1 \circ i \circ \mathrm{T} \\ &= \mathrm{T} \end{aligned}$$

The second condition is idempotence:

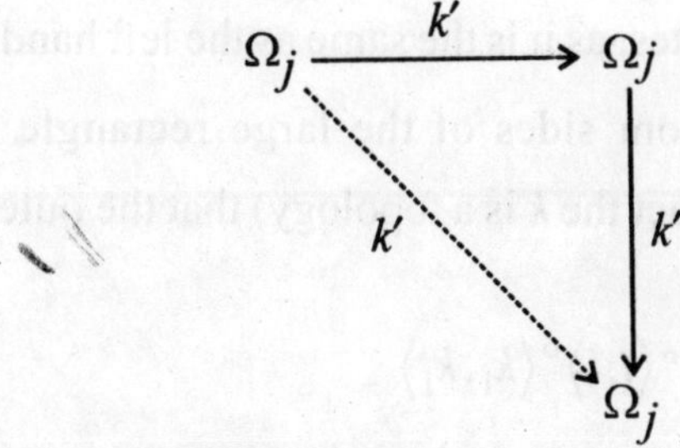

Again, we engage in a diagram chase:

$$\begin{aligned} k' \circ k' &= k_1 \circ i \circ k_1 \circ i \\ &= k_1 \circ k \circ i \\ &= j_1 \circ i \circ k_1 \circ k \circ i \\ &= j_1 \circ k \circ k \circ i \\ &= j_1 \circ k \circ i \\ &= j_1 \circ i \circ k_1 \circ i \\ &= k_1 \circ i \\ &= k' \end{aligned}$$

Finally, we must verify that k' commutes with the meet operator $\wedge_j$ on Ω_j. This will follow if we can show that the outer rectangle in the following diagram commutes:

$$\begin{array}{ccccc}
\Omega_j\times\Omega_j & \xrightarrow{\langle i,i\rangle} & \Omega\times\Omega & \xrightarrow{\langle k_1,k_1\rangle} & \Omega_j\times\Omega_j \\
\downarrow{\scriptstyle\wedge_j} & & \downarrow{\scriptstyle\wedge} & & \downarrow{\scriptstyle\wedge_j} \\
\Omega_j & \xrightarrow[i]{} & \Omega & \xrightarrow[k_1]{} & \Omega_j
\end{array}$$

The fact that the left hand square commutes follows directly from the fact that j is a topology.

To see that the right hand square commutes, consider the following diagram:

$$\begin{array}{ccccc}
\Omega\times\Omega & \xrightarrow{\langle k_1,k_1\rangle} & \Omega_j\times\Omega_j & \xrightarrow{\langle i,i\rangle} & \Omega_j\times\Omega_j \\
\downarrow{\scriptstyle\wedge} & & \downarrow{\scriptstyle\wedge_j} & & \downarrow{\scriptstyle\wedge} \\
\Omega & \xrightarrow[k_1]{} & \Omega_j & \xrightarrow[i]{} & \Omega
\end{array}$$

The right hand square commutes, as it is the same as the left hand square of the previous diagram. Since the top and bottom sides of the large rectangle are just $\langle k,k\rangle$ and k, respectively, it follows (from the fact the k is a topology) that the outer rectangle commutes. Hence:

$$\begin{aligned} i\circ k_1\circ\wedge &= \wedge\circ\langle i,i\rangle\circ\langle k_1,k_1\rangle \\ &= i\circ\wedge_j\circ\langle k_1,k_1\rangle \end{aligned}$$

Since i is a monomorphism, it follows that

$$k_1\circ\wedge = \wedge_j\circ\langle k_1,k_1\rangle$$

But this is just whatweneeded to make the right hand square in the first diagram commute. This completes the proof that k' is a Lawvere-Tierney topology in $\mathrm{Sh}_j(\varepsilon)$.

Proposition: Suppose that ε is an elementary topos, j is a Lawvere-Tierney topology in ε, and k is a Lawvere-Tierney topology in the sheaf topos $\mathrm{Sh}_j(E)$. Let Ω be the subobject classifier in ε, let Ω_j be the subobject classifier in $\mathrm{Sh}_j(\varepsilon)$, let $i\colon \Omega_j \rightarrowtail \Omega$ be the natural inclusion, and let $j_1\colon \Omega\to\Omega_j$ be the closure map of j. Let $\tilde{k} = i\circ k\circ j_1$ denote the following composition:

$$\Omega \xrightarrow{j_1} \Omega_j \xrightarrow{k} \Omega_j \xrightarrow{i} \Omega$$

Then

1. $\tilde{k}$ is a Lawvere-Tierney topology in e.
2. $\tilde{k}$ is a finer topology than j.
3. $\mathrm{Sh}_k\,(\mathrm{Sh}_j(\varepsilon) \simeq Sh_{\tilde{k}}(\varepsilon)$

Proof: 1. As in Proposition, we need to show that j satisfies the required commutative diagrams.

Firstly, to check inflationarity, it is enough to realize that each of the triangles in the following diagram commutes:

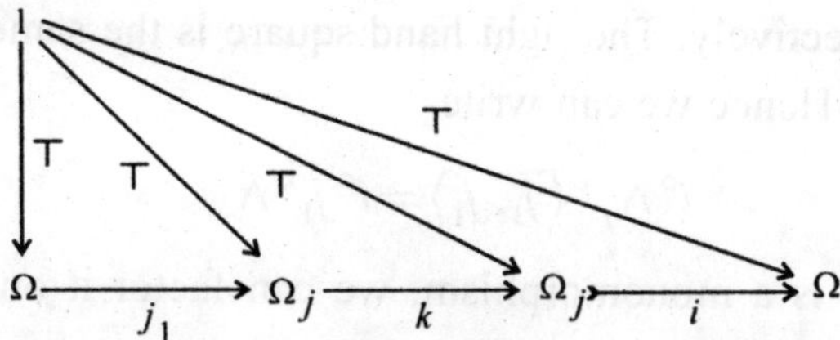

Next, to check that $\tilde{k}$ is idempotent, we use the fact that $j_1 \circ i = id_{\Omega_j}$:

$$\begin{aligned}\tilde{k} \circ \tilde{k} &= i \circ k \circ j_1 \circ i \circ k \circ j_1 \\ &= i \circ k \circ k \circ j_1 \\ &= i \circ k \circ j_1 \\ &= \tilde{k}\end{aligned}$$

Finally, to check that $\tilde{k}$ commutes with products, we need to show that the outer rectangle in the following diagram commutes:

$$\begin{array}{ccccccc}\Omega \times \Omega & \xrightarrow{\langle j_1, j_1 \rangle} & \Omega_j \times \Omega_j & \xrightarrow{\langle k, k \rangle} & \Omega_j \times \Omega_j & \xrightarrow{\langle i, i \rangle} & \Omega \times \Omega \\ \downarrow \wedge & & \downarrow \wedge_j & & \downarrow \wedge_j & & \downarrow \wedge \\ \Omega & \xrightarrow{j_1} & \Omega_j & \xrightarrow{k} & \Omega_j & \xrightarrow{i} & \Omega\end{array}$$

That the right hand square commutes is an immediate consequence of the fact that j is a topology, and that topologies preserve meets.

That the middle square commutes follows from the fact that k is itself a topology in $\mathrm{Sh}_j(\varepsilon)$.

To see that the left hand square commutes, consider the following diagram:

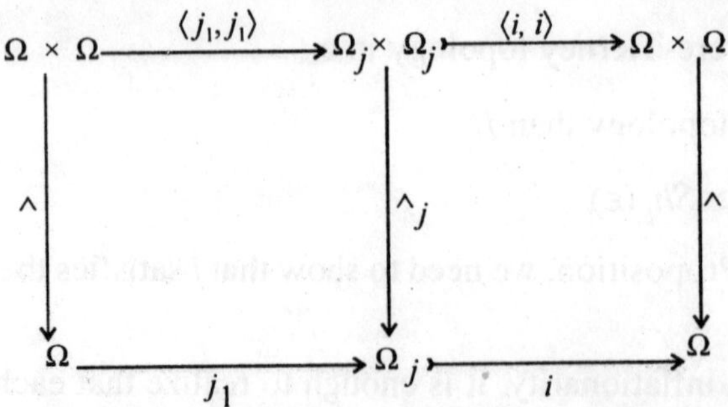

The outer square commutes, since the compositions along the top and the bottom are just $\langle j, j \rangle$ and j, respectively. The right hand square is the same right hand square as in the previous diagram. Hence we can write

$$i^\circ \wedge_j {}^\circ \langle j_1, j_1 \rangle = i^\circ j_1{}^\circ \wedge$$

However, since i is a monomorphism, we can factor it out of the above equation, yielding

$$\wedge_j{}^\circ \langle j_1; j_1 \rangle = j_1{}^\circ \wedge$$

which is exactly what we need to show that the left hand square also commutes.

2. To see that $\tilde{k} \circ j = \tilde{k}$, note that

$$\tilde{k} \circ j = i \circ k \circ j_1 \circ j$$

It will suffice to show that $j_1 \circ j = j_1$. But

$$j_1 \circ j = j_1 \circ i \circ j_1$$

$$= id_{\Omega_j} \circ j_1$$

$$= j_1$$

as required.

3. We show that an object F of ε is both a j-sheaf, and a k-sheaf in $\mathrm{Sh}_j(\varepsilon)$ if and only if F is a $\tilde{k}$-sheaf in ε.

Suppose that F is a j-sheaf, and a k-sheaf in $\mathrm{Sh}_j(\varepsilon)$. Take a pair of objects $A \rightarrowtail E$ in ε, such that A is a $\tilde{k}$-dense subobject of E, with $\chi_E: A \to \Omega$ denoting the characteristic map of A, and let $f: A \to F$ be an arbitrary arrow. Let $A_j \rightarrowtail E$ be the closure of A under the topology j. Then, since F is a j-sheaf, f has a unique extension $\bar{f} : A_j \to F$. Now consider the closure of $A_k \rightarrowtail E$, the closure of A_j under the topology k. We know that $\bar{f}$ must have a unique extension $\tilde{f} : A_k \to F$, since F is a k sheaf.

The characteristic map of A_k is $i \circ k \circ j_1 \circ \chi_A$. But this is just $\tilde{k} \circ \chi_A$. Since we are assuming that A is a $\tilde{k}$ dense subobject of E, it follows that $A_k = E$, and so $\tilde{E} : E \to F$ is the extension of f required to show that F is a $\tilde{k}$ -sheaf.

Now suppose that F is a $\tilde{k}$-sheaf. We first show that F is a j-sheaf. Suppose that $A \mapsto E$ is a j-dense subobject in ε, and let $f: A \to F$ be an arbitrary arrow. Let $\chi_A: E \to \Omega$ denote the characteristic arrow for A. Then $j \circ \chi_A = \text{T}$. But then $\tilde{k} \circ \chi_A = \text{T}$, since

$$\begin{aligned}\tilde{k} \circ \chi_A &= \tilde{k} \circ \text{j} \circ \chi_A \\ &= \tilde{k} \circ \text{T} \\ &= \text{T}\end{aligned}$$

Thus A is a $\tilde{k}$-dense subobject of E. Since F is a $\tilde{k}$-sheaf, it follows that f has a unique extension $\hat{f} : E \to F$. Hence F is a j-sheaf.

Now take a pair of objects $A \mapsto E$ in $\text{Sh}_j(\varepsilon)$ such that A is k-dense, together with an arrow $f: A \to F$. Since A and E are j-sheaves, it follows that when we interpret $A \to E$ in ε, the characteristic map $\chi_A: E \to \Omega$ will factor through i. Hence there is a map $\chi_A^j : E \to \Omega_j$ such that $i \circ \chi_A^j : E \to \Omega$ is just χ_A. χ_A^j (is just the characteristic map of $A \mapsto E$ in $\text{Sh}_j(\varepsilon)$.)

Since A is a k-dense subobject of E, it follows that in $\text{Sh}_j(E)$, $k \circ \chi_A^j = \text{T}$ (or $i \circ k \circ \chi_A^j = \text{T}$ in ε). Since $j_1 \circ i = id_{\Omega_j}$, we get

$$\begin{aligned}\text{T} &= i \circ k \circ \chi_A^j \\ &= i \circ k \circ j_1 \circ i \circ \chi_A^j \\ &= \tilde{k} \circ \chi_A\end{aligned}$$

Hence A is a $\tilde{k}$-dense subobject of E, and so there is a unique $\overline{f} : E \to F$ extending f.

This F is a k-sheaf in $\text{Sh}_j(\varepsilon)$.

The following result connects these two Propositions.

Theorem: The processes in Propositions are inverse to one another. Hence topologies in $\text{Sh}_j(\varepsilon)$ correspond to those topologies in ε which are finer than j.

Proof: We start with a Lawvere-Tierney topology $k: \Omega \to \Omega$ in ε, with k satisfying $j \circ k = k$.

The Lawvere-Tierney topology in $\mathrm{Sh}_j(\varepsilon)$ corresponding to k is $k'\colon \Omega_j \to \Omega_j$, given by

$$k' = k_1 \circ i$$

The Lawvere-Tierney topology in ε corresponding to k' in ε is the arrow $\tilde{k}'\colon \Omega \to \Omega$ given by

$$\tilde{k}' = i \circ k' \circ j_1$$

We must show that $\tilde{k}' = k$.

$$\begin{aligned}(k')' &= i \circ k' \circ j_1 \\ &= i \circ k_1 \circ i \circ j_1 \\ &= k \circ i \circ j_1 \\ &= k \circ j \\ &= k\end{aligned}$$

Now, suppose that $k\colon \Omega_j \to \Omega_j$ is a Lawvere-Tierney topology in $\mathrm{Sh}_j(\varepsilon)$. Then the Lawvere-Tierney topology in ε associated with k is given by

$$\tilde{k} = i \circ k \circ j_1\colon \Omega_j \to \Omega_j$$

To find the Lawvere-Tierney topology in $\mathrm{Sh}_j(\varepsilon)$ associated with $\tilde{k}$, first note that $(\tilde{k})'$ is defined by the following diagram:

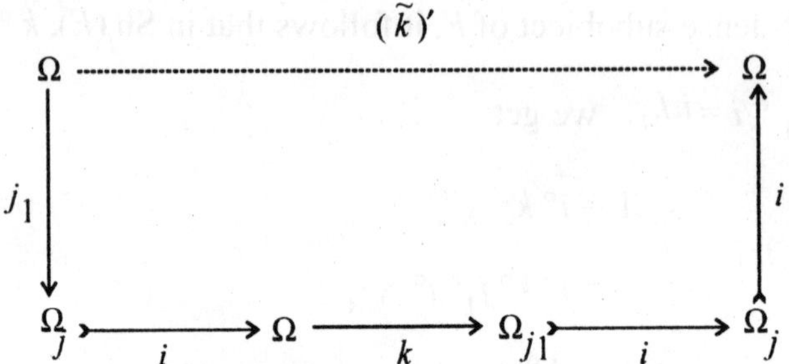

But since $j = i \circ j_1$, it follows that $(\tilde{k})' = j \circ k \circ k$. Applying the fact that k is assumed to be finer than j, this just reduces to k, as required. For most of this Chapter, we will be considering the case where $\varepsilon = \mathrm{Sh}(L)$ is the topos of sheaves on a locale (ie, ε is a localic topos), and j is some topology in ε. In this case, $\mathrm{Sh}_j(L)$ has some useful properties.

Proposition: Let $\varepsilon = \mathrm{Sh}(L)$ be a localic topos, and let k be a Lawvere-Tierney topology in ε. Then

1. There is a Grothendieck topology K on L such that $\mathrm{Sh}(L, K) \simeq \mathrm{Sh}_k(\varepsilon)$
2. $\mathrm{Sh}_k(\varepsilon)$ is a localic topos.

Proof: First note that the second clause follows immediately from the first.

To prove the first clause, let j be the canonical topology on L. Then $\varepsilon \simeq \mathrm{Sh}_j(L)$. Hence, we can apply Proposition to find a topology $\tilde{k}$ on L such that $\mathrm{Sh}_{\tilde{k}}(L) \simeq \mathrm{Sh}_k(\varepsilon)$.

For any elementary topos ε and Lawvere-Tierney topology j, there is a functor i: $\mathrm{Sh}_j(\varepsilon) \to \varepsilon$, embedding the sheaf topos in ε. This functor has a left adjoint a: $\varepsilon \to \mathrm{Sh}_j(\varepsilon)$, called the sheafification, or "associated sheaf" functor. If $\varepsilon = \widehat{\mathbb{C}}$ is a presheaf topos, then the usual method for constructing aP, the associated sheaf of some presheaf P, is to apply the "Grothendieck +" construction twice (once to arrive at a separated presheaf, and then again to arrive at a sheaf). In light of Proposition, this method can clearly be extended to the case where ε is any Grothendieck topos.

However, in the case that ε is a localic topos, the process can be simplified considerably. Note that in this case, a topology j is a closure opeator on L.

Lemma: Let $\varepsilon = \mathrm{Sh}(L)$ be a localic topos, and let j: $\Omega \to \Omega$ be a Lawvere-Tierney topology in ε. Then a sheaf F is a j-sheaf if and only if we have

$$F(A) = F(jA)$$

for any $A \in L$.

Proof: We know already that a sheaf F is a contravariant functor from L to $\mathrm{S_{ET}}$, and hence an object of the presheaf topos $\hat{L}$. We also know that F is a j-sheaf if and only if it is a $\hat{j} = i^{\circ} j^{\circ} c_1$ sheaf in $\hat{L}$ (where c is the cannonical topology on $\hat{L}$). But this just means that for every matching family for some $\hat{j}$-covering family, there is a unique amalgamation. Take $A \in L$, and suppose that C is a $\hat{j}$-cover for A. Since F is a c-sheaf, we know that matching families for C correspond to elements of $F(\curlyvee C)$. Hence F is a $\hat{j}$-sheaf if and only if

$$F(\curlyvee C) = F(A)$$

for any $\hat{j}$-cover C of A.

But this is equivalent to requiring that $F(B) = F(jB)$ for any B. Equipped with this convenient characterization of j-sheaves, we can easily find the associated sheaf $a_j F$ of F:

Theorem: Let ε be a localic topos, and let j be a Lawvere-Tierney topology on ε. Then for any object F of ε, we have

$$a_j F(A) = \coprod_{j(B)=j(A)} F(B) \Big/ \sim$$

where $x_1 \in F(B_1)$ and $x_2 \in F(B_2)$ are related by ~ if

$$\rho^{B_1}_{B_1 \wedge B_2}(x_1) = \rho^{B_2}_{B_1 \wedge B_2}(x_2)$$

Proof: We prove this result by showing firstly that that a_jF, as defined above, is indeed a sheaf, and secondly that every natural transformation $\psi : F \to G$ for some j-sheaf G factors through the evident natural transformation $\eta : F \to a_jF$.

$\eta^A: F(A) \to a_jF(A)$ is given as the following composition:

$$F(A) \mapsto \coprod_{jB=jA} F(B) \to \coprod_{jB=j(A)} F(B) \Big/ \sim$$

To see that a_jF is a sheaf, it suffices to observe that for any $A \in L$, we have

$$a_jF(jA) = a_jF(A)$$

But this is immediate from the definition of a_jF, as $j(B) = j(A) \Leftrightarrow j(A) = j(j(A))$.

Now take an arbitrary j-sheaf G, and an arbitrary natural transformation $\psi : F \to G$. Then since G is a j-sheaf, it follows that $G(A) = G(jA)$ for any A. Thus the following square must commute:

$$\begin{array}{ccc} F(j(A)) & \xrightarrow{\psi^{jA}} & G(jA) \\ \downarrow & & \| \\ F(A) & \xrightarrow[\psi^{A}]{} & G(A) \end{array}$$

For each A, we can extend ψ^A to $\overline{\psi^A} : a_jF(A) \to G(A)$ as follows: Take $x \in a_j(F)(A)$. Then there is a B such that $jB = jA$ and for which $x \in F(B)$. Then let

$$\begin{aligned} \overline{\psi^A}(x) &= \psi^B(x) \\ &\in G(B) \\ &= G(A) \end{aligned}$$

To verify that this operation is independent of the choice of B, it suffices to observe that if $x_1 \sim x_2$, for some $x_1 \in F(B_1)$ and $x_2 \in F(B_2)$ (with $jB_1 = jB_2 = A$), then the following diagram commutes:

$$\begin{array}{ccccc} F(B_1) & \longrightarrow & F(B_1 \wedge B_2) & \longleftarrow & F(B_2) \\ \downarrow \psi^{B_1} & & \downarrow \psi^{B_1 \times B_2} & & \downarrow \psi^{B_2} \\ G(B_1) & = & G(B_1 \wedge B_2) & = & G(B_2) \end{array}$$

It is easy to see that this is the unique extension of ψ to $a_j F$. Thus $a_j F$ is indeed the associated j-sheaf of F, and a_j: $\varepsilon \to \mathrm{Sh}_j(\varepsilon)$ is the sheafification functor.

ALMOST EVERYWHERE COVERS

Suppose that (X, F, μ) is a measure space (where F denotes the σ-algebra $(F, \subseteq, \mathrm{T}, \perp, \neg)$) and that $\varepsilon = \mathrm{Sh}(F)$ is the topos of sheaves on F. Then the notion of formula being true "almost everywhere" induces a topology on Sh(F). Let $N \subseteq F$ be the ideal of μ-negligible sets

$$N = \{A \in F \mid \mu(A) = 0\}$$

Then N is a countably closed sieve, and so corresponds to an arrow

$$\mathrm{N}: 1 \to \Omega$$

in ε.

Define a map $j: \Omega \to \Omega$ by

$$\Omega \xrightarrow{(id_\Omega, N)} \Omega \times \Omega \xrightarrow{\vee} \Omega$$

j is just the "closed topology" induced by N. The idea of j is that it takes a sieve I to the set

$$\{A \in F \mid \exists I \in I \exists N \in N\,(A = I \cup N)\}$$

A sieve $I \in \Omega(A)$ covers A if there is an $I \in I$ such that $\mu(A \cap \neg I) = 0$.

The subtopos of ε induced by this topology is easily seen to be the topos of sheaves on the quotient algebra F/N.

Unfortunately, this conception of an "almost everywhere topology" will not translate to the setting of measures on a locale. The problem is that when we use the expression "almost everywhere", we mean that everything of significance is included. The "closed topology" interpretation above captures the notion that everything that is not included is insignificant. These notions coincide only in certain Boolean settings, in this case because F is a Boolean algebra. Since a locale L is not in general a Boolean algebra, we must find a more direct way to capture the idea of "everything of significance" than "everything that isn't negligible". Such a formulation is the first step towards extending "almost everywhere" equivalence to localic measure theory.

A number of the proofs in this Section make explicit use of sheaf semantics. Hence we assume that we are working in a localic topos $\varepsilon = \mathrm{Sh}(L)$, for some locale $(L, \leq, T, \perp)$. Occasionally, we shall also be referring back to the case where $\varepsilon = \mathrm{Sh}(F)$ is the topos of sheaves on a σ-algebra, as we will want to ensure that we are generalizing the classical notions outlined above.

The first step is to build a notion of "everything significant". To do this, we use the restriction operator ρ.

Definition: Take $\mu \in M$ and $I \in \Omega$. Say that I is dense for μ if

$$\rho(\mu, I) = \mu$$

This logical formula has two free variables (μ and I), and so can be thought of as an arrow

$$[\![\rho(I, \mu) = \mu]\!] : M \times \Omega \to M$$

We can define the collection of μ-dense sieves, by taking the transpose of this arrow:

$$\frac{M \times \Omega \to M}{M \to P\Omega}$$

In fact, this arrow factors through the object $T \rightarrowtail P\Omega$ of topologies.

Theorem: Fix $\mu \in M$. Then the (internal) set $\{I \in \Omega \mid \rho(\mu, I) = \mu\}$ is a topology in ε.

Proof: We need to show that this arrow satisfies the usual algebraic conditions for a topology. We write these conditions as follows:

1. Inflationarity:

$$\rho(\mu, \mathrm{T}) = \mu$$

2. Idempotence:

$$(\rho(\mu, [\![\rho(\mu, I) = \mu]\!]) = \mu) \Leftrightarrow (\rho(\mu, I) = \mu)$$

3. Commutativity with $\wedge$:

$$(\rho(\mu, I \wedge J) = \mu) \Leftrightarrow ((\rho(\mu, I) = \mu) \wedge (\rho(\mu, J) = \mu))$$

The first of these conditions is immediate.

For the two remaining conditions, we will use the fact that if we interpret a measure as a functional on the underlying locale,

$$\rho(\mu, I)(B) = \mu(B \wedge C)$$

whenever $I = \downarrow C$. Note that in Sh(L), all closed sieves are principal.

We simultaneously use the semantics of the topos $\hat{L}$. In particular, we use the fact that the following sequent is a reversible inference in the sheaf semantics:

$$\frac{B \Vdash (\downarrow C) \Rightarrow A}{B \wedge C \Vdash A}$$

Here we are taking advantage of the fact that the formula $\downarrow C$ is a closed sieve, and so a principal sieve. This allows us to view C dually as an element of the base category L (since we can compute $B \wedge C$) and a truth value in E (since we can consider the formula $(\downarrow C) \Rightarrow A$). We use this duality a number of times in the remainder of this proof.

To make this duality more explicit, we adopt the convention of writing $C \Rightarrow A$, rather than $(\downarrow C) \Rightarrow A$.

Likewise, there is no reason to maintain the distinction between the internal conjunction operation "$\wedge$", and the meet operator "$\wedge$" in L. We shall use "$\wedge$" to denote both operations, and reserve "$\wedge$" for external conjunction, and occasionally the join operator in $[0, \infty]$.

Let U be the formula (and hence the element of L) given by

$$U \equiv (\mu, I) = \mu$$

Note that $I \leq U$, and that

$$U \Vdash (I \Rightarrow q \in \mu) \Leftrightarrow (q \in \mu)$$

$$\begin{aligned} q \in \rho(\mu, U)(B) &\leftrightarrow B \Vdash q \in \rho(\mu, U) \\ &\leftrightarrow B \Vdash U \Rightarrow q \in \mu \\ &\leftrightarrow B \wedge U \Vdash q \in \mu \\ &\leftrightarrow B \wedge U \Vdash I \Rightarrow q \in \mu \\ &\leftrightarrow B \wedge U \wedge I \Vdash q \in \mu \\ &\leftrightarrow B \wedge I \Vdash q \in \mu \\ &\leftrightarrow B \Vdash I \Rightarrow q \in \mu \\ &\leftrightarrow B \Vdash q \in \rho(\mu, I) \\ &\leftrightarrow q \in \rho(\mu, I)(B) \end{aligned}$$

For the third condition, we again use the semantic interpretation of $\rho(\mu, B)$. Suppose that $(\mu, I \wedge J) = \mu$. Then for any $B \in L$ we have

$$\begin{aligned} \mu(B) &= \rho(\mu, I \wedge J)(B) \\ &= \mu((I \wedge J) \wedge B) \\ &\leq \mu(I \wedge B) \\ &\leq \mu(B) \end{aligned}$$

Hence $(\mu, I) = \mu$, and likewise for J. Hence

$$\rho(\mu, I \wedge J) = \mu \mid -(\rho(\mu, I) = \mu) \wedge (\rho(\mu, J) = \mu)$$

For the reverse inequality, suppose that $\rho(\mu, I) = \mu = \rho(\mu, J)$. Consequently, we have

$$\mu(B) = (\mu, I \wedge J)$$
$$\leq (\mu, I \vee J)(B)$$
$$\leq \mu(B)$$

Then:

$$\rho(\mu, I \wedge J)(B) = \mu((I \wedge J) \wedge B)$$
$$= \mu((I \wedge B) + (J \wedge B))$$
$$= \mu(I \wedge B) + \mu(J \wedge B) - \mu((I \wedge B) \vee (J \wedge B))$$
$$= \rho(\mu, I)(B) + \rho(\mu, J)(B) - \mu((I \vee J) \wedge B)$$
$$= \mu(B) + \mu(B) - \mu(B)$$
$$= \mu(B)$$

Hence $\rho(\mu, I \wedge J) = \mu$, as required.

Now, we must verify that this topology coincides with the closed topology induced by N, when $\varepsilon \simeq \mathrm{Sh}(F)$ is the topos of sheaves on the -algebra $(F, \subseteq, \mathrm{T}, \perp, \neg)$.

Proposition: Let μ be a measure on F. Then the closed topology on Sh(F) induced by the topology

$$I \mapsto N \vee I$$

coincides with the topology in Definition.

Proof: Viewing Sh(F) as a localic topos Sh(L), the underlying locale is the collection of countably closed sieves in F. There is potential for notational confusion here, as a sieve in L is a sieve of sieves on F. For this reason, for the remainder of this Section, we adpot the following conventions: We use roman script *A, B, C*,... to denote elements of F.

We use the usual calligraphic script, *I, J, K*,... to denote sieves in F (that is, elements of L), and bold face ***A, B, C*** to denote sieves in L. We use $\cap$ and $\cup$ to denote the meet and join in F, $\wedge$ and $\vee$ to denote the meet and join in L (that is, the meet and join in the locale of countably closed sieves in F), and, when needed, $\cap$ and $\cup$ to denote the meet and join in the locale of sieves in L.

In this setting, μ is a measure on F, and $\overline{\mu}$ is the corresponding measure on L, given by

$$\overline{\mu}(I) = \bigvee_{A \in I} \mu(A)$$

Fix $I \in L$. In $\hat{L}$, a sieve on I is some set A of countably complete sieves. However, since we are working in Sh(L), A must be a principal sieve $A = \downarrow J$, for some $J \leq I$.

One element of L is the sieve N of elements of F with measure 0. The traditional "almost everywhere" topology is given by

$$A \mapsto (\downarrow N) \cup A$$

The topology described in Definition, is given by

$$\begin{aligned} A &\mapsto [\![\rho(\overline{\mu}, A) = \overline{\mu}]\!] \\ \downarrow J &\mapsto [\![\rho(\overline{\mu}, J) = \overline{\mu}]\!] \\ &= \{K \in L \mid \overline{\mu}(K \wedge J) = \overline{\mu}(K)\} \\ &= \left\{K \in L \,\middle|\, \left(\bigvee_{\langle J,K\rangle \in J \times K} \mu(K \cap J)\right) = \left(\bigvee_{K \in K} \mu(K)\right)\right\} \end{aligned}$$

Hence a sieve K is in the closure of $\downarrow J$ if and only if for every $K \in K$, and for every natural number n, there is a $J_n \in J$ such that

$$\mu(K \cap J_n) \geq \mu(K) - \frac{1}{n}$$

Since J is countably complete, it follows that there is a $J = \bigcup_n J_n \in J$ such that $\mu(K \cup J) = \mu(K)$. Hence $\mu(K \cap \neg J) = 0$. Thus $K = J \cup N$ for some $N \in N$. Thus the closure of $\downarrow (J)$ is contained in $(\downarrow J) \cup \downarrow (N)$.

The reverse inclusion is immediate.

The following Corollary follows immediately, since the closed sieves of F are just elements of $\uparrow N \subseteq PF$.

Corollary: Let F be a σ algebra, and let μ be a measure on F. Then

$$\mathrm{Sh}_{\mu}(F) \simeq \mathrm{Sh}(F_{\mu})$$

where $F_{\mu} = F/N$ is the -algebra found by taking the quotient of F by the ideal N of μ-negligible sets.

Finally in this Section, we introduce a preorder << on M. This preorder is related to the idea of the "almost everywhere" cover.

Definition: Take $\mu, v \in M$. Then say that μ dominates v, or that $v << \mu$, if

$$\forall I \in \Omega(\rho(\mu, I) = \mu) \Rightarrow (\rho(v, I) = v)$$

This notion of dominance captures the idea of distribution of mass. μ dominates v if every sieve that captures all of μ's mass also captures v's mass. In the case where ε = Sh(F) this can be thought of as saying that v has no mass wherever μ has no mass. Thus $v << \mu$ if and only if $\mu(A) = 0 \Rightarrow v(A) = 0$.

We can formulate this idea internally.

Definition: The map Rull: $M \to \Omega$ is defined in the presheaf topos $\hat{L}$ by

$$Rull(\mu) = [\![\forall q \in Qq \in \mu]\!]$$

Hence $v \ll \mu$ if and only if

$$\text{Rull}(\mu) \subseteq \text{Null}(v)$$

This means that $\ll$ is the pullback of the usual ordering $\leq$ along Null:

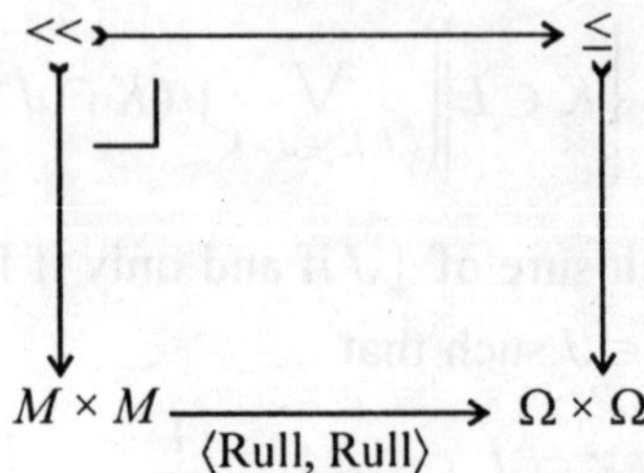

It follows from the semicontinuity condition that Null factors through $\Omega_j \mapsto \Omega$. However, in the case that Sh(L) is a Boolean topos, we can go further.

Proposition: Take $\mu, v \in M$. Then $v \ll \mu$ if and only if Rull $(\mu) \leq$ Rull(v).

Definition therefore restricts to the classical notion of one measure dominating another.

Unsurprisingly, there is a relationship between the dominance relation on measures and the "finer" relation on topologies:

Proposition: Let μ and v be two measures, and let j_μ and j_v be the induced topologies. Then $\mu \ll v$ implies that j_μ is finer than j_v.

Proof: Note that:

$$\mu \ll v \equiv \forall I(\rho(v, I) = v) \Rightarrow (\rho(\mu, I) = \mu)$$

$$\equiv \forall I(\forall q \in \mathbb{Q}(I \Rightarrow q \in v) \Leftrightarrow q \in v) \Rightarrow (\forall q \in \mathbb{Q}(I \Rightarrow q \in \mu) \Leftrightarrow q \in \mu)$$

$$\equiv \forall I j_v(I) \Rightarrow j_\mu(I)$$

Hence we can indicate $\mu \ll v$ by writing $j_v \leq j_\mu$. Recall that j_μ is finer than j_v if anod only if

$$j_m = j_m \circ j_v$$

Since j_μ is a topology, it preserves order. Hence

$$j_\nu \le j_\mu$$
$$j_\mu \circ j_\nu \le j_\mu \circ j_\mu$$
$$j_\mu \circ j_\nu \le j_\mu$$

Likewise,

$$id_\Omega \le j_\nu$$
$$j_\mu \circ id_\Omega \le j_\mu \circ j_\nu$$
$$j_\mu \le j_\mu \circ j_\nu$$

Hence $j_\mu = j_\mu \circ j_\nu$, as required.

ALMOST EVERYWHERE SHEAFIFICATION

In this Section, we investigate some of the important properties of "almost everywhere" topologies. We adopt the following notational conventions in this Section.

$(L, \le, \top, \bot)$ is a locale, $\hat{L}$ is the topos of presheaves on L, and Sh(L) is the topos of sheaves on L (relative to the canonical topology j). M is the sheaf of measures, D is the sheaf of non-negative Dedekind real numbers, Ω is the subobject classifier in $\hat{L}$, and Ω_j is the subobject classifier in Sh(L). $f: D \times M \to M$ is the integration arrow, and $\rho: M \times \Omega_j \to M$ is the restriction arrow. Both f and ρ are arrows in the sheaf topos Sh(L).

Given a measure μ, j_μ is the "almost everywhere" topology associated with μ, and $\mathrm{Sh}_\mu(L)$ is the subtopos of Sh(L) induced by this topology. The sheafification functor is denoted a_μ. $\mathrm{Sh}_\mu(L)$ is itself a localic topos. Let L_μ be the locale of μ closed sieves, so we have

$$L \to L_\mu \mapsto L$$

Since $\mathrm{Sh}_\mu(L)$ is localic, it has its own measure theory. Let M_μ, D_μ, and f_μ denote respectively the sheaf of measures, the sheaf of non-negative Dedekind reals, and the integration arrow in $\mathrm{Sh}_\mu(L)$.

Finally, we consider two subobjects of M. $\downarrow\mu$ is interpreted in Sh(L) as the set

$$\downarrow\mu \equiv \{\nu \in M \mid \nu \ll \mu\}$$

and M^F is interpreted as

$$M^F \equiv \{\nu \in M \mid \exists q \in \mathbb{Q}\mu \ll q\}$$

($\mathbb{Q}$ is the object of positive rationals in Sh(L) and "$\leq$" relation is the associated sheaf of the "element of" relation on $\mathbb{Q} \times M$).

$\downarrow \mu$ is the sheaf of measures that are dominated by μ. M^F is the sheaf of locally finite measures. It is easy to see that if L is the locale of countably closed sieves in some σ-algebra F, then M^F corresponds to the sheaf of -finite measures on F.

The results from this Section can all be displayed in one commutative diagram in Sh(L):

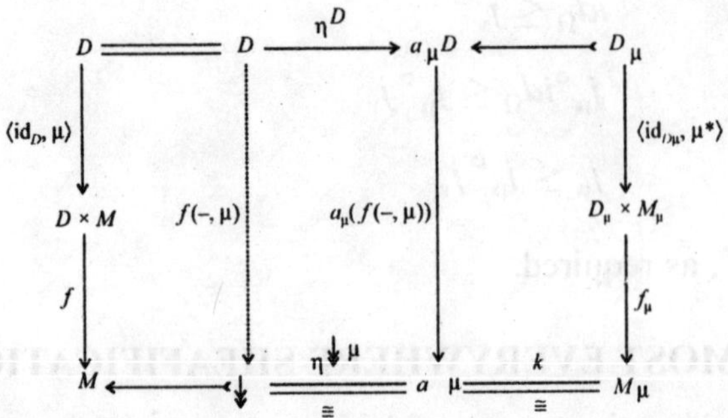

In this diagram, η is the unit of the adjunction $a_\mu \dashv i$, and so $\eta^A: A \to a_\mu A$. The arrow $\mu^*: 1 \to M_\mu$ picks out the measure on the j_μ-closed sieves of L obtained by restricting μ.

Technically $a_\mu A$ is an object of the topos Sh$_j$(L), rather than Sh(L), and $ia_{\mu(\mu)}$ is the corresponding onbect in Sh(L). However, as this whole diagram is assumed to exist in Sh(L), we can drop the is without fear of ambiguity.

We have five results to show:

Theorem:

1. The map $\int \circ \langle id_D, \mu \rangle$ does indeed factor through $\downarrow \mu \mapsto M$.
2. $a_\mu \downarrow \mu \cong M_\mu$
3. There is a natural embedding $D_\mu \mapsto a_\mu D$.
4. The right hand rectangle commutes.
5. The sheaf $\downarrow \mu$ is a μ-sheaf.

Proof: 1. To prove Part 1, we need to show that for any measures μ, and any density f, we have $\int (f, \mu) \ll \mu$.

This can be rewritten as

$$(\rho(\mu, I) = \mu) \Rightarrow \left(\rho\left(\int (f, \mu), I\right) = \int (f, \mu)\right)$$

Start by taking $I \in \Omega_j$ such that $\rho(\mu, I) = \mu$. Then, since we are working in a localic topos, there must be a $B \in L$ such that $I = \downarrow B$. In this case, it follows that for any $A \in L$,

$$\mu(A) = \mu(A \wedge B)$$

We must show that given such a B, we have

$$\int_A f d\mu = \int_{A \wedge B} f d\mu$$

The result is immediate for locally constant $q \in \mathbb{Q}$:

$$\begin{aligned}\int_A q d\mu &= q.\mu(A) \\ &= q.\mu(A \wedge B) \\ &= \int_{A \wedge B} q d\mu\end{aligned}$$

Now, assume that $f = \langle L, U \rangle$ is a Dedekind real. Then

$$\begin{aligned}\int_A f d\mu &= \bigvee_{q \in L} \int_A q d\mu \\ &= \bigvee_{q \in L} \int_{A \wedge B} q d\mu \\ &= \int_{A \wedge B} f d\mu\end{aligned}$$

as required.

2. We know from Theorem that for any $A \in L$,

$$a_\mu \downarrow \mu(A) = \left(\coprod_{j_\mu B = A} \downarrow (B) \right) \Big/ \sim$$

where $v_1 \in \downarrow B_1 \sim v_2 \in \downarrow B_2$ if

$$\rho^{B_1}_{B_1 \cap B_2}(v_1) = \rho^{B_2}_{B_1 \cap B_2}(v_2)$$

We also know that if $A \in L$, then $\mu \in M_\mu(A)$ if μ is a measure on the locale $\downarrow A$.

It is helpful to consider the embedding $\mathrm{Sh}_\mu(L) \rightarrowtail \mathrm{Sh}(L)$ a little more carefully. A functor $F \in \mathrm{Sh}(L)$ is a μ-sheaf if $F(A) = F(B)$ whenever $\overline{A} = \overline{B}$, or equivalently that $F(A) = F(\overline{A})$.

We can view objects in $\mathrm{Sh}_\mu(L)$ in two ways; as functors on L, and as functors on L_μ. A functor on L can be restricted to give the corresponding functor on L_μ. Conversely, a functor on L_μ can be extended to L. If F is a μ-sheaf viewed as a functor on L_μ, we can define $\overline{F}$ by

$$\overline{F}(A) = F(\overline{A})$$

We work in Sh(L). In this setting, we already understand $a_\mu \downarrow \mu$. However, the natural interpretation of M_μ is as a functor on L_μ: For any $B \in L_\mu$, $M_\mu(B)$ is the set of measures on the sublocale $\downarrow B$. In light of the previous remarks, we will consider M_μ to be a functor on L, where for any $A \in L$, $M_\mu(A)$ is the set of measures on the sublocale $\downarrow \overline{A} \subseteq L_\mu$.

We start by taking $v \in M_\mu(A)$. Then is a measure on $\downarrow \overline{A} \subseteq L_\mu$. To extend v to a measure on $\downarrow \overline{A} \subseteq L$, we define $\overline{v}$ by

$$\overline{v}(B) = v(\overline{B})$$

We must verify that $\overline{v}$ is indeed a measure.

1. To see that satisfies the additivity condition, take $U, V \in L$. Then

$$\begin{aligned}\overline{v}(U) + \overline{v}(V) &= v(\overline{U}) + v(\overline{V}) \\ &= v(\overline{U \wedge V}) + v(\overline{U \vee V}) \\ &= \overline{v}(U \wedge V) + \overline{v}(U \vee V)\end{aligned}$$

   as required.

2. To see that satisfies the semicontinuity condition, fix $q \in \mathbb{Q}$. We need to show that the sieve $I = \{U \in L | \overline{v} \leq q\}$ is directed closed (where closure refers to the closure operation j). Take a directed family $D \subseteq I$, and let $D = \curlyvee D$. Let D' be the family $\{B \mid B \in D\}$. Then D' is a directed family in L, and in L_μ.

In fact,

$$v(\overline{D}) = v\left(\curlyvee D'\right) \leq q$$

since for each $D \in D$, we have

$$v(\overline{D}) = \overline{v}\left(D\right) \leq q$$

Finally, we get:

$$\overline{v}(D) = v(\overline{D})$$

as required.

Finally, note that whenever $\overline{C} = \overline{A}$, then $\rho(\overline{v}, C) = v$, and so $\overline{v} \ll \mu$ (since, by definition $\overline{C} = \overline{A} \Leftrightarrow \rho(\mu, Z) = \mu$). Thus $\overline{v}$ is not just an element of $M(A)$, but an element of$(\downarrow\mu)(A)$.

Thus we have built a monomorphism from M_μ to $a_\mu \downarrow \mu$. Composing this with the map $\eta^{\downarrow\mu} : \downarrow\mu \to a_\mu \downarrow \mu$ gives us an arrow in $\mathrm{Sh}_\mu(L)$ from M_μ to $a_\mu \downarrow \mu$.

The reverse direction is simpler. An element of $a_\mu \downarrow \mu(A)$ is a measure $v \ll \mu$ on $\downarrow \overline{A}$. In order to see that such a v restricts to a measure on $\downarrow \overline{A} \subseteq L_\mu$, we just need to verify that $v(B) = v(\overline{B})$ for any $B \leq A$. But this follows immediately from the fact that $v \ll \mu$.

$$a_\mu \downarrow \mu \cong M_\mu$$

These operations are easily seen to be inverse to one another, so we have shown that

$$a_\mu \downarrow \mu \cong M_\mu$$

3. To see that $D_\mu \mapsto a_\mu D$, we consider $a_\mu D$ as a sheaf on L. $a_\mu D(A)$ is the set of equivalence classes of Dedekind reals:

$$\coprod_{\overline{B}=\overline{A}} D(B)/\sim$$

Now consider the sheaf D_μ of Dedekind reals inside $\mathrm{Sh}_\mu(L) \simeq \mathrm{Sh}(L_\mu)$. Since D_μ is a sheaf on L_μ, there is an extension, $\overline{D_\mu}$, a sheaf on L, by

$$\overline{D_\mu}(A) = D_\mu(\overline{A})$$

The elements of $D_\mu(\overline{A})$ are themselves sheaves (in $\mathrm{Sh}(L_\mu)$). These sheaves can be interpreted as sheaves on L. It follows that an element of $D_\mu(\overline{A})$ is a pair $\langle L,U \rangle$ of subsheaves of $\mathbb{Q}$. In order for a pair of sheaves to be a Dedekind real in $\mathrm{Sh}_\mu(L)$, they must satisfy the satisfying the following formulas in $\mathrm{Sh}_\mu(L)$:

1. $\forall q \in \mathbb{Q} \neg(q \in L \wedge q \in U)$
2. $\forall q \in \mathbb{Q} q \in U \Rightarrow \exists r \in \mathbb{Q}(r < q \wedge r \in U)$
3. $\forall q \in \mathbb{Q} q \in L \Rightarrow \exists r \in \mathbb{Q}(r < q \wedge r \in L)$
4. $\forall q \in \mathbb{Q} \forall r \in \mathbb{Q}(q \in U \wedge r > q) \Rightarrow r \in U$
5. $\forall q \in \mathbb{Q} \forall r \in \mathbb{Q}(q \in L \wedge r < q) \Rightarrow r \in L$
6. $\forall q \in \mathbb{Q} \forall r \in \mathbb{Q} q < r \Rightarrow (q \in L \vee r \in U)$
7. $\exists q \in \mathbb{Q} \in U$

Supposing that $\langle L,U \rangle$ satisfies these formulas in $\mathrm{Sh}_\mu(L)$, it is natural to ask what can we say about them in the larger topos Sh(L). In order to determine this, we first consider the nature of the object of rational numbers in $\mathrm{Sh}_\mu(L)$.

It is known that $\mathbb{Q}_\mu$, the object of rational numbers in $\mathrm{Sh}_\mu(L)$ is just $a_\mu\mathbb{Q}$, the associated sheaf of $\mathbb{Q}$, the object of rational numbers in Sh(L). Hence we can consider L and U to be subsheaves of $\mathbb{Q}$, rather than $\mathbb{Q}_\mu$.

The first five conditions taken together imply that for any $A \in L$, $L(A)$ is an open lower set of rationals, that $U(A)$ is an open uper set of rationals, and that $L(A) \cap U(\mathrm{A}) = \phi$. But this is clearly the same as if we had interpreted the formulas in Sh(L). It follows that a pair of μ-sheaves $\langle L,U \rangle$ satisfying the first five conditions will also satisfy them when interpreted as sheaves in Sh(L).

If $\langle L,U \rangle \in D_\mu(A)$ (for some μ-closed sieve A) satisfy the sixth condition for some q, $r \in \mathbb{Q}$, then there must be some $B \in L$ such that $\overline{B} = A$

$$B \Vdash (q \in L) \vee (r \in U)$$

hence $\langle L,U \rangle \in D(B)$. Since

$$D(B) \mapsto \coprod_{\overline{C}=A} D(C) \to \coprod_{\overline{C}=A} D(C) \Big/ \sim \xrightarrow{\cong} a_\mu D$$

Similarly, if$\models \exists q \in \mathbb{Q} q \in U$, then there is some locally constant rational q_0 and a dense $B \in \downarrow A \subseteq L$ such that $B \Vdash q_0 \in U$, or equivalently that $q_0 \in U(B)$. However, since B is dense, and since U is a μ-sheaf, it follows that $q_0 \in U(A)$.

We now have a map from $D_\mu \to a_\mu D$, and need only show that this map is monic. But this follows from the fact that if $\langle L_1,U_1 \rangle$ and $\langle L_2,U_2 \rangle$ are two distinct pairs of μ-sheaves, then they must differ at some μ-closed element of L.

4. The fact that the right hand rectangle commutes follows from the construction of the integral, in Sh(L) and in $\mathrm{Sh}_\mu(L)$. The result is immediate for locally constant q, and the extension to Dedekind reals follows.

5. We know that $\downarrow \mu$ is a sheaf on L. Take $A \in L$. Then a covering family for A is a certain set $C \subseteq \downarrow A \subseteq L$. However, since $\downarrow \mu$ is a sheaf, we know that matching families on C correspond to elements of $\downarrow \mu\,(\curlyvee C)$.

So, take a $B \in L$ such that $\downarrow B$ μ-covers A, and take $v \in \downarrow \mu(B)$. We have to show that v has a unique extension $\overline{v} \in \downarrow \mu(A)$.

For each $D \leq A$, set

$$\overline{v}(D) = v(D \wedge B)$$

It is immediate that $\overline{v}$ is an extension of μ, and that $\overline{v}$ is a measure dominated by μ. Thus we only need to show that $\overline{v}$ is the unique amalgamation of *v*.

Suppose that $\lambda \in \downarrow \mu(A)$ is an arbitrary extension of. Since $\lambda \ll \mu$, it follows that for any *C* we have

$$(\rho(\mu, C) = \mu) \Rightarrow (\rho(\lambda, C) = \lambda)$$

Since *B* is μ-dense, we have $\rho(\mu, B) = \mu$, and hence $\rho(\lambda, B) = \lambda$. Therefore

$$\lambda(D) = \rho(\lambda, B)(D) = \lambda(B \wedge D) = \overline{v}(D)$$

Thus $\lambda = \overline{v}$, as $\overline{v}$ is unique, as required.

These results can also be restricted to the case where we work with M^F, rather than *M*, and interpret $\downarrow \mu$ as a subobject of M^F. There are no significant changes in the proofs.

DIFFERENTIATION IN A BOOLEAN LOCALIC TOPOS

We look at some special properties of the measure theory of a Boolean localic topos. A localic topos is a topos that is equivalent to the topos of sheaves on some locale *L*. The Heyting algebra of truth values in such a topos is just *L*.

As a result, the topos satisfies the law of the excluded middle just in the case that *L* is a complete Boolean algebra, and not merely a Heyting algebra (note that all complete Boolean algebras are locales).

Let *B* be a complete Boolean algebra, let $\widehat{B}$ be the topos of presheaves on *B*, and let Sh(*B*) be the topos of sheaves on *B*, relative to the canonical topology. Note that a measure (that is, as element of *M*(*A*) on $\downarrow A \subseteq B$) is additive for all cardinalities, and not just countable cardinalities. This means that

$$\mu\left(\bigvee_{D \in D}\right) = \bigvee_{D \in D} \mu(D)$$

for any directed set $D \subseteq B$. Likewise:

$$\mu\left(\bigvee_{A \in A}\right) = \sum_{A \in A} \mu(A)$$

for any antichain $A \subseteq B$. This property can be called "complete additivity" (extending the usual measure theoretic terminology of "countable additivity").

Definition: Let ε be a localic topos, and let $\mu: 1 \to M$ be a global element of M. Then we say that μ is differentiable, or has Radon-Nikodym derivatives, if the following arrow in ε has a right inverse, called $\frac{d}{d\mu}$:

$$D \xrightarrow{\langle id_D, \mu\rangle} D \times M \xrightarrow{f} \downarrow \mu^F$$

where $\downarrow_{\mu^F} \mapsto M$ is the sheaf of locally finite measures $v \ll \mu$.

The measure theoretic significance of Boolean localic toposes is the following Theorem:

Radon-Nikodym Theorem

Theorem: Let ε be a Boolean localic topos. Then every locally finite measure in ε is differentiable.

It is possible to view this arrangement in category theoretic terms. We prove that for locally inite μ, the arrow in Definition is the surjective part of the image factorization of

$$D \xrightarrow{\langle id_D, \mu\rangle} D \times M \xrightarrow{f} \downarrow \mu^F$$

The derivative is therefore

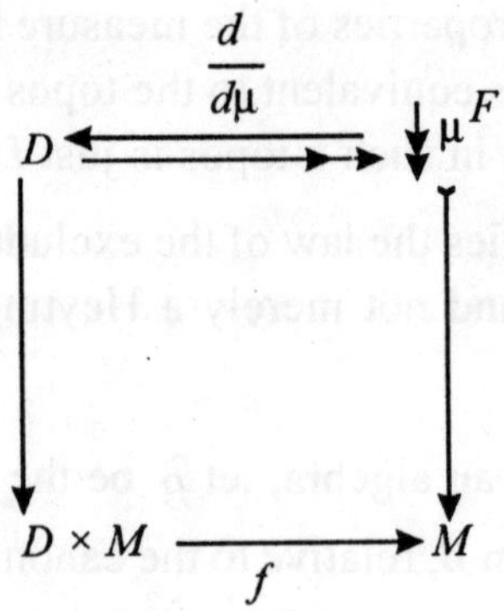

If we can show that the top arrow $D \to \downarrow \mu^F$ is indeed an epimorphism, then the existence of the derivative is an immediate consequence of the fact that epimorphisms split in Boolean toposes.

Before proving this Theorem, some preliminary results are needed.

Definition: Say that two measures μ and are mutually singular if Rull ∨ Rull = T.

Hahn Decomposition Theorem

Proposition: If μ_1 and μ_2 are any two locally finite measures on a complete Boolean algebra B, then there exists $B \in B$ such that the following statements hold:

1. $B \Vdash v \leq \mu$

2. $\neg B \Vdash \mu \leq \nu$

Proof: This proof is based on the usual proof of the Hahn decomposition theorem, with some slight modifications to allow for the fact that we are not using countably additive measures on σ-fields, but rather completely additive measures on complete Boolean algebras.

Start by restricting to a cover $C = \langle C_i \mid i \in I \rangle$ on which both μ and ν are finite. The extension of the result to locally finite μ and ν will be immediate.

Let $\phi: \downarrow \mu C_i \to (-\infty, \infty)$ be given by

$$\phi(B) = \mu(B) - \nu(B)$$

Since $\phi \leq \mu$, it follows that φ is bounded above, and so has a supremum, $\alpha < \infty$.

Suppose that there exists $B \in \downarrow C_i$ such that $\phi(C_i) = \alpha$. Then it follows that for every $D \in B_i$, we have $\phi(D) \geq 0$, or else $\phi(B \wedge \neg D) > \alpha$, which would be a contradiction. Likewise, every $D \in \downarrow (\neg B)$ must have $\phi(D) \leq 0$, for the same reason. Consequently, it will suffice to find such a B.

We know that there must be a sequence $\langle D_n \mid n < \omega \rangle$ such that $\phi(D_n) \uparrow \alpha$. For each n, let $F_n \subseteq \downarrow C_i$ be given by

$$F_n = \left\{ \left(\bigwedge_{i \in S} D_i \right) \wedge \left(\bigwedge_{j \notin S} \neg D_i \right) \middle| S \subseteq \{0, \ldots n\} \right\}$$

In other words, F_n is the set of atoms of the sub-Boolean algebra B_n of $\downarrow C_i$ generated by $\{D_i | i \leq n\}$.

For each n, let $G_n \subseteq F_n$ be given by

$$G_n = \{E \in F_n | \phi(E) \geq 0\}$$

and let $G_n = \curlyvee G_n$. Hence G_n maximizes φ over B_n. Since $D_n \in B_n$, it is obvious that

$$\phi(D_n) \leq \phi(G_n) \leq \alpha$$

Let $H_n = \bigwedge_{i=n}^{\infty} G_i$. We can see that $\phi(H_n)$ is increasing, as

$$\left(\bigcurlyvee_{i=m}^{n} G_i \right) \wedge \neg \left(\bigcurlyvee_{i=m}^{n-1} G_i \right)$$

is the join of elements of F_n.

Finally, let $B = \bigwedge H_n$, so that

$$B = \limsup_n G_n$$

Then $H_n \downarrow B$, and so $\phi(H_n) \to \phi(B)$. Since we already know that $\phi(H_n) \to \alpha$, it follows that $\phi(B) = \alpha$, and we are done.

Corollary: If ε is Boolean, then M^F is (internally) totally ordered.

Proof: We need to show that $\phi \models (\mu \le v) \vee (v \le \mu)$. But we know that

1. $\varepsilon \models B \Rightarrow \le \mu$
2. $\varepsilon \models \neg B \Rightarrow \mu \le v$

Since ε is Boolean, we also have $\varepsilon \models B \vee (\neg B)$, so we are done.

In order to prove the Radon-Nikodym Theorem, we will make some small modifications to the standard proof. The modified proof is based on the fact that a Boolean localic topos is the topos of sheaves on a complete Boolean lattice, since a localic topos is Boolean if and only if the underlying locale is a complete Boolean algebra, and not just a complete Heyting algebra. The main modification between the standard proof and the modified proof is that we do not assume that the Boolean algebra is a field (that is, we do not assume that the elements of B are sets), and so we work with Dedekind cuts rather than with measurable functions.

We first prove the following two Lemmas:

Lemma: Suppose that $v \le \mu \in M^F(A)$. Then there exists $\lambda \in M^F(A)$ such that $v + \lambda = \mu$.

Proof: We know that there is a cover $\langle A_i \mid i \in I \rangle$ for A such that for each $i \in I$ we have $v(A_i) \le \mu(A_1) < \infty$. Given such a cover, there is family of measures $\lambda_i \in M^F(A_i)$. The λ_is form a matching family and so have a unique amalgamation in $M^F(A)$.

First, we must find the λ_is. Let $\lambda_i(B) = \mu(B) - v(B)$. We need to show that λ_i is a measure on $\downarrow A_i$.

It is immediate that if the λ_i's are measures, they are finite (since they are less than or equal to the restrictions of μ). Furthermore, they obviously form a matching family, and have a locally finite amalgamation.

Take $B_1, B_2 \in \downarrow A_i$. Then

$$\begin{aligned}\lambda_i(B_1) + \lambda_i(B_2) &= (\mu_i(B_1) + \mu_i(B_2) - (v_i(B_1) + v_i(B_2)) \\ &= (\mu_i(B_1 \wedge B_2) + \mu_i(B_1 \vee B_2) - (v_i(B_1 \wedge B_2) + v_i(B_1 \vee B_2)) \\ &= \lambda_i(B_1 \wedge B_2) + \lambda_i(B_1 \vee B_2)\end{aligned}$$

Hence λ_i satisfies the additivity condition.

since L is a complete Boolean topos, we can show that λ_i satisfies the semicontinuity condition by showing that for any antichain $A \subseteq \downarrow A_i$ we have

$$\lambda_i(\curlyvee A) = \sum_{B \in A} \lambda(B)$$

Since $\mu(A)$ is finite, we can assume that for all but countably many of the elements of A satisfy

$$\mu(B) = v(B) = \lambda(B) = 0$$

So, without loss of generality, we can assume that A is countable. Write $A = \langle B_i \mid i < \omega \rangle$. By definition, we know that

$$\sum_{i<\omega} \lambda(B_i) = \lim_{n \to \omega} \sum_{i=0}^{n} \lambda(B_i)$$

But for each n we have

$$\sum_{i=0}^{n} \lambda(B_i) = \sum_{i=0}^{n} (\mu(B_i) - v(B_i))$$

$$= \left(\sum_{i=0}^{n} \mu(B_i)\right) - \left(\sum_{i=0}^{n} v(B_i)\right)$$

$$\to \left(\sum_{i<\omega}^{n} \mu(B_i)\right) - \left(\sum_{i=\omega}^{n} v(B_i)\right)$$

$$= \mu(\curlyvee A) - v(\curlyvee A)$$

$$= \lambda(\curlyvee A)$$

so we are done.

Lemma: If $\mu, v \in M^F(A)$, and μ and v are not mutually singular, then there exists $B \in \downarrow A$ and q $\in \mathbb{Q}$ such that $\mu(B) > 0$ and

$$\int (q, \mu) \leq \dot{v}$$

Note that this Lemma can also be stated in the following contrapositive form: If there is no such B and q, then μ and v must be mutually singular.

Proof. We shall simplify our notation by exploiting the fact that integration of a rational number is the same as multiplication, and so we can write $q \times \mu$ for

$$\int(q,\mu)$$

For each $q \in \mathbb{Q}$, apply the Hahn Decomposition Theorem to find B_q such that

$$B_q \Vdash q \times \mu \leq \nu \quad -B_q \Vdash \nu \leq q \times \mu$$

Define B by

$$B = \bigvee_{q \in \mathbb{Q}} B_q$$

Then $\neg B$ is given by

$$-B = \bigwedge_{q \in \mathbb{Q}} -B_q$$

Hence for all $q \in Q$. Since $\mu(-B) < \infty$, it follows that $\nu(\neg B) = 0$.This means that $\neg B \leq$ *Null*(ν). Since μ and ν are not mutually singular, it follows that *Null*(μ) cannot be contain B. Hence $\neg$(*Null*)(μ)^ $B, \perp$. Therefore, $\mu(B) > 0$. But the B_q*'s* form an increasing chain whose join is B. It cannot be the case that all of the B_q*'s* satisfy $\mu(B_q) = 0$, since this would imply that $\mu(B) = 0$. Hence there is a $q \in \mathbb{Q}$ such that $\mu(B_q) > 0$.

This B_q also satisfies (by definition)

$$B_q \Vdash q \times \mu \leq \nu$$

as required.

Now, to prove Theorem

Proof: In light of the comments at the start of this Section, it suffices to show that if μ,ν are locally finite measures on a complete Boolean algebra L, then there is a Dedekind real f on L such that $\int(f,\mu) = \nu$.

Let L be the sheaf of rationals satisfying

$$q \in L \Leftrightarrow \int(q,\mu) \leq \nu$$

Then, since ε is Boolean, D is order complete, and so there is an $f \in D$ such that $f = \vee L$.

All we need to do now is show that

$$\int(f,\mu) = \nu$$

For convenience, we shall let denote the measure $\int(f,\mu)$. It follows from the Monotone Convergence Theorem that

$$\sigma = \int(f,\mu) \vee \nu \int(q,\mu) \leq \nu$$

Applying Lemma, we see that there must be a measure λ such that $\sigma + \lambda = \nu$. (Relative toμ, σ is called the "absolutely continuous" part of ν, and λ is called the "singular" part of ν.) We will show that $\nu \ll \mu\ \lambda \Rightarrow \lambda = 0$.

Suppose that λ is not constantly zero. Then since $\lambda \ll \nu \ll \mu$ and since in a Boolean topos, $(\mu_1 \ll \mu_2) \Leftrightarrow$ $(Null(\mu_1) \geq Null(\mu_2)$, it follows that $Null(\nu)\ Null(\mu)$.

Consequently, μ and λ can only be mutually singular if λ is the zero measure. Suppose that isλ not the zero measure. Then applying Lemma we find that there is a $B \in L$ and $a\ q$ such that $q \times \mu \leq \lambda$ on B. But then we we could add the rational q to f (only locally, at B) and get

$$\int (f+q,\mu) = \int (f,\mu) + q \times \mu \leq \sigma + \lambda \leq \nu$$

But $\int (f,\mu) \leq \int (f + q \times \mu)$, so f is not the supremum claimed in its definition. This is a contradiction, and so λ must be identically zero, whence $\sigma = \nu$ as required.

THE RADON-NIKODYM THEOREM

We saw that locally finite measures are differentiable in a Boolean localic topos. In this Section, we use these derivatives to construct more general Radon-Nikodym derivatives. Rather than requiring $Sh(L)$ to be Boolean, we will consider the case where $Sh_\mu(L)$ is Boolean.

In this case, the derivative of μ in $Sh_\mu(L)$ will be extended to derivatives in $Sh(L)$.

To construct a derivative map, we use a fragment of the diagram from page 91:

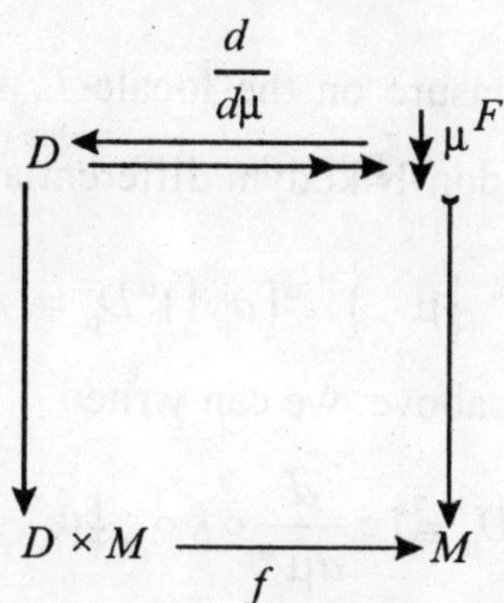

with the added assumption that $\downarrow\mu$ is computed as a subobject ofM^F.

If μ^* is differentiable (in $Sh_\mu(L)$), then the arrow $\int_\mu d\mu : D\mu \to M_\mu^F$ has a section $\frac{d}{d\mu^*} : M_\mu^F \to D_\mu$. We can then write the diagram as follows:

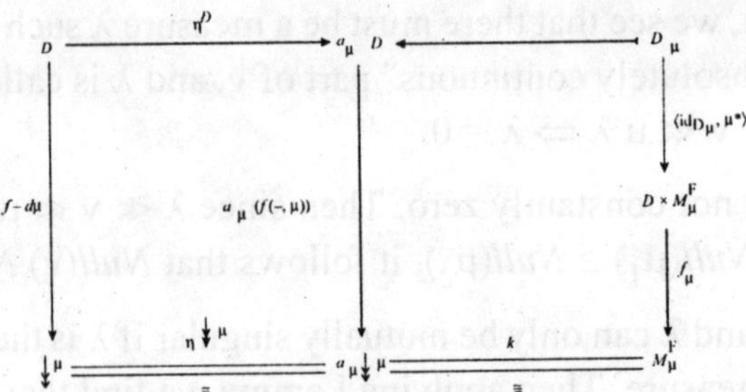

The Radon-Nikodym differentiation map can now be def

Definition: The action of differentiation with respect to μ is given by

$$io\frac{d}{d\mu *}oko\ {}_{\eta}\Downarrow\mu : \Downarrow\mu \to a_{\mu}D$$

This arrow is denoted $D_{\mu} : \Downarrow\mu \to a_{\mu}D$ .it is indicated in the above digram by the dotted arrow.

The differentiation arrow takes a measure dominated by μ to an element of $a_{\mu}D$. Since $a_{\mu}D$ consists of equivalence classes of densities, this is not surprising. In the classical case, the derivative $\frac{d\nu}{d\mu}$ is only defined up to μ-almost everywhere equivalence. We must verify that this notion of derivative is indeed a right inverse to integration.

This task reduces the Radon-Nikodym Theorem to a diagram chase.

Radon-Nikodym Theorem

Theorem: Let μ be a measure on the locale *L*, such that $Sh_{\mu}(L)$ is Boolean. Let $D_{\mu} : \Downarrow\mu \to a_{\mu}D$ denote the Radon-Nikodym differentiation operation.Then

$$\left({}_{\eta}\Downarrow\mu \right)^{-1} \circ \left(a_{\mu} \textstyle\int\right) \circ D_{\mu} = id_{\Downarrow\mu}$$

Proof: Using the diagram above, we can write

$$D_{\mu} = i \circ \frac{d}{d\mu *} \circ k \circ {}_{\eta}\Downarrow\mu$$

The composition in the Theorem can therefore be written:

$$\left({}_{\eta}\Downarrow\mu \right)^{-1} \circ \left(a_{\mu} \textstyle\int\right) \circ ;i \circ \frac{d}{d\mu *} \circ k \circ {}_{\eta}\Downarrow\mu$$

But since ${}_{\eta}\Downarrow\mu$ is an isomorphism, our task reduces to showing that the anticlockwise circuit of the right hand square, starting at ${}_{a\mu}\Downarrow\mu$ is just the identity:

$$\left(a_\mu \int\right)\circ i\circ \frac{d}{d\mu *}\circ k = id_{a\mu} \downarrow\mu$$

Using the fact that

$$\left(a_\mu \int\right)\circ i = k^{-1} \int \mu(-,\mu^*)$$

our composition can be rewritten

$$k^{-1}\circ \int \mu(-,\mu^*)\frac{d}{d\mu *}\circ k$$

Now, using the fact that

$$\int(-,\mu^*)\circ \frac{d}{d\mu *} = id_{M_\mu^F}$$

the composition reduces to

$$k^{-1}\circ k$$

But this is trivially $id_{a\mu} \downarrow\mu$.

The fact that the Radon-Nikodym derivative is not a density, but an equivalence class of densities, captures the idea of "almost everywhere" uniqueness of derivatives. However, there is another interesting feature of this formulation. An element of $a_\mu(D)(A)$ is an equivalence class of densities in

$$\bigcup_{\overline{B}=A} D(B)$$

It is not necessarily the case that such an equivalence class will contain an element of$D(A)$. In this case, the Radon-Nikodym derivative itself is defined only "almost everywhere".

As an example of this phenomenon, let μ be the Lebesgue measure on the locale of open sets of the real line, and let ν be the restriction of this measure to the unit interval I = [0, 1], so that $\nu(A) = \mu(A \cap I)$. Then D is the sheaf of continuous real valued functions. It is clear that there is no continuous function f on $\mathbb{R}$ such that $\int(f,\mu) = \nu$. However, if we let $A = \mathbb{R}\setminus \{0, 1\}$, then there is an element $g \in D(A)$ (that is, a continuous function g: $A \to \mathbb{R}$) such that $\int(g,\mu) = \nu$, namely

$$g(x) = \begin{cases} 1 & \text{if } x \in (0,1) \\ 0 & \text{if } x \in (-\infty,0)\cup(1,\infty) \end{cases}$$

Since A is dense in $\mathbb{R}$, it follows that there is an element of $a_\mu D$ corresponding to g.

Thus g (together with all other continuous functions which agree with g "almost everywhere") is the Radon-Nikodym derivative $\frac{d\mu}{d\nu}$.

It may seem that the requirement that $Sh\mu(L)$ is Boolean is a strong condition to impose. After all, most toposes of interest are not Boolean. However, it turns out that many measures induce Boolean subtoposes.

Lemma: Let $(F, \subseteq, \perp, \perp, \neg)$ be σ-algebra, and let be a σ-finite measure on F. Then the σ-algebra $F_\mu = F/Null(\mu)$ satisfies the countable chain condition.

Proof: For each element $A \in F_\mu$, there is an $A' \in F$ such that

$$A = \{(A' \cup N_1) \cap \neg N_2 \mid N_1, N_2 \in null(\mu)\}$$

We use a contrapositive argument.

Suppose that F_μ does not satisfy the countable chain condition. Then there is an uncountable antichain $A = \{A_\alpha \mid \alpha < k\}$ for some $k \geq \omega_1$. Applying the axiom of choice, there is a corresponding uncountable fami $A = \langle A_\alpha \mid \alpha < k \rangle$. The members of A' are pairwise almost disjoint, meaning that for $A'_\alpha \neq A'_\beta \in A$ we have

$$\mu\left(A'_\alpha \cap A'_\beta\right) = 0$$

We now build a new antichain $\langle B_\alpha \mid \beta < \omega_1 \rangle$ in F:

$$B_\alpha = A_\alpha \cap \neg\left(\bigcup_{\beta<\alpha} B_\beta\right)$$

Note that this recursive expression makes sense only for $\alpha < \omega_1$, as the expression $\left(\bigcup_{\beta<\alpha} B_\beta\right)$ is not necessarily defined if is uncountable.

Then for each $\alpha < \omega_1$, we have $B_\alpha \subseteq A_\alpha$. Furthermore, $\langle B_\alpha \mid \alpha < \omega_1 \rangle$ is an uncountable antichain in F, with

$$\mu(B_\alpha) = \mu(A_\alpha) > 0$$

Hence μ is not σ-finite.

Proposition: Let $(F \subseteq, \perp, \perp, \neg)$ be σ -algebra, and let μ be σ -finite measure on F. Then $Sh\mu(F)$ is a Boolean topos.

Proof: From Corollary, we know that L_μ is the quotient of F by the ideal $Null(\mu)$. If μ is σ-finite, then this quotient algebra, F_μ,satisfies the countable chain condition.

Lemma: The countable chain condition here has two important consequences:

1. F_μ is a complete Boolean algebra
2. If $A \subseteq F_\mu$, then there is a some countable $A_0 \subseteq A$, such that

$$\bigcup A = \bigcup A_0$$

As a result, it follows that $Sh_\mu(F)$ is the topos of sheaves on a complete Boolean algebra, and is hence a Boolean topos.

To see that F_μ has these properties:

1. Take $I \subseteq F_\mu$ Each $A \in I$ consists of an equivalence class of elements of F, where

$$B \sim C \Leftrightarrow \mu((B \cap \neg C) \cup (\neg B \cap C)) = 0$$

Assume without loss of generality that I is a countably complete sieve in F (that is, an element of Ω_j). We will show that I contains a maximal element.

Use the axiom of choice to well order the elements of I. Define a sequence $\langle G_\alpha \mid \alpha \triangleleft |I| \rangle$ in I by

$$G_\alpha \bigcup_{\beta<\alpha} F_\alpha$$

In order to make sure that this is well defined, we must show that G_α exists when α has uncountable cofinality.

Since $\langle G_\gamma \mid \alpha \rangle$ is a chain, it must follow that $\langle \mu(G_\gamma) \mid \gamma < \alpha \rangle$ is also a chain, and in fact must be increasing. However, every increasing sequence with uncountable cofinality of elements of $[0,\propto]$ must terminate, and so the sequence $\langle \mu(G_\gamma) \mid \gamma < \alpha \rangle$ must have countable cofinality. Since $U \not\subseteq V \Rightarrow \mu(U) \leq \mu(V)$ in F_μ, it follows that $\langle G_\gamma \mid \gamma < \alpha \rangle$ also terminates, and so has a supremum.

2. So now, given $I \subseteq F_\mu$, we know that $\cup I$ exists.

First consider the case where I is a countable complete sieve. The argument used above shows that the sequence $\langle G_\alpha \mid \alpha < |I| \rangle \uparrow \cap I$ has countable cofinality, and so there is a countable subsequence $\langle G_{\alpha_i} \mid i < \omega \rangle$ in I whose supremum is $\bigcup I$.

If I is an not a countably complete sieve, then we have shown that there is a countable subset of $\overline{\downarrow I}$ whose join is $\bigcup I$. But each member of $\overline{\downarrow I}$ is the countable join of members of F, and so we are done.

Proposition: If $Sh(L)$ is the topos of sheaves on a locale L, and μ is a measure on L satisfying$\mu(A) = \mu(\neg\neg A)$, then $Sh_\mu(L)$ is Boolean. (Such a measure is called"continuous".)

Note that the Lebesgue measure, λ, is a continuous measure, as for every open set U, we have $\lambda(\vartheta U) = 0$.

Proof: In the case where L is a locale, and μ has the property that $\mu(A) = \mu(\neg\neg A)$ for all $A \in L$, it follows that

$$\rho(A, \mu) = \rho(\neg\neg A, \mu)$$

Hence $j_\mu(A) = \perp$ if and only if $j_\mu(\neg\neg A) = \perp$.

It follows that j_μ factors through the double negation topology $\neg\neg$, and so is a topology in Sh$\neg\neg$ (Sh(L)). But this sheaf topos is itself a localic topos, and is in fact the topos of sheaves on the complete Boolean algebra of $\neg\neg$-stable elements of L. (In the event that L is spatial, and so is the algebra of open sets in topological space, the $\neg\neg$-stable elements of L are just the regular open sets.) In this topos, we can apply Wendt's argument to see that the topology corresponding to j_μ is just the closed topology induced by the sieve of ($\neg\neg$-stable) elements A of L satisfying $\mu(A) = 0$.

But the resulting locale of closed sieves is just the quotient of a complete Boolean algebra by a complete ideal and is hence a complete Boolean algebra. Thus the topos $Sh_\mu(L)$ is equivalent to the topos of sheaves on a complete Boolean algebra, and so is a Boolean topos, as required.

The following Corollary justifies calling a $\neg\neg$ stable measure "continuous".

Corollary: Let (X,τ) be a topological space, let $\nu \ll \mu$ be two locally finite continuous measures on τ. Then there is a continuous function f which is defined on a μ–dense open set X_0 such that

$$\nu = \int f d\mu$$

POSSIBILITIES FOR FURTHER WORK

This dissertation lays the groundwork for a locale based measure theory, and a topos based measure theory. There are several immediate possibilities for further research leading out of these ideas.

SLICING OVER *M*

Rather than fixing a single measure μ, and sheafifying with respect to that measure, it would be desirable to study differentiation of all measures simultaneously. The natural way to approach this is to work in the slice topos ε/M. An object of this topos is an arrow (of ε) with codomain$M$. An arrow between objects $f: A \to M$ and $g: B \to M$ is simply an arrow $h: A \to$ B such that $g \circ h = f$.

We can build the following diagram, consisting of arrows over M:

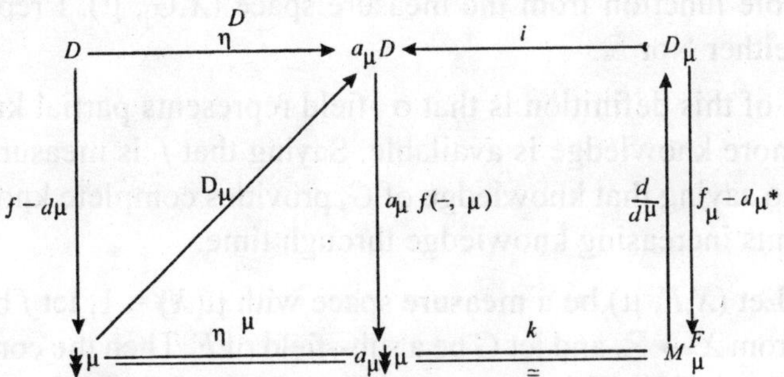

Differentiation would therefore be a section to the arrow $M \times D \to \ll$, taking a pair of measures $\langle \nu, \mu \rangle$ to a pair $\langle f, \mu \rangle$ such that $\int \langle f, \mu \rangle = \nu$.

Further study of this topic would start with an investigation of the μ-sheafification operation in the slice topos. This should result in a single Lawvere-Tierney topology in ε/M. The existence of derivatives would then be conditional on the law of the excluded middle.

EXAMPLES OF LOCALIC MEASURE THEORY

We have studied the essentials of measure theory in an arbitrary localic topos. It would be natural to investigate particular localic toposes, and determine which additional measure theoretic results hold.

Aparticular case that has practical implications is the case where B is a Boolean algebra, and J is the finite join topology on B. Carrying out the construction of M in $\widehat{B}$ yields the sheaf of finitely additive measures. The Radon-Nikodym theorem is much harder to apply to finitely additive measures, so it would be interesting to approach this problem using the sheaf theoretic approach.

The first steps would be to verify that the sheaf of measures in *Sh*(B) is equivalent to the sheaf of measures in *Sh*(L), where L is the locale of ideals (closed sieves) in B. The next step would be to determine the nature of the μ-almost everywhere topologies, and determine which measures induce Boolean subtoposes.

STOCHASTIC PROCESSES AND MARTINGALES

We start with some definitions from probability theory:

Definition: Let (X, F, μ) be a measure space with $\mu(X) = 1$.

1. A stochastic process is a sequence $f = \langle f_i \mid i \in I \rangle$ of measurable real valued functions, for some ordered index set I
2. A filtration is an in increasing family $G = \langle G_i \mid i \in I \rangle$ of sub--fields of F.

3. A stochastic process f is adapted to the filtration G if for every $i \in I$ f_i is a measurable function from the measure space (X, G_i, μ). I represents time, and is usually either $\mathbb{N}$ or $\mathbb{R}$.

The essence of this definition is that σ -field represents partial knowledge. The finer the σ-field, the more knowledge is available. Saying that f_i is measurable with respect to G_i is equivalent to saying that knowledge of G_i provides complete knowledge of f_i. Thus a filtration represents increasing knowledge through time.

Definition: Let (X, F, μ) be a measure space with $\mu(X) = 1$, let f be a measurable real valued function from $X \to \mathbb{R}$, and let G be a sub--field of F. Then the conditional expectation of f given G, denoted $E[f \| G]$ is a measurable real valued function g such that

1. g is measurable relative to G
2. For any $G \in G$,

$$\int_G g d\mu = \int_G f d\mu$$

Note that unconditional expectation is just conditional expectation, with the conditioning done relative to the trivial σ-field $\{\theta, X\}$. The existence of g is a consequence of the Radon-Nikodym theorem: Let ν be the measure on (X, G, μ^*) (where μ^* is the obvious restriction of μ to G) given by

$$\nu(G) = \int_G f d\mu$$

It is clear that $\nu >> \mu^*$, and so there is a Radon-Nikodym derivative $\frac{d\nu}{d\mu^*}$ on $(XG\mu^*)$. This derivative is g.

Definition: A martingale consists of a stochastic process $f = \langle f_i \mid i \in I \rangle$ and a filtration $G = \langle G_i \mid i \in I \rangle$ such that

1. f is adapted to G
2. For every $i < j$ we have

$$f_i = E[f_j \| G_i]$$

The machinery presented in this dissertation provides for capturing all the ingredients necessary to study martingales, except for sub-σ-fields, and filtrations. However, these objects can be embedded into the topos theoretic framework. The topos of sheaves on G is a subtopos of the topos of sheaves on G. The topology J that induces this subtopos is given by saying that for any $C \in F$,

$$J(C) = \{I \in \Omega(C) \mid \{G \in G \mid G \subseteq C\} \subseteq I\}$$

Such a topology has a very special property. For any given C, there is a smallest sieve that covers C, namely the sieve

$$\{D \in F \mid \exists G \in G D \subseteq G \subseteq C\} = \bigcup \{\downarrow G \mid G \in G \cap C \downarrow\}$$

This means that the closure arrow j: $\Omega \to \Omega$ has an internal left adjoint m: $\Omega \to \Omega$. Furthermore, $m \perp = \perp$.

Definition: Say that a Lawvere-Tierney topology $j < \Omega \to \Omega$ is nice if it has a left adjoint m: $\Omega \to \Omega$ such that $m \perp = \perp$. These "niceness" properties allow for the following important proposition:

Proposition: Let ε be a topos, and let j: $\Omega \to \Omega$ be a nice Lawvere-Tierney topology, with a left adjoint m: $\Omega \to \Omega$. Let $\mathbb{R}_j$ be the Dedekind real numbers object in $Sh_j(\varepsilon)$, and let $\mathbb{R}$ be the Dedekind real numbers object on ε. Then $\mathbb{R}j\ \mathbb{R}$.

Proof: Let $\langle L,U \rangle$ be an element of $\mathbb{R}_j$. We must show that $\langle L,U \rangle$ is a Dedekind real in ε. Let $\mathbb{R}_j$ be the Dedekind real numbers object in $Shj(\varepsilon)$, let $\mathbb{R}_\varepsilon$ be the Dedekind real numbers object in ε, and let $\mathbb{R}$ denote the "internal" set of Dedekind reals.

$$\langle L,U \rangle \in \mathbb{R}_j$$

$$\leftrightarrow sh_j(\varepsilon) \models \langle L,U \rangle \in \mathbb{R}$$

$$\leftrightarrow \varepsilon \models J^\circ(\langle L,U \rangle \in \mathbb{R})$$

$$\leftrightarrow \varepsilon \models \perp \Rightarrow J^\circ(\langle L,U \rangle \in \mathbb{R})$$

$$\leftrightarrow \varepsilon \models m^\circ \perp \Rightarrow (\langle L,U \rangle \in \mathbb{R})$$

$$\leftrightarrow \varepsilon \models \perp \Rightarrow (\langle L,U \rangle \in \mathbb{R})$$

$$\leftrightarrow \varepsilon \models \langle L,U \rangle \in \mathbb{R}$$

$$\leftrightarrow \langle L,U \rangle \in \mathbb{R}_\varepsilon$$

This proposition allows us to define a filtration as an increasing sequence of Lawvere-Tierney topologies with left adjoint's, and a stochastic process adapted to that filtration asa sequence of Dedekind reals, in the chain

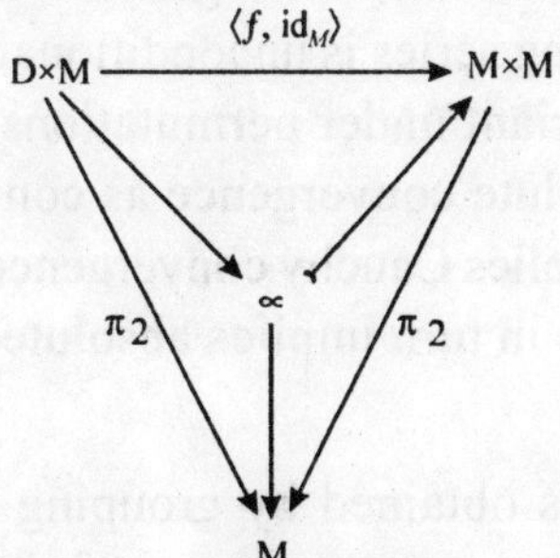

The next step in the construction would be the expression of the martingale property.

Chapter 4

Modes of Convergence

In mathematics, there are many senses in which a sequence or a series is said to be convergent. The aim of this article is to serve as an index for various modes (senses) of convergence, organized according to the settings in which they are defined. It must be noted that each of the following objects is a special case of the types preceding it: sets, topological spaces, uniform spaces, TAGs (topological abelian groups), normed spaces, Euclidean spaces, and the real/complex numbers.

Also, note that any metric space is a uniform space. Convergence can be defined in terms of sequences in first-countable spaces. Nets are a generalization of sequences that is useful in spaces which are not first countable. Filters further generalize the concept of convergence. In metric spaces, one can define Cauchy sequences.

Cauchy nets and filters are generalizations to uniform spaces. Even more generally, Cauchy spaces are spaces in which Cauchy filters may be defined. Convergence implies "Cauchy-convergence", and Cauchy-convergence, together with the existence of a convergent subsequence implies convergence. The concept of completeness of metric spaces, and its generalizations is defined in terms of Cauchy sequences.

SERIES OF ELEMENTS IN A TOPOLOGICAL ABELIAN GROUP

Convergence of a series is defined as convergence of the sequence of partial sums. An important concept when considering series is unconditional convergence, which guarantees that the limit of the series is invariant under permutations of the summands. In a normed vector space, one can define absolute convergence as convergence of the series of norms ($\Sigma| b_k |$). Absolute convergence implies Cauchy convergence of the sequence of partial sums (by the triangle inequality), which in turn implies absolute-convergence of some grouping (not reordering).

The sequence of partial sums obtained by grouping is a subsequence of the partial sums of the original series. The norm convergence of absolutely convergent series is an equivalent condition for a normed linear space to be Banach (i.e.: complete). Absolute convergence and convergence together imply unconditional convergence, but unconditional convergence does not imply absolute convergence in general, even if the space is Banach, although the implication holds in $\mathbb{R}^d$.

CONVERGENCE OF SEQUENCE OF FUNCTIONS TO A TOPOLOGICAL SPACE

The most basic type of convergence for a sequence of functions (in particular, it does not assume any topological structure on the domain of the functions) is pointwise convergence.

It is defined as convergence of the sequence of values of the functions at every point. If the functions take their values in a uniform space, then one can define pointwise Cauchy convergence, uniform convergence, and uniform Cauchy convergence of the sequence. Pointwise convergence implies pointwise Cauchy-convergence, and the converse holds if the space in which the functions take their values is complete.

Uniform convergence implies pointwise convergence and uniform Cauchy convergence. Uniform Cauchy convergence and pointwise convergence of a subsequence imply uniform convergence of the sequence, and if the codomain is complete, then uniform Cauchy convergence implies uniform convergence.

If the domain of the functions is a topological space, local uniform convergence (i.e. uniform convergence on a neighborhood of each point) and compact (uniform) convergence (i.e. uniform convergence on all compact subsets) may be defined.

Note that "compact convergence" is always short for "compact uniform convergence," since "compact pointwise convergence" would mean the same thing as "pointwise convergence" (points are always compact).

Uniform convergence implies both local uniform convergence and compact convergence, since both are local notions while uniform convergence is global. If X is locally compact (even in the weakest sense: every point has compact neighborhood), then local uniform convergence is equivalent to compact (uniform) convergence. Roughly speaking, this is because "local" and "compact" connote the same thing.

SERIES OF FUNCTIONS TO A TOPOLOGICAL ABELIAN GROUP

Pointwise and uniform convergence of series of functions are defined in terms of convergence of the sequence of partial sums.

For functions taking values in a normed linear space, absolute convergence refers to convergence of the series of positive, real-valued functions $\Sigma \mid g_k \mid$. "Pointwise absolute convergence" is then simply pointwise convergence of $\Sigma| g_k |$. Normal convergence is uniform convergence of the series of non-negative, real-valued functions obtained by taking the (pointwise) norm of each function in the series (uniform convergence of $\Sigma \mid g_k \mid$).

In Banach spaces, pointwise absolute convergence implies pointwise convergence, and normal convergence implies uniform convergence. For functions defined on a topological space, one can define (as above) local uniform convergence and compact (uniform) convergence in terms of the partial sums of the series.

If, in addition, the functions take values in a normed linear space, then local normal convergence (local, uniform, absolute convergence) and compact normal convergence (absolute convergence on compact sets).

Normal convergence implies both local normal convergence and compact normal convergence. And if the domain is locally compact (even in the weakest sense), then local normal convergence implies compact normal convergence.

FUNCTIONS DEFINED ON A MEASURE SPACE

If one considers sequences of measurable functions, then several modes of convergence that depend on measure-theoretic, rather than solely topological properties, arise. This includes pointwise convergence almost-everywhere, convergence in *p*-mean and convergence in measure. These are of particular interest in probability theory.

MODES OF CONVERGENCE

Real analysis studies four basic modes of convergence for a sequence of functions f_n on a measure space Ω with measure μ.

1. Almost everywhere (AE). The sequence f_n converges almost everywhere to f if $f_n(x)$converges to f(x) except possible for a set of x values of measure zero.
2. Almost uniform (AU). The sequence f_n converges almost uniformly if for every $\varepsilon > 0$ there exists a corresponding set A_ε of measure less than ε such that f_n converges uniformly to f on the complement of A_ε.
3. L^p. The sequence f_n converges to f in L^p norm if

$$\lim_{n\to\infty}\left(\int_\Omega |f_n - f|^p \, d\mu\right)^{1/p} = 0.$$

4. In measure (M). The sequence f_n converges to f in measure if for every $\varepsilon > 0$,

$$\lim_{n\to\infty} \mu\{x : | f_n(x) - f(x)\} > \varepsilon\} = 0.$$

This page summarizes and diagrams the relations between these four modes of convergence in general, in a finite measure space, and under dominated convergence. Then counterexamples are given that show no more relationships exist in general.

GENERAL MEASURE SPACES

The first diagram shows the general relationships between the four modes of convergence.

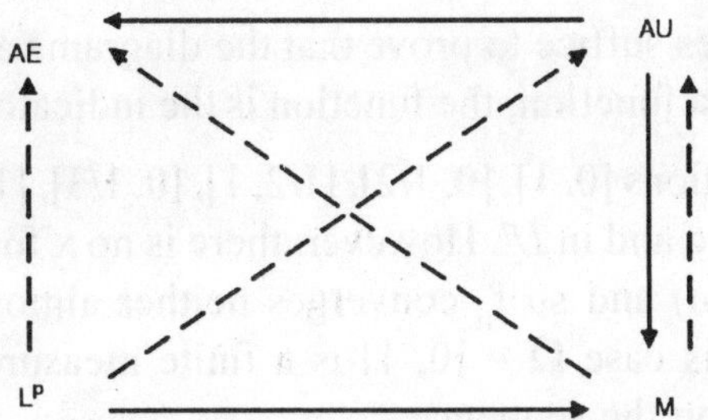

FINITE MEASURE SPACES

Now suppose μ(Ω). For finite measure spaces, almost everywhere and almost uniform convergence are equivalent. Convergence in measure is the weakest form of convergence since it is implied by the other forms. The following diagram summarizes the relationships between the four modes of convergence for finite measure spaces.

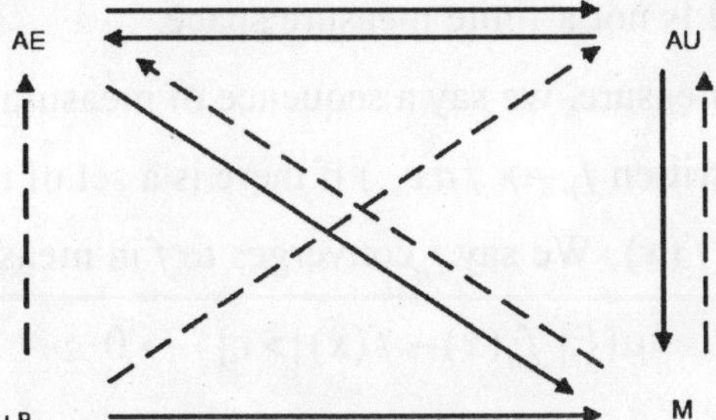

DOMINATED CONVERGENCE

If Ω is a general measure space, but the sequence f_n is uniformly dominated by an L^p function g, then more relationships exist, as summarized in the diagram below.

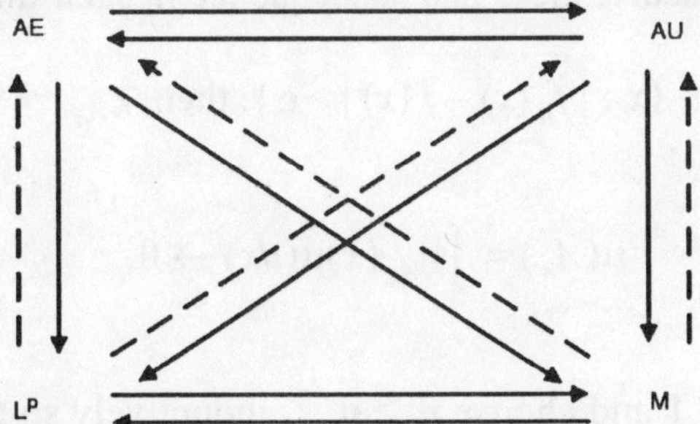

Note that the even though we do not require Ω to be a finite measu2re space, all the convergence relationships for finite measure spaces continue to hold, as well as two new ones.

In this setting, almost everywhere and almost uniform convergence are equivalent. Also, L^p convergence and convergence in measure are equivalent.

COUNTEREXAMPLES

Three counter examples suffice to prove that the diagrams above are complete. When an interval is referred to as a function, the function is the indicator function of that interval.

First, consider the functions [0, 1], [0, 1/2], [1/2, 1], [0, 1/3], [1/3, 2/3], etc. The sequence f_n converges to 0 in measure and in L^p. However, there is no x for which $f_n(x)$ converges to 0 ($f_n(x) = 1$ infinitely often) and so f_n converges neither almost everywhere nor almost uniformly. Note that in this case $\Omega = [0, 1]$ is a finite measure space, and the constant function 1 is an L^p bound on the sequence.

Next, consider the functions $f_n = n\ [1/n, 2/n]$. The sequence f_n converges pointwise to 0 everywhere. It converges almost uniformly and converges in measure. However, the L^p norm of f_n is 1 for all n and so no subsequence converges to 0 in L^p norm. Note again $\Omega = [0, 1]$ is a finite measure space in this example.

Finally, let f_n be [n, n+1]. Then f_n converges to 0 everywhere, but f_n does not converge in measure. In this case Ω is not a finite measure space.

Definition: If μ is a measure, we say a sequence of measurable function f_n converges to f almost everywhere (written $f_n \to f a.e.$,) if there is a set of measure 0and for x not in this set we have $f_n(x) \to f(x)$. We say $f_n$converges to f in measure if for each $\varepsilon > 0$

$$\mu(\{:|f_n(x) - f(x)| > \varepsilon\}) \to 0$$

as $n \to \infty$

Proposition: Suppose μ is a finite measure.

(a) If $f_n \to f$,a.e., then f_n converges to f in measure.

(b) If $f_n \to f$ in measure, there is a subsequence n_j such that $f_{nj} \to f$, a.e.

Proof: Let $\varepsilon > 0$. If $= \{x : |f_n(x) - f(x)| > \varepsilon\}$. then $\chi_{A_n} \to 0$ a.e., and by dominated convergence,

$$\mu(A_n) = \int \chi_{A_n}(x)\mu(dx) \to 0.$$

This proves (a).

To prove (b) let $n_1 = 1$ and choose $n_j > n_{j-1}$ inductively so that

$$\mu(\{x : |f_{nj}(x) - f(x)| > 1/j\}) \le 2^{-j}.$$

Let $A_j = \{x : | f_{nj}(x) - f(x)| > 1/j\})$. Then $\mu(A_j >) \le 2^{-j}$, and

$$A = \bigcap_{k=1}^{\infty} \bigcup_{j=1}^{\infty} A_j$$

has measure less $\bigcup_{j=k} A_j$ than for every k, hence less than 2^{-k+1} for every k. Therfore A has measure 0. If $x \notin A$, then $x \notin \bigcup_{j=1}^{\infty} A_j$ for some k, so $| f_{nj}(x) - f(x) \le 1/j$ for $j \ge k$. which means $f_{nj} \to f$ a.e. on A^c.

Example:

1. Part (a) of the above proposition is not true if $\mu(X) = \infty$. Let $X = \mathbb{R}$ and let

   $f_n = \chi_{(n,n+1)}$.

2. For an example where $f_n \to f$ in the measure but not almost everywhere. Let

   $X = [0, 1]$, let μ be lebesgue measure and let $f_{n(x)} = \chi_{(F_n)}(x)$, where

   $$F_n = \left\{ y : \left(\sum_{j=1}^{n} 1/j \right) (\text{mod} 1) \le y \le \left(\sum_{j=1}^{n+1} 1/j \right) (\text{mod} 1) \right\} z (\text{mod} 1)$$

   is defined as the fractional part of z (where the largest integer less than z is subtracted from z). Let $f(x = 0$ for all x.

Then $\mu(F_n) \le 1/n \to 0$, so $f_n \to f$ in measure. But any x will be in infintely many F_n ' so f_n does not converge to $f(x)$ at any point. The following is known as Egoroff's theorem.

Theorem: If μ is a finite measure $\varepsilon > 0$, and $f_n \to f$ then there exists a measurable set A such that $\mu(A) < \varepsilon$ and $f_n \to f$ uniformly on A^c.

This type of convergence is sometimes known as almost uniform convergence.

Proof: Let

$$E_{nk} \cup_{m=n}^{\infty} \{x : | f_m(x) - f(x)| > 1/k\}.$$

for fixed k, E_{nk} decreases as n increases and the intersection $\bigcap_n E_{nk}$ has measure 0. so $\mu(E_{nk}) \to 0$. Then there exists an interger n_k such that $\mu(E_{nk^k}) < \varepsilon 2^{-k}$. Let $E = \bigcup_{k=1}^{\infty} E_{nk^k}$. Then $\mu(E) < \varepsilon$, and if $x \notin E$ and $n > n_k$, then $| f_n(x) - f(x)| \le 1/k$. Thus $f_n \to f$ uniformly on E^c.

PRODUCT MEASURES

If $A_1 \subset A_2 \subset ....$ and $A = \bigcup_{i=1}^{\infty} A_i$, we write $A_i \uparrow A$ .If $A_1 \supset A_2 \supset ....$ and $A = \bigcap_{i=1}^{\infty} A_i$, we write $A_i \downarrow A$.

Definition: *M* is a monotone class is *M* is a collection of subsets of *X* such that

1. If $A_i \uparrow A$ and each $A_i \in M$, then $A \in M$;

2. If $A_i \downarrow A$ and each $A_i \in M$, then A $\in M$.

The intersection of montone classes is a monotone class, and the intersecrtion of all monotone classes containing a given collection of sets is the smallest monotone class containing that collection.

The next theorem, the monotone class lemma, is rather techinical, but very useful.

Theorem: Suppose A_0 is as algebra, *A* is the smallest $\sigma-$algebra containing A_0 and *M* is the smallest monotone class containing A_0. Then $M = A$.

Proof: *A* $\sigma-$algebra is clearlly a monotone class, so $M \subset A$. We must show $A \subset M$.

Let $N_1 = \{A \in M : A^c \in M\}$. Note N_1 is contained in *M*, contains A_0, and is a monotone class. So $N_1 = M$, and therefore *M* is closed under the operation of taking complements.

Let ddd for all $B \in A_0\}.N_2$ is contained in *M*; N_2 contains A_0, contains A_0 because A_0 is an algebra; N_2 is a monotone class because $\left(\bigcup_{i=1}^{\infty} A_n\right) \cap \mathrm{B} = \bigcup_{i=1}^{\infty} (A_i \cap B)$, and similarly for intersections. Therefore $N_2 = M$; in other words, if $B \in A_0$ and $A \in M$ then $A \cap B \in M$ ·

Let $N_3 = \{A \in M : A \cap B \in M$ for all $B \in M\}$. As in *j*the preceding paragraph, N_3 is a monotone class contained in *M*. By the last sentences of the preceding paragraph, N_3 contains A_0. Hence, $N_3 = M$. We thus have that *M* is a monotone class closed under the operations of taking complements and taking intersection. This shows *M* is a $\sigma-$algebra, and so $A \subset M$.

Suppose (X, A, μ) and (Y, B, ν) are two measure spaces i.e., *A* and *B* are $\sigma-$algebra on *X*and *Y*, resp., and μ and $\upsilon$are measures on *A* and *B* resp. A rectangle is a set of the from $A \times B$, where $A \in A$ and $B \in B$. Define a set function $\mu \times \nu$ on rectangle by

$$\nu \times \mu(A \times B) = \mu(A)\nu(B).$$

Lemma: Suppose $A \times B = \bigcup_{i=1}^{\infty} A_i \times b_i$, where $A, A_{i,} \in A$ and $B, B_i \in B$ and the $A_i \times B_i$ are disjoint. Then

$$\mu \times \nu(A \times B) = \sum_{i=1}^{\infty} \mu \times \nu(A_i \times B_i).$$

Proof: We have

$$\chi_{A \times \mathrm{B}}(x, y) = \sum_{i=1}^{\infty} \chi_{A_i \times B_i}(x, y),$$

and so

$$\chi_A(x)\chi_B(y) = \sum_{i=1}^{\infty} \chi_{A_i}(x)\chi_{B_i}(y).$$

Holding x fixed and integtating over y with respect to ν.

$$\chi_A(x)\nu(B) = \sum_{i=1}^{\infty} \chi_{A_i}(x)\nu(\mathrm{B}_i).$$

Now use again and integrate over x with respect to μ obtain the result.

Let $C_0 = \{\text{finite unions of rectangles}\}$. It is clear that C_0 is an algebra. By Lemma and linearity, we see that $\mu \times \nu$ is a measure on C_0. Let $A \times B$ be the smallest $\sigma-$ algebra containing C_0; this is called the product $\sigma-$algebra.By the Caratheodory extension theorem, $\mu \times \nu$ can be extend to a measure on $A \times B$.

We will nee the following observation. Suppose a μ measure is $\sigma-$finite So there exist E_i which have finite μ measure and whose union is X. If we let $F_n = \bigcup_{i=1}^{\infty} E_i$, then $F_i \uparrow X$ and $\mu(F_n)$ is finite for each n.

If μ and ν are both $\sigma-$finite say with $F_i \uparrow X$ and $G_i \uparrow Y$, then $\mu \times \nu$ will be $\sigma-$finite,using the sets $F_i \times G_i$.

The main result of this section is Fubinis's theorem, which allows one to interchange the order on integration.

Theorem: Suppose $f : X \times Y \to \mathbb{R}$ is measurable with respect to $A \times B$. If f is nonnegative of $\int |f(x, y)| \, d(\mu \times \nu)(x, y) < \infty$, then

(a) The function $g(x) = \int f(x, y)\nu(dy)$ is measurable with respect to A;

(b) The function $h(y) = \int f(x, y)\mu(dx)$ is measurable with respect to B;

(c) We have

$$\int f(x,y)d(\mu\times\nu)(x,y) = \int\left(\int f(x,y)d\mu(x)\right)d\nu(y)$$
$$= \int\left(\int f(x,y)d\nu(y)\right)\mu(dx).$$

Proof: First suppose μ and ν are measure. if f is the characteristic function of a rectangle, then (a) – (c) are obvious. By linearity, (a) – (c) hold if f is the characteristic function of a set $C_{0,}$ the set of finite unions of rectangle.

Let M be the collection of sets C such that (a)–(c) hold for χ_C. If $C_i \uparrow C$ and $C_i \in M$ then (c) holds for χ_C by dominted convergence. If $C_i \downarrow C$, then (c) holds for χ_C by dominated convergence. (a) and (b) are easy. So M is a monotone class containing A_0, so $M = A \times B$.

If μ and ν are σ – finite, applying monotone convergence to $C \cap (F_n \times G_n)$ for suitable F_n and G_n and monotone convergence, we see that (a) –(c) holds for the charateristic fuctions so sets in $A \times B$ In this case as well. By linearity, (a) – (c) holds for nonnegative simple functions. By monotone convergence, (a) – (c) hold for nonegative functions. In case $\int |f| < \infty$, writting f = f + – f⁻ and using linearity proves (a)–(c) for this case, too,

THE RADON -NIKODYM THEOREM

Suppose f is nnnegative, system, measurable, and interable with respect to μ. If we define ν by

$$\nu(A) = \int_A f d\mu,$$

then ν is a measure. The only part that needs thought is the countable additivity, and this follows from applied to the functions $f_{\chi_{A_i}}$. Moreover, $\nu(A)$ zero whenever $\mu(A)$ is. We sometimes write.

Definition: *A* measure is v is called absolutely continuous with respect to a measure This is frequently wreten μ if $\nu(A) = 0$ whenever $\mu(A) = 0$. This is frequently written $d\nu \ll \mu$.

Proposition: A finite measure ν is absolutely continuous with respect to μ if and only if for all $\in$ there exists δ such that $\mu(A)$ implies $\nu(A) < \varepsilon$.

Proof: If the condition given in the statement of the proposition holds, it is clear that $\nu \ll \mu$. Suppose now that $\nu \ll \mu$. If the condition does not hold, there exists E_k such that $\mu(E_k) < 2^{-k}$ but $\nu(E_k) \geq \varepsilon$. Let $F = \bigcap_{n=1}^{\infty} \bigcup_{k=n}^{\infty} E_k$.

Then

$$\mu(F) = \lim_{n\to\infty} \mu\left(\bigcup_{k=n}^{\infty} E_k\right) \le \lim_{n\to\infty} \sum_{k=n}^{\infty} 2^{-k} = 0,$$

but

$$\nu(F) = \lim_{n\to\infty} \nu\left(\bigcup_{k=n}^{\infty} E_k\right) \ge \varepsilon;$$

This contradicts the absolute continuity.

Definition: A function $\mu : A \to (-\infty, \infty]$ is called a signed measure if $\mu(\theta) = 0$ and $\mu\left(\bigcup_{i=1}^{\infty} A_i\right) = \sum_{i=n}^{\infty} \mu(A_i)$ whenever the A_i are disjoint and all the A_i are in A.

Definition: Let μ be a signed measure. A set $A \in A$ is called a positive set for μ if $\mu(B) \ge 0$ whenever $B \subset A$ and $B \in A$. We define a negative set similarly. A null set A is one where $\mu(B) = 0$ whenever $B \subset C$ is measurable.

Example: Suppose m is Lebesgue measure and $\int_A f\, dm$ for some integrable f. if we let $P = \{x : f(x) \ge 0\}$, then P is easily seen to be a positive set, and if then $N = \{x : f(x) < 0\}$, then N is a negative one. The Hahn decomposition which we give below is a decomposition of our space (in this case $\mathbb{R}$) into positive and negative sets. This decomposition is unique, except that $C = \{x : f(x) = 0\}$ could be included in N instead of P, or apportioned partially to P and partially to N. Note however, that C is a null set. The Jordan decomposition below is decomposition of μ into μ^+ and μ^-, where $\mu^+(A) \int_A f^+ dm$, and similary $\mu^-(A) = \int_A f^- d\mu$.

Note that if μ is a signed measure, then $\mu\left(\bigcup_{i=1}^{\infty} A_i\right) = \lim_{n\to\infty} \mu\left(\bigcup_{i=1}^{n} A_i\right)$. The proof is the same as in the case of positive measure.

Proposition: Let μ be a signed measure taking values in $(-\infty, \infty]$. Let E be measurable with $\mu(E) < 0$. Then there exists a subset F of E that is a negative set with $\mu(F) < 0$.

Proof: If E is a negative set, we are done. If note, there exists a subset with postive measure. Let n_1 be the smallest positive interger such that there exists $E_1 \subset E$ with $\mu(E_1) \ge 1/n$. Let $k \ge 2$. If $F_k = E - (E_1 \cup .... \cup E_{k-1})$ is negative, we are done. If not, let n_k be the smallest positive integer such that there exists $E_k \subset F_k$ with $\mu(E_k \ge 1/n_k$. We continue.

If the construction stops after a finite numnber of sets, we are done. If not, let $F = \bigcap_k F_k = E - \left(\bigcup_k E_k\right)$. Since $0 > (E) > -\infty$ and $\mu(E_k) \geq 0$, then

$$\mu(E) = \mu(F) + \sum_{k=1}^{\infty} \mu(E_k).$$

Then $\mu(F) \leq \mu(E) < 0$, so the sum converges. If $G \subset F$ is measurable with $\mu(G) > 0$, then $\mu(G) \geq 1/N$ for some *N*, which contradicts the construction Therefore *F* must be a negative set.

We write $A \Delta B$ for $(A-B) \cup (B-A)$. The following is known as the Hahn decomposition theorem.

Theorem: Let μ be a signed measure taking values in $(-\infty, \infty]$. There exists set *E* and *F* in *A* that are disjoint whose is *X* and such that *E* is a negative set and *F* is a positive set. If *E'* and *F'* are another such pair, then $E \Delta E' = F \Delta F'$ is a null set with respect to μ.

Proof: Let L = inf { $\mu(A)$: *A* is negative set}. Choose negative sets A_n such $\mu(A_n) \to$ L. Let $E = \bigcup_{n=1}^{\infty} A_n$. Let $B_n = A_n - (B_1 \cup \ldots \cup B_{n-1})$ for each *n*. Since A_n is a negative set, so is each that B_n. Also the B_n are disjoint. If $C \subset E$, then

$$\mu(C) = \lim_{n\to\infty} \mu\left(C \cap \left(\bigcup_{i=1}^{n} B_i\right)\right) = \lim_{n\to\infty} \sum_{i=1}^{n} \mu(C \cap B_i) \leq 0.$$

So *E* is negative set.

Since *E* is negative,

$$\mu(E) = \mu(A_n) + \mu(E - A_n) \leq \mu(A_n).$$

Letting $n \to \infty$, we obtain $\mu(E) = L$.

Let $F = E^c$. If *F* were not a positive set, there would exist $B \subset F$ with μ(B) < 0.By proposition there exists a negitive set *C* contained in *B* with $\mu(C) < 0$. But then $E \cup C$ would be a negative set with $\mu(E \cup C) < \mu(E) = L$, a contradiction.

To prove uniqueness, if *E'*, *F'* and another such pair of sets and $A \subset E - E' \subset E$, then $\mu(A) \leq 0$. But $A \subset E - E' = F' - F \subset F'$, so $\mu(A) \leq 0$. Therefore $\mu(A) = 0$. The same argument work if $A \subset E' - E$ and any subset of $E \,\Delta\, E'$ can be written as the union of A_1 and A_2 whre $A \subset E - E'$ and $A_2 \subset E' - E$.

Let us say two measures μ and ν are mutually singular if there exists two disjoint sets *E* and *F* in *A* whose union is *X* With $\mu(E) = \nu(F) = 0$ This often written $\mu \perp \nu$.

Example:

1. If is Lebesgue measure restricted to [0, 1/2], $\mu(A) = m(A \cap [0,1/2])$, and ν is Lebesgue measure restricted to [1/2, 1], then μ and ν are mutually singular. We let E = [0, 1/2] and F (1/2, 1]. This example works because the Lebesgue measure of {1/2} is 0.
2. A more interesting example is the following. Let *f* be the Cantor-Lebesgue function and let ν be the Lebesgue -Stieltjes measure associated with *f*. Let μ be Lebesgue measure restricted to [0, 1]. Then $\mu \perp \nu$. To see this, we let $E = C$, where C is the Cantor set, and F = [0, 1] – C. We already know that $m(E)$ = 0and we nee to show $\nu(F) = 0$. To so that. we need to show $\nu(I) = 0$ for every open interval contained in F. This will follow if we show $\nu(J) = 0$ for every interval of the form $J = (a, b]$ contained in F. But f is constanant on every such interval, so $f(b) = f(a)$, and therefore $\nu(J) = f(b) - f(a) = 0$.

The Jordan Decompostion Theorem

Theorem: If μ is a signed measure, there exists μ^+ and ν^- such that u = $\mu^+ - \mu^-$ and $\mu^+$and μ^- are mutually singular. This decomposition is unique.

Proof: Let E and F be positive and negative sets for μ and let $\mu^+(A) = \mu(E \cap A)$, $\mu^-(A) = -\mu(A \cap F)$. This gives the desired decomposition.

If $\mu = \nu^+ - \nu^-$ is another such decomposition with μ^+, ν^- mutually singular, let E' and F'` be the sets in the definition of mutually singular. Then $X = E' \cup F'$ gives another Hahn decompostion, hence $E \Delta E'$ is a null set with respect to μ. Then for any $A \in A$

$$\nu^+(A) = \mu(A \cap E') = \mu(A \cap E) = \mu^+(A),$$

and similarly for ν^-, μ, μ^-.

The measure $\mu^+ + \mu^+$ is called the total variation measure and is writtin $|\mu|$.

The Radon–Nikodym Theorem

Theorem: Suppose μ is a σ-finite measure and ν is a finite measure such that ν is absolutely continuous with respect to μ. There exists a μ- integrable nonegative function f such that $\nu(A) = \int_A f\, d\mu$ for all $A \in A$. Moreover, if g is another such function then $f = g$ almost everywhere with respect to μ.

Proof: Let us first prove the uniqueness assertion. For every set A we have

$$\int_A (f - g)d\mu = \nu(A) - \nu(A) = 0.$$

By proposition we have $f-g=0$ a.e. with respect to μ.

Since μ is $\sigma-$finite, there exist $F_i \uparrow X$ such that $\mu(F_i)<\infty$ for each *i*. Let μ_i be the restriction of μ to F_i that is, $\mu_i(A)=\mu(A\cap F_i)$. Define ν_i the restriction of ν to. F_i, similarly. If f_i is a function such that $\nu_i(A)=\int_A(A)\int_i f_i du_i$ for all *A*, the argument of the first paragraph shows that $f_i=f_j$ on F_i if $i\leq j$. If we define *f* by $f(x=f_i(x)$ if $x\in F_i$, we see that *f* wil be the desired function. So it suffices to restrict attention to the case where μ is finite

Let $$F=\left\{g:0\leq g,\int_A gd\mu\leq \nu(A)\text{for all } A\in A\right\}.$$

F is not empty because $0\in F$ Let $L=\sup\left\{\int gd\mu : g\in F\right\}$, and let g_n be a sequence in *F* such that $\int gn\, d\mu \to L$. Let $h_n \max(g_1,..., g_n)$.

If g_1 amd g_2 are in *F*, then $h_2 = \max(g_1, g_2)$ is also in *F*. To see this, let $B=\{x: g_i(x)\geq g_2(x)\}$, and write

$$\int_A h_2\mu = \int_{A\cap B} h_2 d\mu + \int_{A\cap B^c} h_2 d\mu$$

$$= \int_{A\cap B} g_1 d\mu + \int_{A\cap B^c} g_2 d\mu$$

$$\leq \nu(A\cap B)+\nu(A\cap B^c) = \nu(A).$$

By an induction arguments, h_n is in *F*.

The h_n increases, say to *f*. By monotone convergence $\int fd\mu = L$ and

$$\int_A f\, d\mu \leq (A)$$

for all *A*.

Let *A* be a set where there is strict inequality; let ε be chosen sufficiently small so that if π is defined by

$$\pi(B)=\nu(B)-\int_B f\, d\mu - \varepsilon\mu(B),$$

then $\pi(A)>0.\pi$ is a signed measure; let *F* be the positive set as constructed in Theorem. In particular, $\pi(F)>0$. So for every *B*

$$\int_{B\cap F} fd\mu + \varepsilon\mu(B\cap F)\leq \nu(B\cap F).$$

We then have, that

$$\int_B (f+\varepsilon\chi_F)d\mu = \int_{B\cap F} fd\mu + \varepsilon\mu(B\cap F)$$

$$= \int_{B\cap F^c} fd\mu + \int_{B\cap F} fd\mu + \varepsilon\mu(B\cap F)$$

$$\leq \nu(B\cap F^c) + (B\cap F) = \nu(B).$$

This says that $f+\varepsilon\chi_F \in F$. However,

$$L \geq \int (f+\varepsilon\chi_F)d\mu = \int (fd\mu + \varepsilon\mu(F) = L + \varepsilon\mu(F),$$

which implies μ $(F) = 0$. But then $v(F) = 0$, and hence $\pi(F) = 0$, contradicting the fact that F is a positive set for F with $\pi(F) > 0$.

Lebesegue Decomposition Theorem

Theorem: Suppose μ and ν are two finite measures. There exists λ, p measures such that $\nu = \lambda + p$, p is absolutely continuous with respect to μ, and λ and μ are mutually singular.

Proof: Define Fand Land construct f as proof of the Randon-Nikodym theorem Let $p(A) = \int_A fd\mu$ and let $\lambda = \nu - p$. We have $\int_A fd\mu \leq \nu(A)$, so $\lambda(A) \geq 0$ for all A. To keep things straight, we record that we have $f = dp/d\mu$ and $p + \lambda = \nu$. We need to show μ and λ are mutually singular. Suppose not. Then there exist $F \in A$ with $\mu(F) > 0$ and $\lambda(F) > 0$, and so $v(F) \geq \lambda(F) > 0$. Note that ε for small enough $(\gamma - \varepsilon\mu)(F) > 0$. We claim that there exist $\varepsilon > 0$ and $E \subset F$ such that E is a positive set with respect to $\gamma - \varepsilon\mu$ and $\mu(E) > 0$. Given the claim, if $A \in A$,

$$\varepsilon \int_A \chi_E d\mu = \varepsilon\mu\nu(A\cap E) \leq \nu(A\cap E)$$

$$\leq \lambda(A\cap E) \leq \lambda(A)$$

$$= \nu(A) - \int_A fd\mu.$$

This says that

$$\int_A (f+\varepsilon\chi_E)d\mu \leq \nu(A)$$

For all $A \in A$, or $f + \varepsilon\chi_E \in F$. But

$$\int (f+\varepsilon\chi_E)d\mu = \int (fd\mu + \varepsilon\mu(E) > L,$$

a contradiction to the definition of L.

It remains to prove the claim. Let $F = P_n \cup N_n$ be a Hahn decomposition for the measure $\lambda - \frac{1}{n}\mu$ restricted to *F*, let $F = \cup P_n$, and $N = \cap N_n = P$. Then *N* is negative set for $\lambda - \frac{1}{n}\mu$ for each *n* or $0 \le \lambda(N) \le -\frac{1}{n}\mu(N)$; implies $\lambda(N) = 0$. If $\mu(p) = 0$, then $\mu \perp \lambda$, and we are supposing that is not the case. Therefore $\mu(p) > 0$, hence $\mu(p_n) > 0$ for some *n*, and P_n is a positive set for $\lambda - \frac{1}{n}\mu$. Now take $\varepsilon = 1/n$ and $E = P_n$.

DIFFERENTIATION OF REAL-VALUED FUNCTIONS

In this section we want to look at when $f : \mathbb{R} \to \mathbb{R}$ is defferentiable and when the fundamental theorem of calculus holds. Briefly,

1. Function of bounded variation are differenntiabl;
2. The derivative of $\int_a^x f(y)dy$ is equal to *f* a.e. if *f* is integrable;
3. $\int_a^x f(y)dy = f(b) - f(a)$ if *f* ' is absolutely continuous.

Let $E \subset \mathbb{R}$ be a measurable set let *O* be a collection of interval. We say *O* is a Vitali cover of *E* if for each $x \in E$ each there exists $\varepsilon > 0$ there exist an interval $G \in O$ containing *x* whose length is less than ε. *m* will denote Lebesgue measure.

Lemma: Let *E* have finite measure and let *O* be Vitali cover of *E*. $\varepsilon > 0$ Given there exists a finite subcollection of disjoint intervals $I_1 .. I_n$ such that $m\left(E - \bigcup_{i=1}^{n} I_n\right) < \varepsilon$.

Proof: We may replace each interval in *O* by closed one, since the set of endpoints of a finite subcollection will have measure 0.

Let *O* be an open set of finite measure containing *E*. Since *O* is a Vitali cover, we may suppose without loss of generality that each set of is contained in *O*. Let $a_0 = \sup\{m(I) :\in O\}$.

Let I_1 be any element of *O* with $m(I_1) \ge a_0 / 2$. Let $a_1 = \sup\{m(I) : 1 \in O, I$ disjoint from $I_1\}$. And choose $I_2 \in O$ disjoint from I_1 such that $m(I_2) \ge a_1 / 2$. Continue in this way choosing I_{n+1} disjoint from $I_1 ..... I_2$ and in *O* with length at least one half as large as any other such interval in *O* that is disjoint from $I_1 ..... I_2$ If the process stops at some finite stage, we are done. If not, we generate a sequence of disjoint intervals $I_1, I_2, ...$ Since they are disjoint and all contained in O, then $\sum_{i=1}^{\infty} m(I_i) \leq m(O) < \infty$. So there exists N such that $\sum_{i=N+1}^{\infty} m(I_i) < \varepsilon / 5$.

Let R = E $-\bigcup_{i=1}^{N} I_i$; we will show $m(R) < \varepsilon$. Let J_n be the interval with the same center as I_n but five times the lenght. Let $x \in R$. There exists an interval $I \in O$ containing x with I disjoint from $I_1,..., I_N$.

Since $\sum m(I_n) < \infty$, then $\sum a_n \leq 2\sum m(I_n) < \infty$, and $a_n \to 0$. So I must either be one of the I_n for some $n > N$ or at least intersect it, for otherwise we would have chosen I at some stage. Let n be the smallest integer such that I intersects I_n; note $n > N$. We have $m(I) \leq a_{n-1} \leq 2m(I_n)$. Since x is in $m(I) + m(I_n)/2 \leq (5/2)\, m(I_n)$. Since x is in I and I intersects I_n, the distance from x to the midpoint of I_n is at most $m(I) + m(I_n)/ 2 \leq (5/2)\, m(I_n)$. Therefore $x \in J_n$.

Let $R \subset \bigcup_{i=N+1}^{\infty} J_n$, so $m(R) \leq \sum_{i=N+1}^{\infty} m(J_n) = 5\sum_{i=N+1}^{\infty} m(I_n) < \varepsilon$.

Given a function f, we define the derivates of f at x by

$$D^{+} f(x) = \limsup_{h \to 0+} \frac{f(x+h) - f(x)}{h}, \qquad D^{-} f(x) = \limsup_{h \to 0-} \frac{f(x+h) - f(x)}{h}$$

$$D_{+} f(x) = \limsup_{h \to 0+} \frac{f(x+h) - f(x)}{h}, \qquad D_{-} f(x) = \limsup_{h \to 0-} \frac{f(x+h) - f(x)}{h}$$

If all the derivates are equal, we say that f is differentiable at x and define $f'(x)$ to be the common values.

Theorem: Suppose f is nondecreasing on [a, b]. Then f is diffrentiable almost everywhere, f' is integrable, and $\int_a^b f'(x)dx \leq f(b) - f(a)$.

Proof: We will show that the set where any two derivates are unequal has measure zero. We consider the set E where $D^{+} f(x) > D_{-} f(X)$, the other sets being similar. Let $E_{u,v} = \{x : D^{+} f(x) > u > v > D_{-} f(x)\}$. If we show $m(E_{u,v}) = 0$, then taking the union of all pairs of rationals with $u > v$ shows $m(E) = 0$.

Let $s = m(E_{u,v})$, let $\varepsilon > 0$, and choose an open set O such that $E_{u,v} \subset O$ and $m(O) < s + \varepsilon$. For each $x \in E_{u,v}$ there exists an arbitrarily small interval $[x - h, x]$ contained in O such that $f(x) - f(x - h) < vh$.

Use Lemma to choose $I_1,..., I_n$ which are disjoint and whose interiors cover a subset A of $E_{u,v}$ of measure greater than $x - \varepsilon$. Suppose In = $[x_n - h_n, x_n]$. Summing over these intervals.

$$\sum_{n=1}^{N}[f(x_n)-f(x_n-h_n)]<v\sum_{n=1}^{n}h_n<vm(O)<v(s+\varepsilon).$$

Each point $y \in A$ is the left endpoint of an arbitrarily small interval $(y, y + k)$ that is contained in some I_n and for which $f(y + k) - f(y) > uk$. Using Lemma again, we pick out a finite collection $J_1,.., J_M$ whose union contains a subset of A of measure larger than $s - 2\varepsilon$. Summing over these intervals yields

$$\sum_{i=1}^{M}[f(y_i+k_i)-f(y_i)]<u\sum k_i>u(s-2\varepsilon).$$

Each interval J_i is contained in some interval I_n, and if we sum over those i for which $J_i \subset I_n$ we find

$$\sum[f(y_i+k_i)-f(y_i)]\le f(x_n)-f(x_n-h_n),$$

since f is increasing. Thus

$$\sum_{n=1}^{N}[f(x_n)-f(x_n-h_n)]\ge\sum_{i=1}^{M}[f(y_i+k_i)-f(y_i)],$$

and so $v(s + \varepsilon) > u(s - 2\varepsilon)$. This is true for each ε, for $v_s \ge u_s$. Since $u > v$, this implies $s = 0$.

This shows that

$$g(x)=\lim_{h\to 0}\frac{f(x+h)-f(x)}{h}$$

is defined almost everywhere and that f is differentiable wherever g is finite. Define $f(x) = f(b)$ if $x \ge b$. Let $g_n(x) = n[f(x + 1/n) - f(x)]$. Then $g_n(x) \to g(x)$ for almost all x, and so g is measurable. Since f is increasing $g_n \ge 0$. By Fatou's lemma

$$\int_a^b g \le \liminf \int_a^b g_n = \liminf n\int_a^b [f(x+1/n)-f(x)]dx$$

$$=\liminf\left[n\int_b^{b+1/n} f-n\int_a^{a+1/n} f\right]=\liminf\left[f(b)-n\int_a^{a+1/n} f\right]$$

$$\le f(b)-f(a).$$

For the last inequality, we use the fact that f is increasing. This shows that g is integrable and hence finite almost everywhere.

A function is of bounded variation if $\sup\sup\left\{\sum_{i=1}^{k}|f(x_i)-f(x_{i-1})|\right\}$ is finite, where the supremum is over all partitions $a = x_0 < x_1 <... < x_k = b$ of $[a, b]$.

Lemma: If f is of bounded variation on [a, b], then f can be written as the differece of two nondecreasing functions on [a, b].

Proof. Define

$$P(y)\sup\left\{\sum_{i=1`}^{k}[f(x_i)-f(x_{i-1})]^{+}\right\}, N(y)\sup\left\{\sum_{i=1`}^{k}[f(x_i)-f(x_{i-1})]^{-}\right\},$$

where the supremum is over all partitions a = $x_0 < x_1 < ..... < x_k$ = y for $y \in [a, b]$. *P* and *N* are measurable since they are both increasing. Since

$$\sum_{i=1`}^{k}[f(x_i)-f(x_{i-1})]^{+} = \sum_{i=1`}^{k}[f(x_i)-f(x_{i-1})]^{-} + f(y) - f(a),$$

taking the supremum over all partitions of *[a, y]* yields

$$P(y) = N(y) + f(y) - f(a).$$

Clearly *P* and *N* are nonedecreaing in *y*, ande the result follows by solven jfor *f*(*y*).

From this lemma, we see that functions of bounded variation we differentialble a.e. But the function (1/*x*)defined on (0, 1] is defferentiable everwhere, but is not of bounded variation.

Next we look at when the derivative of $\int_a^x f(t)dt$ is equal to *f*(*x*) a.e. Define the indefintite integraol of an intregrable *f* by

$$F(x)\int_a^x f(t)dt.$$

Lemma: If *f* is integrable, then F is continuous and of bounded variation.

Proof: The continuity follows from the dominated convergence theorem. The bounded variation follows from

$$\sum_{i=1`}^{k}[F(x_i)-F(x_{i-1})| = \sum_{i=1`}^{k}|\int_{x_{i-1}}^{x_i} f(t)dt| \leq = \sum_{i=1`}^{k}\int_{x_{i-1}}^{x_i}|f(t)dt \leq \int_a^b|f(t)dt$$

for all partitions.

Lemma: If *f* is integrable and *F*(*x*) = 0 for all *x*, then *f* = 0 a.e.

Proof: For any interval, $\int_c^d f = \int_a^d f - \int_a^c f = 0$. By dominated convergence and the fact that any open set is the countable union of disjoint open intervals, $\int_O f = 0$ for any open set *O*.

If E is any measurable set, take O_n open that such that χ_{O_n} decreases to χ_E By dominated convergence,

$$\int_E f = \int f_{\chi_E} = \lim \int f\chi_{O_n} = \lim \int_{O_n} f = 0.$$

Proposition: If f is bounded and measurable, then $F'(x) = f(x)$ for almost every x.

Proof: By Lemma F is continuous and of bounded varition, and so F' exists a.e. Let K be a bounded for $|f|$. If

$$f_n(x) = \frac{F(x+1/n) - F(x)}{1/n},$$

then

$$f_n(x) = n\int_x^{x+1/n} f(t)dt,$$

so $|f_n|$ is also bounded by K. Since $f_n \to f'$ then by dominated convergence,

$$\int_a^c F'(x)dx = \lim \int_a^c f_n(x)dx = \lim n \int_a^c F(x+1/n) - F(x)]dx$$

$$= \lim\left[n \int_c^{c+1/n} F(x)dx - n = \int_a^{a+c} F(x)\,dx\right]$$

$$F(c) - F(a) = \int_a^c F(x)dx,$$

using the fact F is continuous. So $\int_a^c [F'(x) - f(x)]dx = 0$ for all c, which implies $F' = f$.

Theorem: If f is integrable, then $F' = f$ almost everywhere.

Proof: Without loss of generality we may assume $f \geq 0$. Let $f_n(x) = f(x)$ if $f(x) \leq n$ and let $f_n(x) = n$ if $f(x) > n$. Then $f - f_n \geq 0$. If $G_n(x) = \int_a^x [F' - f_n]$, then G_n is nondecreasing, and hence has a derivative almost everywhere. By Proposition, we know the derivative of $\int_a^x F` - f_n$ is equal to f_n almost everywhere. Therefore

$$F'(c) = G_n'(x) + \left[\int_a^c f_n\right]' \geq f_n(x)$$

a.e Since n is arbitrary, $F' \geq f$ a.e. So $\int_a^b F' \geq \int_a^b f. = f(b) - F(a)$. On the other hand, by Theorem, $\int_a^b F'(x)dx \leq F(b) - f(a) = \int_a^b f$. We conclude that $\int_a^b [F' - f] = 0$; since $F' - f \geq 0$, this tells us that $F' = f$ a.e.

Finally, we look at when $\int_a^b F' \geq \int_a^b F'(y)dy = F(b) - F(a).$

A function is absolutely continuous on $[a, b]$ if given ε there exists δ such that $\sum_{i=1}^{k} | f(x_i') | < \varepsilon$ whenever $\{(x_i, x'_i)\}$ is a finite collection of nonoverlaping intervals with $\sum_{i=1}^{k} | x_i' - x_i) | < \delta.$

It is to see that absolutely continuous functions are continuous and that the Cantor - Lebesgue fuction is not absolutely continuous.

Lemma: If $F(x) = \int_a^x f(t)dt$ for f integrable on $[a, b]$ then F is absolutely continuous.

Proof: Let $\varepsilon > 0$. Choose a simple function s such that $\int_a^b | f - s | < \varepsilon / 2$. Let K be a bound for $|s|$ and let $\delta = \varepsilon > / 2K$. If $\{(x_i, x'_i)\}$ is a collection of nonoverlapping intervals, the sum of whose lengths is less δ then set $A = \bigcup_{i=1}^{k} (x_i, x_i')$ and note $\int_A | f - s | < \varepsilon / 2. =$ and $\int_A s < K\delta = \varepsilon / 2.$

Lemma: If f is absolutely continuous, then it is of bounded variation.

Proof: Let δ correspond to $\varepsilon = 1$ in the definition of absolute continuity. Given a partition, add point if necessay so that subinterval has length at most δ. We can then group the subintervals into at most K collections, each of total length less than δ, where K is an integer larger tht $(1 + b - a)/\delta$. So the total varition is then less than K.

Lemma: If f is absolutely continuous on $[a, b]$ and $f'(x) = 0$ a.e., then f is contant.

The Cantor- Lebesgue function is an example to show that we need *j*the absolute continuity.

Proof: Let $c \in [a, b]$, let $E = \{x \in [a, c] : f'(x) = 0\}$, and let $\varepsilon > 0$. For each point $x \in E$ there exists arbitrarily small intervals $[x, x + h] \subset [a, b]$ such that $|f(x + h) - f(x)| < \varepsilon h$. By lemma we can find a finite collection where δ is the δ in the definition of absolute continuity. If the intervals are $[x_i, y_i]$ with $x_i < y_i \leq x_{i+1}$, then $\sum | f(x_{i+1}) - f(y_i) | < \varepsilon$ by the definition of absolute continuity, while $\sum | f(y_i) - f(x_i) | < \varepsilon \sum (y_i - x_i) \leq \varepsilon(c - a)$. So adding these two inequalties together,

$$| f(c) - f(a) | = | \sum | f(x_{i+1})] + \sum | f(y_i) - f(x_i)] \leq \varepsilon + \varepsilon(c - a).$$

since ε is arbitrary, then $f(c) = f(a)$, which implies that f is contant.

Theorem: If F is absolutely continuous, then

Proof: Suppose F is absolutely continuous on $[a, b]$. Then is F is of bounded variation, so $F = F_1 - F_2$ where F_1 and F_2 are nondecreasing, and F' exists a.e. Since $|F'(x) \leq |F'_1(x) + F'_2(x,)$, then $\int |F'(x)|\,dx \leq F_1(b) + F_2(b) - F_1(a) - F_2(a)$, and hence F' is intergrable. If $G(x) - \int_a^x F'(t)dt$, then G is absolutely continuous by Lemma, so $F - G$ is absolutely continuous.Then $(F - G)' = 0$ a.e., and therefore $F - G$ is absolutely continuous. Thus $F(x) = \int_a^x F'(t)dt + F(a)$. If we set $x = b$, we get out result.

L^p SPACES

We assume throughout this section that the measure is σ – finite. For $1 \leq p < \infty$, define the L^p norm of f by

$$\| f \|_p \left(\int | f(x) |^p \, d\mu \right)^{1/p}.$$

For $p = \infty$ define the L^∞ norm of f by

$$\| f \|_\infty = \inf\{M : \mu(\{x : | f(x) | \geq M\}) = 0\}.$$

For $1 \leq p \leq \infty$ the space L^p space is the set $\{f : \| f \|_p < \infty\}$.

The L^∞ norm of a function f is the supremum of f provided we disregard sets of measure 0.

Holder's Inequality

Proposition: If $1 < p, q < \infty$ and $p^- + q^- = 1$,

Then $\int | f(x)g(x) | \, d\mu \leq \| f \|_p \| g \|_q$.

This also holds if $P = \infty$ and $g = 1$.

Proof: If $M = \| f \|_\infty$, then $\int fg \leq M \int | g |$ and the case p = q = 1 follows. So let us assume $1 < p, q < \infty$. If $\| f \|_p = 0$ then $f = 0$ a.e. and $\int fgo = 0$, so the result is clear if $\| f \|_p = 0$. Let $F(x) = | f(x)| / \| f \|_p$ and $G(x) = |g(x)[/\|g \|_q$. Note $\|F\|_p =$ 1and $\|G\|_q = 1$, and it suffices to show that $\int FG \leq 1$. The second derivative of the function e^x is again e^x, which is postive and so e^x is convex.Therefore if $0 \leq \lambda \leq 1$, we have

$$e^{\lambda n + (1-\lambda)b} \leq \lambda e^a + (1-\lambda)e^b.$$

If $F(x), G(x) \neq 0$, let $a = p \log F(x), b = q \log G(x), \lambda = 1/p$, and $1 - \lambda = 1/p$. we then obtain

$$F(x)G(x) \leq \frac{F(x)^p}{p} + \frac{G(x)^q}{q}.$$

Clearly this inequality also holds if $F(x) = 0$ or $G(x) = 0$. Integrating,

$$\int FG \leq \frac{\| F \|_p^p}{p} + \frac{\| G \|_q^q}{q} = \frac{1}{p} + \frac{1}{q} = 1.$$

One application of Holder's inequality is to prove Minkowski's inequality' which is simply the triangle inequality for L^p.

Lemma: If $a, b > 0$ and $1 \leq p < \infty$, then

$$(a+b)^p \leq 2^{p-1} a^p + 2^{p-1} b^p.$$

Proof: To this, we may without loss of generality assume $a \leq b$. The case $a = 0$ is obvious, so we assume $a > 0$. Dividing both sides by a and letting $x = b/a$, the innequality we want is equivalent to

$$(1+x)^p \leq 2^{p-1} + 2^{p-1} x^p, \quad x \geq 1.$$

clearly t]his inequality is valid for $x = 1$.So to proove it suffices to show that the derivative of

$$(1+x)^p - 2^{p-1} - 2^{p-1} x^p$$

is less than equal to 0, or

$$p(1+x)^{p-1} \leq 2^{p-1} p x^{p-1}, \quad x \geq 1.$$

This last inequality hods because $2x \geq 1 + x$ when $x \geq 1$.

Minkowski's Inequality

Proposition: If $1 \leq p \leq \infty$, then

$$\| f + g \|_p \leq \| f \|_p + \| g \|_p.$$

Proof: Since $|(f+g)|(x)| \leq |f(x)| + |g(x)|$, integrating gives the case when $p = 1$. The case $p = \infty$ is also easy. So let us suppose $1 < p < \infty$. If $\| f \|_p$ or $\| g \|_p$. The case P is infinite, the result is obnvious, so we may asume boy are finite. The inequality Lemma with yields, after an integration,

$$\int |(f+g)(x)|^p \, d\mu \leq 2^p \int |f(x)|^p \, d\mu + 2^p \int |g(x)|^p \, d\mu.$$

So we have $\| f+g \|_p < \infty$. Clearly we may assume $\| f+g \|_p > 0$.

Now write

$$| f+g |^p \leq | f \| f+g |^{p-1} + | g \| f+g |^{p-1}$$

and apply Holder's inequality with $q = \left(1-\frac{1}{p}\right)^{-1}$, We obtain

$$\int | (f+g |^p \leq \| f \|_p \left(\int | f+g |^{(p-1)q}\right)^{1/q} + \| g \|_p \left(\int | f+g |^{(p-1)q}\right)^{1/q}.$$

Since $p^{-1}+q^{-1}=1$ then $(p-1)\,q = p$, so we have

$$\| f+g \|_p^p \leq \left(\| f \|_p + \| g \|_p\right) \| f+g \|_p^{p/q}.$$

Dividing both sides by $\| f+g \|_p^{p/q}$ and using the fact that $p-(p/q) = 1$ gives us our result.

Minkowski's inequality says that L^p is a normed linear space, proveded we identify fuctions that are equal a.e.

We say f_n converges to f in L^P if $\|f_n - f\|_p \to 0$ as $n \to \infty$. The next proposition compares convergece in L^p to convergence in measure. Before we prove this, we prove an easy preliminary result known as Chebyshev's inequality.

Lemma: If $1 \leq p < \infty$,

$$\mu(\{x : | f(x) | \geq a\}) \leq \frac{\| f \|_p^p}{a^p}.$$

Proof: If $A = \{x : | f(x) | \geq a\}$, then

$$\mu(A) \leq \int_A \frac{| f(x) |^p}{a^p} d\mu \leq \frac{1}{a^p} \int | f |^p \, d\mu.$$

Proposition: If f_n converges to f in L^p then it converges in measure.

Proof: $\varepsilon > 0$, if by Chebyshev's inequality

$$\mu(\{x : | f_n(x) - f(x) | > \varepsilon\}) = \mu(\{x : | f_n(x) - f(x) |^p > \varepsilon^p\}) \leq \frac{\| f_n - f \|_p^p}{\varepsilon^p} \to 0.$$

Letting $f_n = n^2 \chi_{(0,1/n)}$ with on [0,1] the measure being Lebesgue measure give an example where f_n converges to 0 a.e. and in measure, but does not converge in L^p. We next show that L^p is complete. This is often phrased as saying that L^p is Banach space i.e. a complete normed linear space.

Proposition: If $1 \le p \le \infty$, then L^p complete.

Proof: We do only the case $p > \infty$; the case $p = \infty$ is easy. Suppose f_n is a Cauchy sequence in L_p.Given $\varepsilon = 2^{-(j+1)}$, there exists n_j such that if $n, m \le n_j$, then $\| f_n - f_m \| \varepsilon = 2^{-(j+1)}$. Without loss of generallity we may assum $n_j \ge n_{j-1}$ for each j.

Set $n_0 = 0$ and define $f_0 \equiv 0$. If $A_j \equiv \left\{ x : | f_{n_j}(x) - f_{n_{j-1}}(x) | > 2^{-j/2} \right\}$, then from Lemma, $\mu(A_j) \le 2^{-jp/2}$ We have

$$\mu\left(\bigcap_{j=1}^{\infty} \bigcup_{m=j}^{\infty} A_m\right) = \lim_{j \to \infty} \mu\left(\bigcup_{m=j}^{\infty} A_m\right) \le \sum_{m=j}^{\infty} \mu(A_m) = 0.$$

So except for a set of measure 0, for each x there is a last j for which $x \in \bigcup_{m=j}^{\infty} A_m$, hence a last j for which $x \in A_j$. So for each x (except for the null set) there is a j_0 (depending on x) such that if $j \ge j_0$, then $| f_{n_j}(x) - f_{n_{j-1}}(x) | \le 2^{-j}$.

Set

$$g_j(x) = \sum_{m=1}^{j} | f_{n_m}(x) - f_{n_{m-1}}(x) |.$$

$g_j(x)$ increases for each x, and the limit is finite for almost every x by the preceding paragraph. Let us call the limit $g(x)$. We have

$$\| g_j \|_p \le \sum_{m=1}^{j} 2^{-j} + \| f_{n_1} \|_p \le 2 + \| f_{n_1} \|_p$$

by Minkowski's inequality, and so by Fatou's lemma, $\| g \|_p \le \| f_{n_1} \|_p < \infty$.

We have

$$f_{n_j}(x) = \sum_{m=1}^{j} (f_{n_m}(x) - f_{n_{m-1}}(x)).$$

Suppose x is not in the null set where $g(x)$ is infinite. Since $| f_{n_j}(x) - f_{n_k}(x) | \le | g_{n_j}(x) - g_{n_k}(x) | \to 0$ as $j, k \to \infty$, then $f_{n_j}(x)$ is a Cauchy series (in $\mathbb{R}$), and hence converges, say to f(x). We have $\| f - f_{n_j} \|_p = \lim_{n \to \infty} \| f_{n_m} - f_{n_j} \|_p$;

this follows by dominated convergence with the function g defined above as the dominating function.

We have thus shown that $\| f - f_{n_j} \|_p \to 0$. Given $\varepsilon = 2^{-(j+1)}$, if $m \geq n_j$, then $\| f - f_m \|_p \leq \| f - f_{n_j} \|_p + \| f_m - f_{n_j} \|_p$. This shows that fm converges to f in L^p norm.

Next we show:

Proposition: The set of continuous functions with compact support is dense in $L^p(\mathbb{R}^d)$

Proof. Suppose $f \in L^p$. By dominated covergence $\int | f - f\chi_{-[n,n]} |^p \to 0$ as $n \to \infty$, the dominating function being $| f |^p$. So we may suppose f has compact support. By writing $f = f^+ - f^-$ we may suppose $f \geq 0$. By taking simple functions s_m increasing to f, we have $\int | f - s_m |^p \to 0$ by dominated convergence, so it suffices to consider simple functions. By linearitry, it suffices to consider characteristic functions with compact support. Given such a χ_E and $\varepsilon > 0$ we showed in Proposition that there exist g continuous with compact supporter and with values in [0,1] such that $\int | g - \chi_E | < \varepsilon$. Since $| g - \chi_E | \leq 1$, then $\int | g - \chi_E |^p \leq \int | g - \chi_E | < \varepsilon$.

The following in very useful.

Proposition: For $1 < p < \infty$ and $p^{-1} + q^{-1} = 1$,

$$\| f \|_p = \sup\left\{ \int f_g : \| g \|_q \leq 1 \right\}$$

When $p = 1$ holds if we take $q = \infty$, and if $p = \infty$ holds if we take $q = 1$.

Proof: The right hand side of is less than the left hand side by Holder's inequality. So we need only show that the right hand side is greater than the left hand side.

First suppose $p = 1$. Take $g(x) = \operatorname{sgn} f(x)$, where sgn α is > 0, is 0 if $\alpha = 0$, and is -1 if $\alpha < 0$. Then g is bounded by 1 and $f_g = | f |$. This takes care of the case $p = 1$.

Next suppose $p = \infty$. Since μ is σ-finite, there exist sets F_n increasing up to X such that $\mu(F_n) < \infty$ for each n. If $M = \| f \|_\infty$, let a be any finite real less than M. By the definition of L^∞ norm, the measure of $A = \{x \in F_n : | f(x)| > a\}$ must be positive if n is sufficiently large. Let $g(x) = (\operatorname{sgn} f(x))\chi_A(x)/\mu(A)$. Then the L^1 norm of g is 1 and $\int fg = \int_A | f | / \mu(A) \geq \alpha$. Since a is arbitrary, the spermium on the right hand side must be M.

Now suppose $1 < p < \infty$. We may suppose $\|f\|_p > 0$. Let q_n be a sequence of nonnegative simple functions increasing to f^+, r_n a sequence of nonnegative simple functions increasing to f^-, and $s_n(x) = (q_n(x) - r_n(x))\chi_{Fn}(x)$. Then $S_n(x) \to f(x)$ for each x, $|s_n(x)| \le |f(x)|$ for each x, s_n is a simple function, and $\|s_n\|_p < \infty$ for each n. If $f \in L^p$, then $\|s_n\|_p \to \|s_n\|_p$ by dominated convergence. If $\int |f|^p = \infty$, then $\int |s_n|^p \to \infty$ by monotone convergence. For n sufficiently large, $\|s_n\|_p > 0$.

Let

$$g_n(x) = (\operatorname{sgn} f(x)) \frac{|s_n(x)|^{p=1}}{\|Sn\|_p^{p/q}}$$

Since $(p-1)q = p$, then

$$\|g_n\|_q = \frac{\left(\int |s_n|^{(p-1)q}\right)^{1/q}}{\|Sn\|_p^{p/q}} = \frac{\|S_n\|_p^{p/q}}{\|S_n\|_p^{p/q}} = 1.$$

On the other hand, since $|f| \ge |s_n|$,

$$\int fg_n = \frac{f|f||s_n|^{p-1}}{\|Sn\|_p^{p/q}} \ge \frac{|S_n|^p}{\|Sn\|_p^{p/q}} = \|Sn\|_p^{p-(p/q)}$$

Since $p - (q/q) = 1$, then $\int fg_n \ge \|s_n|_p$,, which tends to $\|f\|_p$.

The above proof also establishes

Corollary: For $1 < p < \infty$ and $p^{-1} + q^{-1} = 1$,

$$\|f\|_p = \sup \int fg = \|g\|_q \le 1, g\ \{simple\}.$$

The space L^P is a normed linear space. We can thus talk about its dual, namely, the set of bounded linear functional on L^P. The dual of a space Y is denoted Y^*. If H is a bounded linear functional on L^P, we define the normk of H to be $\|H\| = \sup\{H(f) : \|f\|_P \le 1\}$.

Theorem: If $1 < p < \infty$ *and* $p^{-1} + q^{-1} = 1$, then $(L^p)^* = L^q$.

What this means is that is H is a bounded linear functional on L^P, then there exists g $\in L^q$ such that $H(f) = \int fg$ and that if g $\in L^q$, then $H(f) = \int fg$ is a bounded linear functional on L^p.

Proof: If g $\in L^q$, then setting $H(f) = \int fg$ for $f \in L^p$ yields a bounded linear functional; the boundedness follows form Holder's inequality. Moreover, from holder's inequality and Proposition we see that $\|H\| = \|g\|_q$.

Now suppose we are given a bounded linear function linear functional H on L^p and we must show there exists $g \in L^q$ such that $H(f) = \int fg$. First suppose $\mu(X) < \infty$. Define $v(A) = H(\chi_A)$. If A and B are disjoint, then

$$v(A \cup B) = H(\chi_{A \cup B}) = H(\chi_A + \chi_B) = H(\chi_A) + H(\chi_B) = v(A) + v(B).$$

To show v is countably additive, it suffices to show that if $A_n \uparrow A$,then $v(A_n) \to v(A)$ But if $A_n \uparrow A$, then $\chi A_n \to \chi_A$ in L^p, and so $v(A_n) = H(\chi A_n) \to H(\chi_A) = (A)$; we use here the fact that $\mu(X) < \infty$.Therefore v is a countably additive signed measure. Moreover, if $\mu(A) = 0$, then $\chi_A = 0$ a.e., hence $v(A) = H(\chi_A) = 0$. By writing $v = v^+ - v^-$ and using the radon-Nikodym theorem for both the positive and negative parts, we see there exist an integrable g such that $v(A) = \int_A g$ for all sets A. If $s = \sum a_i \chi A_i$ is a simple function, by linearity we have

$$H(s) = \sum a_i H(\chi A_i) = \sum a_i v(A_i) = \sum a_i \int g\chi A_i = \int gs.$$

By Corollary,

$$\| g \|_q = \sup\left\{\int gs : \| s \|_p \leq 1, s \textit{ simple}\right\} \leq \sup \{H(s) : \| s \|_p \leq 1 \} \leq \| H \|$$

If s_n are simple function tending to f in L^p, then $(S_n) \to H(f)$, while by Holder's inequality $\int s_n g \to \int fg$. We thus have $H(f) = \int fg$ for all $f \in L^p$, and $\| g \|_p \leq \| H \|$. By Holder's inequality, $\| H \| \leq \| g \|_p$.

In the case where μ is σ-finite, but not finite, let $n \uparrow X$ be such that $\mu(F_n) < \infty$ for each n. Define function is H_n by $H_n(f) = H(f\chi F_n)$. Clearly each H_n by $H_n(f) = H(f\chi F_n)$. Clearly each H_n is a bounded linear functional on L^p. Applying the above argument, we see there exit g_n such that $H_n(f) = \int fg_n$ and $\| g_n \|_p = \| H_n \| \leq \| H \|$. It is easy to see that g_n is 0 if $x \notin F_n$. Moreover, by the uniqueness part of the Radon-Nikodym theorem, if $n > m$, then $g_n = g_m$ on F_m. Define g by setting $g(x) = g_n(x)$ if $x \in F_n$.

Then g is well defined. By Faou's lemma, g is in L^q with a norm bounded by $\|H\|$. Since $f\chi_{Fn} \to f$ in L^P By dominated convergence, then $H_n(f) = H(f\chi F_n) \to H(f)$, since H is a bounded linear functional on L^p. On the other hand $H_n(f) = \int_{F_n} fg_n = \int_{F_n} fg \to \int fg$ by dominated covergence. So $H(f) = fg$. Again by holder's inequality $\| H \| \leq \| g \|_p$.

FOURIER TRANSFORMS

Fourier transforms give a representation of a function in terms of frequencies. We give the basic properties here.

If $f \in L^1(\mathbb{R}^n)$, define the Fourier transform $\widehat{f}$ by

$$\widehat{f}(u) = \int_{\mathbb{R}^n} e^{iu.x} f(x)dx, \quad u \in \mathbb{R}^n.$$

We are using $u.\ x$ for the standard inner product is $\mathbb{R}^n$. Various books have slightly different definitions. Some a negative sign before the $iu.\ x$, some have a 2π either in front of the integral or in the exponent. Then basic theory is the same in any case.

Some basic properties of the Fourier transform are given by

Proposition: Suppose f and g are in L^1. Then

(a) $\widehat{f}$ is bounded and continuous;

(b) $\left(\widehat{f+g}\right)(u) = \widehat{f}(u) + \widehat{g}(u)$; $(\widehat{af})(u) = a\widehat{f}(u)$;

(c) if $f_a(x) = f(x+a)$, then $\widehat{f}_a(u) = e^{-iu.a}\widehat{f}(u)$;

(d) if $g_a(x) = e^{ia.x} g(x)$, then $\widehat{g}_a(u) = \widehat{f}(u+a)$;

(e) if $h_a(x) = f(ax)$, then $\widehat{h}_a(u) = a^{-n}\widehat{f}(u/a)$.

Proof: (a) $\widehat{f}$ is bounded because $f \in L^1$ and $|e^{iu.x}| = 1$. We have

$$\widehat{f}(u+h) - \widehat{f}(u) = (e^{i(u+h).x} - e^{iu.x}) f(x)dx.$$

So

$$|\widehat{f}(u+h) - \widehat{f}(u)| \leq \int |e^{iu.x}|\, e^{ih.x} - 1|\, |f(x)|\, dx.$$

The integrand is bounded by $2|f(x)$, which is integrable, and $e^{ih.x} - 1 \to 0$ as $h \to 0$, as and thus the continuity followed by dominate convergence.

(b) is obvious. (c) follows because

$$\widehat{f}_a(u) = \int e^{iu.x} f(x+a)dx = \int e^{iu.(x-a)} f(x)dx = e^{-iu.a}\widehat{f}(u)$$

by a change of variable. For (d),

$$\widehat{g}_a(u) = \int e^{iu.x} e^{ia.x} f(x)dx = \int e^{i(u+a).x} f(x)dx = \widehat{f}(u+a).$$

Finally for (e), by a change of variables,

$$\widehat{h}_a(u) = \int e^{iu.x} f(ax)dx = a^{-n} \int e^{iu.(y/a)} f(y)dy$$

$$= a^{-n} \int e^{i(u/a).y} f(y)dy = a^{-n}\widehat{f}(u/a).$$

One reason for the usefulness of Fourier transforms is that they relate derivatives and multiplication.

Proposition: Suppose $f \in L^1$ and $x_j f(x) \in L^1$, where x_j is the jth coordinate of x. Then

$$\frac{\partial \widehat{f}}{\partial u_j}(u) = i \int e^{iu.x} x_j f(x) dx.$$

Proof: Let e_j be the unit vector in the jth direction. Then

$$\frac{\widehat{f}(u+he_j) - \widehat{f}(u)}{h} = \frac{1}{h}\int (e^{i(u+he_j).x} - e^{iu.x}) f(x) dx$$

$$= \int e^{iu.x}\left(\frac{e^{ihx_j} - 1}{h}\right) f(x) dx.$$

Since

$$\left|\frac{1}{h}\left(e^{ihx_j} - 1\right)\right| \le |x_j|$$

and $x_j f(x) \in L^1$, the right hand side coverages to $\int e^{iu.x} ix_j f(x) dx$ by dominated covergence. Therefore the left hand side converges. Of course, the limit is $\partial \widehat{f} / \partial u_j$.

The convolution of f and g is defined by

$$f*g(x) = \int f(x-y)\, g(y) g(y)\, dy.$$

By a change of variables, this is the same as $\int f(y) g(x-y)\, dy$, so $f * g = g * f$.

Proposition: (a) If $f, g \in L^1$, then $f * g$ is in L^1 and $\| f*g \|_1 \le \| f \|_1 \| g \|_1$.

(b) The Fourier transform of $f * g$ is $\widehat{f}(u)\ \widehat{g}(u)$.

Proof: (a) We will show $f * g(x) dx < \infty$.

$$\int | f * g(x) g | dx \le \int\int | f(x-y) \| g(y) | dydx.$$

Since the integrand is nonnegative, we can apply Fubini and the right hand side is equal to

$$\int\int | f(x-y) | dx\, | g(y) dy$$

$$= \int\int | f(x) | dx\, | g(y) | dy = \| f \|_1 \| g \|_1.$$

The first equality here follows by a change of variables. To verify that we can do a change of variables, we reduce to simple functions and then characteristic functions, and then use the translation invariance of Lebesgue measure.

(b) We have

$$\widehat{f*g}(u) = \int e^{iu.x} \int f(x-y)g(y)dydx$$

$$= \int\int e^{iu.x(x-y)} f(x-y)dxe^{iu.y} g(y)dy$$

$$= \int \widehat{f}(u)e^{iu.y} g(y)dy = \widehat{f}(u)\widehat{g}(u).$$

We applied fubini in the first equality; equality; this is valid because as we saw in (a), the absolute value of the integrand in integrable.

We want to give a formula for recovering f from $\widehat{f}$. First we need to calculate the Fourier transform of a particular function.

Proposition: (a) Suppose $f_1 : \mathbb{R} \to \mathbb{R}$ is defined by

$$f_1(x) = \frac{1}{\sqrt{2\pi}} e^{-x^{2/2}}.$$

Then $\widehat{f}_n(u) = e^{-|u|^2/2}$.

Proof: (a) may be proved using contour integration, but let's give a real variable proof. Let $g(u) = \int e^{iux}e^{-x^2/2}dx$. Differentiate with respect to u. We may differentiate under the integral sign because $(e^{i(u+h)x} - e^{iux})/h$ is bounded in absolute value by $|x|\,e^{-x^2/2}dx$ is integrable; therefore dominated convergence applies. We then obtain

$$g'(u) = i\int e^{iux}xe^{-x^2/2}dx.$$

By integration by parts this is equal to

$$-u\int e^{iux}xe^{-x^2/2}dx = -ug(u).$$

Solving the differential equation $g'(u) = -ug(u)$, we have

$$[\log g(u)]' = \frac{g'(u)}{g(u)} = -u,$$

So $\log g(u) = -u^2/2 + c_1$, and so then

$$g(u) = c_2e^{-u^2}.$$

Since $g(0) = \int e^{-x^2/2} dx = \sqrt{2\pi}, c_2 = \sqrt{2\pi}$.

Substituting this value of c_2 in (14.2) and dividing both sides by $\sqrt{2\pi}$ proves (a).

For (b), since $f_n(x) = f_1(x_1)...f_1(x_n)$ if $x_1 = (x_1, ..., x_n)$,

$$\widehat{f}_n(u) = \int ... \int e^{i\sum_j u_j x_j} f_1(x_1)...f_1(x_n) dx_1...dx_n$$

$$\widehat{f}_1(u_1)...\widehat{f}_1(u_n) = e^{-|u|^2/2}.$$

One more preliminary before providing the inversion theorem.

Proposition: Suppose j is in L^1 and $\int \varphi(x)dx = 1$. Let $\varphi_A(x) = A^{-n}\varphi(x/A)$.

(a) Then $\| f * j_A - F \|_1 \to 0$ as $A \to 0$

(b) If f is continuous with compact support, then $f * \varphi_A$ converges to f pointwise.

Proof: (a) Let $\varepsilon > 0$. Choose g continuous with compact support so that $\| f - g \|_1 < \varepsilon$. Let $h = f - g$. A change of variables shows that $\|\varphi_A\|_1 = \| \varphi \|_1$.

Observe

$$\|f * \varphi_A - f\|_1 \leq \| g * \varphi_A - g \|_1 + \| h * \varphi_A - h\|_1$$

and $$\| h * \varphi_A - h \|_1 \leq \| h \|_1 + \| h * \varphi_A \|_1 \leq \| h \|_1 + \| h \|_1 \| \varphi_A \|_1 < \varepsilon(1 + \| \varphi \|_1).$$

So since ε is arbitrary, it suces to show that $g * \varphi_A \to g$ in L^1.

We start by writing

$$g * \varphi_A(x) - g(x) = \int g(x-y)\varphi_A(y)dy - g(x)$$

$$= \int g(x - Ay)\varphi(y)dy - g(x)$$

$$= \int [g(x - Ay) - g(x)]\varphi(y)dy$$

We used a change of variables and the fact that $\int \varphi(y)dy = 1$. Because g is continuous with compact support, then g is bounded, and the integral on the right goes to 0 by dominated convergence, the dominating function being $\| g \|_\infty |\varphi(y)|$. Therefore $g * \varphi_A(x)$ converges to $g(x)$ pointwise. To show the convergence in L^1, we have

$$\int | g * \varphi_A(x) - g(x) | dx \leq \int \int | g(x - Ay) - g(x) || \varphi(y) | dydx$$

$$= \int \int | g(x - Ay) - g(x) || \varphi(y) | dxdy.$$

Since g is continuous with compact support and hence bounded, for each y

$$G_A(y) = \int |g(x - Ay) - g(x)|\, dx$$

converges to 0 as $A \to 0$ by dominated convergence. Also

$$G_A(y) \le \int |g(x - Ay)|\, dx + \int |g(x)|\, dx \le 2\| g \|_1 < \infty.$$

Then

$$\int G_A(y)\, |\varphi(y)|\, dy$$

converges to 0 as $A \to 0$ by dominated convergence, the dominating function being $2\| g \|_1 | \varphi(y) |$.

(b) This follows from the argument we used for g above.

Now we are ready to give the inversion formula. The proof seems longer than it might be, but there is no avoiding the introduction of the function H_a or some similar function.

Plancherel Theorem

Theorem: Suppose $f, \widehat{f} \in L^1$. Then

$$f(y) = \frac{1}{(2\pi)^n} \int e^{-iu.y} \widehat{f}(u)du, \quad a.e.$$

Proof: If $g(x) = a^{-n} f(x/a)$, then its Fourier transform is $\widehat{f}(au)$. So the Fourier transform of

$$\frac{1}{a^n} \frac{1}{(2\pi)^{n/2}} e^{-x^2/2a^2}$$

is $e^{-a^2u^2/2}$. Therefore if we let

$$H_a(x) = \frac{1}{(2\pi)^n} e^{-|x|^2/2a^2},$$

We have

$$\widehat{H}_a(u) = (2\pi)^{-n/2} a^n e^{-a^2|u^2|/2}.$$

We have

$$\begin{aligned}
\int \widehat{f}(u) e^{-iu.y} H_a(u)du &= \int\int e^{-iu.x} f(x) e^{-iu.y} H_a(u)dxdu \\
&= \int\int e^{-iu.(x-y)} H_a(u)du\ f(x)dx \\
&= \int \widehat{H}_a(x - y) f(x)dx.
\end{aligned}$$

We can interchange the order of integration because

$$\int\int |f(x)||H_a(u)|\,dxdu < \infty$$

The left hand side of the first line of converges to $(2\pi)^{-n}\int \widehat{f}(u)e^{-iu.y}dy$ as $a \to \infty$ by dominated convergence and the fact that $\widehat{f} \in L^1$. The last line of is equal to

$$\int \widehat{H}_a(y-x)f(x)dx = f * \widehat{H}_a(y)$$

since $\widehat{H}_a$ is symmetric. But by Proposition, $f * \widehat{H}_a$ converges to f in L^1 as $a \to \infty$.

Theorem: (a) Suppose f is continuous with compact support. Then $\widehat{f} \in L^2$ and

$$\| f \|_2 = (2\pi)^{-n/2} \| \widehat{f} \|_2 .$$

(b) We can use the result in (a) to define $\widehat{f}$ when $f \in L^2$ and so that holds.

Proof: (a) Let $g(x) = \overline{f(-x)}$. Note

$$\widehat{g}(u) = \int e^{iu.x}\overline{f(-x)}\,dx = \overline{\int e^{-iu.x}f(-x)dx} = \overline{\int e^{-iu.x}f(-x)dx} = \overline{\widehat{f}(u)}.$$

with $y = 0$

$$f * g * \widehat{H}_a(0) = \int \widehat{f * g}(u)H_a(u)du.$$

Since $\widehat{f*g}\,(u) = \widehat{f}\,(u)\,\widehat{g}\,(u) = |\,\widehat{f}\,(u)|^2$, the right hand side of converges by monotone convergence to $(2\pi)^{-n}\int |\widehat{f}(u)|^2\,du$ as a π 1. Since f and g are continuous with compact support, then it is easy to see that $f * g$ is also, and so the left hand side of converges to $f * g(0) = \int f(y)g(-y)dy = \int |f(y)|^2\,dy$ by Proposition.

(b) The set of continuous functions with compact support is dense in L^2. Given a function f in L^2, choose a sequence of continuous functions with compact support $\{f_m\}$ such that $f_m \to f$ in L^2. By the result in (a), $\{\widehat{f}_m\}$ is a Cauchy sequence in L^2, and therefore converges to a function in L^2, which we call $\widehat{f}$. If $\{f_m'\}$ is another sequence of continuous functions with compact support converging to f in L^2, then $\{f_m - f_m'\}$ is a sequence of continuous functions with compact support converging to 0 in L^2; by the result in (a), $\widehat{f}_m - \widehat{f}_m'$ converges to 0 in L^2, and therefore bf is defined uniquely up to almost everywhere equivalence.

Chapter 5

Product Spaces

A product space is the cartesian product of a family of topological spaces equipped with a natural topology called the product topology. This topology differs from another, perhaps more obvious, topology called the box topology, which can also be given to a product space and which agrees with the product topology when the product is over only finitely many spaces. However, the product topology is "correct" in that it makes the product space a pullback of its factors, whereas the box topology is too fine; this is the sense in which the product topology is "natural".

Definition: Let I be a (possibly infinite) index set and suppose X_i is a topological space for every i in I. Set $X = \Pi X_i$, the Cartesian product of the sets X_i. For every i in I, we have a canonical projection $p_i : X \to X_i$. The product topology on X is defined to be the coarsest topology (i.e. the topology with the fewest open sets) for which all the projections p_i are continuous. The product topology is sometimes called the Tychonoff topology.

Explicitly, the product topology on X can be described as the topology generated by sets of the form $p_i^{-1}(U)$, where i in I and U is an open subset of X_i. In other words, the sets $\{p_i^{-1}(U)\}$ form a subbase for the topology on X. A subset of X is open if and only if it is a union of (possibly infinitely many) intersections of finitely many sets of the form $p_i^{-1}(U)$. The $p_i^{-1}(U)$ are sometimes called open cylinders, and their intersections are cylinder sets.

We can describe a basis for the product topology using bases of the constituting spaces X_i. A basis consists of sets, where for cofinitely many (all but finitely many) i, $U_i = X_i$ (it's the whole space), and otherwise it's a basic open set of X_i.

In particular, for a finite product (in particular, for the product of two topological spaces), the products of base elements of the X_i gives a basis for the product.

In general, the product of the topologies of each X_i forms a basis for what is called the box topology on X. In general, the box topology is finer than the product topology, but for finite products they coincide.

Examples: If one starts with the standard topology on the real line **R** and defines a topology on the product of n copies of **R** in this fashion, one obtains the ordinary Euclidean topology on $\mathbf{R}^n$.

The Cantor set is homeomorphic to the product of countably many copies of the discrete space {0,1} and the space of irrational numbers is homeomorphic to the product of countably many copies of the natural numbers, where again each copy carries the discrete topology.

Several additional examples are given in the article on the initial topology.

Properties

The product space X, together with the canonical projections, can be characterized by the following universal property: If Y is a topological space, and for every i in I, $f_i : Y' \to X_i$ is a continuous map, then there exists *precisely one* continuous map $f : Y' \to X$ such that for each i in I the following diagram commutes:

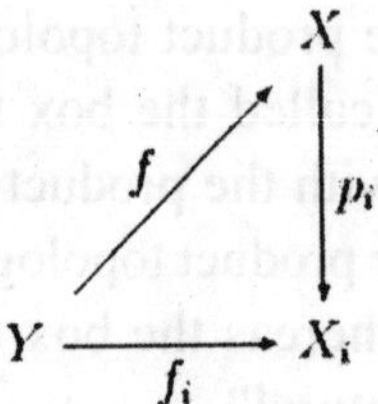

This shows that the product space is a product in the category of topological spaces. If follows from the above universal property that a map $f : Y' \to X$ is continuous iff $f_i = p_i \circ f$ is continuous for all i in I. In many cases it is often easier to check that the component functions f_i are continuous. Checking whether a map $g : X' \to Z$ is continuous is usually more difficult; one tries to use the fact that the p_i are continuous in some way.

In addition to being continuous, the canonical projections $p_i : X' \to X_i$ are open maps. This means that any open subset of the product space remains open when projected down to the X_i. The converse is not true: if W is a subspace of the product space whose projections down to all the X_i are open, then W need not be open in X. (Consider for instance $W = \mathrm{R}^2 \backslash (0,1)^2$.) The canonical projections are not generally closed maps (consider for example the closed set whose projections onto both axes are $\mathrm{R} \backslash \{0\}$).

The product topology is also called the *topology of pointwise convergence* because of the following fact: a sequence (or net) in X converges if and only if all its projections to the spaces X_i converge. In particular, if one considers the space $X = \mathrm{R}^I$ of all real valued functions on I, convergence in the product topology is the same as pointwise convergence of functions.

Any product of closed subsets of X_i is a closed set in X.

An important theorem about the product topology is Tychonoff's theorem: any product of compact spaces is compact. This is easy to show for finite products, while the general statement is equivalent to the axiom of choice.

FUBINI'S THEOREM

Let $\{X_j, M_j, \mu_j\}$, j = 1, 2 be two measure spaces, and consider the σ algebra M is subsets of $X = X_1 \times X_2$ generated by rectangles of the form $E_1 \times E_2$, where $E_j \in M_j$. This σ algebra is called the product algebra and is denoted by $M = M_1 \otimes M_2$. We shall denote point in X_1 and X_2 by x and y respectively.

A product measure on x is constructed as follows. Denote the algebra of sets generated by finite disjoint unions of rectangle by A. Define μ' on rectangle by $\mu' (E_1 \times E_2) = \mu_1 (E_1)\mu_2(E_2)$. Then μ' is a pre-measure on A.

Using μ', we construct an outer measure μ^* on X; by Caratheodory's theorem we obtain a complete measure μ on X whsoe σ-algebra of measurable ets contains the product algebra $M = M_1 \otimes M_2$. If μ_j are σ-finite, then so is μ.

Clearly this construction of product algebras and measure extends to any finite number of factors. Moreover, this definition of product algebras and measures is associative.

In this section we prove the following theorems of Tonelli and Fubini:

Theorem: Let $\{X_j, M_j, \mu_j\}$be σ-finite measure space, and let the product measure space be denoted by $\{X, M, d\mu\}$.

Tonelli: Let $f \in L^+(d\mu)$. Then the functions $g(x) = f_{X_2} f(x,y)d\mu_2(y)$ and $h(y) = f_{X_1} f(x,y)d\mu_1(x)$ are $L^+(d\mu_1)$ and $L^+(\mu_2)$ respectively; and

$$\int\int_X f d\mu = \int\left[\int f(x,y)d\mu_2(y)\right]d\mu_1(x) = \int\left[\int f(x,y)d\mu_1(x)\right]d\mu_2(y).$$

These three integrals may assume the value $+\infty$

Fubini: Let $f \in L^1(d\mu)$. Then

$$\int_{x_2} f(x,y)d\mu(y) \in L^1(d\mu_1),\ \int_{x_1} f(x,y)d\mu(x) \in L^1(d\mu_2)$$

and holdes, that is, the double and each of the iterated integrals are equal.

The strategy of the proof is to begin by proving the result for characteristic functions of rectangles, then simple functions, and ultimately extend the result to general measurable functions on X.

The x and y sections of a set $E \subset X$ are defined to be the subsets.

$$E_x = \{y : y \in X_2, (x,y) \in E\},\ x \in X_1$$

$$E^y = \{x : x \in X_1, (x,y) \in E\},\ y \in X_2$$

Similarly, if $f : X \mapsto \mathbb{R}$, then the x and y sections of f are the mappihngs $f_x : X_2 \mapsto \mathbb{R}$ and $f^y : X_1 \mapsto \mathbb{R}$ defined by $f_x(y) = f(x,y)$ and $f^y(x) = f(x,y)$.

Proposition: *If* $E \in M$, *then the sections* E_x *and* E_y *respecitvely belong to* M_2 *for each* $x \in X_1$, *and* M_1 *for each* $y \in X_2$.

If f is measurable with respect to the product algebra $M_1 \otimes M_2$, *then its sections* f_x *and* f_y *are measurable with respect to the factors* M_2 *and* M_1 *respectively.*

Proof: Let R be the collection of all subsets $E \subset X$ such that $E_x \in M_2$ for all $x \in X_1$ and $E^y \in M_1$ for all $y \in X_2$. Then

$$(A \times B)_x = \begin{cases} B & x \in A \\ \phi & x \in A^\circ \end{cases}$$

and similiary for the section $(A \times B)^y$. Hence R contains all rectangles. Moroever, R is a σ algebra, since

$$\left(\bigcup_1^{\infty} E_j\right)_x = \bigcup_1^{\infty} (E_j)_x \,, \; (E_x)^c = (E^c)_x$$

and similiary for y-sections. Therefore $R \supset M_1 \otimes M_2$. The measurability of f^x and f^y follows from the first statement and the relationship

$$(f_x)^{-1}(B) = (f^{-1}(B))_x, \; (f^y)^{-1}(B) = (f^{-1}(B))^y$$

A monotone class on a space X is a collection of subsets $C \subset P(X)$ that is closed under countable increasing unions and countable decreasing intersection.

Every σ algebra is a montone class; and the intersection of monotone classes is a monotone class. Thus for any collection $\varepsilon \subset P(X)$, there is a unique smallest monotone class containing ε, called the monotone class generated by ε.

Proposition: If A is an algebra of subsets of X then the monotone class generated by A coincides with the σ algebra generated by A. Give a measure μ we denote the non-negative μ measurable functions by $L^+(\mu)$.

Lemma: Let (X_j, M_j, μ_j) be σ*-finite measure spaces for* $j = 1, 2$, *and let* M *and* μ *be the product algebras and measures on* $X = X_1 \times X_2$. *Given* $E \in M$, *the sections* $(\chi_E)_x$ *and* $(\chi_E)^y$ *are in* $L^+(d\mu_2)$ *and* $L^+(d\mu_1)$ *respectively; and*

$$\mu(E) = \iint_X \chi_E d\mu = \int_{X_1}\left\{\int_{X_2} (\chi_E)_x d\mu_2(y)\right\} d\mu_1(x)$$

$$= \int_{X_2}\left\{\int_{X_1} (\chi_E)^y d\mu_1(x)\right\} d\mu_2(y).$$

Proof: We shall establish the lemma for the case in which the u_i are finite meaures; the σ - finite case is left as an exercise. Let C be the class of sets in M for which the lemma holds. When E is a rectangle, $E = A \times B$ then $(\chi_E)_x(y) = \chi_A(x)\chi_B(y) = (\chi_E)^y(x)$ and all three integrals above are equal to $\mu_1(A)\,\mu_2(B)$. Thus $E \subset C$. By finite additivity of the integral C contains finite disjoint unions of rectangles. It therefore suffices to show that C is a monotone class.

Let $E = U_n E_n$ where $\{E_n\}$ is an increasing sequence of sets. Then the sequence of functions χ_{E_n} increasing montonically to χ_E; and their sections increasing montonically to the corresponding sections of χ_E; follows from the montone convergence theorem; and $E \in C$. Similarly let $\{E_n\}$ be a decreasing sequence of sets in C and put $E = \bigcap_n E_n$. The characteristic functions χE_n then converge pointwise to χE; and their sections converge pointwise to the correspnding sections of χE. Since $\mu(E_1) < \infty$ the conclusion follows from the dominated convergence theorem. Thus C is a monotone class and is therefore equal to the σ algebra M.

Given $f \in L^+(X, d\mu)$ we approximate f from below by simple functions and pass to the limit using the monotone convergence theorem. All three integrals are either finite and equal, or all equal to infinity. Give $f \in L^1(d\mu)$ we decompose $f = f_+ - f_-$ and apply Tonelli's theorem to each of the components.

Lebesgue Measures in $\mathbb{R}^n$

A specific exmple of a product measure, obtained by the procedure is n dimensional Lebesgue measure on $\mathbb{R}^n$, it is obtained by the premeasure defined on 'rectangles' in $\mathbb{R}^n$ defiend by $R = \prod_j I_j$, where the $\{I_j]$ are intervals (a_j, b_j) on the real axis. We have

$$m_0(R) = \mu\left(\prod_{j=1}^{n} I_j\right) = \prod_{j=1}^{n}(b_j - a_j)$$

Using m_0 we obtain an outer measure on $\mathbb{R}^n$ and by Caratheodory's theorem we obtain a measure, called the Lebesgue measure on $\mathbb{R}^n$. We denote it by m when the dimension is understood.

As on $\mathbb{R}$, Lebesgue measure on $\mathbb{R}^n$ has has a number of regularity properties

Theorem: *Suppose* $E \in £^n$. *Then (a)* $m(E) = \inf\{m(U): E \subset U, U \text{ open}\} = \sup\{m(K): K \subset E, K \text{ compact}\}$; *(b)* $E = A_1 \cup N_1 = A_2 \backslash N_2$, *where* N_1 *and* N_2 *are sets of measure 0,* $A_1 \in F_\sigma$ *(the class of countable unions of closed sets); and* $A_2 \in F_\delta$ *(the class of countable intersections of open sets); and (c) If* $m(E) < \infty$ *then for any* $\varepsilon > 0$ *there is a finite collection of disjoint rectangles whose union R satisfies* $\mu(E \Delta R) < \varepsilon$

Theorem: *The class of simple functions* $f = \sum_j f_j XR_j$, *where* $R_j = \coprod_k I_{jk}$ *is a product of intervals* $I_{jk} \subset \mathbb{R}$, *is dense in* $L^1(\mathbb{R}^n)$. *Furthermore the class of continuous functions with compact support in* $\mathbb{R}^n$ *is dense in* $L^1(\mathbb{R}^n)$.

Proof: The proof is essentially the same as the proof of the corresponding theorem for Borel measures on $\mathbb{R}$.

Theorem: *Lebesgue measure on* $\mathbb{R}^n$ *is invariant under the group of rigid motions.*

Proof: It suffices to establish that Lebesgue outer measure is invariant under the group of rigid motions, that is, under the group of translations and rotations in $\mathbb{R}^n$. We prove the result in $\mathbb{R}^2$ and leave its extension to $\mathbb{R}^n$ as an exercise.

Lebesgue outer measure was defined in terms of coverings of rectangles with sides parallel to the axes. Thus the outer measure is invariant under translations and reflections in the axes. We therefore have two set functions, area and outer measure, that are boy additivie and invariant under translations and reflections. Any theorem on area in classical geometry that uses only these three properties is therefore valid for outer measure.

To show that the outer measure is invariant under rotations, we must prove that the outer measure o a rectangle in general position is equal to the area of that rectangle. A rectangle in general position may be written as the union of two triangles and a parallelogram (degenerate if one of the diagonals is parallel to the x axis) with bases parallel to the x-axis; so by additivity of both the area and the outer measure, it suffices to prove that the two functionals agree on triangles and parallelograms with base parallel to one of the axes.

Give a parallelogram with base parallel to the x axis, trnaslate one of its triangular caps parallel to the x axis to the other end, forming a rectnalge with base parallel to the x axis. Both the outer measure and the area of the parallelogram are invariant under this operation; and they coincide on the rectangle. Hence the outer measure and area of the parallelogram coincide.

By similar arguments from planet geometry, using only additivity and invariance under reflections and translations, one can prove that the outer measure of any triangle is equal to its area. A second approach is combine a translation of the upper triangle with the lower triangle to produce a second parallelogram with base parallel to the x axis, and then apply the previous argument showing the measure of a parallelogram is equal to its area.

Now let $T \in GL(n, R)$ and assume that $\det T \neq 0$. We define an invertible transformation acting on the Lebesgue measurable functions $f \in £(\mathbb{R}^n)$ by $f_T = foT^{-1}$. We define an action of $GL(n, R)$ on n dimensional Lebesgue measure by $m_T(E) = m(TE)$, where E is a Lebesgue measurable subset of $\mathbb{R}^n$ and TE denotes its image under T. For a rectangle $R \in R^n$ we have $m(TR) = \det Tm(R)$.

Theorem: *For* $f \in L^1(\mathbb{R}^n)$ *and any invertible linear transformation* $T \in GL(n, R)$ *we have*

$$\iint f_T dm = \iint f dm_T = |\det T| \iint f dm$$

Proof: It is sufficient to prove the result for characteristic functions. It then follow for simple functions by additivity, and for the general case by the dominated convergence theorem.

First, note that for rectangles, $A(TR) = \det TA(R)$, hence on rectangles, $m_T(R) = m(TR) = |\det T| m(R)$. By continuity, this identity extends to all Lebesgue measurable set E. Next, for any set E we observe that $\chi_E(T^{-1}x) = \chi_{TE}(x) = (\chi_E)_T$; hence

$$\iint (\chi_E)_T dm = \iint \chi_{TE} dm = m(TE)$$

$$= |\det T| m(E) = \det T \iint \chi_E dm = \iint \chi_E dm_T$$

The volume $V(E)$, where $E \subset \mathbb{R}^n$, is initially defined on rectangles, and then extended to polygons by its invriance under translation and reflection. The extension of the volume functional to bounded subsets in $\mathbb{R}^n$ is defined by the method of exhaustion, a reincarnation of which is known as Jordan content: Let Q_k denote the collection of cubes with sides of length 2^{-k} and vertices at the points $2^{-k}\mathbb{Z}^n$ in $\mathbb{R}^n$. The volume of each cube in Q_k is 2^{-nk}, and the collection Q_{k+1} is obtained by bisecting the sides of Q_k.

For $E \subset \mathbb{R}^n$ we define the inner and outer dyadic approximations of E to be the open sets

$$I_k(E) = \bigcup\{Q \in Q_k : Q \subset E\},$$

$$O_k(E) = \bigcup\{Q \in Q_k : Q \cap E \neq \emptyset\}$$

For each k, $I_k(E) \subset E \subset O_k(E)$. The $I_k(E)$ form an increasing sequence of open sets, and the $O_k(E)$ a decreasing sequence. Moreover, if E is bounded, then the volumes of these sets are finite and respectively increasing and decreasing. Denoting their limits respectively by $\underline{k}(E)$ and $\overline{k}(E)$, we have

$$\underline{k}(E) = \lim_k V(I_k(E)) \leq \lim_k V(O_k(E)) = \overline{k}(E)$$

If these two quantities are equal, then the volume of E, denoted by $V(E)$, is defined to be their common value. In Jordan's theory, $\underline{k}(E)$ and $\overline{k}(E)$ are know as the inner and outer contents of E.

Theorem: *If the volume of a set $E \subset \mathbb{R}^n$ is defined in the above sense, then E is Lebesgue measurable, and its Lebesgue measure is equal to the volume.*

Proof: Let

$$\underline{A}(E) = \bigcup_{k=1}^{\infty} I_k(E), \quad \overline{A}(E) = \bigcup_{k=1}^{\infty} O_k(E)$$

Then $\underline{A}(E) \subset E \subset \overline{A}(E), \underline{A}(E)$ and $\overline{A}(E)$ are Borel sets, and by the continuity o Lebesgue measure,

$$m(\underline{A}(E)) = \underline{k}(E), \;\; m(\underline{A}(E)) = \overline{k}(E)$$

L^p SPACES

Let (X, M, μ) be a measure space and for $1 \leq p < \infty$ define

$$L^p(X, d\mu) = \{f : \| f \|_p < \infty\}, \| f \|_p = \left(\int | f |^p \, d\mu\right)^{1/p}.$$

Technically L^p consists of equivalence classes of functions which are equal almost everywhere; but normally we shall not refer to this matter explicitly.

A *Banach space* is a complete normed linear vector space. A norm on a vector space B is defined to be a positive functional, generally denoted by $\|.\|$, with the properties

1. $\| f \| \geq 0$; $\| f \| = 0$ iff $f = 0$.
2. $\| \lambda f \| = | \lambda | \, \| f \|$;
3. $\| f + g \| \leq \| f \| + \| g \|$.

We shall show that the spaces $L^p(X, d\mu)$ are Banach spaces for $1 \leq p < \infty$ with the norm $\|\cdot\|_p$ given as above. The case $p = \infty$ is also included, and will be discussed below. By complete we mean that any Cauchy sequence $\{f_n\} \in B$ converges to a point $f \in B$.

The first property of the norm is clear, since $\|f\|_p = 0$ implies that $f = 0$ a.e. The second property is also easy to prove. Hence the only item of substance is the triangle inequality.

Holder's Inequality

Theorem: For $1 < p < \infty$ and $p^{-1} + q^{-1} = 1$; we have $\|fg\|_1 \leq \|f\|_p \|g\|_q$ for all $f \in L^p$ and $g \in L^q$. Equality holds iff f and g are scalar multiples of one another.

Proof: The result is trivial if either f or g is zero, or if $\|f\|_p$ or $\|g\|_q$ is infinite. In the general case we may rescale f and g so that $\|f\|_p = \|g\|_q = 1$. Then, by the concavity of the logarithm,

$$\ln a^{1/p} b^{1/q} = \frac{1}{p}\ln a + \frac{1}{q}\ln b \leq \ln\left(\frac{a}{p} + \frac{b}{q}\right).$$

Exponentiating this inequality, and taking $a = |f|^p$, $b = |g|^q$, we obtain

$$|f||g| \leq \frac{|f|^p}{p} + \frac{|g|^q}{q}.$$

Integrating this inequality over X we obtain

$$\|fg\|_1 \leq \frac{1}{p}\|f\|^p + \frac{1}{q}\|g\|^q = 1 = \|f\|_p \|g\|_q .$$

The result follows by reintroducing the original scaling.

The indices p and q are called *conjugate indices.* Unless otherwise stated we shall take p and q to be conjugate indices in the following.

Minkowski's Inequality

Theorem: If $1 \leq p < \infty$ and $f, g \in L^p$ then

$$\|f + g\|_p \leq \|f\|_p + \|g\|_p.$$

Proof: The result is immediate if $p = 1$ or $f + g = 0$ a.e. Otherwise, note the inequality $|f + g|^p \leq (|f| + |g|)|f + g|^{p-1}$; and integrate over X:

$$\|f+g\|_p^p = \int |f+g|^p \, d\mu \leq \int (|f| + |g|)|f+g|^{p-1} d\mu$$

$$\leq \int |f||f+g|^{p-1} + \int |g||f+g|^{p-1} d\mu$$

$$\leq (\| f \|_p + \| g \|_p)\left(\int | f + g |^{(p-1)}\, d\mu \right)^{1/q}$$

$$\leq (\| f \|_p + \| g \|_p)\left(\int | f + g |^{p}\, d\mu \right)^{p/q}$$

$$= (\| f \|_p + \| g \|_p) \| f + g \|_p^{p-1},$$

and the inequality follows by dividing both sides by $\| f + g \|_p^{p-1}$.

Theorem: The L^p spaces are Banach spaces for $1 \leq p < \infty$.

Proof: Let $\{ f_j \}$ be a Cauchy sequence in L^p and choose a subsequence n_j so that $\| f_{n_{j+1}} - f_{n_j} \|_p < 2^{-j}$.

Putting $G_n = \sum_{j=1}^{n} | f_{n_{j+1}} - f_{n_j} |$, we see that $\| G_n \|_p < \sum_j 2^{-j} = 1$. Since $\{G_n\}$ is a monotone increasing sequence, we have, by the monotone convergence theorem, that

$$\int | G |^p \, d\mu = \lim_{n \to \infty} \int | G_n |^p \, d\mu < 1.$$

Therefore $G = \sum_{j=1}^{\infty} | f_{n_{j+1}} - f_{n_j} | < \infty$ a.e. and the series $\sum_j (f_{n_{j+1}} - f_{n_j})$ converges pointwise a.e. The limit $f = \lim_j f_{n_j}$ therefore exists and is finite a.e.; and, moreover,

$$\| f - f_{n_k} \|_p = \left\| \sum_{j=k+1}^{\infty} f_{n_{j+1}} - f_{n_j} \right\|_p \leq \sum_{j=k+1}^{\infty} \| f_{n_{j+1}} - f_{n_j} \|_p < 2^{-k},$$

$$\| f \|_p \leq \| f_{n_k} \|_p + \| f - f_{n_k} \|_p.$$

It follows that $f_{n_k} - f$ in the L^p norm, and $f \in L^p$. Moreover, the original sequence $f_n \to f$, since

$$\| f - f_n \|_p \leq \| f - f_{n_j} \|_p + \| f_{n_j} - f_n \|_p < 2^{-j} + 2^{-j} = 2^{1-j}, \; n > n_j.$$

We conclude by treating the case $p = \infty$. We define

$$\| f \|_\infty = \inf\{a\colon a \geq 0,\ \mu(\{| f | > a\}) = 0\}.$$

The space $L^\infty = L^\infty(X, M, \mu)$ is the vector space of all measurable functions f for which $\| f \|_\infty$. The set $E = \{| f | > \| f \|_\infty\}$ necessarily has measure 0. Putting $g = f\chi_E$ we see that $f = g$ a.e. and $\| f \|_\infty = \sup | g |$. Hence $\|.\|_\infty$ is a norm. Moreover, L^∞ is a Banach space, since any Cauchy sequence in norm is a uniform Cauchy sequence, and so converges pointwise a.e.

Theorem: If $0 < p < q < r \leq \infty$ then $L^p \cap L^r \subset L^q$ and

$$\| f \|_q \leq \| f \|_p^{\lambda} \| f \|_r^{1-\lambda}, \frac{1}{q} = \frac{\lambda}{p} + \frac{1-\lambda}{r}, 0 < \lambda < 1.$$

Theorem: Let $1 \leq pj, r \leq \infty$ and suppose that $f_j \in L^{pj}$ for $j = 1,..., n$. Then

$$\| f_1 ... f_n \| r \leq \prod_{j=1}^{n} \| f_j \|_{pj}, \text{ where } \frac{1}{r} = \sum_{j=1}^{n} \frac{1}{pj}.$$

The proof is left as an exercise.

DUALITY

A *bounded linear functional* on a Banach space B is a mapping $F: B \mapsto C$ such that $|F(x)| \leq C\| x \|$, where $\|.\|$ is the norm on B. The collection of all bounded linear functionals on a Banach space B is itself a normed vector space with norm

$$\| F \| = \sup_{x \in B} \frac{| F(x) |}{\| x \|} = \inf\{A : | F(x) | \leq A \| x \|\}.$$

The space of linear functionals on B is called the *dual* space and is denoted by B^*.

Theorem: B is itself a Banach space.

Proof: We leave as an exercise the proof that the expression $\| F \|$ defined above is space of functionals defined above is in fact a norm. Suppose $\{F_n\}$ is a Cauchy sequence of functionals in the dual norm. Then for each $x \in B$, $\{F_n(x)\}$ is a Cauchy sequence of complex numbers, hence converges to a limit, denoted by $F(x)$. Since the limit is a linear operation, $F(x)$ is linear in x, hence F is a linear functional. Moreover, $|F(x)| \leq \lim_n |F_n(x)| \leq \lim_n \|F_n\| \| x \|$. In particular, $\|F\| \leq \lim_n \| F_n \|$.

Given $\varepsilon > 0$, choose N so that $\| F_n - F_m \| < \varepsilon$ for all $m, n > N$. Then

$$| F(x) - F_n(x)| = \sup_{m \to \infty} | F_m(x) - F_n(x) |$$
$$\leq \lim_{m \to \infty} \| F_m - F_n \| \| x \| < \varepsilon \| x \|, \text{ for all } n \geq N.$$

It follows that $\|F - F_n\| \leq \varepsilon$, hence $\lim_{n \to \infty} \| F - F_n \| = 0$. This shows that every Cauchy sequence in B^* has a limit in B^*.

Given a function g, measurable on X, consider the functional F_g defined by $F_g(f) = \int fgd\mu$. By Holder's inequality, $|F_g(f)| \leq \| f \|_p \| \| g \|_q$; thus if $g \in L^q$ then F_g is a linear functional on L^p, i.e. $Fg \in L^{p*}$. In other words, $L^q \to (L^p)^*$; more precisely, every element $q \in L^q$ is mapped into $(L^p)^*$ via the mapping $g \mapsto F_g$.

A measure μ is *semi-finite* if every set B of positive μ-measure has a subset of finite μ measure.

Theorem: For $1 \leq p < \infty$ the injection $L^q \to (L^p)^*$ defined by $g \mapsto F_g$ is an isometric injection of L^q into $(L^p)^*$; that is $\| F_g \| = \| g \|_q$. The same holds for $p = \infty$ when μ is semi-finite.

Proof: We need to show that

$$\| g \|_q = \sup_f \left\{ \int fgd\mu, \| f \|_p = 1 \right\}.$$

For $1 < p < \infty$, $\| F_g \| \leq \| g \|_q$ by Holder's inequality. If $g \neq 0$ let

$$f = \frac{|g|^{q-1} \operatorname{sgn}(g)}{|g|_q^{q-1}}.$$

Then $\| f \|_p = 1$ and

$$F_g(f) = \int fgd\mu = \int \frac{|g|^q}{|g|_q^{q-1}} d\mu = \frac{\| g \|_q^q}{\| g \|_q^{q-1}} = \| g \|_q .$$

This shows that the supremum is attained by this choice of f.

For $p = 1$ take $f = \operatorname{sgn} g$. Then $\| f \|_\infty = 1$ while $F_g(f) = \int fgd\mu = \int |g| \, d\mu = \|g\|_1$. Therefore

$$\| g \|_1 \leq \sup_{\|f\|_\infty = 1} F_g(f) = \| F_g \|.$$

On the other hand,

$$| F_g(f) | \leq \left| \int fgd\mu \right| \leq \int | fg | \, d\mu \leq \| g \|_1 \| f \|_\infty ,$$

so $\| F_g \| \leq \| g \|_1$.

Now consider the case $p = \infty$ and suppose μ is semi-finite. Given $\varepsilon > 0$ choose $A_\varepsilon = \{x : | g(x) | > \| g \|_\infty - \varepsilon\}$. If $\mu(A_\varepsilon) > 0$ we choose a subset $B \subset A_\varepsilon$ with finite measure, and put

$$f = \frac{\chi_B}{\mu(B)} \operatorname{sgn} g.$$

Then $\| f \|_1 = 1$ so

$$\| F_g \| \geq | F_g(f) | = \left| \int fgd\mu \right| = \int_B \frac{|g|}{\mu(B)} d\mu \geq \| g \|_\infty - \varepsilon.$$

Since this holds for all $\varepsilon > 0$ we have $\| g \|_\infty \leq \| F_g \|$. On the other hand, $|F_g(f)| \leq \int | fg \,| d\mu \leq \|g\|_\infty \, \|f\|_\infty$; $\| F_g \| \leq \| g \|_\infty$; and therefore $\| F_g \| = \|g\|_\infty$.

Theorem: Let $1 \leq p < \infty$, let q be the conjugate index; and let μ be σ-finite. Then the embedding $g \mapsto F_g$ is a surjective mapping of $L^q(d\mu)$ onto $L^p(d\mu)^*$.

Proof: Given a functional $F \in L^p(d\mu)9^*$, let $\nu_F(E) = F(\chi_E)$ for all measurable sets E. Then ν_F is a countably additive σ-finite measure, and is absolutely continuous with respect to μ. By the Radon-Nikodym theorem, there is a measurable function $g \in L^p(d\mu)$ such that $\nu_F(E) = \int g d\mu$; and by linearity the result also holds for all simple functions ϕ: $F(\phi) = \nu_F(\phi) = \phi$. Since the simple functions are dense in L^p, the result immediately extends to all measurable functions $f \in L^p$.

DISTRIBUTION FUNCTIONS

Given a real valued function f on a measure space (X, M, μ) its distribution function λ_f is defined by

$$\lambda_f(\alpha) = \mu(\{x: |f(x)| > \alpha\}).$$

Proposition: The distribution function is decreasing and right continuous in α, and increasing in $|f|$. If $f_n \uparrow f$ then $\lambda_{f_n} \uparrow \lambda_f$. Finally,

$$\lambda_{f+g}(\alpha) \leq \lambda_f(\alpha/2) + \lambda_g(\lambda/2).$$

Proof: The proof is left to the reader.

If $\lambda_f(\alpha)$ is ¯nite for all $\alpha > 0$, then the distribution function defines a negative Borel measure on $\mathbb{R}^+$ via the premeasure $\lambda_f((a, b]) = \lambda_f(b) - \lambda_f(a)$. We denote the associated

Proposition: For any Borel measurable function ϕ defined on $\mathbb{R}^+$ we have

$$\int \phi(|f|) d\mu = -\int_0^\infty \phi(\alpha) d\lambda_f(\alpha),$$

where the integral on the right is the Lebesgue-Stieltjes integral with respect to the measure $d\lambda_f$.

In particular,

$$\int |f|^p = -\int_0^\infty \alpha^p d\lambda_f(\alpha).$$

Proof: The result is immediate for $\phi = \chi_{(\alpha,\infty)}$. The result extends immediately to characteristic functions for any Borel set, then to simple functions, then to integrable functions by the monotone convergence theorem.

LINEAR TRANSFORMATIONS

A bounded linear transformation from a Banach space B to a Banach space B' is a linear transformation $L: B \mapsto B'$ such that $\| Lu \|_{B'} \leq C \| u \|_B$ for some fixed positive constant C. The best possible such constant is the *norm* of L, denoted by $\|L\|$, and defined by

$$| L |= \sup_{u \in B} \frac{\| Lu \|_{B'}}{\| u \|_B}.$$

An important class of linear transformations on Banach spaces is the class of linear integral operators mapping $L^p(Y, d\mu)$ to $L^q(X, dv)$ defined by

$$(Tf)(x) = \int K(x, y) f(y) d\mu(y),$$

where $K(x, y)$, called the *kernel* is a measurable function on the product space $(X \times Y, M \otimes N)$.

The question of when an integral operator is a bounded transformation from L^p to L^q is of fundamental interest in applications; and a variety of techniques exist. For example, the Fourier transformation

$$Ff = \frac{1}{\sqrt{2\pi}} \int e^{-ikx} f(y) dy,$$

is an isometry from $L^2(\mathbb{R})$ to itself, though this fact is not immediately obvious.

An interesting class of linear integral operators is the class of *convolution* operators, which typically arise in connection with the Fourier transform and partial differential equations with constant coeffcients. The convolution of two functions f and g on the line is defined by

$$(f * g)(x) = \int f(x-y) g(y) dy.$$

By Fubini's theorem,

$$\| f * g \|_1 = \int \left| \int f(x-y) g(y) dy \right| dx \leq \int \int | f(x-y) g(y) | dy dx$$

$$= \int | g(y) | \left\{ \int | f(x-y) | dx \right\} dy = \int | g(y) | \| f \|_1 dy = \| f \|_1 \| g \|_1 .$$

This calculation shows that the linear transformation defined by $L_g f = f * g$ is a bounded transformation on $L^1(\mathbb{R})$ into itself, and that $\| L_g \| \leq \| g \|_1$. Similarly, $\| f * g \|_\infty \leq \| g \|_1 \| f \|_\infty$.

More generally, we have

Generalized Young Inequality

Theorem: *Let* $f \in L^p(\mathbb{R})$, $g \in L^q(\mathbb{R})$. Then

$$\| f * g \|_r \le \| f \|_p \| g \|_q, \frac{1}{p} + \frac{1}{q} = 1 + \frac{1}{r},$$

hence the convolution operator L_g is a bounded linear transformation from L^p to L^r with norm $\| L_g \| \le \| g \|_q$.

This theorem is easily proved using an interpolation argument.

Riesz-Thorin Interpolation Theorem

***Theorem**: Let* (X, M, μ) and (Y, N, v) be measure spaces and $1 \le p_0, p_1, q_0, q_1 \le \mu$. Suppose T is a bounded linear transformation from $L^{p_j}(d\mu)$ to $L^{q_j}(dv)$ for $j = 0; 1$ with respective norms M_0 and M_1.

$$\| Tf \|_{q_0} \le \| f \|_{p_0}, \| Tf \|_{q_1} \le M_1 \| f \|_{p_1}.$$

Let

$$\frac{1}{p\lambda} = \frac{1-\lambda}{p_0} + \frac{\lambda}{p_1}, \frac{1}{q\lambda} = \frac{1-\lambda}{q_0} + \frac{\lambda}{q_1}, 0 < \lambda < 1.$$

Then T is a bounded linear transformation from $L^{p\lambda}$ to $L^{q\lambda}$ and

$$\| Tf \|_{q\lambda} \le M_0^{1-\lambda} M_1^{\lambda} \| f \|_{p\lambda}.$$

We apply the interpolation theorem in two stages, first establishing for the case $q = 1$. By the argument above, we know that L_g is a bounded operator from L^1 to L^1 and from L^∞ to L^∞, in both cases with norm $\| g \|_1$.

We apply the Interpolation theorem with

$$q_0 = p_0 = 1, q_1 = p_1 = \infty$$

$$q_\lambda = p_\lambda = \frac{1}{1-\lambda}, M_0 = M_1 = \| g \|_1.$$

Hence by the interpolation theorem, ($p = p_\lambda$ for $1 < p < \infty$),

$$\| L_g f \|_p \le \|g\|_1 \|f\|_p, 1 \le p \le \infty.$$

This inequality is known as *Young's inequality*.

Next, suppose $1 < q$, let $g \in L^q$, and consider the linear transformation $L_g f = f * g$. For $p = 1$ and $r = q$ we have $\| L_g f \|_q \le \| g \|_q \|f\|_1$ by Young's inequality; and for $r = \infty$ and $p = q/(q-1)$ we have $\| L_g f \|_\infty \le \| g \|q \|f\|_p$ by Holder's inequality. Thus

$$\| L_g f \|_{r_0} \leq M_0 \| f \|_{p_0}, p_0 = 1, r_0 = q;$$

$$\| L_g f \|_{r_1} \leq M_1 \| f \|_{p_1}, p_1 = \frac{q}{q-1}, r_1 = \infty,$$

and $M_0 = M_1 = \| g \|_q$. By Theorem,

$$\| f * g \|_{r_\lambda} = \| L_g f \|_{r_\lambda} \leq \| g \|_q \| f \|_{p_\lambda},$$

where

$$\frac{1}{p_\lambda} = \frac{1-\lambda}{p_0} + \frac{\lambda}{p_1}, \frac{1}{r_\lambda} = \frac{1-\lambda}{q}.$$

The result then follows from the identity

$$\frac{1}{p_\lambda} + \frac{1}{q} = 1 + \frac{1}{r_\lambda}.$$

If $\{v_j\}$ is a collection of vectors in a Banach space, then by a repeated application of the triangle inequality, $\left\| \sum_j v_j \right\| \leq \sum_j \| v_j \|$. This result extends to integrals of a family of vectors $\{v(t)\}$. Namely,

$$\left\| \int v(t) d\mu(t) \right\| \leq \int \| v(t) \| d\mu(t).$$

This result is established by approximating the integral by integrals of simple, vector valued functions $s(t) = \sum_j v_j \chi_{E_j}(t)$. In that case the inequality is a direct consequence of the triangle inequality for the norm.

Theorem: Let $(X, M, d\mu)$, (Y, N, dv) be measure spaces and $K(x, y)$ be measurable with respect to $M \otimes N$. Suppose that $\| K(., y) \|_1 \leq C$ and $\| K(x,.) \|_1 \leq C$ for some fixed constant C, where $\| K(., y) \|_1$ denotes the L^1 norm of $K(x, y)$ in $(X, d\mu)$ for fixed y, etc. Then the integral operator T defined is a bounded linear transformation from $L^p(Y, dv)$ to $L^p(X, d\mu)$ with norm less than or equal to C.

Proof: For the cases $p = 1$, μ the proofs are straightforward and require only the hypotheses $\| K(x,.) \|_1 \leq C$, respectively $\| K(., y) \|_1 \leq C$.

In the case $1 < p < \infty$, let q be the conjugate index and apply Holder's inequality:

$$\int | K(x,y) f(y) | dv(y) \leq \left[\int | K(x,y) | dv(y) | \right]^{1/q}$$

$$\left[\int | K(x,y) \| f(y) |^p dv(y) | \right]^{1/p}$$

$$\leq C^{1/q}\left[\int |K(x,y)||f(y)|^p\,dv(y)\right]^{1/p}, \text{ a.e.}$$

Hence, by the Tonelli-Fubini theorem

$$\int\left[\int |K(x,y)||f(y)|^p\,dv(y)\right]^p d\mu(x)$$

$$\leq C^{p/q}\iint |K(x,y)||f(y)|^p\,dv(y)d\mu(x)$$

$$\leq C^{(p/q)+1}\int f(y)|^p\,dv(y).$$

This implies that the left side is finite; and, taking *pth* roots, we obtain

$$\| Tf \|_p \leq \| f \|_p.$$

Theorem: Let $K(x, y)$ be Lebesgue measurable in $\mathbb{R}^+ \times \mathbb{R}^+$ and suppose that $K(\lambda x, \lambda y) = \lambda^{-1} K(x, y)$ for all $\lambda > 0$. Suppose also that

$$\int_0^\infty |K(x,1)|x^{-1/p}dx = C < \infty, 1 \leq p < \infty.$$

Let q be the conjugate index to p. For $f \in L^p(\mathbb{R}^+)$ and $g \in L^q(\mathbb{R}^+)$ let

$$Tf(x) = \int_0^\infty K(y,x)f(y)dy, Sg(x) = \int_0^\infty K(x,y)g(y)dy.$$

Then T and S are bounded linear transformation, with norm less than or equal to C, on $L^p(\mathbb{R}^+)$ and $L^q(\mathbb{R}^+)$ respectively.

Proof: We have

$$\int_0^\infty K(y,x)f(y)dy = \int_0^\infty K(xz,x)f(xz)xz = \int_0^\infty K(z,1)f_z(x)dz,$$

where $f_z(x) = f(xz)$. Now $\| f_z \|_p = z^{-1/p} \| f \|_p$; so by,

$$\| Tf \|_p \leq \int_0^\infty |K(z,1)| \| f_z \|_p\,dz = \| f \|_p \int_0^\infty |K(z,1)|z^{-1/p}dz = C\| f \|_p\,.$$

The proof of the boundedness of the operator S on L^q is similar; we note that

$$\int_0^\infty |K(1,y)|y^{-1/q}dy = \int_0^\infty |K(u,1)|u^{-1-1/q}dy$$

$$= \int_0^\infty |K(u,1)|u^{-1/p}du = C.$$

Chapter 6

Complex Analysis

Complex analysis is the study of complex numbers together with their derivatives, manipulation, and other properties. Complex analysis is an extremely powerful tool with an unexpectedly large number of practical applications to the solution of physical problems. Contour integration, for example, provides a method of computing difficult integrals by investigating the singularities of the function in regions of the complex plane near and between the limits of integration. The key result in complex analysis is the Cauchy integral theorem, which is the reason that single-variable complex analysis has so many nice results.

A single example of the unexpected power of complex analysis is Picard's great theorem, which states that an analytic function assumes every complex number, with possibly one exception, infinitely often in any neighborhood of an essential singularity! A fundamental result of complex analysis is the Cauchy-Riemann equations, which give the conditions a function must satisfy in order for a complex generalization of the derivative, the so-called complex derivative, to exist. When the complex derivative is defined "everywhere," the function is said to be analytic.

Complex analysis, traditionally known as the theory of functions of a complex variable, is the branch of mathematics investigating functions of complex numbers. It is useful in many branches of mathematics, including number theory and applied mathematics, and in physics. Complex analysis is particularly concerned with the analytic functions of complex variables, which are commonly divided into two main classes: the holomorphic functions and the meromorphic functions. Because the separable real and imaginary parts of any analytic function must satisfy Laplace's equation, complex analysis is widely applicable to two-dimensional problems in physics.

One central tool in complex analysis is the line integral. The integral around a closed path of a function which is holomorphic everywhere inside the area bounded by the closed path is always zero; this is the Cauchy integral theorem. The values of a holomorphic function inside a disk can be computed by a certain path integral on the disk's boundary.

Path integrals in the complex plane are often used to determine complicated real integrals, and here the theory of residues among others is useful. If a function has a *pole* or *singularity* at some point, that is, at that point its values "blow up" and have no finite value, then one can compute the function's residue at that pole, and these residues can be used to compute path integrals involving the function; this is the content of the powerful residue theorem.

The remarkable behavior of holomorphic functions near essential singularities is described by the Weierstrass-Casorati theorem. Functions which have only poles but no essential singularities are called meromorphic. Laurent series are similar to Taylor series but can be used to study the behavior of functions near singularities. A bounded function which is holomorphic in the entire complex plane must be constant; this is Liouville's theorem. It can be used to provide a natural and short proof for the fundamental theorem of algebra which states that the field of complex numbers is algebraically closed.

An important property of holomorphic functions is that if a function is holomorphic throughout a simply connected domain then its values are fully determined by its values on any smaller subdomain. The function on the larger domain is said to be analytically continued from its values on the smaller domain. This allows the extension of the definition of functions such as the Riemann zeta function which are initially defined in terms of infinite sums that converge only on limited domains to almost the entire complex plane. Sometimes, as in the case of the natural logarithm, it is impossible to analytically continue a holomorphic function to a non-simply connected domain in the complex plane but it is possible to extend it to a holomorphic function on a closely related surface known as a Riemann surface.

All this refers to complex analysis in one variable. There is also a very rich theory of complex analysis in more than one complex dimension where the analytic properties such as power series expansion still remain true whereas most of the geometric properties of holomorphic functions in one complex dimension are no longer true. The Riemann mapping theorem about the conformal relationship of certain domains in the complex plane, which may be the most important result in the one-dimensional theory, fails dramatically in higher dimensions. It is also applied in many subjects throughout engineering, particularly in power engineering.

Principle of the Point of Accumulation

In the rigorous discussion of the fundamentals of analysis, the leading part is played by Weierstrass' principle of the point of accumulation.

The principle follows: *If infinitely many numbers are given in a finite interval, these numbers possess at least one point of accumulation, that is, there is at least one point x such that in every interval about the point x, however small, there lie infinitely many of the given numbers.*

Proof of Weierstrass' Principle of the Point of Accumulation

In order to prove this principle arithmetically, we assume to begin with that the given interval is the interval [0, 1]. We now subdivide this interval into ten equal parts by means of the points 0.1, 0.2,..., 0.9. At least one of these subintervals must contain infinitely many points. Let the interval beginning with the number $0.a_1$ be that interval (or one of those intervals if there are several). We now subdivide this interval into ten parts by means of the points of subdivision $0.a_1 1, 0.a_1 2,..., 0.a_1 9$. Again, it is true that at least one of these subintervals must contain infinitely many points; let it be the subinterval beginning with the number $0.a_1a_2$. We again subdivide into ten parts, note that one of these parts must contain infinitely many points and continue the process. We thus arrive at a sequence of digits $a_1, a_2, a_3,...,$ each having one of the values 0, 1, 2, ..., 9. We now consider the decimal

$$\xi = 0.a_1a_2a_3 \ldots$$

It is clear that this is a point of accumulation of our set of numbers. In fact, every interval, no matter how small, in the interior of which the point ξ lies, contains the subintervals of our system of subdivision from a certain degree of fineness onward, and these subintervals contain infinitely many numbers of the set.

If the interval under consideration instead of being the interval from 0 to 1 is, say, the interval from a to $a + A$, nothing essential in the above argument is changed. The point of accumulation is then represented simply by a number of the form

$$a + h < 0.a_1a_2a_3 \ldots$$

Limits of Sequences

These considerations throw new light upon the concept of the limit of an infinite sequence of numbers $a_1, a_2, a_3, \ldots, a_n, \ldots$. We first consider the exceptional case in which infinitely many numbers of the sequence are equal to one another and extend our definition by applying the name point of accumulation also to this point (or these points). If there are infinitely many different numbers in the sequence and if we assume that the numbers a_n of this sequence are bounded, i.e., that there is a number M such that $| a_n | < M$ for all values of n, the numbers of the sequence form an infinite set of numbers in a finite interval, since they all lie between - M and M. Hence they must possess at least one point of accumulation (ξ).

If there is only one point of accumulation, it is easy to show that the sequence converges and that its limit is the number ξ. In fact, let us mark off any small interval about the point ξ. If infinitely many points of the sequence were outside this interval, they would have a limit point other than ξ, contrary to the hypothesis, whence only a finite number of the numbers of the sequence are exterior to the interval and, by definition, the sequence approaches ξ. On the other hand, if there are several points of accumulation, the sequence approaches no limit. The existence of a limit and the uniqueness of the point of accumulation of a bounded sequence of numbers are therefore equivalent ideas.

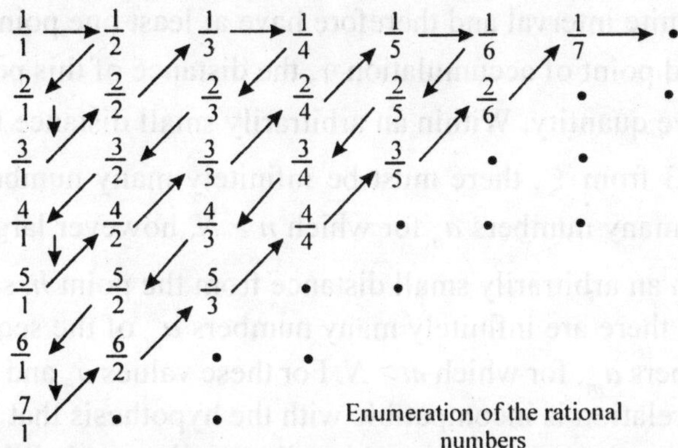

Enumeration of the rational numbers

The case of the non-existence of a limit is to be regarded as the rule rather than the exception. For example, the sequence with the terms

$$a_{2n} = \frac{1}{n},\ a_{2n-1} = 1-,\ (n=1, 2, \ldots)$$

has the two points of accumulation 0 and 1.

The aggregate of the positive rational numbers may be regarded as a sequence of numbers, in which the ordering by magnitude is, of course, completely destroyed. We arrive most easily at such an arrangement in a sequence by first writing down the rational numbers as shown and then running through this array as shown by the arrows, disregarding those numbers which have already been encountered (such as 2/4).

The system of rational numbers obviously has all rational and irrational points as points of accumulation. It therefore forms a simple example of a sequence with an infinite number of points of accumulation.

By means of the concept of convergence, we can state the principle of the point of accumulation in a remarkable form which is often convenient for applications.

From every bounded infinite set of numbers, it is possible to choose an infinite sequence $a_1, a_2, a_3, \ldots$ which converges to a definite limit ξ. For this purpose, we have only to take a point of accumulation ξ of the given set of numbers, then select a number a_1 of the set the distance of which from ξ is less than, then a second number a_2 of the set the distance of which from x is less than, then a third number a_3, the distance of which from ξ is less than, and so on. We see at once that this sequence actually converges to the limit ξ.

Cauchy's Convergence Test

Let us now return to convergent sequences, i.e., to bounded sequences with only one point of accumulation. Cauchy's convergence test now reduces almost to a triviality. In fact, when m and n are sufficiently large, let $| a_m - a_n |$ be arbitrarily small. Then all the numbers a_n lie in a finite interval and therefore have at least one point of accumulation ξ. If there were a second point of accumulation η, the distance of this point from x would be $|\xi - \eta| = a$, a positive quantity. Within an arbitrarily small distance from x, say, within a distance less than $a/3$ from ξ, there must be infinitely many numbers a_n and hence, in particular, infinitely many numbers a_n for which $n > N$, however large N is chosen.

Similarly, within an arbitrarily small distance from the point h, say, within a distance less than $a/3$ from η, there are infinitely many numbers a_m of the sequence; in particular, infinitely many numbers a_m, for which $m > N$. For these values a_n and a_m, it is true that $| a_m - a_n | > a/3$, and this relation is incompatible with the hypothesis that for sufficiently large values of N the difference $| a_n - a_m |$ is arbitrarily small provided that n and m are both greater than N. Hence there are not two distinct points of accumulation and Cauchy's test has been proved.

The Existence of Limits of Bounded Monotonic Sequences

It is equally easy to see that a bounded monotonic increasing or monotonic decreasing sequence of numbers must possess a limit. In fact, let the sequence be monotonic increasing and ξ be a point of accumulation of the sequence; such a point of accumulation must certainly exist. Then x must be greater than any number of the sequence. For if a number a_l of the sequence were equal to or greater than ξ, every number a_n for which $n > l + 1$ would satisfy the inequality $a_n > a_{l+1} > a_l ? \xi$. Hence all numbers of the sequence, except at most the first $(l + 1)$, would lie outside the interval of length $2(a_{l+1} - \xi)$ the mid-point of which is at the point ξ.

However, this contradicts the assumption that ξ is a point of accumulation. Hence no numbers of the sequence, and, a no points of accumulation lie above x. Thus, if another point of accumulation h exists, we must have $\eta < \xi$. But if we repeat the above argument with η in place of ξ, we obtain $\xi < \eta$, which is a contradiction. Hence only one point of accumulation can exist and the convergence is proved. Naturally, an argument exactly analogous to this one applies to monotonic decreasing sequences.

We can extend our statements about monotonic sequences by including the limiting case in which successive numbers of the sequence are equal, to one another. It is in this case better to speak of monotonic non-decreasing and monotonic non-increasing sequences, respectively. The theorem about the existence of a limit remains valid for such sequences,

Upper and Lower Points of Accumulation

Here we are going to discuss about upper and lower bounds and their properties.

Upper and Lower Bounds of a Set of Numbers

In the construction of subintervals, which led us to a point of accumulation ξ, we must at each step choose a subinterval containing infinitely many points of the set. Had we always chosen the last subinterval which contained an infinite number of points, we should have been led to a certain definite point of accumulation β.

This point of accumulation β is called the upper point of accumulation or the upper limit of the set of numbers and is denoted by $\overline{\lim}$. It is that point of accumulation of the sequence which lies furthest to the right, i.e., it is quite possible that an infinite number of points of the sequence lie above β, but no matter how small the positive number ε may be, there are not an infinite number above $\beta+\varepsilon$.

If we had always chosen in the construction of the first of the intervals containing an infinite number of points of the set, we should again have arrived at a certain definite point of accumulation ξ. This point ξ is called the lower point of accumulation or lower limit of the set and is denoted by $\underline{\lim}$. There may be infinitely many numbers of the set below α, but, no matter how small is the positive number *e*, there are only a finite number below $\alpha - \varepsilon$. The proofs of this facts is left for the students.

Neither the upper limit β nor the lower limit α need belong to the set. For example, for the set of numbers a_{2n} = l/*n*, a_{2n-1} = 2 –, these limits are α = 0 and β = 2, respectively, but the numbers 0 and 2 do not themselves occur in the set.

There is in this example no number of the set above β = 2. In this case, we say that β = 2 is also the upper bound M of the set, according to the following definition: M is called the least upper bound or simply the upper bound of a set of numbers, if: (1) there is no number of the set greater than M, but (2) there is a number of the set greater than M – ε for every positive number ε.

The upper bound may coincide with the upper bound, as in the example above. But the set $a_n = 1 + \frac{1}{n}$ (n = 1, 2,...) shows that this is not necessarily the case. Here M = 2 and β = 1. Every bounded set of numbers has a least upper bound. In fact, let β be the upper limit of the set. Either there are no numbers of the set larger than β or there are such numbers. In the first case, β is the least upper bound, since no numbers are above β and there are numbers arbitrarily close to β below it.

In the second case, let α be a number of the set greater than β. There are only a finite number of numbers of the set equal to or greater than α, since otherwise there would be a point of accumulation above β, which is impossible. We therefore need only choose the greatest of these numbers; it will be the upper bound of the set.

We see that in any case M > β and we recognize the fact: If the upper bound of a set does not coincide with the upper limit, it must belong to the set and is an isolated point of the set. Corresponding statements hold for the lower bound m; it is always equal to or less than a, and if m and a do not coincide, m belongs to the set and is an isolated point of the set.

THEOREMS ON CONTINBUOUS FUNCTIONS

Greatest and Least Values of Continuous functions

A bounded, infinite set of numbers must possess a least upper bound *M* and a largest lower bound *m*. However, as we have seen, these numbers *M* and *m* do not necessarily belong to the set; as we say: The set does not necessarily have a greatest or a least value.

In view of this fact, the following theorem on continuous functions is by no means as obvious as it appears to be to simple intuition: Every function $f(x)$, which is continuous in a closed interval $a \leq x \leq b$, assumes at least once a greatest and a least value or, as we say, it possesses a greatest and a least value.

This may easily be proved as follows: The values assumed by the continuous function $f(x)$ in the interval *a* ? *x* ? *b* form a bounded set of numbers and therefore possess a least upper bound *M*. Otherwise, a sequence of numbers ξ_1,, ξ_2,..., ξ_n,... in our interval would exist for which $f(\xi)$ increases beyond all bounds. This sequence would have at least one point of accumulation $\bar{\xi}$ in the interval. In that case, arbitrarily near to ξ, there would always be numbers ξ_n of our sequence for which the expression $|f(\xi) - f(\xi_n|$ exceeds 1 (and, in fact, is arbitrarily large), that is, the function would be discontinuous at the point $\bar{\xi}$. Thus, a least upper bound *M* exists and hence either there is a point ξ such that $f(\xi) = M$, which would prove the statement, or there is a sequence of numbers x_1, x_2, ???, x_n, ??? in the interval for which

$$\lim_{n\to\infty} f(x_n) = M.$$

According to the formulated principle of the point of accumulation, we can select a subsequence of the numbers ξ_n,, which converges to a limit ξ. Let us call this subsequence ξ_1, ξ_2,..., ξ_n,, so that

$$\lim_{n\to\infty} \xi_n = \xi$$

It is then certain that

$$\lim_{n\to\infty} f(\xi_n) = M.$$

Ott the other hand, the function has been assumed to be continuous in the interval and, in particular, at ξ, whence

$$\lim_{n\to\infty} f(\xi_n) = f(\xi).$$

Hence $f(\xi) = M$. The value M is therefore assumed by the function at a definite point ξ in the interior or on the boundary of the interval, as stated. An exactly similar discussion applies to the least value.

In general, the theorem about the greatest and least values of continuous functions does not remain true unless we expressly assume the interval to be closed, that is, unless we make the hypothesis of continuity also refer to the end-points. For example, the function $y = 1/x$ is continuous in the open interval $0 < x < \xi$. However, it has no greatest value, but has arbitrarily large values near $x = 0$. Similarly, it has no least value, but comes arbitrarily near 0 for sufficiently large values of x without ever assuming the value 0.

The Uniformity of Continuity

As we have already seen and as we shall further see, the continuity of a function $f(x)$ in a closed interval a ? x ? b leaves room for a variety of possibilities, which do not suggest themselves intuitively.

For this reason, we shall give logically rigorous proofs of certain consequences of the idea of continuity which, from a na point of view, seem to be quite obvious. The definition of continuity simply states that there follows from the relation

$$\lim_{n\to\infty} \xi_n = \xi \Rightarrow \lim_{n\to\infty} f(\xi_n) = f(\xi).$$

We can also express this as follows: There correspond to every $\varepsilon > 0$ for each point ξ a number $\delta > 0$ such that $| f(x) - f(\xi) | < \varepsilon$ whenever $| x - \xi | < \delta$, provided that all the numbers x considered lie in the interval $a < x < b$.

For example, in the case of the function $y = cx$ (where $c > 0$), such a number δ is given by the relation $\delta = \varepsilon | c |$. For the function $y = f(x)$, we can find such a number as follows: We assume that $a = 0$ and $b = 1$, and ask ourselves how near to ξ the number x must lie in order that the expression $| x^2 - \xi^2 |$ may be less than ε. For this purpose, we write

$$| x^2 - \xi^2 | = | x - \xi | | x + \xi | \leq | x - \xi | (1 + \xi).$$

Hence, if we choose $\delta < \varepsilon/(1 + \xi)$, we can be sure that $| x^2 - \xi^2 | < \varepsilon$. We see in this example that the number δ found in this way depends not only on ε, but also on the point of the interval at which we are investigating the continuity of the function.

However, if we give up the attempt to make the best possible choice of δ for each ξ, we can eliminate this dependence of δ on ξ. In fact, we need only replace ξ on the right hand side by the number 1 and thus obtain for δ the expression $\varepsilon/2$, which is smaller than the previous expression for δ, but serves equally well for all points ξ.

There arises now the question whether something similar does not hold for every function which is continuous in a closed interval, that is, we enquire whether it may not be possible to determine for each *e a d* = *d* (*e*) which depends only on *e* and not on *x* such that the inequality

$$|f(x) - f(\xi)| < \in$$

holds, provided $|x - \xi| < \delta$ for all values of ξ at the same time (or, better expressed, uniformly with respect to ξ). As a matter of fact, this is possible merely as a consequence of the general definition of continuity without any additional assumptions. This fact, which first attracted attention late in the Nineteenth Century, is called the Theorem of the Uniform Continuity of Continuous Functions.

We shall prove this theorem indirectly, that is, we shall show that the assumption that a function $f(x)$ exists, which in a closed interval $a \leq x \leq b$ is continuous and yet not uniformly continuous, leads us to a contradiction. Uniform continuity means that, if we wish to make the difference $|f(u) - f(v)|$ less than an arbitrarily chosen positive number ε, the numbers *u* and *v* being chosen in the closed interval $a \leq x \leq b$, we need only choose *u* and *v* near enough to one another, namely, at a distance apart which is less than $\delta = \delta(\varepsilon)$; it is immaterial where in the interval the pair of numbers *u*, *v* is chosen.

Now, if $f(x)$ were not uniformly continuous, there would exist a positive (perhaps very small) number *a* with the property: There corresponds to every number δ_n of an arbitrary sequence $\delta_1, \delta_2, \ldots$ of positive numbers tending to zero a pair of values u_n, v_n of the interval for which $|u_n - v_n| < \delta_n$ and $|f(u_n) - f(v_n)| > \alpha$. According to the principle of the point of accumulation, the numbers u_n must have a point of accumulation ξ and the numbers v_n must have the same point of accumulation. If we select an arbitrarily small interval $|x - \xi| < \delta$ about this point ξ, an infinite number of the pairs u_n, v_n will lie in this interval. But this contradicts the assumed continuity of $f(x)$ at the point ξ; in fact, by Cauchy's convergence test, this requires that the points x_2 and x_1 are near enough to ξ.

$$|f(x) - f(x_2)| < a.$$

The uniformity of the continuity is thus proved. The theorem of uniformity of continuity does not hold for intervals which are not closed.

1. Otherwise the point of accumulation need not belong to the interval.

For example, the function l/x is continuous in the half-open interval $0 < x < 1$, but it is not uniformly continuous, because no matter how small the length $\delta(< l)$ of an interval is chosen, the function will take values differing by any fixed number, say 1, in the interval, if only the interval lies near enough to the origin, say, $\delta/2 < x < 3\delta/2$. Of course, the non-uniformity of continuity is due to the fact that in the closed interval $0 \leq x \leq 1$ the function possesses at the origin a discontinuity.

Intermediate Value Theorem

A function $f(x)$, continuous in a closed interval $a \leq x \leq b$, which is negative for $x = a$ and positive for $x = b$ (or conversely), assumes the value 0 at least once in the interval. Geometrically speaking, this theorem is trivial, since it merely states that a curve, which begins below the x-axis and ends above it, must cut the axis somewhere in between. Analytically speaking, the theorem is very easily proved. There are in the interval an infinite number of points for which $f(x) < 0$; in fact, due to the continuity of the function, this is true for an entire interval beginning at the point a. The set consisting of those points x for which $f(x) < 0$ has a least upper bound x which is greater than a.

Since there are points in every neighbourhood of x for which $f(x) < 0$, we must have $f(x) < 0$ (whence, in particular, x ? b).

However, it is impossible that $f(\xi) < 0$, (whence, in particular, $\xi < \beta$). However, it is impossible that $f(\xi) < 0$, because then $f(\xi)$ would be negative in a sufficiently small neighbourhood of ξ, including values of $x > \xi$, in contradiction to the assumption that ξ is the upper bound of the values of ξ for which $f(x) < 0$. Hence $f(\xi) = 0$ and our assertion is proved. A slight generalization of our theorem is: If we assume that $f(a) = \alpha$ and $f(b) = \beta$ and if μ is any value between α and β, the continuous function $f(x)$ assumes the value $f(\mu)$ at least once in the interval, because the continuous function

$$\phi(x) = f(x) - \mu$$

will have different signs at the two ends of the interval and will therefore assume the value 0 somewhere inside it.

The Inverse of a Continuous Monotonic Function

If the continuous function $y = f(x)$ is monotonic in the interval $a < x < b$, it will assume each value between $f(a)$ and $f(b)$ once and only once; hence, if y describes the closed interval between the values $\alpha = f(a)$ and $\beta = f(b)$, there will correspond to each value of y exactly one value of x. We can therefore think of x as a single-valued function of y in this interval, i.e., the function $y = f(x)$ has a unique inverse. We assert that this inverse function $x = \phi(y)$ is also a continuous, monotonic function of y, as y varies within the interval between α and β.

The monotonic character of the inverse function $x = \phi(y)$ is obvious. In order to prove its continuity, we observe that it follows from the monotonic character of $f(x)$ that

$$|f(x_2) - f(x_1)| = |y_2 - y_1)| > 0,$$

provided that x_1 and x_2 are distinct numbers of the interval. If h is a positive number less than $b - a$, the function

$$|f(x+h) - f(x)|$$

is continuous in the closed interval a ? x ? $b - h$, whence it has a least value

$$|(\xi+h)-f(\xi)|=a(h),$$

which by our preceding remark is not zero.* We conclude from this that, if x_1 and x_2 are two points in the interval for which $|x_1 - x_2| < h$, then

$$|f(x_1)-f(x_2)|\geq a(h).$$

However, this implies the continuity of the inverse function, because, if $|y_1 - y_2|$ falls below the positive number $\alpha(h)$, then we must have $|x_1 - x_2| < h$, whence, if a positive number e is given, we need only choose δ equal to $\alpha(\varepsilon)$ in order to ensure that for all values y for which $|y_1 - y_2| < \delta$ it is also true that $|\phi(y_1) - \phi(y_2)| < \varepsilon$.

* On account of the continuity of $f(x)$, naturally, $\alpha(h)$ itself tends to 0 with h.

Hence we have established the theorem: If the function $y = f(x)$ is continuous and monotonic in the interval $a < x < b$ and $f(a) = \alpha$, $f(b) = \beta$, then it has a single-valued inverse function $x = \phi(y)$, $\alpha < y < \beta$ and this inverse function is also continuous and monotonic.

Further Theorems on Continuous Functions

We leave it to the reader to prove the following almost trivial fact: A continuous function of a continuous function is itself continuous, i.e., if $\phi(x)$ is a function, continuous in the interval $a < x < b$, its functional values lie in the interval $\alpha < \phi < \beta$ and, in addition, if $f(\phi)$ is a continuous function of ϕ in this last interval, then $f(\phi(x))$ is a continuous function of x in the interval $a < x < b$.

It has already been mentioned that the sum, difference and product of continuous functions are themselves continuous and that the quotient of continuous functions is continuous, provided that the denominator remains different from zero.

Polar Co-ordinates

Functions can be represented functions geometrically by means of curves. However, it is useful to recall that analytical geometry follows the reverse procedure; it begins with a curve given by some geometrical property and represents this curve by a function, for example, by a function which expresses one of the coordinates of a point of the curve in terms of the other coordinate.

This point of view naturally leads us to consider, apart from rectangular coördinates, other systems of coordinates which may be better suited for the representation of curves which are given geometrically,

The most important example is that of the polar coordinates r, θ, connected with the rectangular co-ordinates x, y of a point P by the equations

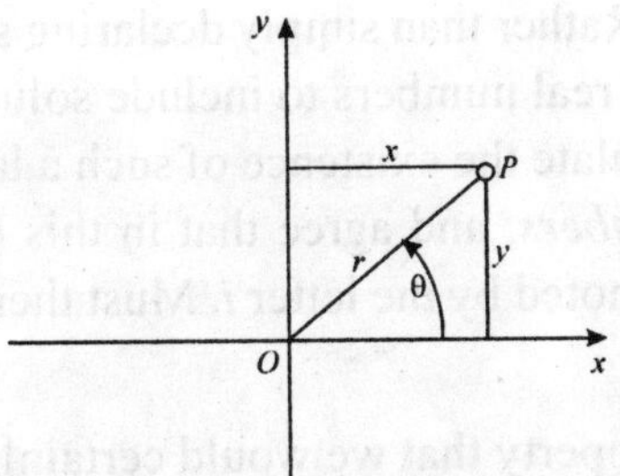

Fig. Polar Coordinates

$$x = r\cos\theta, y = r\sin\theta, r^2 = x^2 + y^2, \tan\theta = \frac{y}{x},$$

and the geometrical interpretation of which follows from Figure 1.

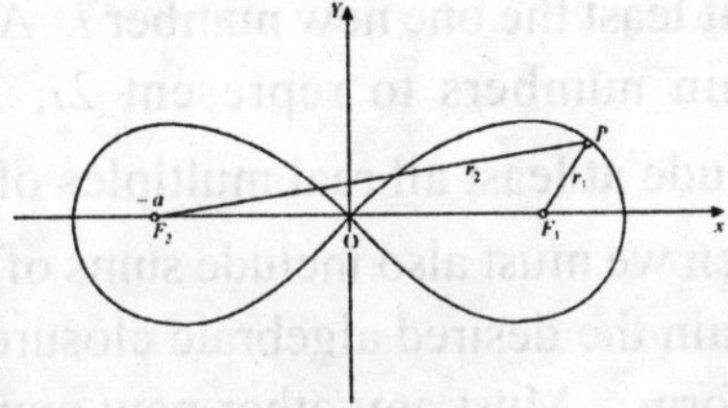

Fig. Graph

Consider as an example the lemniscate. It is geometrically defined as the locus of all points for which the product of the distances r_1 and r_2 from the fixed points P_1 and P_2 with the rectangular co-ordinates $x = a$, $y = 0$ and $x = -a$, $y = 0$, respectively, has the constant value a. Since

$$r_1^2 = (x - a)^2 + y^2, r_2^2 = (x + a)^2 + y^2,$$

a simple calculation yields the equation of the lemniscate in the form

$$(x^2 + y^2)^2 - 2a^2 (x^2 - y^2) = 0.$$

If we now introduce polar coordinates, we obtain

$$r^4 - 2a^2r^2 (\cos^2\theta - \sin^2\theta) = 0;$$

now, division by r and use of a simple trigonometric formula yields

$$r^2 = 2a^2 \cos 2\theta.$$

Thus, we see that the equation of the lemniscate in polar coordinates is simpler than in rectangular coordinates.

COMPLEX NUMBERS

Does the process of creation of new "numbers" end with the real numbers? In spite of creating these successively larger number systems in order to have solutions to more and more equations, there are still equations that have no solution, the simplest being the

innocent-looking $x^2 + 1 = 0$. Rather than simply declaring such equations to be unsolvable, why not extend the system of real numbers to include solutions to equations of this type? In order to do this, let us postulate the existence of such a larger system of numbers, which we will call the *complex numbers*, and agree that in this larger system, a solution to the equation $x^2 + 1 = 0$ will be denoted by the letter i. Must there be any other numbers in such a number system?

Algebraic closure is a property that we would certainly expect any set that calls itself a number system to have. This means that adding and multiplying any two such numbers should result in another such number.

In particular, then, since this new number system is an extension of (that is, a more inclusive set than) the real numbers, our new complex number system must contain all the real numbers, in addition to at least the one new number i. Algebraic closure then requires that our new system contain numbers to represent $2i$, $3i$, $4.783i$, $-8.\dot{6}i$, and even $\sqrt{2}i$ and πi. So we must include at least all real multiples of i. (These are called the *pure imaginary* numbers.) But then we must also include sums of real numbers with the various multiples of i, in order to retain the desired algebraic closure. That is, we must include at least the "numbers" of the form . Must any other new numbers be included in our new system of complex numbers?

Consider a simple quadratic equation $x^2 + 1 = 0$. There is no real number which satisfies this equation. So there was a need to find a system which could answer to this problem. Euler used the symbol 'i' to denote $\sqrt{-1}$ to solve the above equation. Complex number system consists of the set of all ordered pairs of real numbers (a, b) denoted by $a + ib$, where $i = \sqrt{-1}$.

IMAGINARY NUMBER

Square root of a negative number is known as an imaginary number. $\sqrt{-a}$, $a > 0$ is an imaginary number.

or

A number whose square is negative is known as an imaginary number.

$$\sqrt{-1}, \sqrt{-2}, \sqrt{-3}, \ldots$$

The symbol i,

$$i = \sqrt{-1}$$

We write,

$$\sqrt{-3} = \sqrt{3}\sqrt{-1} = \sqrt{3}i = i\sqrt{3}$$

$$\sqrt{-4} = 2i = i2$$

Powers of i,

$$i = \sqrt{-1}$$
$$i^2 = -1$$
$$i^3 = (i)^2\, i = -i$$
$$i^4 = (-1)^2 = 1$$

Let $p > 0$ be a positive integer such that $p > 4$. Let q be the quotient and r be the remainder, when p is divided by 4.

i.e., $p = 4q + r$ where $0 \leq r < 4$.

Hence $i^p = i^{4q+r} = i^{4q}.i^r = (i^4)^q.i^r = 1^q.i^r = i^r \qquad (\because\ i^4 = 1)$

General Form of Complex Number

If x and y are real numbers, then $x + iy$ is called a complex number.

x is called the real part and y is called the imaginary part.

The complex number $x + iy$ is also written as an ordered pair (x, y) and is denoted by z.

i.e., $z = x + iy$

The positive value $\sqrt{x^2 + y^2}$ is called the modulus of Z and is denoted by $| Z |$.

i.e., $Z = x + iy$, then $| Z | = \sqrt{x^2 + y^2}$

In $Z = x + iy$, if $y = 0$, then Z is purely real and if $x = 0$, then Z is imaginary.

Example: $2 + 4i$, $6 - 5i$, $3 + \sqrt{2i}, \sqrt{3} - 2i$ are complex numbers.

Note: Set of real numbers is a proper subset of the set of Complex numbers.

Equality of Complex Numbers

Two complex numbers are equal iff their corresponding real parts and imaginary parts are separately equal.

i.e., $a + ib = x + iy \Leftrightarrow a = x$ and $b = y$

Sum of two Complex Numbers

If $Z_1 = a + ib$ and $Z_2 = x + iy$, then we define their sum as

$Z_1 + Z_2 = (a + ib) + (x + iy)$

$= (a + x) + i(b + y)$ which is a complex number.

Negative of a Complex Number

If $Z = a + ib$, then Z is called the negative of Z.

$$-Z = -(a + ib)$$
$$= -a - ib$$

Additive Identity of the Complex Number

The complex number $0 + i0$ is the additive identity for the set of complex numbers.

$0 + i0$ is called the additive identity for the complex number.

$$Z + (0 + i\,0) = a + i\,b + (0 + i\,0)$$
$$= (a + 0) + i(b + 0)$$
$$= a + i\,b$$

Additive Inverse of a Complex Number

Let $Z = a + i\,b$ and $Z' = x + iy$ be the additive inverse of Z, then

$$(a + ib) + (x + iy) = 0 + i0$$
$$(a + x) + i(b + y) = 0 + i0$$
$$a + x = 0 \text{ and } b + iy = 0$$
$$\Rightarrow x = -a \text{ and } y = -b$$

$\therefore$ Additive inverse of $a + ib$ is $-a - ib$.

Product of Two Complex Numbers

Let $Z_1 = a + ib$ and $Z_2 = c + id$, then

$$Z_1Z_2 = (a + ib)(c + id)$$
$$= ac + iad + ibc + i^2db$$
$$= ac + i(ad + bc) - bd \because (i^2 = -1)$$
$$= (ac - bd) + i(ad + bc) \text{ which is a}$$

Multiplicative Identity of Complex Numbers

Let $Z = a + i\,b$ and $Z' = x + iy$, then

$$(a + ib)(x + iy) = a + ib$$
$$(ax - by) + i(ay + bx) = a + ib$$
$$ax - by = a$$

and
$$ay + bx = b$$

Solving (*i*) and (*ii*), we have

$$x = 1, y = 0$$

Multiplicative identity is $1 + i0$.

Conjugate Complex Numbers

If $Z = a + ib$ is a complex number, then $a - ib$ is called the complex conjugate of $a + ib$ and the conjugate is denoted by $\overline{Z}$.

Remark: The sum and product of two conjugate numbers is always real.

For $Z = a + ib, \overline{Z} = a - ib$

$$Z + \overline{Z} = (a + ib) + (a - ib)$$

$$= 2a + i(0) = 2a$$

$$Z.\overline{Z} = (a + ib)(a - ib)$$

$$= a^2 - (ib)^2$$

$$= a^2 - (-1)b^2 = a^2 + b^2$$

Hence $Z + \overline{Z}$ and $Z.\overline{Z}$ are both real.

Quotient of two non-zero Complex numbers

If $Z_1 = a + ib$ and $Z_2 = c + id$ are two complex numbers, then $\frac{Z_1}{Z_2}$, quotient is defined as

$$\frac{Z_1}{Z_2} = \frac{a+ib}{c+id} = \frac{a+ib}{(c+id)} \cdot \frac{c-id}{(c-id)}$$

$$\frac{(ac+bd)+i(bc-ad)}{c^2+d^2}$$

$$\left(\frac{ac+bd}{c^2+d^2}\right) + i\left(\frac{bc-ad}{c^2+d^2}\right)$$

If $Z_1 = 0 + i0$, then quotient is also $0 + i0$ if $Z_2 \neq 0$.

Reciprocal of a non-zero complex number or multiplicative inverse of a non-zero complex number

Let $Z = a + ib$, then $\frac{1}{Z} = \frac{1}{a+ib}$

$$\frac{1}{Z} = \frac{1}{a+ib}\frac{(a-ib)}{(a-ib)}$$

$$= \frac{a-ib}{a^2-b^2}$$

$$= \left(\frac{a}{a^2+b^2} - \frac{ib}{a^2+b^2}\right)$$

Properties of Complex Numbers

If Z, Z_1 and Z_2 are complex numbers, then

(i) $|Z| = |\overline{Z}|$

$$Z = x+iy \Rightarrow |Z| = \sqrt{x^2+y^2}$$

$$\overline{Z} = x-iy \Rightarrow |\overline{Z}| = \sqrt{x^2+(-y)^2} = \sqrt{x^2+y^2}$$

(ii) $Z.\overline{Z} = |Z|^2$

$$(x + iy)(x - iy) = x^2 + y^2$$

$$= \left[\sqrt{x^2+y^2}\right]^2$$

$$= |Z|^2$$

(iii) $|Z_1.Z_2| = |Z_1|.|Z_2|$

Let $Z = a + ib$, $Z_2 = c + id$

so that $|Z_1| = \sqrt{a^2+b^2}, |Z_2| = \sqrt{c^2+d^2}$

$$Z_1.Z_2 = (a + ib)(c + id)$$

$$= (ac - bd) + i(bc + ad)$$

$$|Z_1 Z_2| = \sqrt{(ac - bd)^2 + (bc + ad)^2}$$

$$= \sqrt{a^2c^2+b^2d^2+b^2c^2+a^2d^2}$$

$$= \sqrt{a^2(c^2+d^2)+b^2(c^2+d^2)}$$

$$= \sqrt{a^2+b^2}\sqrt{c^2+d^2}$$

$$= |Z_1|.|Z_2|$$

(iv) $|kZ| = |k||Z| \forall k \in R$

Let $Z = a + ib$

$$k|Z| = k(a + ib) = ka + ikb$$

$$k|Z| = \sqrt{k^2a^2 + k^2b^2}$$

$$= |k|\sqrt{a^2 + b^2}$$

$$= |k||Z|$$

(v) $Z + \overline{Z} = 2R_e(Z), Z - \overline{Z} = a - ib.$

Let Z = a + ib then $\overline{Z}$ = a – ib.

$$Z + \overline{Z} = (a + ib) + (a - ib) Z - \overline{Z} = (a + ib) - (a - ib)$$

$$= 2a + i0 = (a - a) + i(b + b)$$

$$= 2a = 0 + i(2b)$$

$$= 2R_e(Z) = 2I_m(Z)$$

(vi) $|R_e(Z)| \leq |Z|, |I_m(\overline{Z})| \leq |Z|$

Let $Z = a + ib$, then $|Z| = \sqrt{a^2 + b^2}$

$$R_e(Z) = a$$

$a \leq \sqrt{a^2 + b^2}$ $\because$ (Both terms in the radical sign are positive)

Hence $R_e(Z) \leq |Z|$

$$I_m(Z) = b$$

$$b \leq \sqrt{a^2 + b^2}$$

$$I_m(Z) \leq |Z|$$

(vii) $|Z_1 + Z_2|^2 = +|Z_1|^2 + |Z_2|^2 + 2R_e(Z_1.\overline{Z}_2)$

LHS = $(Z_1 + Z_2)(\overline{Z_1 + Z_2})$

$$= (Z_1 + Z_2)(\overline{Z}_1 + \overline{Z}_2)$$

$$= Z_1\overline{Z}_1 + Z_1\overline{Z}_2 + Z_2\overline{Z}_1 + Z_2\overline{Z}_2$$

$$= Z_1\overline{Z}_1 + Z_2\overline{Z}_2 + Z_1\overline{Z}_2 + Z_2\overline{Z}_1$$

$$= |Z_1|^2 + |Z_2|^2 + 2R_e(Z_1.\overline{Z}_2)$$

= RHS

(viii) $|Z_1 + Z_2|^2 + |Z_1 - Z_2|^2$

$$= 2(|Z_1|^2 + |Z_2|^2)\,|Z_1 + Z_2|^2 + |Z_1 - Z_2|^2$$

$$= |Z_1|^2 + |Z_2|^2 + 2R_e(Z_1.\overline{Z}_2) + |Z_1|^2 + |Z_2|^2 + 2R_e(Z_1(-\overline{Z}_2))$$

$$= 2|Z_1|^2 + 2|Z_2|^2 + 2R_e(Z_1.\overline{Z}_2) - 2R_e(Z_1\overline{Z}_2)$$

$$= 2|Z_1|^2 + 2|Z_2|^2$$

(ix) $(\overline{Z}) = Z$

Let $Z = a + ib$, then

$$\overline{Z} = \overline{(a - ib)} = (a + ib) = Z$$

(x) $(Z - \overline{Z})$ is zero or imaginary.

Let $Z = a + ib$ then $\overline{Z} = a - ib$

$$Z - \overline{Z} = (a + ib) - \overline{(a - ib)}$$

$$= (a + ib) - a - ib)$$

$$= 2ib$$

which is zero when $b = 0$ and purely imaginary when $b \neq 0$.

(xi) $\overline{Z_1 + Z_2} = (\overline{Z}_1 + \overline{Z}_2)$

Let $Z_1 = a + ib$, $Z_2 = c + id$

So that $\overline{Z_1} = a - ib$ and $\overline{Z_2} = c - id$

$$Z_1 + Z_2 = (a + c) + i(b + d)$$

$$\overline{Z_1 + Z_2} = (a + c) - i(b + d)$$

$$= (a - ib) + (c - id)$$

$$= \overline{Z_1} + \overline{Z_2}$$

(xii) $\overline{Z_1.Z_2} = \overline{Z}_1.\overline{Z}_2$

Let $Z_1 = a + ib$, $Z_2 = c + id$

So that $\overline{Z}_1 = a - ib$ and $\overline{Z}_2 = c - id$

$$Z_1.Z_2 = (ac - bd) + i(ad + bc)$$

$$\overline{Z_1.Z_2} = (ac - bd) - i(ad + bc)$$

$$= c(a - ib) - id(a - ib)$$

$$= (a - ib)(c - id)$$

$$= \overline{Z}_1.\overline{Z}_2$$

(xiii) $|Z_1 + Z_2| \leq |Z_1| + |Z_2|$

$$= |Z_1 + Z_2|^2 = (Z_2 + Z_2)(\overline{Z_1 + Z_2}) \quad (\because |Z|^2 = Z.\overline{Z})$$

$$= (Z_1 + Z_2)(\overline{Z}_1 + \overline{Z}_2) \quad (\because \overline{Z_1 + Z_2} = \overline{Z}_1 + \overline{Z}_2)$$

$$= Z_1.\overline{Z}_1 + Z_1.\overline{Z}_2 + Z_2.\overline{Z}_1 + Z_2.\overline{Z}_2$$

$$= |Z_1|^2 + |Z_2|^2 + Z_1.\overline{Z}_2 + (\overline{Z_1 \overline{Z}_2})$$

$$= |Z_1|^2 + |Z_2|^2 + 2R_e(Z_1.\overline{Z}_2) \quad (\because Z + \overline{Z} = 2R_e(Z))$$

$$\leq |Z_1|^2 + |Z_2|^2 + 2|Z_1.\overline{Z}_2| \quad (\because R_e(Z) \leq |Z|)$$

$$= |Z_1|^2 + |Z_2|^2 + 2|Z_1|.|\overline{Z}_2|$$

$$= |Z_1|^2 + |Z_2|^2 + 2|Z_1|.|\overline{Z}_2| \quad (\because |\overline{Z}_2| = |Z_2|)$$

$$= [|Z_1| + |Z_2|]^2$$

$$\therefore \quad |Z_1 + Z_2|^2 \leq [|Z_1| + |Z_2|]^2$$

$$|Z_1 + Z_2| \leq |Z_1| + |Z_2|$$

(xiv) $|Z_1 - Z_2| \geq |Z_1| - |Z_2|$

$$|Z_1| = |Z_1 - Z_2 + Z_2| \leq |Z_1 - Z_2| + |Z_2|$$

$$|Z_1| - |Z_2| \leq |Z_1| - |Z_2|$$

$$|Z_1 - Z_2| \geq |Z_1| - |Z_2|$$

Graphical Representation of Complex Numbers

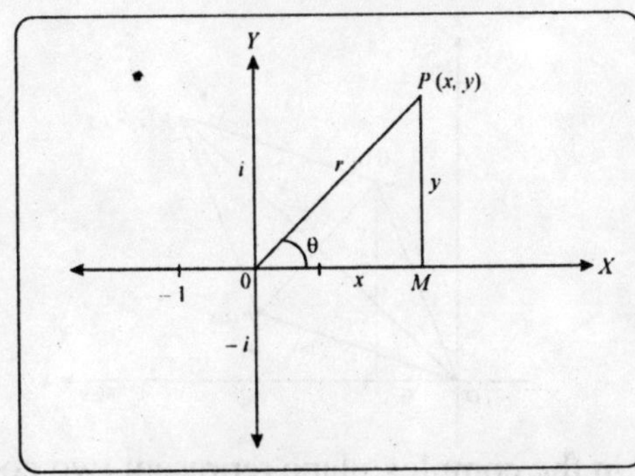

The complex number $Z = x + iy$ may be represented graphically by the point P whose rectangular co-ordinates are (x, y). Thus each point in the plane is associated with a complex number. In the figure, P defines $Z = x + iy$. It is customary to choose x-axis as real axis and y-axis as imaginary axis.

Draw PM perpendicular to x-axis.

In ΔPOM,

$$P\hat{O}M = \theta, OM = x, PM = y, OP = r$$

$$r = \sqrt{x^2 + y^2}, x = r\cos\theta, y = r\sin\theta, \tan\theta = \frac{y}{x}$$

Here the positive value of r is called the modulus of Z or absolute value of Z. ? is called the amplitude or argument of Z.

$$\therefore \quad r = |Z| = |x + iy| = \sqrt{x^2 + y^2}$$

$$\theta = ampZ = \arg Z = \tan^{-1}\frac{y}{x}$$

Geometrical Representation of a Complex Number, Argand Diagram

Since every complex number $z = x + iy$ is an order pair of real numbers (x, y), it can therefore be represented by a point $P(x, y)$ in the xy plane by taking the real part along the x-axis and the imaginary part along the y-axis.

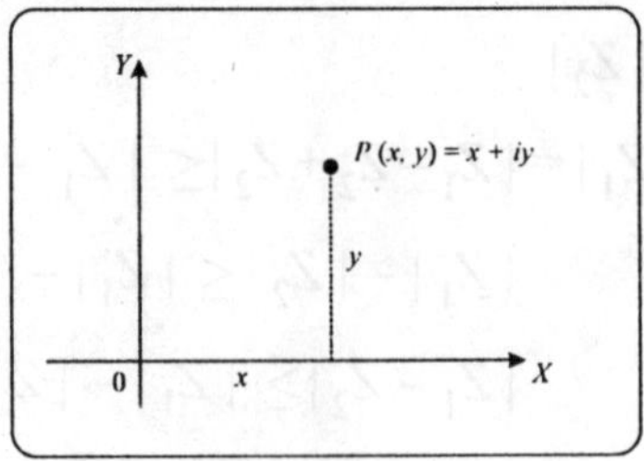

This representation of a complex number in a plane is called the argand diagram.

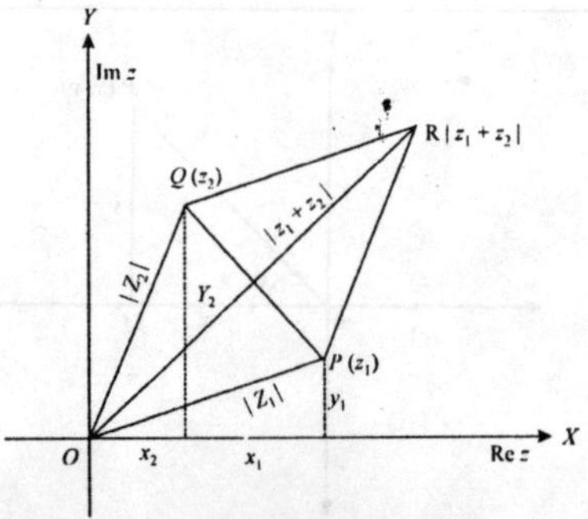

Let points P and Q in the complex plane represent two complex numbers $z_1 = x_1 + iy_1$ and $z_2 = x_2 + iy_2$, respectively, as shown in figure.

Join the origin O with the points P and Q and complete the parallelogram $OPRQ$.

It is clear from the figure that the coordinates of R are $(x_1 + x_2, y_1 + y_2)$ and it represents the complex number $(x_1 + x_2) + i\,(y_1 + y_2)$, i.e., $z_1 + z_2$.

The absolute values of z_1, z_2 and $z_1 + z_2$ are geometrically given by

$$|Z_1| = OP, |Z_2| = OQ = PR \text{ and } |Z_1 + Z_2| OR.$$

We know that the sum of any two sides of a triangle is greater than the third. Hence, in ?ORP, we have

$$OR \leq OP + PR \text{ implies } |Z_1 + Z_2| \leq |Z_1| + |Z_2|$$

with the equality holding only when O, P, Q are lying in a straight line. That is why this inequality for the absolute values of complex numbers is called the triangle inequality.

The triangle inequality can be extended to n complex numbers by finite induction, i.e., for any n complex numbers $z_1, z_2 \ldots\ldots, z_n$, we obtain

$$|Z_1 + Z_2 + \ldots + Z_n| \leq |Z_1| + |Z_2| + \ldots + |Z_n|$$

Polar form of a Complex Number

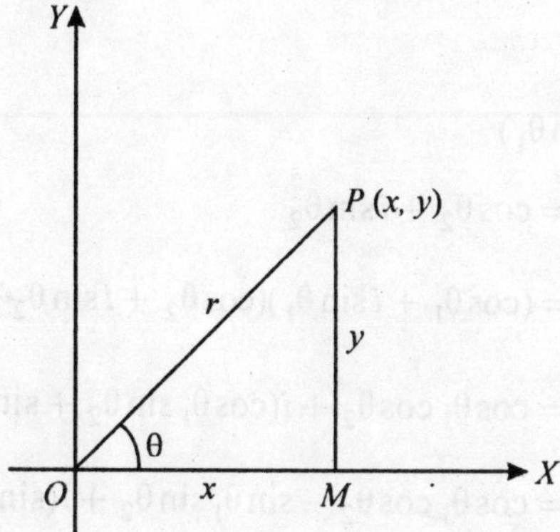

Let $z = x + iy$ denote the point $P(x, y)$ in the xy plane.

Let $OP = r$ and $M\hat{O}P = \theta$.

Then from the figure, we have

$$\cos\theta = \frac{x}{r} \text{ and } \sin\theta = \frac{y}{r}$$

$$\therefore\ z = x + iy = r\cos\theta + ir\sin\theta$$

$$z = r(\cos\theta + i\sin\theta)$$

$$r^2 = x^2 + y^2 \Rightarrow r = \sqrt{x^2 + y^2}$$

$z = r\{\cos\theta + i\sin\theta\}$ is called the polar form of the complex number. $x + iy$ (which is in the cartesian form) where $r = \sqrt{x^2 + y^2}$ is called the modulus of the complex number z denoted by $|z|$ and $= \tan^{-1} y/x$ is called the amplitude or argument of the complex number z denoted by amp (z) or arg(z).

The value of θ is such that $-\pi < \theta \leq \pi$, is called the principal value of the amplitude.

The general value of the amplitude is $\theta + 2n\pi$, where n is positive or negative integer and is the principal value of the amplitude.

To find the principal value of (amplitude) from the equation $\cos\theta = x/r$ and $\sin\theta = y/r$, the following table is useful.

cos θ	sin θ	θ
+ ve	+ ve	α (say)
+ ve	– ve	– α (say)
– ve	+ ve	(π – α)
– ve	– ve	– (π – α)

De Moivre's Theorem

(i) Let $Z_1 = (\cos\theta_1 + i\sin\theta_1)$

$$Z_2 = \cos\theta_2 + i\sin\theta_2$$

$$Z_1.Z_2 = (\cos\theta_1 + i\sin\theta_1)(\cos\theta_2 + i\sin\theta_2)$$

$$= \cos\theta_1\cos\theta_2 + i(\cos\theta_1\sin\theta_2 + \sin\theta_1\cos\theta_2) + i^2\sin\theta_1\sin\theta_2$$

$$= \cos\theta_1\cos\theta_2 - \sin\theta_1\sin\theta_2 + i(\sin\theta_1\cos\theta_2 + \cos\theta_1\sin\theta_2)$$

$$= \cos(\theta_1 + \theta_2) + i\sin(\theta_1 + \theta_2)$$

(ii) Let $Z_1 = (\cos\theta_1 + i\sin\theta_1)$

$$Z_2 = (\cos\theta_2 - i\sin\theta_2)$$

$$Z_1.Z_2 = (\cos\theta_1 + i\sin\theta_1)(\cos\theta_2 - i\sin\theta_2)$$

$$= (\cos\theta_1\cos\theta_2 + \sin\theta_1\sin\theta_2) + i(\sin\theta_1\cos\sin\theta_2 - \cos\theta_1\sin\theta_2)$$

$$= \cos(\theta_1 - \theta_2) + i\sin(\theta_1 - \theta_2)$$

(iii) Let $Z_1 = (\cos\theta_1 - i\sin\theta_1)$

$$Z_1.Z_2 = \cos\theta_1\cos\theta_2 - \sin\theta_1\sin\theta_2 - i(\sin\theta_1\cos\theta_2 + \cos\theta_1\sin\theta_2)$$

$$= \cos(\theta_1+\theta_2) - i\sin(\theta_1+\theta_2)$$

Statement of DeMoivre's Theorem

If 'n' be any rational number, positive or negative, then

$$[\cos\theta + i\sin\theta]^n = \cos n\theta + i\sin n\theta$$

Proof:

Part I: When 'n' is any integer, positive or negative then

$$[\cos\theta + i\sin\theta]^n = \cos n\theta + i\sin n\theta$$

Part II: When 'n' is a fraction.

Part I:

Case 1: When n is a positive integer.

By actual multiplication, we have

$$(\cos\theta_1 + i\sin\theta_1)(\cos\theta_2 + i\sin\theta_2) = \cos\overline{\theta_1+\theta_2} + i\sin\overline{\theta_1+\theta_2}$$

Again $(\cos\theta_1 + i\sin\theta_1)(\cos\theta_2 + i\sin\theta_2)(\cos\theta_3 + i\sin\theta_3)$

$$= [\cos(\theta_1+\theta_2) + i\sin(\theta_1+\theta_2)][\cos\theta_3 + i\sin\theta_3]$$

$$= \cos(\theta_1+\theta_2+\theta_3) + i\sin(\theta_1+\theta_2+\theta_3)$$

Similarly by the method of induction,

$$(\cos\theta_1 + i\sin\theta_1)(\cos\theta_2 + i\sin\theta_2) \,.......\, (\cos\theta_n + i\sin\theta_n)$$

$$= \cos(\theta_1+\theta_2+\theta_3+...+\theta_n) + i\sin(\theta_1+\theta_2+\theta_3+...+\theta_n)$$

In this result, if we put $\theta_1 = \theta_2 = \theta_3 = ... = \theta_n = \theta$,

then we obtain, $(\cos\theta + i\sin\theta)^n = \cos n\theta + i\sin n\theta$

Case 2: When 'n' is a negative integer.

Let $n = -m$ and $m > 0$

Now $(\cos\theta + i\sin\theta)^n = (\cos\theta + i\sin\theta)^{-m}$

$$= \frac{1}{(\cos\theta + i\sin\theta)^m}$$

$$= \frac{1}{\cos m\theta + i\sin m\theta} \text{ (by case 1)}$$

$$= \frac{\cos m\theta - i\sin m\theta}{(\cos m\theta - i\sin m\theta)(\cos m\theta - i\sin m\theta)}$$

$$= \frac{\cos m\theta - i\sin m\theta}{\cos^2 m\theta + \sin^2 m\theta}$$

$$= \cos m\theta - i\sin m\theta$$
$$= \cos(-m\theta) + i\sin(-m\theta)$$

$$= \cos n\theta + i\sin n\theta$$

Part II:

Case 3: When 'n' is a fraction.

Let $n = \frac{p}{q}$

where 'p' is any positive integer and 'q' is any integer.

From Part I,

$$(\cos\theta + i\sin\theta)^p = (\cos p\theta + i\sin p\theta)$$

$$= \cos\frac{p}{q}.q\theta + i\sin\frac{p}{q}.q\theta$$

$$= \left[\cos\frac{p}{q}\theta + i\sin\frac{q}{p}\theta\right]^q$$

Now taking q^{th} root of both sides,

$$(\cos\theta + i\sin\theta)^{\frac{p}{q}} \text{ has one of its value } = \cos\frac{p}{q}\theta + i\sin\frac{p}{q}\theta$$

Now writing $\frac{p}{q} = n$,

$$= (\cos\theta + i\sin\theta)^n = \cos n\theta + i\sin n\theta$$

Hence the theorem.

Corollary:

$$(\cos\theta + i\sin\theta)^{-m} = \cos(-m\theta) + i\sin(-m\theta)$$

$$= \cos m(-\theta) + i\sin m(-\theta)$$

$$= [\cos(-\theta) + i\sin(-\theta)]^m =.$$

$$= (\cos\theta - i\sin\theta)^m$$

Corollary:

$$(\cos\theta + i\sin\theta)^{mn} = \cos mn\theta + i\sin mn\theta$$

$$= (\cos m\theta + i\sin m\theta)^n$$

$$= (\cos n\theta + i\sin n\theta)^m$$

Corollary:

$$\frac{1}{\cos\theta + i\sin\theta} = (\cos\theta + i\sin\theta)^{-1} = \cos\theta - i\sin\theta$$

Find the qth Roots of a Complex Number

One of the most important applications of De Moivre's theorem is to find the qth roots of a complex number.

Let $z = x + iy$ be a complex number.

Let $z = r\{\cos\theta + i\sin\theta\}$ be its polar form.

And $z^{\frac{1}{q}}$ be the qth root of z.

We have $z = r\{\cos(2n\pi + \theta) + i\sin(2n\pi + \theta)\}$

[2n? + ? is the general amplitude]

$$\therefore\ z^{\frac{1}{q}} = [r\{\cos(2n\pi + \theta) + i\sin(2n\pi + \theta)\}]^{\frac{1}{q}}$$

$$r^{\frac{1}{q}}\left\{\cos\left(\frac{2n\pi + \theta}{q}\right) + i\sin\left(\frac{2n\pi + \theta}{q}\right)\right\}$$ (using De Moivre's theorem

where $n = 0, 1, 2, 3, \ldots\ldots\ldots, q - 1, \ldots.$

Since we are finding the qth roots of a complex number, $z^{1/q}$ should satisfy the equation $x^q = z$ which is an algebraic equation of degree q and should have exactly q roots.

$$\therefore\ z^{\frac{1}{q}} = r^{\frac{1}{q}}\left\{\cos\left(\frac{2n\pi + \theta}{q}\right) + i\sin\left(\frac{2n\pi + \theta}{q}\right)\right\}$$

where $n = 0,1,2,3,\ldots\ldots\ldots, q - 1$

By giving values for $n = 0, 1, 2, 3, \ldots\ldots\ldots, q - 1$, we get q distinct q^{th} roots of z.

Cube Roots of Unity

Let $z^3 = 1 = 1\{\cos 0 + i\sin 0\}$

$$z^3 = 1\{\cos(2n\pi + 0) + i\sin(2n\pi + 0)\}$$

where 2*n*? + 0 is the general amplitude.

$$z = 1^{\frac{1}{3}}\{\cos 2n\pi + i\sin 2n\pi\}^{\frac{1}{3}}$$

$$z = \cos\frac{2n\pi}{3} + i\sin\frac{2n\pi}{3}, \text{ where n = 0, 1, 2...}$$

when $n = 0$, $z = \cos 0 + i \sin 0 = 1$

$$n = 1, z = \cos\frac{2\pi}{3} + i\sin\frac{2\pi}{3} = \frac{-1}{2} + i\frac{\sqrt{3}}{2}$$

$$n = 2, z = \cos\frac{4\pi}{3} + i\sin\frac{4\pi}{3} = -\frac{1}{2} - i\frac{\sqrt{3}}{2}$$

$\therefore$ The cube roots of unity are $1, \frac{-1}{2} + i\frac{\sqrt{3}}{2}, \frac{-1}{2} - \frac{i\sqrt{3}}{2}$ which are usually denoted by 1, ?

Note 1:

$$\omega^3 = 1, (\omega^2)^3 = 1$$

Note 2:

$$1 = \omega^3 = \omega^6 = \omega^9 = \omega^{12} = ...$$

$$\omega = \omega^4 = \omega^7 = \omega^{10} = \omega^{13} = ...$$

$$\omega^2 = \omega^5 = \omega^8 = \omega^{11} = \omega^{14} = ...$$

Note 3:

We have $\omega^3 = 1$

$\Rightarrow \quad (\omega - 1)(\omega^2 + \omega + 1) = 0$

$\Rightarrow \quad \omega^2 + \omega + 1 = 0$

Note 4:

$$\frac{1}{\omega} = \omega^2 \text{ and } \frac{1}{\omega^2} = \omega$$

Fourth roots of unity

Let $z^4 = 1 = 1\ \{\cos 0 + i \sin 0\}$

Replacing the amplitude 0 by general amplitude 2*n*? + 0, we get

$$z^4 = \cos 2n\pi + i\sin 2n\pi$$

$$z = (\cos 2n\pi + i \sin 2n\pi)^{\frac{1}{4}}$$

$$z = \cos\frac{n\pi}{2} + i\sin\frac{n\pi}{2}, n = 0,1,2,3...$$

when $n = 0$, $z = \cos 0 + i \sin 0 = 1$

$$n = 1, z = \cos\frac{\pi}{2} + i\sin\frac{\pi}{2} = i$$

$$n = 2, z = \cos\pi + i\sin\pi = -1$$

$$n = 3, z = \cos\frac{3\pi}{2} + i\sin\frac{3\pi}{2} = -i$$

$\therefore$ The fourth roots of unity are $\pm\ 1, \pm\ i$.

Note:

$$1 = 1^4 = i^8 = i^{12} = ...$$

$$i = i^5 = i^9 = i^3 = ...$$

$$-1 = i^2 = i^6 = i^{10} = ...$$

$$-i = i^3 = i^7 = i^{11} = ...$$

Argand diagram of the qth roots of a Complex number: All the q-th roots of z lie on a circle centred at the origin O and having radius equal to the real, positive q^{th} root of r. One of them has amplitude $\frac{\theta}{q}$ and others are uniformly spaced around the circle separated from others by an angle $\frac{2\pi}{q}$.

Representation of Cube Roots of Unity

radius = 1

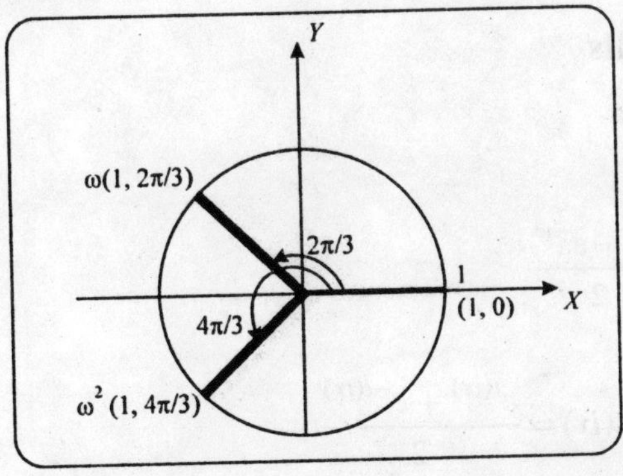

Representation of Fourth Roots of Unity

radius = 1

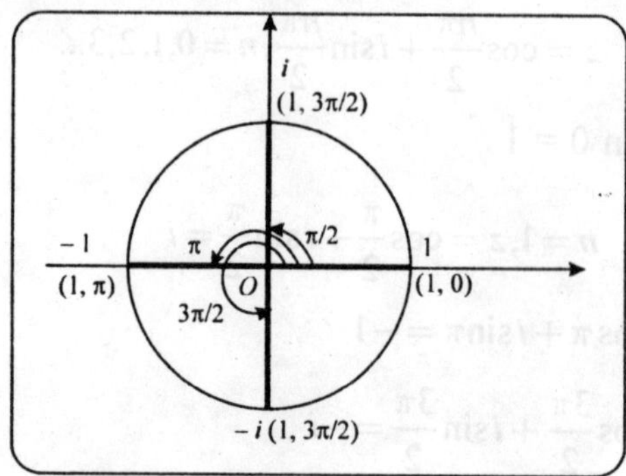

Exponential form of a Complex Number

If $z = x + iy$

Then $e^z = e^{x+iy}$

$$= e^x e^{iy}$$

$e^z = e^x \{\cos y + i \sin y\}$ is called the exponential form of the complex number.

We have,

$$e^{i\theta} = \cos\theta + i\sin\theta$$

and $e^{-i\theta} = \cos\theta - i\sin\theta$

$\therefore$ We get,

$$\cos\theta = \frac{\theta^{i\theta} + e^{-i\theta}}{2}$$

and
$$\sin\theta = \frac{e^{i\theta} - e^{-i\theta}}{2i}$$

Some Important Results

(i) $\cos(ix) = \cos hx$

Proof:

We have $\cos\theta = \dfrac{e^{i\theta} + e^{-i\theta}}{2}$

Replacing θ by ix,

$$\cos(ix) = \frac{e^{i(ix)} + e^{-i(ix)}}{2}$$

$$= \frac{e^{-x} + e^{x}}{2}$$

$$= \cos hx$$

(ii) sin (*ix*) = *i* sin *hx*

Proof: We have,

$$\sin\theta = \frac{e^{i\theta} - e^{-i\theta}}{2i}$$

Replacing θ by *ix*,

$$\sin(ix) = \frac{e^{i(ix)} - e^{-i(ix)}}{2i}$$

$$= \frac{1}{i}\left(\frac{e^{-x} - e^{x}}{2}\right)$$

$$= -\frac{1}{i}\sin hx$$

$$= i \sin hx$$

(iii) sin ($x \pm iy$) = sin *x* cos $hy \pm i$ sin *hy* cos *x*

Proof:

$$\sin(x + iy) = \sin x \cos(iy) + \cos x \sin(iy)$$

$$= \sin x \cos hy + \cos xi \sin hy$$

$$= \sin x \cos hy + i \cos x \sin hy$$

Similarly,

$$\sin(x - iy) = \sin x \cos hy - i \cos x \sin hy$$

(iv) cos ($x \pm iy$) = cos *x* cos $hy \pm i$ sin *hy* sin *x*

Proof:

$$\cos(x + iy) = \cos x \cos(iy) - \sin x \sin(iy)$$

$$= \cos x . \cos hy - \sin x . i \sin hy$$

$$= \cos x \cos hy - i \sin x \sin hy$$

Similarly,

$$\cos(x - iy) = \cos x \cos hy + i \sin x \sin hy$$

To find the Logarithm of a Complex Number: Let $z = x + iy$ be a complex number.

Let $z = re^{i\theta}$ where $r = \sqrt{x^2 + y^2}$ and $\theta = \tan^{-1}\left(\frac{y}{x}\right)$

Taking log on both sides,

$$\log_e z = \log(re^{i\theta})$$

$$= \log r + \log(e^{i\theta})$$

$$= \log_e z = \log(r + i\theta)$$

$$\therefore \log_e(x + iy) = \log\sqrt{x^2 + y^2} + i\tan^{-1}\frac{y}{x}$$

Note 1: The above formula is useful for finding the logarithm of negative numbers also.

Note 2:

$$\log(-1) = \log(1 + i\pi)$$

$$\log(-1) = i\pi$$

Chapter 7

Measurable Functions

We always denote by X our universe, i.e. all the sets we shall consider are subsets of X. Recall some standard notation. 2^X everywhere denotes the set of all subsets of a given set X. If $A \cap B = \phi$ then we often write $A \cup B$ rather than $A \cup B$, to underline the disjointness. The complement (in X) of a set A is denoted by A^c.

By $A \Delta B$ the symmetric difference of A and B is denoted, i.e. $A \Delta B = (A \setminus B) \cup (B \setminus A)$. Letters i, j, k always denote positive integers. The sign is used for restriction of a function (operator etc.) to a subset (subspace).

THE RIEMANN INTEGRAL

Recall how to construct the Riemannian integral. Let $f: [a, b] \to \mathbb{R}$. Consider a partition π of $[a, b]$:

$$a = x_0 < x_1 < x_2 < ... < x_{n-1} < x_n = b$$

and set $\Delta x_k = x_{k+1} - x_k$, $|\pi| = \max \{\Delta x_k: k = 0, 1, ..., n-1\}$, $m_k = \inf\{f(x): x \in [x_k, x_{k+1}]\}$, $M_k = \sup\{f(x): x \in [x_k, x_{k+1}]\}$. Define the upper and lower Riemann–Darboux sums

$$\underline{s}(f, \pi) = \sum_{k=0}^{n-1} m_k \Delta x_k, \overline{s}(f, \pi) = \sum_{k=0}^{n-1} M_k \Delta x_k.$$

One can show that the following limits exist

$$\lim_{|\pi| \to 0} \underline{s}(f, \pi) = \sup_{\pi} \underline{s}(f, \pi) \underline{\int_a^b} f dx$$

$$\lim_{|\pi| \to 0} \overline{s}(f, \pi) = \inf_{\pi} \overline{s}(f, \pi) \overline{\int_a^b} f dx$$

Clearly,

$$\underline{s}(f,\pi) \le \underline{\int_a^b} f dx \le \overline{\int_a^b} f dx \le \overline{s}(f,\pi)$$

for any partition π.

The function *f* is said to be Riemann integrable on [*a, b*] if the upper and lower integrals are equal. The common value is called Riemann integral of *f* on [*a, b*].

The functions cannot have a large set of points of discontinuity. More presicely this will be stated further.

THE LEBESGUE INTEGRAL

It allows to integrate functions from a much more general class. First, consider a very useful example. For $f, g \in C[a, b]$, two continuous functions on the segment $[a, b] = \{x \in \mathbb{R} : a \le x \le b\}$ put

$$\rho_1(f,g) \lim_{a \le x \le b} | f(x) - g(x) |,$$

$$\rho_2(f,g) = \int_a^b | f(x) - g(x) | dx.$$

Then $(C[a, b], \rho_1)$ is a complete metric space, when $(C[a, b], \rho_2)$ is not. To prove the latter statement, consider a family of functions $\{\varphi_n\}_{n=1}^{\infty}$ as drawn on Figure. This is a Cauchy sequence with respect to ρ_2. However, the limit does not belong to $C[a, b]$.

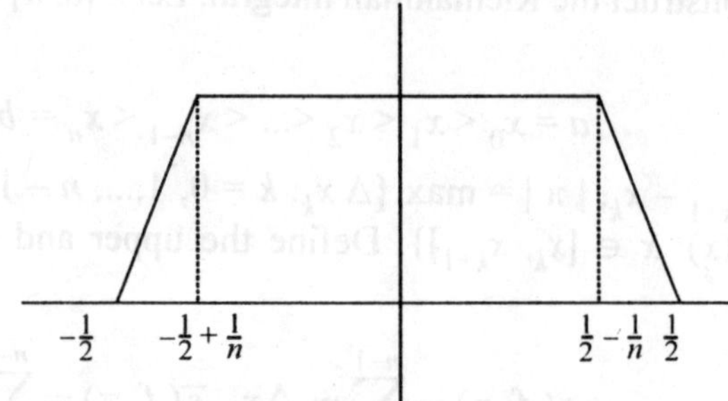

Fig. The Function φ_n.

SYSTEMS OF SETS

Definition: A ring of sets is a non-empty subset in 2^X which is closed with respect to the operations ∪.

Proposition: Let $\mathfrak{R}$ be a ring of sets. Then $\phi \in \mathfrak{R}$.

Proof: Since $\mathfrak{R} \ne \phi$, there exists $A \in \mathfrak{R}$. Since $\mathfrak{R}$ contains the difference of every two its elements, one has $A \setminus A = \phi \in \mathfrak{R}$.

Examples:

1. The two extreme cases are $\Re = \{\phi\}$ and $\Re = 2^X$.
2. Let $X = \mathbb{R}$ and denote by $\Re$ all finite unions of semi-segments $[a, b)$.

Definition: A semi-ring is a collection of sets $B \subset 2^X$ with the following properties:

1. If $A, B \in B$ then $A \cap B \in B$;
2. For every $A, B \in B$ there exists a finite disjoint collection $(C_j)\, j = 1, 2, \ldots, n$ of sets (i.e. $C_i \cap C_j = \phi$ if $i \neq j$) such that

$$A \setminus B = \bigcup_{j=1}^{n} C_j.$$

Example: Let $X = \mathbb{R}$, then the set of all semi-segments, $[a, b)$, forms a semi-ring.

Definition: An algebra (of sets) is a ring of sets containing $X \in 2^X$.

Examples:

1. $\{\phi, X\}$ and 2^X are the two extreme cases (note that they are different from the corresponding cases for rings of sets).
2. Let $X = [a, b)$ be a fixed interval on $\mathbb{R}$. Then the system of finite unions of sub intervals $[\alpha, \beta) \subset [a, b)$ forms an algebra.
3. The system of all bounded subsets of the real axis is a ring (not an algebra).

Remark: U is algebra if (i) $A, B \in A \Rightarrow A \cup B \in U$, (ii) $A \in U \Rightarrow A^c \in U$. Indeed, 1) $A \cap B = (A^c \cap B^c)^c$; 2) $A \setminus B = A \cap B^c$.

Definition: A σ-ring (a σ-algebra) is a ring (an algebra) of sets which is closed with respect to all countable unions.

Definition: A ring (an algebra, a σ-algebra) of sets, $k(U)$ generated by a collection of sets $U \subset 2^X$ is the minimal ring (algebra, σ-algebra) of sets containing U. In other words, it is the intersection of all rings (algebras, σ-algebras) of sets containing U.

MEASURES

Let X be a set, u an algebra on X.

Definition: A function $\mu: A \to \mathbb{R}_+ \cup \{\infty\}$ is called a measure if

1. $\mu(A) \geq 0$ for any $A \in u$ and $\mu(\phi) = 0$;
2. if $(A_i)_{i\geq 1}$ is a disjoint family of sets in u ($A_i \cap A_j = \phi$ for any $i \neq j$) such that $\bigcup_{i=1}^{\infty} A_i \in u$, then

$$\mu\left(\bigcup_{i=1}^{\infty} A_i\right) = \sum_{i=1}^{\infty} \mu(A_i).$$

The latter important property, is called countable additivity or σ-additivity of the measure μ.

Let us state now some elementary properties of a measure. Below till the end of this section u is an algebra of sets and μ is a measure on it.

1. *Monotonicity of m*: If $A, B \in u$ and $B \subset A$ then $m(B) \leq m(A)$.

Proof: $A = (A \setminus B) \cup B$ implies that

$$\mu(A) = \mu(A \setminus B) + \mu(B).$$

Since $\mu(A \setminus B) \geq 0$ it follows that $\mu(A) \geq \mu(B)$.

2. *Subtractivity of μ*: If $A, B \in u$ and $B \subset A$ and $\mu(B) < \infty$ then $\mu(A \setminus B) = \mu(A) - \mu(B)$.

Proof: In 1) we proved that

$$\mu(A) = \mu(A \setminus B) + \mu(B).$$

If $\mu(B) < \infty$ then

$$\mu(A) - \mu(B) = \mu(A \setminus B).$$

3. If $A, B \in A$ and $\mu(A \cap B) < \infty$ then $\mu(A \cup B) = \mu(A) + \mu(B) - \mu(A \cap B)$.

Proof: $A \cap B \subset A$, $A \cap B \subset B$, therefore

$$A \cup B = (A \setminus (A \cap B)) \cup B.$$

Since $\mu(A \cap B) < \infty$, one has

$$\mu(A \cup B) = (\mu(A) - \mu(A \setminus B)) + \mu(B).$$

4. *Semi-additivity of μ*: If $(A_i)_{i \geq 1} \subset u$ such that $\bigcup_{i=1}^{\infty} A_i \in u$ then

$$\mu\left(\bigcup_{i=1}^{\infty} A_i\right) \leq \sum_{i=1}^{\infty} \mu(A_i).$$

Proof: First let us proove that

$$\mu\left(\bigcup_{i=1}^{n} A_i\right) \leq \sum_{i=1}^{n} \mu(A_i).$$

Note that the family of sets

$$B_1 = A_1$$

$$B_2 = A_2 \setminus A_1$$
$$B_3 = A_3 \setminus (A_1 \cup A_2)$$
$$\dots$$
$$B_n = A_n \setminus \bigcup_{i=1}^{n-1} A_i$$

is disjoint and $\bigcup_{i=1}^{n} B_i = \bigcup_{i=1}^{n} A_i$. Moreover, since $B_i \subset A_i$, we see that $\mu(B_i) \leq \mu(A_i)$. Then

$$\mu\left(\bigcup_{i=1}^{n} A_i\right) = \mu\left(\bigcup_{i=1}^{n} B_i\right) = \sum_{i=1}^{n} \mu(B_i) \leq \sum_{i=1}^{n} \mu(A_i).$$

Now we can repeat the argument for the infinite family using σ-additivity of the measure.

CONTINUITY OF A MEASURE

Theorem: Let u be an algebra, $(A_i)_{i\geq 1} \subset u$ a monotonically increasing sequence of sets $(A_i \subset A_{i+1})$ such that $\bigcup_{i\geq 1} \in u$. Then

$$\mu\left(\bigcup_{i=1}^{\infty} A_i\right) = \lim_{n\to\infty} \mu(A_n).$$

Proof: 1. If for some n_0 $\mu(A_{n_0}) = +\infty$ then $\mu(A_n) = +\infty \; \forall n \geq n_0$ and $\mu\left(\bigcup_{i=1}^{\infty} A_i\right) = +\infty$.

2. Let now $m(A_i) < \infty \; \forall i \geq 1$.

Then

$$\mu\left(\bigcup_{i=1}^{\infty} A_i\right) = \mu \; (A_1 \cup (A_2 \setminus A_1) \cup \dots \cup (A_n \setminus A_{n-1}) \cup \dots)$$

$$= \mu(A_1) + \sum_{k=2}^{\infty} \mu(A_k \setminus A_{k-1})$$

$$= \mu(A_1) + \lim_{n\to\infty} \sum_{k=2}^{n} \mu(A_k) - \mu(A_{k-1}) = \lim_{n\to\infty} \mu(A_n).$$

OUTER MEASURE

Let a be an algebra of subsets of X and μ a measure on it. Our purpose now is to extend μ to as many elements of 2^X as possible. An arbitrary set $A \subset X$ can be always covered by sets from u, i.e. one can always find $E_1, E_2, ... \in u$ such that $\bigcup_{i=1}^{\infty} E_i \supset A$. For instance, $E_1 = X$, $E_2 = E_3 = ... = \phi$.

Definition: For $A \subset X$ its outer measure is defined by

$$\mu^*(A) = \inf \sum_{i=1}^{\infty} \mu(E_i)$$

where the infimum is taken over all u-coverings of the set A, i.e. all collections (E_i), $E_i \in u$ with $\bigcup_i E_i \supset A$.

Remark: The outer measure always exists since $\mu(A) \geq 0$ for every $A \in u$.

Example: Let $X = \mathbb{R}^2$, $u = \Re(B)$, -σ-algebra generated by B, $B = \{[a, b) \times \mathbb{R}^1\}$. Thus u consists of countable unions of strips like one drawn on the picture. Put $\mu([a, b) \times \mathbb{R}^1) = b - a$. Then, clearly, the outer measure of the unit disc $x^2 + y^2 \leq 1$ is equal to 2.

The same value is for the square $|x| \leq 1$, $|y| \leq 1$.

Theorem: For $A \in u$ one has $\mu(A) = \mu(A)$.

In other words, μ* is an extension of μ.

Proof: 1. A is its own covering. This implies $\mu^*(A) \leq \mu(A)$.

2. By definition of infimum, for any $e > 0$ there exists a u-covering (E_i) of A such that $\sum_i \mu(E_i) < m^*(A) + e$. Note that

$$A = A \cap \left(\bigcup_i E_i \right) = \bigcup_i (A \cap E_i)$$

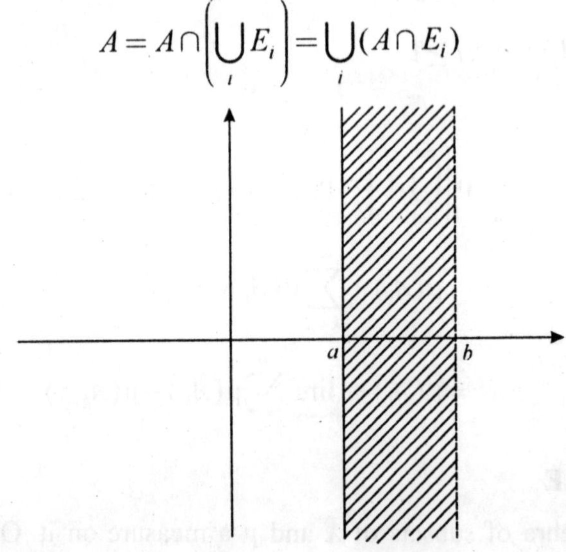

Using consequently σ-semiadditivity and monotonicity of μ, one obtains:

$$\mu(A) \le \sum_i \mu(A \cap E_i) \le \sum_i \mu(E_i) < \mu^*(A) + \varepsilon.$$

Since ε is arbitrary, we conclude that $\mu(A) \le \mu^*(A)$.

It is evident that $\mu^*(A) \ge 0$, $\mu^*(\phi) = 0$ (Check !).

Lemma. Let u be an algebra of sets (not necessary σ-algebra), μ a measure on u. If there exists a set $A \in u$ such that $\mu(A) < \infty$, then $\mu(\phi) = 0$.

Proof: $\mu(A \setminus A) = \mu(A) - \mu(A) = 0$.

Therefore the property $\mu(\phi) = 0$ can be substituted with the existence in u of a set with a finite measure.

Monotonicity of Outer Measure

Theorem: If $A \subset B$ then $\mu^*(A) \le \mu^*(B)$.

Proof: Any covering of B is a covering of A.

S-semiadditivity of m

Theorem: $\mu^*\left(\bigcup_{j=1}^{\infty} A_j\right) \le \sum_{j=1}^{\infty} \mu^*(A_j)$..

Proof: If the series in the right-hand side diverges, there is nothing to prove. So assume that it is convergent. By the definition of outer measur for any $\varepsilon > 0$ and for any j there exists an u-covering $\bigcup_k E_{kj} \supset A_j$ such that

$$\sum_{k=1}^{\infty} \mu(E_{kj}) < \mu^*(A_j) + \frac{\varepsilon}{2^j}.$$

Since

$$\bigcup_{j,k=1}^{\infty} E_{kj} \supset \bigcup_{j=1}^{\infty} A_j,$$

the definition of μ^* implies

$$\mu^*\left(\bigcup_{j=1}^{\infty} A_j\right) \le \sum_{j,k=1}^{\infty} \mu(E_{kj})$$

and therefore

$$\mu^*\left(\bigcup_{j=1}^{\infty} A_j\right) < \sum_{j=1}^{\infty} \mu^*(A_j) + \varepsilon.$$

MEASURABLE SETS

Let u be an algebra of subsets of X, μ a measure on it, $\mu*$ the outer measure defined in the previous section.

Definition: $A \subset X$ is called a measurable set (by Carath'eodory) if for any $E \subset X$ the following relation holds:

$$\mu*(E) = \mu*(E \cap A) + \mu*(E \cap A^c).$$

Denote by $\tilde{u}$ the collection of all set which are measurable by Carath'eodory and set $\tilde{\mu} = \mu* \mid \tilde{u}$.

Remark: Since $E = (E \cap A) \cup (\mathbf{E} \cap A^c)$, due to semiadditivity of the outer measure

$$\mu*(E) \leq \mu*(E \cap A) + \mu(E \cap A^c).$$

Theorem: $\tilde{u}$ is a σ-algebra containing u, and $\tilde{\mu}$ is a measure on $\tilde{u}$.

Proof: We devide the proof into several steps.

1. If $A, B \in \tilde{u}$ then $A \cup B \in \tilde{u}$.

By the definition one has

$$\mu*(E) = \mu*(E \cap B) + \mu*(E \cap B^c).$$

Take $E \cap A$ instead of E:

$$\mu*(E \cap A) = \mu*(E \cap A \cap B) + \mu*(E \cap A \cap B^c).$$

Then put $E \cap A^c$ in (1) instead of E

$$\mu(E \cap A^c) = \mu*(E \cap A^c \cap B) + \mu*(E \cap A^c \cap B^c).$$

Add (2) and (3):

$$\mu*(E) = \mu*(E \cap A \cap B) + \mu*(E \cap A \cap B^c) + \mu*(E \cap A^c \cap B) + \mu*(E \cap A^c \cap B^c).$$

Substitute $E \cap (A \cup B)$ in (4) instead of E. Note that

1. $E \cap (A \cup B) \cap A \cap B = E \cap A \cap B$
2. $E \cap (A \cup B) \cap A^c \cap B = E \cap A^c \cap B$
3. $E \cap (A \cup B) \cap A \cap B^c = E \cap A \cap B^c$
4. $E \cap (A \cup B) \cap A^c \cap B^c = \phi$.

One has

$$\mu*(E \cap (A \cup B)) = \mu*(E \cap A \cap B) + \mu*(E \cap A^c \cap B) + \mu*(E \cap A \cap B^c).$$

From (4) and (5) we have

$$\mu*(E) = \mu*(E \cap (A \cup B)) + \mu*(E \cap (A \cup B)^c).$$

2. If $A \in \tilde{u}$ then $A^c \in \tilde{u}$.

The definition of measurable set is symmetric with respect to A and A^c. Therefore $\tilde{u}$ is an algebra of sets.

3. Let $A, B \in u$, $A \cap B = \phi$. From (5)

$$\mu*(E\cap(A \cup B)) = \mu*(E \cap A^c \cap B) + \mu*(E \cap A \cap B^c) = \mu*(E \cap B) + \mu*(E \cap A)$$

4. $\tilde{u}$ is a σ-algebra.

From the previous step, by induction, for any finite disjoint collection (B_j) of sets:

$$\mu*\left(E\cap\left(\bigcup_{j=1}^{n} B_j\right)\right)=\sum_{j=1}^{n}\mu*(E\cap B_j).$$

Let $A=\bigcup_{j=1}^{\infty} A_j, A_j \in u$. Then $A=\bigcup_{j=1}^{\infty} B_j$, $B_j = A_j \setminus \bigcup_{k=1}^{j-1} A_k$ and $B_i \cap B_j = \phi\,(i \neq j)$. It suffices to prove that

$$\mu*(E)\geq\mu*\left(E\cap\left(\bigcup_{j=1}^{\infty} B_j\right)\right)+\mu*\left(E\cap\left(\bigcup_{j=1}^{\infty} B_j\right)^c\right).$$

Indeed, we have already proved that $\mu*$ is σ-semi-additive.

Since $\tilde{u}$ is an algebra, it follows that $\bigcup_{j=1}^{n} B_j \in \tilde{u}(\forall n \in \mathbb{N})$ and the following inequality holds for every n:

$$\mu*(E)\geq\mu*\left(E\cap\left(\bigcup_{j=1}^{n} B_j\right)\right)+\mu*\left(E\cap\left(\bigcup_{j=1}^{n} B_j\right)^c\right).$$

Since $E\cap\left(\bigcup_{j=1}^{\infty} B_j\right)^c \subset E\cap\left(\bigcup_{j=1}^{n} B_j\right)^c$, by monotonicity of the mesasure and (8)

$$\mu*(E)\geq\sum_{j=1}^{n}\mu*(E\cap B_j)+\mu*(E\cap A^c).$$

Passing to the limit we get

$$\mu*(E)\geq\sum_{j=1}^{\infty}\mu*(E\cap B_j)+\mu*(E\cap A^c).$$

Due to semiadditivity

$$\mu^*(E\cap A)=\mu^*\left(E\cap\left(\bigcup_{j=1}^{\infty}B_j\right)\right)=\mu^*\left(\bigcup_{j=1}^{\infty}(E\cap B_j)\right)\leq\sum_{j=1}^{\infty}\mu^*(E\cap B_j)$$

Compare this

$$\mu^*(E) \geq \mu^*(E \cap A) + \mu^*(E \cap A^c).$$

Thus, $A \in \tilde{u}$, which means that $\tilde{u}$ is a σ-algebra.

5. $\tilde{\mu} = \mu \,|\, \tilde{u}$ is a measure.

We need to prove only σ-additivity. Let $E = \bigcup_{j=1}^{\infty} A_j$. We get

$$\mu^*\left(\bigcup_{j=1}^{\infty}A_j\right)\geq\sum_{j=1}^{\infty}\mu^*(A_j).$$

The oposite inequality follows from -semiadditivity of μ^*.

6. $\tilde{u} \supset u$.

Let $A \in u$, $E \subset X$. We need to prove:

$$\mu^*(E) > \mu^*(E \cap A) + \mu^*(E \cap A^c).$$

If $E \in u$ then is clear since $E \cap A$ and $E \cap A^c$ are disjoint and both belong to u where $\mu^* = \mu$ and so is additive.

For $E \subset X$ for $\forall \varepsilon > 0$ there exists a u-covering (E_j) of E such that

$$\mu^*(E) + \varepsilon > \sum_{j=1}^{\infty}\mu(E_j).$$

Now, since $E_j = (E_j \cap A) \cup (E_j \cap A^c)$, one has

$$\mu^*(E_j) = \mu(E_j \cap A) + \mu(E_j \cap A)$$

and also

$$E\cap A\subset\bigcup_{j=1}^{\infty}(E_j\cap A)$$

$$E\cap A^c\subset\bigcup_{j=1}^{\infty}(E_j\cap A^c)$$

By monotonicity and σ-semiadditivity

$$\mu^*(E\cap A)\leq\bigcup_{j=1}^{\infty}\mu(E_j\cap A)$$

$$\mu^*(E \cap A^c) \le \sum_{j=1}^{\infty} \mu(E_j \cap A^c)$$

Adding the last two inequalities we obtain

$$\mu^*(E \cap A) + \mu^*(E \cap A^c) \le \sum_{j=1}^{\infty} \mu^*(E_j) < \mu^*(E) + \varepsilon .$$

Since $\varepsilon > 0$ is arbitrary, is proved.

The following theorem is a direct consequence of the previous one.

Theorem: Let u be an algebra of subsets of X and μ be a measure on it. Then there exists a σ-algebra $u_1 \supset u$ and a measure μ_1 on u_1 such that $\mu_1 \mid u = \mu$.

Remark: Consider again an algebra u of subsets of X. Denot by u_σ the generated σ-algebra and construct the extension μ_σ of μ on u_σ. This extension is called minimal extension of measure.

Since $\tilde{u} \supset u$ therefore $u_\sigma \subset \tilde{u}$. Hence one can set $\mu = \tilde{\mu} \mid u_\sigma$. Obviously μ_σ is a minimal extension of μ. It always exists. On can also show that this extension is unique.

Theorem: Let μ be a measure on an algebra u of subsets of X, μ^* the corresponding outer measure. If $\mu^*(A) = 0$ for a set $A \subset X$ then $A \in \tilde{u}$ and $\tilde{\mu}(A) = 0$.

Proof: Clearly, it suffices to prove that $A \in \tilde{u}$. Further, it suffices to prove that $\mu^*(E) \ge \mu^*(E \cap A) + \mu^*(E \cap A^c)$. The latter statement follows from monotonicity of μ^*. Indeed, one has $\mu^*(E \cap A) \le \mu^*(A) = 0$ and $\mu^*(E \cap A^c) \le \mu^*(E)$.

Definition: A measure μ on an algebra of sets u is called complete if conditions $B \subset A$, $A \in u$, $\mu(A) = 0$ imply $B \in u$ and $\mu(B) = 0$.

Corollary: $\tilde{\mu}$ is a complete measure.

Definition: A measure μ on an algebra u is called finite if $\mu(X) < \infty$. It is called σ-finite if the is an increasing sequence $(F_j)_{j \ge 1} \subset u$ such that $X = \bigcup_j F_j$ and $\mu(F_j) < \infty \, \forall j$.

Theorem: Let μ be a σ-finite measure on an algebra u. Then there exist a unique extension of μ to a measure on $\tilde{u}$.

Proof: It suffices to sjow uniqueness. Let v be another extension of μ ($v \mid u = \mu \mid u$).

First, let μ (and therefore v, μ^*) be finite. Let $A \in \tilde{u}$. Let $(E_j) \subset u$ such that $A \subset \bigcup_j E_j$.

We have

$$v(A) \le v\left(\bigcup_{j=1}^{\infty} E_j\right) \le \sum_{j=1}^{\infty} v(E_j) = \sum_{j=1}^{\infty} \mu(E_j).$$

Therefore

$$v(A) \le \mu^*(A) \quad \forall A \in \tilde{u}$$

Since μ^* and v are additive (on $\tilde{u}$) it follows that

$$\mu^*(A) + \mu^*(A^c) = v(A) + v(A^c).$$

The terms in the RHS are finite and $v(A) \le \mu^*(A)$, $v(A^c) \le \mu^*(A^c)$. From this we infer that

$$v(A) = \mu^*(A) \; \forall A \in \tilde{u}.$$

Now let μ be σ-finite, (F_j) be an increasing sequence of sets from u such that $\mu(F_j) < \infty \, \forall j$ and $X = \bigcup_{j=1}^{\infty} F_j$. From what we have already proved it follows that

$$\mu^*(A \cap F_j) = v(A \cap F_j) \; \forall A \in \tilde{u}.$$

Therefore

$$\mu^*(A) = \lim_j \mu^*(A \cap F_j) = \lim_j v(A \cap F_j) = v(A).$$

Continuity of Measure

Theorem: Let u be a σ-algebra with a measure μ, $\{A_j\} \subset u$ a monotonically increasing sequence of sets. Then

$$\mu\left(\bigcup_{j=1}^{\infty} A_j\right) = \lim_{j \to \infty} \mu(A_j).$$

Proof: One has:

$$A = \bigcup_{j=1}^{\infty} A_j = \bigcup_{j=1}^{\infty} (A_{j+1} \setminus A_j) \cup A_1.$$

Using σ-additivity and subtractivity of μ,

$$\mu(A) = \bigcup_{j=1}^{\infty} (\mu A_{j+1}) - \mu(A_j)) + \mu(A_1) = \lim_{j \to \infty} \mu(A_j)$$

Similar assertions for a decreasing sequence of sets in u can be proved using de Morgan formulas.

Theorem: Let $A \in \tilde{u}$. Then for any $\varepsilon > 0$ there exists $A_\varepsilon \in u$ such that $\mu^*(A \,\Delta\, A_\varepsilon) < \varepsilon$.

Proof: 1. For any $\varepsilon > 0$ there exists an u cover $\bigcup E_j \supset A$ such that

$$\sum_j \mu(E_j) < \mu^*(A) + \frac{\varepsilon}{2} = \tilde{u}(A) + \frac{\varepsilon}{2}.$$

On the other hand,

$$\sum_j \mu(E_j) \geq \tilde{\mu}\left(\bigcup_j E_J\right).$$

The monotonicity of $\tilde{u}$ implies

$$\tilde{\mu}\left(\bigcup_{j=1}^{\infty} E_J\right) = \lim_{n\to\infty} \tilde{\mu}\left(\bigcup_{j=1}^{n} E_j\right),$$

hence there exists a positive integer N such that

$$\tilde{\mu}\left(\bigcup_{j=1}^{\infty} E_J\right) - \tilde{\mu}\left(\bigcup_{j=1}^{N} E_j\right) < \frac{\varepsilon}{2}.$$

2. Now, put

$$A_\varepsilon = \bigcup_{j=1}^{N} E_j$$

and prove that $\mu^*(A \,\Delta\, A_\varepsilon) < \varepsilon$.

(a) Since

$$A \subset \bigcup_{j=1}^{\infty} E_j,$$

one has

$$A \setminus A_\varepsilon \subset \bigcup_{j=1}^{\infty} E_j \setminus A_\varepsilon.$$

Since

$$A_\varepsilon \subset \bigcup_{j=1}^{\infty} E_j,$$

one can use the monotonicity and subtractivity of $\tilde{\mu}$. Together with estimate, this gives

$$\tilde{\mu}(A \setminus A_\varepsilon) \leq \tilde{\mu}\left(\bigcup_{j=1}^{\infty} E_j \setminus A_\varepsilon\right) < \frac{\varepsilon}{2}$$

(b) The inclusion

$$A_\varepsilon \setminus A \subset \bigcup_{j=1}^{\infty} E_j \setminus A$$

implies

$$\tilde{\mu}(A_\varepsilon \setminus A) \leq \left(\bigcup_{j=1}^{\infty} E_j \setminus A\right) = \tilde{\mu}\left(\bigcup_{j=1}^{\infty} E_j\right) - \mu(A) < \frac{\varepsilon}{2}$$

Here we used the same properties of $\tilde{\mu}$ as above and the choice of the cover (E_j).

3. Finally,

$$\tilde{\mu}(A \Delta A_\varepsilon) \leq \tilde{\mu}(A \setminus A_\varepsilon) + \tilde{\mu}(A_\varepsilon \setminus A).$$

MONOTONE CLASSES AND UNIQUENESS OF EXTENSION OF MEASURE

Definition: A collection of sets, M is called a monotone class if together with any monotone sequence of sets M contains the limit of this sequence.

Example: Any σ-ring.

Theorem: Let k be a ring of sets, k_σ the σ-ring generated by k. Then $m(k) = k_\sigma$.

Proof: Clearly, $m(k) \subset k_\sigma$. Now, it suffices to prove that $m(k)$ is a ring. This follows from the minimality of k_σ. $m(k)$ is a ring.

For $B \subset X$, set

$$k_B = \{A \subset X: A \cup B,\ A \cap B,\ A \setminus B,\ B \setminus A \in m(k)\}.$$

This definition is symmetric with respect to A and B, therefore $A \in k_B$ implies $B \in k_A$.
b. k_B is a monotone class.

Let $(A_j) \subset K_B$ be a monotonically increasing sequence. Prove that the union, $A = \bigcup A_j$ belongs to k_B.

Since $A_j \in k_B$, one has $A_j \cup B \in k_B$, and so

$$A \cup B = \bigcup_{j=1}^{\infty} (A_j \cup B) \in m(k).$$

In the same way,

$$A \setminus B = \left(\bigcup_{j=1}^{\infty} A_j\right) \setminus B = \bigcup_{j=1}^{\infty} (A_j \setminus B) \in m(k);$$

$$B \setminus A = B \setminus \left(\bigcup_{j=1}^{\infty} A_j\right) = \bigcap_{j=1}^{\infty} (B \setminus A_j) \in m(k).$$

Similar proof is for the case of decreasing sequence (A_j).

If $B \in k$ then $m(k) \subset k_B$.

Obviously, $k \subset k_B$. Together with minimality of $m(k)$, this implies $m(k) \subset k_B$.

If $B \in m(k)$ then $m(k) \subset k_B$.

Let $A \in k$. Then $m(k) \subset k_A$. Thus if $B \in m(k)$, one has $B \in k_A$, so $A \in k_B$.

Hence what we have proved is $k \subset k_B$. This implies $m(k) \subset k_B$.

It follows from 2a. – 2d. that if $A, B \in m(k)$ then $A \in k_B$ and so $A \cup B, A \cap B, A \setminus B$ and $B \setminus A$ all belong to $m(k)$.

Theorem: Let u be an algebra of sets, μ and v two measures defined on the σ-algebra u_σ generated by u. Then $\mu \mid u = v \mid u$ implies $\mu = v$.

Proof: Choose $A \in u_\sigma$, then $A = \lim_{n\to\infty} A_n$, $A_n \in u$, for $u_\sigma = m(u)$. Using continuity of measure, one has

$$\mu(A) = \lim_{n\to\infty} \mu(A_n) = \lim_{n\to\infty} v(A_n) = v(A).$$

Theorem: Let u be an algebra of sets, $B \subset X$ such that for any $\varepsilon > 0$ there exists $A_\varepsilon \in u$ with $\mu^*(B \,\Delta\, A_\varepsilon) < \varepsilon$. Then $B \in \tilde{u}$.

Proof: Since any outer measure is semi-additive, it suffices to prove that for any $E \subset X$ one has

$$\mu^*(E) \geq \mu^*(E \cap B) + \mu^*(E \cap B^c).$$

Since $u \subset \tilde{u}$, one has

$$\mu^*(E \cap A_\varepsilon) + \mu^*(E \cap A_\varepsilon^c) \leq \mu^*(E)$$

Since $A \subset B \cup (A \,\Delta\, B)$ and since the outer measure μ^* is monotone and semiadditive, there is an estimate $\mid \mu^*(A) - \mu^*(B) \mid \leq \mu^*(A \,\Delta\, B)$ for any $A, B \subset X$.

It follows from the monotonicity of μ^* that

$$\mid \mu^*(E \cap A_\varepsilon) - \mu^*(E \cap B) \mid \leq \mu^*((E \cap A_\varepsilon) \,\Delta\, (E \cap B)) \leq \mu(A_\varepsilon \cap B) < \varepsilon.$$

Therefore, $\mu^*(E \cap A_\varepsilon) > \mu^*(E \cap B) - \varepsilon$.

In the same manner, $\mu^*(E \cap A_\varepsilon^c) > \mu^*(E \cap B^c) - \varepsilon$.

Using, one obtains

$$\mu^*(E) > \mu^*(E \cap B) > \mu^*(E \cap B^c) - 2\varepsilon.$$

THE LEBESGUE MEASURE ON THE REAL LINE $\mathbb{R}^1$

THE LEBESGUE MEASURE OF BOUNDED SETS OF $\mathbb{R}^1$

Put u for the algebra of all finite unions of semi-segments (semi-intervals) on $\mathbb{R}^1$, i.e. all sets of the form

$$A = \bigcup_{j=1}^{k} [a_j, b_j).$$

Define a mapping $\mu: u \to \mathbb{R}$ by:

$$\mu(A) = \sum_{j=1}^{k} (b_j - a_j).$$

Theorem: μ is a measure.

Proof: All properties including the (finite) additivity are obvious. The only thing to be proved is the σ-additivity.

Let $(A_j) \subset u$ be such a countable disjoint family that

$$A = \bigcup_{j=1}^{\infty} A_j \in u.$$

The condition $A \in u$ means that $\bigcup A_j$ is a finite union of intervals.

For any positive integer n,

$$\bigcup_{j=1}^{n} A_j \subset A,$$

hence

$$\sum_{j=1}^{n} \mu(A_j) \le \mu(A),$$

and

$$\sum_{j=1}^{\infty} \mu(A_j) = \lim_{n\to\infty} \sum_{j=1}^{n} \mu(A_j) \le \mu(A).$$

Now, let A^ε a set obtained from A by the following construction. Take a connected component of A. It is a semi-segment of the form $[s, t)$. Shift slightly on the left its right-hand end, to obtain a (closed) segment. Do it with all components of A, in such a way that

$$\mu(A) < \mu(A^\varepsilon) + \varepsilon.$$

Apply a similar procedure to each semi-segment shifting their left end point to the left $A_j = [a_j, b_j)$, and obtain (open) intervals, A_j^ε with

$$\mu\left(A_j^\varepsilon\right) < \mu(A_j) + \frac{\varepsilon}{2^j}.$$

By the construction, A^ε is a compact set and (A_j^ε) its open cover. Hence, there exists a positive integer n such that

$$\bigcup_{j=1}^{n} A_j^\varepsilon \supset A^\varepsilon.$$

Thus

$$\mu(A^\varepsilon) \le \sum_{j=1}^{n} \mu(A_j^\varepsilon).$$

The formulas imply

$$\mu(A) < \sum_{j=1}^{n} \mu(A_j^\varepsilon) + \varepsilon \le \sum_{j=1}^{n} \mu(A_j) + \sum_{j=1}^{n} \frac{\varepsilon}{2^j} + \varepsilon,$$

thus

$$\mu(A) < \sum_{j=1}^{\infty} \mu(A_j) + 2\varepsilon.$$

Now, one can apply the Carath'eodory's scheme developed above, and obtain the measure space $(\tilde{u}, \tilde{\mu})$.

The result of this extension is called the Lebesgue measure. We shall denote the Lebesgue measure on $\mathbb{R}^1$ by m.

Definition: Borel algebra of sets, B on the real line $\mathbb{R}^1$ is a σ-algebra generated by all open sets on $\mathbb{R}^1$. Any element of B is called a Borel set.

Theorem: Let $E \subset \mathbb{R}^1$ be a Lebesgue measurable set. Then for any $\varepsilon > 0$ there exists an open set $G \supset E$ such that $m(G \setminus E) < \varepsilon$.

Proof: Since E is measurable, $m^*(E) = m(E)$. According the definition of an outer measure, for any $\varepsilon > 0$ there exists a cover $A = \bigcup [a_k, b_k) \supset E$ such that

$$m(A) < m(E) + \frac{\varepsilon}{2}.$$

Now, put

$$G = \bigcup \left(a_k - \frac{\varepsilon}{2^{k+1}}, b_k \right).$$

Corollary: For any $\varepsilon > 0$ there exist an open set G and a compact set F such that $G \supset E \supset F$ and $m(G \setminus F) < \varepsilon$.

Such measures are called regular.

THE LEBESGUE MEASURE ON THE REAL LINE $\mathbb{R}^1$

We now abolish the condition of boundness.

Definition: A set A on the real numbers line $\mathbb{R}^1$ is Lebesgue measurable if for any positive integer n the bounded set $A \cap [-n, n)$ is a Lebesgue measurable set.

Definition: The Lebesgue measure on $\mathbb{R}^1$ is

$$m(A) = \lim_{n \to \infty} m(A \cap [-n, n)).$$

Definition: A measure is called σ-finite if any measurable set can be represented as a countable union of subsets each has a finite measure.

Thus the Lebesgue measure m is σ-finite.

THE LEBESGUE MEASURE IN $\mathbb{R}^d$

Definition: We call a d-dimensional rectangle in $\mathbb{R}^d$ any set of the form

$$\{x: x \in \mathbb{R}^d: a_i \leq x_i < b_i\}.$$

Using rectangles, one can construct the Lebesque measure in $\mathbb{R}^d$ in the same fashion as we deed for the $\mathbb{R}^1$ case.

MEASURABLE FUNCTIONS

Let X be a set, u a σ-algebra on X.

Definition: A pair (X, u) is called a measurable space.

Definition: Let f be a function defined on a measurable space (X, u), with values in the extended real number system. The function f is called measurable if the set

$$\{x: f(x) > a\}$$

is measurable for every real a.

Theorem: The following conditions are equivalent

$\{x: f(x) > a\}$ is measurable for every real a.

$\{x: f(x) \geq a\}$ is measurable for every real a.

$\{x: f(x) < a\}$ is measurable for every real a.

$\{x: f(x) \leq a\}$ is measurable for every real a.

Proof: The statement follows from the equalities

$$\{x: f(x) \geq a\} = \bigcap_{n=1}^{\infty} \left\{x : f(x) > a - \frac{1}{n}\right\},$$

$$\{x: f(x) < a\} = X \setminus \{x: f(x) \geq a\},$$

$$\{x: f(x) \leq a\} = \bigcap_{n=1}^{\infty} \left\{x : f(x) < a + \frac{1}{n}\right\},$$

$$\{x: f(x) > a\} = X \setminus \{x: f(x) \leq a\}$$

Theorem: Let (f_n) be a sequence of measurable functions. For $x \in X$ set

$$g(x) = \sup_n f_n(x) (n \in \mathbb{N})$$

$$h(x) = \limsup_{n \to \infty} f_n(x)$$

Then g and h are measurable.

Proof:

$$\{x: g(x) \leq a\} = \bigcap_{n=1}^{\infty} \{x : f_n(x) \leq a\},$$

Since the LHS is measurable it follows that the RHS is measurable too. The same proof works for inf.

Now

$$h(x) = \inf g_m(x),$$

where

$$g_m(x) = \sup_{n \geq m} f_n(x).$$

Theorem: Let f and g be measurable real-valued functions defined on X. Let F be real and continuous function on $\mathbb{R}^2$. Put

$$h(x) = F(f(x), g(x))\ (x \in X).$$

Then h is measurable.

Proof: Let $G_a = \{(u, v): F(u, v) > a\}$. Then G_a is an open subset of $\mathbb{R}^2$, and thus

$$G_a = \bigcup_{n=1}^{\infty} I_n$$

where (I_n) is a sequence of open intervals

$$I_n = \{(u, v): a_n < u < b_n, c_n < v < d_n\}.$$

The set $\{x: a_n < f(x) < b_n\}$ is measurable and so is the set

$$\{x: (f(x), g(x)) \in I_n\} = \{x: a_n < f(x) < b_n\} \cap \{x: c_n < g(x) < d_n\}.$$

Hence the same is true for

$$\{x: h(x) > a\} = \{x: (f(x), g(x)) \in G_a\} = \bigcup_{n=1}^{\infty}\{x:(f(x), g(x)) \in I_n\}.$$

Corollories: Let f and g be measurable. Then the following functions are measurable

(i) $f + g$

(ii) $f \cdot g$

(iii) $|f|$

(iv) $\frac{f}{g}$ (if $g \ne 0$)

(v) $\max\{f, g\}$, $\min\{f, g\}$ since $\max\{f, g\} = 1/2\ (f + g + |f - g|)$, $\min\{f, g\} = 1/2\ (f + g - |f - g|)$.

STEP FUNCTIONS

Simple Functions

Definition: A real valued function $f: X \to \mathbb{R}$ is called simple function if it takes only a finite number of distinct values.

We will use below the following notation

$$\chi_E(x) = \begin{cases} 1 & \text{if } x \in E \\ 0 & \text{otherwise} \end{cases}$$

Theorem: A simple function $f = \sum_{j=1}^{n} c_j \chi_{E_j}$ is measurable if and only if all the sets E_j are measurable.

Theorem: Let f be real valued. There exists a sequence (f_n) of simple functions such that $f_n(x) \to f(x)$ as $n \to \infty$, for every $x \in X$. If f is measurable, (f_n) may be chosen to be a sequence of measurable functions. If $f \geq 0$, (f_n) may be chosen monotonically increasing.

Proof: If $f \geq 0$ set

$$f_n(x) = \sum_{i=1}^{n.2^n} \frac{i-1}{2^n} \chi E_{n_i} + n\chi F_n$$

where

$$E_{n_i} = \left\{x : \frac{i-1}{2^n} \leq f(x) < \frac{i}{2^n}\right\}, F_n = \{x : f(x) \geq n\}.$$

The sequence (f_n) is monotonically increasing, f_n is a simple function. If $f(x) < \infty$ then $f(x) < n$ for a sufficiently large n and $|f_n(x) - f(x)| < 1/2^n$. Therefore $f_n(x) \to f(x)$. If $f(x) = +\infty$ then $f_n(x) = n$ and again $f_n(x) \to f(x)$.

In the general case $f = f^+ - f^-$, where

$f^+(x) := \max\{f(x), 0\}, f^-(x) := -\min\{f(x), 0\}$.

Note that if f is bounded then $f_n \to f$ uniformly.

INTEGRATION

Definition: A triple (X, u, μ), where u is a σ-algebra of subsets of X and μ is a measure on it, is called a measure space.

Let (X, u, μ) be a measure space. Let $f: X \mapsto \mathbb{R}$ be a simple measurable function.

$$f(x) = \sum_{i=1}^{n} c_i \chi_{E_i}(x)$$

and

$$\bigcup_{i=1}^{n} E_i = X, E_i \cap E_j = \phi \ (i \neq j).$$

There are different representations of f. Let us choose the representation such that all c_i are distinct.

Definition: Define the quantity

$$I(f) = \sum_{i=1}^{n} c_i \mu(E_i).$$

First, we derive some properties of $I(f)$.

Theorem: Let f be a simple measurable function. If $X = \bigcup_{j=1}^{k} F_j$ and f takes the constant value b_j on F_j then

$$I(f) = \sum_{j=1}^{k} b_j \mu(F_j).$$

Proof: Clearly, $E_i = \bigcup_{i:b_j=c_i} F_j$.

$$\sum_i c_i \mu(E_i) = \sum_{i=1}^{n} c_i \mu\left(\bigcup_{j:b_j=c_i} F_j\right) = \sum_{i=1}^{n} c_i \sum_{j:b_j=c_i} \mu(F_j) = \sum_{j=1}^{k} b_j \mu(F_j).$$

This show that the quantity $I(f)$ is well defined.

Theorem: If f and g are measurable simple functions then

$$I(\alpha f + \beta g) = \alpha I(f) + \beta I(g).$$

Proof: Let $f(x) = \sum_{j=1}^{n} b_j \chi_{F_j}(x), X = \bigcup_{j=1}^{n} F_j, g(x) = \sum_{k=1}^{m} c_k \chi_{G_k}(x), X = \bigcup_{k=1}^{n} G_k$. Then

$$\alpha f + \beta g = \sum_{j=1}^{n} \sum_{k=1}^{m} (\alpha b_j + \beta c_k) \chi_{E_{jk}}(x)$$

where $E_{jk} = F_j \cap G_k$.

Theorem: Let f and g be simple measurable functions. Suppose that $f \leq g$ everywhere except for a set of measure zero. Then

$$I(f) \leq I(g).$$

Proof: If $f \leq g$ everywhere then in the notation of the previous proof $b_j \leq c_k$ on E_{jk} and $I(f) \leq I(g)$ follows.

Otherwise we can assume that $f \leq g + \phi$ where is non-negative measurable simple function which is zero every exept for a set N of measure zero. Then $I(\phi) = 0$ and

$$I(f) \leq I(g + \phi) = I(f) + I(\phi) = I(g).$$

Definition: If $f: X \mapsto \mathbb{R}^1$ is a non-negative measurable function, we define the Lebesgue integral of f by

$$\int f d\mu := \sup I(\phi)$$

where sup is taken over the set of all simple functions ϕ such that $\phi \leq f$.

Theorem: If f is a simple measurable function then $\int f d\mu = I(f)$.

Proof: Since $f \leq f$ it follows that $\int f d\mu \geq I(f)$.

On the other hand, if $\phi \leq f$ then $I(\phi) \leq I(f)$ and also

$$\sup_{\phi \leq f} I(\phi) \leq I(f)$$

which leads to the inequality

$$\int f d\mu \leq I(f).$$

Definition: If A is a measurable subset of X ($A \in u$)and f is a non-negative measurable function then we define

$$\int_A f d\mu = \int f \chi_A d\mu.$$

$$\int f d\mu = \int f^+ d\mu - \int f^- d\mu$$

if at least one of the terms in RHS is finite. If both are finite we call f integrable.

Remark: Finiteness of the integrals $\int f^+ d\mu -$ and $\int f^- d\mu$ is equivalent to the finitenes of the integral

$$\int |f| d\mu.$$

If it is the case we write $f \in L^1(X, \mu)$ or simply $f \in L^1$ if there is no ambiguity.

The following properties of the Lebesgue integral are simple consequences of the definition. The proofs are left to the reader.

1. If f is measurable and bounded on A and $\mu(A) < \infty$ then f is integrable on A.

2. If $a \leq f(x) \leq b$ ($x \in A$)), $\mu(A) < \infty$ then

$$a\mu(A) \leq \int_A f d\mu \leq b\mu(a).$$

3. If $f(x) \leq g(x)$ for all $x \in A$ then

$$\int_A f d\mu \leq \int_A g d\mu.$$

4. Prove that if $\mu(A) = 0$ and f is measurable then

$$\int_A f d\mu = 0.$$

The next theorem expresses an important property of the Lebesgue integral. As a consequence we obtain convergence theorems which give the main advantage of the Lebesgue approach to integration in comparison with Riemann integration.

Theorem: Let f be measurable on X. For $A \in u$ define

$$\phi(A) = \int_A f d\mu$$

Then ϕ is countably additive on u.

Proof: It is enough to consider the case $f \geq 0$. The general case follows from the decomposition $f = f^+ - f^-$.

If $f = \chi_E$ for some $E \in u$ then

$$\mu(A \cap E) = \int_A \chi_E d\mu$$

and σ-additivity of ϕ is the same as this property of μ.

Let $f(x) = \sum_{k=1}^{n} c_k \chi_{E_k}(x), \bigcup_{k=1}^{n} E_k = X$. Then for $A = \bigcup_{i=1}^{n} A_i, A_i \in u$ we have

$$\phi(A) = \int_A f d\mu = \int f \chi A d\mu = \sum_{k=1}^{n} c_k \mu(E_k \cap A)$$

$$= \sum_{k=1}^{n} c_k \mu\left(E_k \cap \left(\bigcup_{i=1}^{\infty} A_i\right)\right) = \sum_{k=1}^{n} c_k \mu\left(\bigcup_{i=1}^{\infty} (E_k \cap A_i)\right)$$

$$= \sum_{k=1}^{n} c_k \sum_{i=1}^{\infty} \mu(E_k \cap A_i) = \sum_{i=1}^{\infty} \sum_{k=1}^{n} c_k \mu(E_k \cap A_i)$$

(the series of positive numbers)

$$= \sum_{i=1}^{\infty} \int_{A_i} f d\mu = \sum_{i=1}^{\infty} \phi(A_i).$$

Now consider general positive f's. Let φ be a simple measurable function and $\varphi \leq f$. Then

$$\int_A \varphi d\mu = \sum_{i=1}^{\infty} \int_{A_i} \varphi d\mu \leq \sum_{i=1}^{\infty} \phi(A_i)$$

Therefore the same inequality holds for sup, hence

$$\phi(A) \leq \sum_{i=1}^{\infty} \phi(A_i).$$

Now if for some i $\phi(A_i) = +\infty$ then $\phi(A) = +\infty$ since $\phi(A) \geq \phi(A_n)$. So assume that $\phi(A_i) < \infty$ $\forall i$. Given $\varepsilon > 0$ choose a measurable simple function φ such that $\varphi \leq f$ and

$$\int_{A_1} \varphi d\mu \geq \int_{A_1} f d\mu - \varepsilon, \int_{A_2} \varphi d\mu \geq \int_{A_2} f - \varepsilon$$

Hence

$$\phi(A_1 \cup A_2) \geq \int_{A_1 \cup A_2} \varphi d\mu = \int_{A_1} + \int_{A_2} \varphi d\mu \geq \phi(A_1) + \phi(A_2) - 2\varepsilon,$$

so that $\phi(A_1 \cup A_2) \geq \phi(A_1) + \phi(A_2)$.

By induction

$$\phi\left(\bigcup_{i=1}^{n} A_i\right) \geq \sum_{i=1}^{n} \phi(A_i).$$

Since $A \supset \bigcup_{i=1}^{n} A_i$ we have that

$$\phi(A) \geq \sum_{i=1}^{n} \phi(A_i).$$

Passing to the limit $n \to \infty$ in the RHS we obtain

$$\phi(A) \geq \sum_{i=1}^{\infty} \phi(A_i).$$

This completes the proof.

Corollary: If $A \in u$, $B \subset A$ and $\mu(A \setminus B) = 0$ then

$$\int_A f d\mu = \int_B f d\mu .$$

Proof:

$$\int_A f d\mu = \int_B f d\mu + \int_{A \setminus B} f d\mu = \int_B f d\mu + 0..$$

Definition: f and g are called equivalent ($f \sim g$ in writing) if $\mu(\{x: f(x) \neq g(x)\}) = 0$.

It is not hard to see that $f \sim g$ is relation of equivalence.

(i) $f \sim f$, (ii) $f \sim g$, $g \sim h \Rightarrow f \sim h$, (iii) $f \sim g \Leftrightarrow g \sim f$.

Theorem: If $f \in L^1$ then $|f| \in L^1$ and

$$\left|\int_A f d\mu\right| \leq \int_A |f| d\mu$$

Proof:

$$-|f| \leq f \leq |f|$$

Monotone Convergence Theorem

Theorem: Let (f_n) be nondecreasing sequence of nonnegative measurable functions with limit f. Then

$$\int_A f d\mu = \lim_{n\to\infty} \int_A f_n d\mu, A \in u$$

Proof: First, note that $f_n(x) \leq f(x)$ so that

$$\lim_n \int_A f_n d\mu \leq \int f d\mu$$

It is remained to prove the opposite inequality.

For this it is enough to show that for any simple φ such that $0 \leq \varphi \leq f$ the following inequality holds

$$\int_A \varphi d\mu \leq \lim_n \int_A f_n d\mu$$

Take $0 < c < 1$. Define

$$A_n = \{x \in A: f_n(x) \geq c\varphi(x)\}$$

then $A_n \subset A_{n+1}$ and $A = \bigcup_{n=1}^{\infty} A_n$.

Now observe

$$c\int_A \varphi d\mu = \int_A c\varphi d\mu = \lim_{n\to\infty} \int_{A_n} c\varphi d\mu \leq$$

(this is a consequence of σ-additivity of ϕ proved above)

$$\leq \lim_{n\to\infty} \int_{A_n} f_n d\mu \leq \lim_{n\to\infty} \int_A f_n \varphi d\mu$$

Pass to the limit $c \to 1$.

Theorem: Let $f = f_1 + f_2, f_1, f_2 \in L^1(\mu)$. Then $f \in L^1(\mu)$ and

$$\int f d\mu = \int f_1 d\mu + \int f_2 d\mu$$

Proof: First, let $f_1, f_2 \geq 0$. If they are simple then the result is trivial. Otherwise, choose monotonically increasing sequences $(\varphi_{n,1})$, $(\varphi_{n,2})$ such that $\varphi_{n,1} \to f_1$ and $\varphi_{n,2} \to f_2$.

Then for $\varphi_n = \varphi_{n,1} + \varphi_{n,2}$

$$\int \varphi_n d\mu = \int \varphi_{n,1} d\mu + \int \varphi_{n,2} d\mu$$

and the result follows from the previous theorem.

If $f_1 \geq 0$ and $f_2 \leq 0$ put

$$A = \{x: f(x) \geq 0\},\ B = \{x: f(x) < 0\}$$

Then f, f_1 and $-f_2$ are non-negative on A.

Hence $\int_A f_1 = \int_A f d\mu + \int_A (-f_2) d\mu$

Similarly

$$\int_B (-f_2) d\mu = \int_B f_1 d\mu + \int_B (-f) d\mu$$

The result follows from the additivity of integral.

Theorem: Let $A \in u$, (f_n) be a sequence of non-negative measurable functions and

$$f(x) = \sum_{n=1}^{\infty} f_n(x), x \in A$$

Then

$$\int_A f d\mu = \sum_{n=1}^{\infty} \int_A f_n d\mu$$

Fatou's lemma

Theorem: If (f_n) is a sequence of non-negative measurable functions defined a.e. and

$$f(x) = \underline{\lim}_{n\to\infty} f_n(x)$$

then

$$\int_A f d\mu \leq \underline{\lim}_{n\to\infty} \int_A f_n d\mu$$

$$A \in u$$

Proof: Put $g_n(x) = \inf_{i \geq n} f_i(x)$

Then by definition of the lower limit $\lim_{n\to\infty} g_n(x) = f(x)$.

Moreover, $g_n \leq g_{n+1}$, $g_n \leq f_n$. By the monotone convergence theorem

$$\int_A f d\mu = \lim_n \int_A g_n d\mu = \underline{\lim}_n \int_A g_n d\mu \leq \underline{\lim}_n \int_A f_n d\mu.$$

Lebesgue's Dominated Convergence Theorem

Theorem: Let $A \in u$, (f_n) be a sequence of measurable functions such that $f_n(x) \to f(x)$ $(x \in A.)$ Suppose there exists a function $g \in L^1(\mu)$ on A such that

$$|f_n(x)| \leq g(x)$$

Then

$$\lim_n \int_A f_n d\mu = \int_A f d\mu$$

Proof: From $|f_n(x)| \leq g(x)$ it follows that $f_n \in L_1(\mu)$. Sinnce $f_n + g \geq 0$ and $f + g \geq 0$, by Fatou's lemma it follows

$$\int_A (f+g)d\mu \leq \underline{\lim}_n \int_A (f_n + g)$$

or

$$\int_A f d\mu \leq \underline{\lim}_n \int_A f_n d\mu.$$

Since $g - f_n \geq 0$ we have similarly

$$\int_A (g-f)d\mu \leq \underline{\lim}_n \int_A (g - f_n)d\mu$$

so that

$$-\int_A f d\mu \leq -\underline{\lim}_n \int_A f_n d\mu$$

which is the same as

$$\int_A f d\mu \geq \overline{\lim}_n \int_A f_n d\mu$$

This proves that

$$\underline{\lim}_n \int_A f_n d\mu = \overline{\lim}_n \int_A f_n d\mu = \int_A f d\mu.$$

COMPARISON OF THE RIEMANN AND THE LEBESGUE INTEGRAL

To distinguish we denote the Riemann integral by $(R)\int_a^b f(x)dx$ and the Lebesgue integral by $(L)\int_a^b f(x)dx$.

Theorem: If a function f is Riemann integrable on $[a, b]$ then it is also Lebesgue integrable on $[a, b]$ and

$$(L)\int_a^b f(x)dx = (R)\int_a^b f(x)dx.$$

Proof: Boundedness of a function is a necessary condition of being Riemann integrable. On the other hand, every bounded measurable function is Lebesgue integarble.

So it is enough to prove that if a function f is Riemann integrable then it is measurable.

Consider a partition π_m of $[a, b]$ on $n = 2^m$ equal parts by points $a = x_0 < x_1 <... < x_{n-1} < x_n = b$ and set

$$\underline{f}_m(x) = \sum_{k=0}^{2^m-1} m_k \chi_k(x), \overline{f}_m(x) = \sum_{k=0}^{2^m-1} M_k \chi_k(x),$$

where χ_k is a charactersitic function of $[x_k, x_{k+1})$ clearly,

$$\underline{f}_1(x) \le \underline{f}_2(x) \le ... \le f(x),$$

$$\overline{f}_1(x) \ge \overline{f}_2(x) \ge ... \ge f(x).$$

Therefore the limits

$$\underline{f}(x) = \lim_{m\to\infty} \underline{f}_m(x), \overline{f}(x) = \lim_{m\to\infty} \overline{f}_m(x)$$

exist and are measurable. Note that $\underline{f}(x) \le f(x) \le \overline{f}(x)$. Since $\underline{f}_m$ and $\overline{f}_m$ are simple measurable functions, we have

$$(L)\int_a^b \underline{f}_m(x)dx \le (L)\int_a^b \underline{f}(x)dx \le (L)\int_a^b \overline{f}(x)dx \le (L)\int_a^b \overline{f}_m(x)dx$$

Moreover,

$$(L)\int_a^b \underline{f}_m(x)dx = \sum_{k=0}^{2^m-1} m_k \Delta x_k = \underline{s}(f, \pi_m)$$

and similarly

$$(L)\int_a^b \overline{f}_m(x) = \overline{s}(f, \pi_m).$$

So

$$\underline{s}(f, \pi_m) \le (L)\int_a^b \underline{f}(x)dx \le (L)\int_a^b \overline{f}(x)dx \le \overline{s}(f, \pi_m)$$

Since f is Riemann integrable,

$$\lim_{m\to\infty} \underline{s}(f, \pi_m) = \lim_{m\to\infty} \overline{s}(f, \pi_m) = (R)\int_a^b f(x)dx.$$

Therefore

$$(L)\int_a^b (\overline{f}(x)-\underline{f}(x))dx=0$$

and since $\overline{f} \geq \underline{f}$ we conclude that

$$f=\overline{f}=\underline{f} \text{ almost everywhere.}$$

From this measurability of f follows.

L^p-SPACES

Let (X, u, μ) be a measure space. In this section we study $L^p(X, u, \mu)$-spaces which occur frequently in analysis.

AUXILIARY FACTS

Lemma: Let p and q be real numbers such that $p > 1$, $\frac{1}{p}+\frac{1}{q}=1$ (this numbers are called conjugate). Then for any $a > 0$, $b > 0$ the inequality

$$ab \leq \frac{a^p}{p}+\frac{b^q}{q}.$$

holds.

Proof: Note that $\varphi(t):=\frac{t^p}{p}+\frac{1}{q}-t$ with $t \geq 0$ has the only minimum at $t = 1$. It follows that

$$t \leq \frac{t^p}{p}+\frac{1}{q}.$$

Then letting $t=ab^{-\frac{1}{p-1}}$ we obtain

$$\frac{a^p b^{-q}}{p}+\frac{1}{q} \geq ab^{-\frac{1}{p-1}},$$

and the result follows.

Lemma: Let $p \geq 1$, $a, b \in \mathbb{R}$. Then the inequality $|a+b|^p \leq 2^{p-1}(|a|^p+|b|^p)$ holds.

Proof: For $p = 1$ the statement is obvious. For $p > 1$ the function $y = x^p$, $x \geq 0$ is convex since $y'' \geq 0$. Therefore

$$\left(\frac{|a|+|b|}{2}\right)^p \leq \frac{|a|^p+|b|^p}{2}$$

THE SPACES L^p, $1 \le P < \infty$. DEFINITION

Recall that two measurable functions are said to be equivalent (with respect to the measure μ) if they are equal μ-a;most everywhere.

The space $L^p = L^p(X, u, \mu)$ consists of all μ-equivalence classes of *u*-measurable functions *f* such that $|f|^p$ has finite integral over *X* with respect to μ.

We set

$$\| f \|_p := \left(\int_X | f |^p \, d\mu \right)^{1/p}.$$

HOLDER'S INEQUALITY

Theorem: Let $p > 1, \frac{1}{p} + \frac{1}{q} = 1$. Let *f* and *g* be measurable functions, $|f|^p$ and $|g|^q$ be integrable. Then *fg* is integrable and the inequality

$$\int_X | fg | \, d\mu \le \left(\int_X | f |^p \, d\mu \right)^{1/p} \left(\int_X | g |^q \, d\mu \right)^{1/q}.$$

Proof: It suffices to consider the case

$$\| f \|_p > 0, \| g \|_q > 0.$$

Let

$$a = | f(x) | \, \| f \|_p^{-1}, b = | g(x) | \, \| g \|_q^{-1}.$$

By Lemma

$$\frac{| f(x)g(x) |}{\| f \|_p \| g \|_q} \le \frac{| f(x) |^p}{p \| f \|_p^p} + \frac{| g(x) |^q}{q \| g \|_q^q}.$$

After integration we obtain

$$\| f \|_p^{-1} \| g \|_q^{-1} \int_X | fg | \, d\mu \le \frac{1}{p} + \frac{1}{q} = 1.$$

MINKOWSKI'S INEQUALITY

Theorem: If $f, g \in L^p$, $p \ge 1$, then $f + g \in L^p$ and

$$\| f + g \|_p \le \| f \|_p + \| g \|_p.$$

Proof: If $\| f \|_p$ and $\| g \|_p$ are finite then by Lemma $| f + g |_p$ is integrable and $\| f + g \|_p$ is well-defined.

$$|f(x)+g(x)|^p = |f(x)+g(x)||f(x)+g(x)|^{p-1}$$
$$\leq |f(x)||f(x)+g(x)|^{p-1} + |g(x)||f(x)+g(x)|^{p-1}.$$

Integratin the last inequality and using H¨older's inequality we obtain

$$\int_X |f+g|^p \, d\mu \leq \left(\int_X |f|^p \, d\mu\right)^{1/p} \left(\int_X |f+g|^{(p-1)q} \, d\mu\right)^{1/q}$$
$$+\left(\int_X |g|^p \, d\mu\right)^{1/p} \left(\int_X |f+g|^{(p-1)q} \, d\mu\right)^{1/q}.$$

The result follows.

L^p, $1 \leq P < \infty$, IS A BANACH SPACE

It is readily seen from the properties of an integral and Theorem that L^p, $1 \leq p < \infty$, is a vector space. We introduced the quantity $\|f\|_p$. Let us show that it defines a norm on L^p, $1 \leq p < \infty$,. Indeed,

1. By the definition $\|f\|_p \geq 0$.
2. $\|f\|_p = 0 \Rightarrow f(x) = 0$ for μ-almost all $x \in X$. Since L^p consists of μ-eqivalence classes, it follows that $f \sim 0$.
3. Obviously, $\|\alpha f\|_p = \|\alpha\| \|f\|_p$.
4. From Minkowski's inequality it follows that $\|f+g\|_p \leq \|f\|_p + \|g\|_p$.

So L^p, $1 \leq p < \infty$, is a normed space.

Theorem: L^p, $1 \leq p < \infty$, is a Banach space.

Proof: It remains to prove the completeness.

Let (f_n) be a Cauchy sequence in L^p. Then there exists a subsequence $(f_{n_k})(k \in \mathbb{N})$ with n_k increasing such that

$$\|f_m - f_{n_k}\| < \frac{1}{2^k} \; \forall m \geq n_k.$$

Let

$$g_k := |f_{n_1}| + |f_{n_2} - f_{n_1}| + \ldots + |f_{n_{k+1}} - f_{n_k}|$$

Then g_k is monotonocally increasing. Using Minkowski's inequality we have

$$\|g_k^p\|_1 = \|g_k\|_p^p \leq \left(\|f_{n_1}\|_p + \sum_{i=1}^{k} \|f_{n_{i+1}} - f_{n_i}\|_p\right)^p < \left(\|f_{n_1}\|_p + 1\right)^p.$$

Put

$$g(x) := \lim_k g_k(x).$$

By the monotone convergence theorem

$$\lim_k \int_X g_k^p d\mu = \int_A g^p d\mu.$$

Moreover, the limit is finite since $\| g_k^p \|_1 \le C = \left(\| f_{n_1} \|_p + 1\right)^p$.

Therefore

$|f_{n_1}| + \sum_{j=1}^{\infty} |f_{n_{j+1}} - f_{n_j}|$ converges almost everywhere and so does

$f_{n_1} + \sum_{j=1}^{\infty} (f_{n_{j+1}} - f_{n_j})$, which means that

$f_{n_1} + \sum_{j=1}^{N} (f_{n_{j+1}} - f_{n_j}) = f_{n_{N+1}}$ converges almost everywhere as $N \to \infty$.

Define

$$f(x) := \lim_{k \to \infty} f_{n_k}(x)$$

where the limit exists and zero on the complement. So f is measurable.

Let $\varepsilon > 0$ be such that for $n, m > N$

$$\| f_n - f_m \|_p^p = \int_X | f_n - f_m |^p d\mu < \varepsilon / 2.$$

Then by Fatou's lemma

$$\int_X | f - f_m |^p d\mu = \int_X \lim_k | f_{n_k} - f_m |^p d\mu \le \underline{\lim}_k \int_X | f_{n_k} - f_m |^p d\mu$$

which is less than ε for $m > N$. This proves that

$$\| f - f_m \|_p \to 0 \text{ as } m \to \infty.$$

Chapter 8

Gamma and Beta Functions

GAMMA FUNCTION

Many important functions in applied sciences are defined via improper integrals. Maybe the most famous among them is the Gamma Function. This is why we thought it would be a good idea to have a page on this function with its basic properties. You may consult any library for more information on this function.

Historically the search for a function generalizing the factorial expression for the natural numbers was on. In dealing with this problem one will come upon the well-known formula

$$\int_0^{+\infty} e^{-t} t^x dt$$

A very quick approach to this problem suggests to replace n by x in the improper integral to generate the function

$$f(x) = \int_0^{+\infty} e^{-t} t^x dt$$

Clearly this definition requires a close look in order to determine the domain of $f(x)$. The only possible bad points are 0 and $+\infty$. Let us look at the point 0. Since $e^{-i} \sim 1$ when $t \approx 0$, then we have

$$e^{-i} \sim t^x = \frac{1}{t^{-x}}$$

when $t \approx 0$. The p-test implies that we have convergence around 0 if and only if $-x < 1$ (or equivalently $x > -1$). On the other hand, it is quite easy to show that the improper integral is convergent at $+\infty$ regardless of the value of x. So the domain of $f(x)$ is $(-1, +\infty)$. If we like to have $(0, +\infty)$ as a domain, we will need to translate the x-axis to get the new function

$$\Gamma(x) = f(x-1) = \int_0^{+\infty} e^{-i} t^{x-1} dt$$

which explains somehow the awkward term $x - 1$ in the power of t. Now the domain of this new function (called the Gamma Function) is $(0, +\infty)$. The above formula is also known as Euler's second integral (if you wonder about Euler's first integral, it is coming a little later).

BASIC PROPERTIES OF Γ(x)

1. First, from the remarks above we have

$$\Gamma(n) = (n - 1)!\ n = 1, 2, \ldots$$

2. One of the most important formulas satisfied by the Gamma function is

$$\Gamma(x+1) = x\Gamma(x)$$

for any $x > 0$. In order to show this formula from the definition of $\Gamma(x)$, we will use the following identity

$$\int_a^b e^{-i} t^x dt = \left[-e^{-i} t^x\right]_a^b + x\int_a^b e^{-i} t^{x-1} dt$$

(this is just an integration by parts). If we let a goe to 0 and b goe to $+\infty$, we get the desired identity.

In particular, we get

$$\Gamma(x+n) = (x+1-1)(x+n-2)\ldots(x+1)x\Gamma(x)$$

for any $x > 0$ and any integer $n \geq 1$. This formula makes it possible for the function $\Gamma(x)$ to be extended to $(-\infty, 0)$ (except for the negative integers). In particular, it is enough to know $\Gamma(x)$ on the interval $(0, 1]$ to know the function for any $x > 0$. Note that since

$$\int_0^{+\infty} e^{-i} dt = 1$$

we get $\Gamma(1) = 1$. Combined with the above identity, we get what we expected before :

$$\Gamma(n) = (n - 1)!\ n = 1, 2, \ldots$$

1. A careful analysis of the Gamma function (especially if we notice that $\ln(\Gamma(x))$ is a convex function) yields the inequality

$$\frac{n^x n!}{x(x+1)\ldots(x+n)} \leq \Gamma(x) \leq \frac{n^x n!}{x(x+1)\ldots(x+n)} \frac{x+n}{n}$$

or equivalently

$$\Gamma(x)\frac{n}{x+n} \leq \frac{n^x n!}{x(x+1)\ldots(x+n)} \leq \Gamma(x)$$

for every $n \geq 1$ and $x > 0$. If we let n go to $+\infty$, we obtain the identity

$$\Gamma(x) = \lim_{n\to+\infty} \frac{n^x n!}{x(x+1)...(x+n)}$$

Note that this formula identifies the Gamma function in a unique fashion.

Weierstrass identity: A simple algebraic manipulation gives

$$\frac{n^x n!}{x(x+1)...(x+n)} = e^{x(\ln(n)-1-1/2-...-1/n)} \frac{1}{x} \frac{e^{x/1}}{1+x/1} \frac{e^{x/2}}{+x/2} ... \frac{e^{x/n}}{1+x/n}$$

Knowing that the sequence $(\ln(n) - 1 - 1/2 - ... - 1/n)$ converges to the constant $-C$, where

$$C = \lim_{n\to+\infty} 1 + \frac{1}{2} + \frac{1}{3} + ... + \frac{1}{n} - \ln(n)$$

is the Euler's constant. We get

$$\Gamma(x) = e^{-Cx} \frac{1}{x} \lim_{n\to+\infty} \prod_{k=1}^{k=n} \frac{e^{x/k}}{1+x/k}$$

or

$$\Gamma(x) = e^{-Cx} \frac{1}{x} \prod_{n=1}^{+\infty} \frac{e^{x/n}}{1+x/n}.$$

The logarithmic derivative of the Gamma function: Since $\Gamma(x) > 0$ for any $x > 0$, we can take the logarithm of the above expression to get

$$\ln(\Gamma(x)) = -Cx - \ln(x) + \sum_{n=1}^{+\infty} \left(\frac{x}{n} - \ln\left(1 + \frac{x}{n}\right) \right)$$

If we take the derivative we get

$$\frac{d}{dx}(\ln(\Gamma(x))) = -C - \frac{1}{x} + \sum_{n=1}^{+\infty} \left(\frac{1}{n} - \frac{1}{x+n} \right)$$

or

$$\frac{\Gamma'(x)}{\Gamma(x)} = -C - \frac{1}{x} + \sum_{n=1}^{+\infty} \frac{x}{(x+n)}$$

In fact, one can differentiate the Gamma function infinitely often. In "analysis" language we say that $\Gamma(x)$ is of C^{∞}-class. Below you will find the graph of the Gamma function.

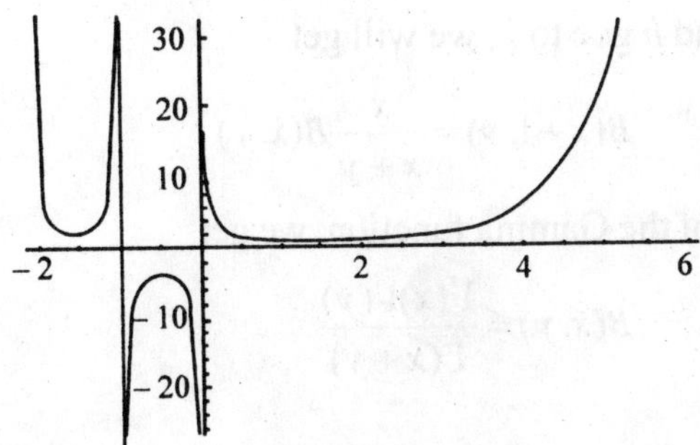

BETA FUNCTION

Euler's first integral or the Beta function: In studying the Gamma function, Euler discovered another function, called the **Beta function**, which is closely related to $\Gamma(x)$. Indeed, consider the function

$$B(x, y) = \int_0^1 t^{x-1}(1-t)^{y-1}\, dt$$

It is defined for two variables *x* and *y*. This is an improper integral of Type I, where the potential bad points are 0 and 1. First we split the integral and write

$$\int_0^1 t^{x-1}(1-t)^{y-1}\, dt = \int_0^{1/2} t^{x-1}(1-t)^{y-1}\, dt + \int_{1/2}^1 t^{x-1}(1-t)^{y-1}\, dt$$

When $t \approx 0$, we have

$$t^{x-1}(1-t)^{y-1} \sim t^{x-1}$$

and when $t \approx 0$, we have

$$t^{x-1}(1-t)^{y-1} \sim (1-t)^{y-1}$$

So we have convergence if and only if $x > 0$ and $y > 0$ (this is done via the *p*-test). Therefore the domain of $B(x, y)$ is $x > 0$ and $y > 0$. Note that we have

$$B(x+1, y) = \int_0^1 (1-t)^{x+y-1}\left(\frac{t}{1-t}\right)^x dt$$

Let *a* and *b* such that $0 < a \le b < 1$, we have (via an integration by parts)

$$\int_a^b (1-t)^{x+y-1}\left(\frac{t}{1-t}\right)^x dt = \left[-\frac{(1-t)^{x+y}}{x+y}\left(\frac{t}{1-t}\right)^x\right]_a^b +$$

$$\int_a^b \frac{x}{x+y}(1-t)^{x+y}\left(\frac{t}{1-t}\right)^{x-1} \frac{1}{(1-t)^2}$$

If we let a goe to 0 and b goe to 1, we will get

$$B(x+1,y)=\frac{x}{x+y}B(x,y)$$

Using the properties of the Gamma function, we get

$$B(x,y)=\frac{\Gamma(x)\Gamma(y)}{\Gamma(x+y)}$$

or

$$\frac{\Gamma(x)\Gamma(y)}{\Gamma(x+y)}=\int_0^1 t^{x-1}(1-t)^{y-1}dt$$

In particular, if we let $x=y=1/2$, we get

$$\Gamma^2\left(\frac{1}{2}\right)=\int_0^1 t^{-1/2}(1-t)^{-1/2}dt$$

If we set $t=\sin^2(\theta)$ or equivalently $\sqrt{t}=\sin(\theta)$, then the technique of substitution implies

$$\int_0^1 t^{-1/2}(1-t)^{-1/2}dt=2\int_0^{\pi/2}d\theta=\pi$$

Hence we have

$$\Gamma^2\left(\frac{1}{2}\right)=\pi$$

or

$$\Gamma\left(\frac{1}{2}\right)=\sqrt{\pi}$$

Using this formula, we can now easily calculate the value of $\Gamma\left(n+\frac{1}{2}\right)$.

OTHER IMPORTANT FORMULAS:

The following formulas are given without detailed proofs. We hope they will be of some interest.

Asymptotic behavior of the Gamma function when x is large: We have

$$\Gamma(x)=\sqrt{2\pi}x^{x-1/2}e^{-x+\mu(x)}$$

where

$$\mu(x)=\sum_{n=0}^{+\infty}\left(x+n+\frac{1}{2}\right)\ln\left(1+\frac{1}{x+n}\right)-1=\frac{\theta}{12x},0<\theta<1$$

If we take, $x = n$, we get after multiplying by n

$$n! = \sqrt{2\pi}n^{n+1/2}e^{-n+\theta/12n}$$

This is a well known result, called Stirling's formula. So for large n, we have

$$n! \sim \sqrt{2n\pi}n^n e^{-n}$$

The connection with sin(x): For any $x > 0$, we have

$$\Gamma(x)\Gamma(1-x) = \frac{\pi}{\sin(\pi x)}$$

which implies

$$\sin(\pi x) = \frac{\pi}{\Gamma(x)\Gamma(1-x)} = -\frac{\pi}{x\Gamma(x)\Gamma(-x)}$$

Using the Weierstrass product formula (for $\Gamma(x)$ and $\Gamma(-x)$), we get

$$\sin(\pi x) = \pi x \prod_{n=1}^{+\infty}\left(1 - \frac{x^2}{n^2}\right).$$

If we use the Beta function ($B(x, y)$), we get the following formulas:

$$\int_0^1 t^{x-1}(1-t)^{-x}\,dt = \frac{\pi}{\sin(\pi x)}, 0 < x < 1,$$

$$\int_0^{+\infty} \frac{t^{x-1}}{1+t}\,dt = \frac{\pi}{\sin(\pi x)}, 0 < x < 1,$$

$$2\int_0^{\pi/2} \tan^{2x-1}(\alpha)\,d\alpha = \frac{\pi}{\sin(\pi x)}, 0 < x < 1.$$

CLASSES OF SUBSETS

TOPOLOGY

Definition: A Topological space (X, T) consists of a non-empty set X together with a collection T of subsets of X such that

(T1) $X, \phi \in T$,

(T2) If $A_1, \ldots, A_n \in T$ then $\bigcap_{i=1}^{n} A_i \in T$,

(T3) If $A_i \in T$, $i \in I$ for some index set I then $\bigcup_{i \in I} A_i \in T$.

The elements of T are called the open sets.

Note Every set has at least two topologies on it, namely the trivial ones $T = \{X, \phi\}$ and $T = P(X)$, i.e. the collection of all subsets of X.

Lemma: If $T_j, j \in J$ are topologies on X then $\bigcap_{j\in J} T_j$ is a topology on X.

Proof:

(T1) $X, \phi \in T_j$ for all j so $X, \phi \in \bigcap_{j\in J} T_j$.

(T2) Take any finite collection of sets $A_1, \ldots A_n \in \bigcap_{j\in J} T_j$. Then $A_1, \ldots A_n \in T_j$ for each j and so $\bigcap_{i=1}^{n} A_i \in T_j$ for each j and so $\bigcap_{i=1}^{n} A_i \in \bigcap_{j\in J} T_j$.

(T3) Take any collection of sets $A_i \in \bigcap_{j\in J} T_j$, $i \in I$. Then $A_i \in T_j$, for all $i \in I$ and all $j \in J$ and so $\bigcup_{i\in I} A_i \in T_j$ for each j and thus $\bigcup_{i\in I} A_i \in T_j \in \bigcap_{j\in J} T_j$.

Corollary: There exists a topology u on $\mathbb{R}$ containing all intervals (a, b) and such that if T_0 is any other topology containing all such intervals then $u \subseteq T_0$.

Proof: There exists at least one topology on $\mathbb{R}$ containing all (a, b), namely $T_0 = P(\mathbb{R})$, the power set of $\mathbb{R}$. Set $u = \bigcap T$, the intersection over all topologies containing all the intervals (a, b). This is a topology by Lemma and the minimality property is immediate from the definition.

Definition: We say that u is the usual topology on $\mathbb{R}$.

Definition: Given any collection A of subsets of X, the intersection of all topologies containing A is said to be the topology *generated* by A.

Example: The usual topology on $\mathbb{R}$ is the topology *generated* by the intervals (a, b).

There are other topologies on $\mathbb{R}$, such as

Example: The co-finite topology on $\mathbb{R}$ is defined by $A \in$ co-finite if, and only if, either $A = \phi$ or $|A^c| < \infty$. (That is, the complement is finite.)

For the following definition we require the idea of *pre-image* of a map $f: X \to Y$. So if $A \subseteq Y$ then the pre-image of A is given by

$$f^{-1}(A) = \{x \in X : f(x) \in A\}.$$

Definition: Given topological spaces (X, T_X), (Y, T_Y) and a map $f: X \to Y$ we say that f is continuous with respect to T_X, T_Y if

$$U \in T_Y \Rightarrow f^{-1}(U) \in T_X.$$

Heine-Borel

Theorem: If $[a, b] \subseteq \mathbb{R}$ is covered by a collection of (c_i, d_i), so $[a, b] \subseteq \bigcup_{i \in I}(c_i, d_i)$, then there exists a finite sub-collection of the (c_i, d_i), which can relabeled as $1 \leq i \leq N$ such that $[a, b] \subseteq \bigcup_{i=1}^{N}(c_i, d_i)$.

Lindel of's Theorem

Theorem: If $G = \{I_\alpha : \alpha \in A\}$ is a collection of intervals $(a, b) \subseteq \mathbb{R}$, possibly an uncountable collection, then there exists a countable subcollection $\{I_i : i \geq 1\} \subseteq G$ such that

$$\bigcup_{\alpha \in A} I_\alpha = \bigcup_{i=1}^{\infty} I_i .$$

Corollary: Let S be the set of all countable unions of intervals $(a, b) \subseteq \mathbb{R}$. Then $S = u$, the usual topology.

Proof: Let $S \in S$ so $S = \bigcup_{i=1}^{\infty}(a_i, b_i)$ for some (a_i, b_i) But $(a_i, b_i) \in u$ for all i, and u is a topology so $\bigcup_{i=1}^{\infty}(a_i, b_i) \in U$. Thus $S \in u$ and so $S \subseteq u$.

We now show that S is a topology by verifying the properties T_1, T_2 and T_3.

(T1) $\mathbb{R} = \bigcup_{n \in \mathbb{N}}(-n, n) \in S$ and $\phi = (0, 0) \in S$.

(T2) If $S_1, S_2, S_3, \ldots S_n \in S$ then each

$$S_i = \bigcup_{j_i=1}^{\infty} I_i \text{ for some } I_{j_i} = \left(a_{j_i}, b_{j_i}\right).$$

Then

$$\bigcap_{i=1}^{n} S_i = \bigcap_{i=1}^{n} \bigcup_{j_i=1}^{\infty} I_{j_i}$$

$$= \left(\bigcup_{j_1=1}^{\infty} I_{j_1}\right) \cap \left(\bigcup_{j_2=1}^{\infty} I_{j_2}\right) \cap \ldots \cap \left(\bigcup_{j_n=1}^{\infty} I_{j_n}\right)$$

$$= \bigcup_{j_1=1}^{\infty} \bigcup_{j_2=1}^{\infty} \ldots \bigcup_{j_n=1}^{\infty} \left(I_{j_1} \cap I_{j_2} \cap \ldots \cap I_{j_n}\right).$$

Importantly we have a finite intersection of open intervals which is therefore an open interval. (It is possible that an infinite intersection of open intervals is closed.) Thus $\bigcap_{i=1}^{n} S_i$ is a countable union of open intervals, hence it is an element of S.

(T3) If $S_k \in S$ for $k \in K$, perhaps an uncountable collection, then $\bigcup_{k\in K} S_k = \bigcup_{k\in K}\bigcup_{jk=1}^{\infty} I_{jk}$ is a, possibly uncountable, union of intervals $\left(A_{jk}, b_{jk}\right)$. But by Lindelof's Theorem this can be written as a countable union and so $\bigcup_{k\in K} S_k \in S$.

Hence S is a topology containing the intervals (a, b). But u is the minimal topology containing these intervals. Hence $u \subseteq S$. Thus $u = S$.

RINGS

Definition: A collection, S, of subsets of the non-empty set X is a semi-ring if

(i) $\phi \in S$,

(ii) $A, B \in S \Rightarrow A \cap B \in S$,

(iii) $A, B \in S \Rightarrow A \backslash B = \bigcup_{i=1}^{N} E_i$, a finite disjoint union of $E_i \in S$.

Example: The collection, P, of all finite intervals of the form $(a, b] \subseteq \mathbb{R}$ form a semi-ring.

This is the most important example of a semi-ring we shall study. It should be compared with the collection of all intervals of the form (a, b) that does not form a semi-ring.

Definition: A non-empty collection, R, of subsets of X is a *ring* if

(i) $A, B \in R \Rightarrow A \cup B \in R$,

(ii) $A, B \in R \Rightarrow A \backslash B \in R$.

Note (i) $R \neq \phi$ implies that there exists a set A in R and so $\phi = A \backslash A \in R$.

*(ii) Such a collection, with the operations

$$A + B = A \,\Delta\, B \text{ and } A \,.\, B = A \cap B$$

is a ring as defined in the second year course. (Here $A \,\Delta\, B = (A\backslash B) \cup (B\backslash A)$ is the symmetric difference.)

Note: The collection P is not a ring, for instance $(0, 3]$ and $(1, 2] \in P$ but $(0, 3]\backslash(1, 2] = (0, 1] \cup (2, 3] \notin P$.

Definition: The collection, ε, of all finite unions of disjoint members of P, is called the set elementary figures in $\mathbb{R}$.

Example: The collection ε is a ring. Verification.

Let $A, B \in \varepsilon$, so $A = \bigcup_{i=1}^{m} A_i$ and $B = \bigcup_{j=1}^{n} B_j$ where $A_i, B_j \in P$, disjoint unions. Then

$$A \setminus B = A \cap B^c = \left(\bigcup_{i=1}^{m} A_i\right) \cap B^c$$

a disjoint union. Writing $A_i = (a, b]$ and $B = \bigcup_{j=1}^{n} (a_j, b_j]$ we see that

$$A_i \cap B^c = (a, b] \cap \left\{(-\infty, a_1] \cup \bigcup_{j=1}^{n-1} (b_j, a_{j+1}] \cup (b_n, +\infty]\right\}$$

which, on applying the distributive law, is seen to be in ε. Then, since the A_i are disjoint we see that $A \setminus B$ is a disjoint union of sets from ε and so $A \setminus B \in$ ε, i.e. condition (ii) holds.

Also $A \cup B = (A \setminus B) \cup (A \cap B) \cup (B \setminus A)$, a disjoint union. By what we have already proved it suffices to show that $A \cap B \in$ ε. But $A \cap B = \bigcup_{i=1}^{m} \bigcup_{j=1}^{n} (A_i \cap B_j)$ a disjoint union of sets from P and so in ε. Thus condition (i) is satisfied. Hence ε is a ring.

Definition: A ring R is a *σ-ring* if it is closed under countable unions. That is, given $A_n \in R$, $n \geq 1$ then $\bigcup_{n\geq 1} A_n \in R$.

Note: Given $A_n \in R$ a σ-ring, write $A = \bigcup_{n\geq 1} A_n$. It is trivial that $\bigcup_{n\geq 1} A_n \subseteq A$ but this observation means that we can write

$$\bigcap_{n\geq 1} A_n = A \setminus \left(A \setminus \bigcap_{n\geq 1} A_n\right).$$

Here

$$A \setminus \bigcap_{n\geq 1} A_n = A \cap \left(\bigcap_{n\geq 1} A_n\right)^c = A \cap \left(\bigcup_{n\geq 1} A_n^c\right)$$

$$= \bigcup_{n\geq 1} (A \cap A_n^c) = \bigcup_{n\geq 1} (A \setminus A_n).$$

Since A and $A_n \in \mathbb{R}$ for all $n \geq 1$ we have $A \setminus A_n \in R$, and so $\bigcup_{n\geq 1} (A \setminus A_n) \in R$ and thus $A \setminus \left(A \setminus \bigcap_{n\geq 1} A_n\right)$ or $\bigcap_{n\geq 1} A_n \in R$. Hence σ-rings are closed under countable intersections.

FIELDS

Definition: A non-empty collection F of subsets of a non-empty set X is a field (or algebra) if

(i) $X \in F$,

(ii) F is a ring.

Further X is a σ-field (or σ-algebra) if

(i) $X \in F$,

(ii) F is a σ-ring.

Note A field can be defined to satisfy

(i) $X \in F$,

(ii) If $A \in F$ then $A^c \in F$,

(iii) If $A, B \in F$ then $A \cup B \in F$.

Similarly for σ-fields.

Verification that (i) $X \in F$, (ii) $A, B \in R \Rightarrow A \cup B \in R$, and (iii) $A, B \in R \Rightarrow A \setminus B \in R$ are equivalent to (i') $X \in F$, (ii) $A, B \in R \Rightarrow A \cup B \in R$, and (iii') if $A \in F$ then $A^c \in F$.

Assume (i), (ii) and (iii). Then (i') and (ii') obviously holds while (iii') follows from (iii) and (i) with $X, A \in F$ implying $X \setminus A \in F$, i.e. $A^c \in F$.

Assume (i'), (ii') and (iii'). Then (i) and (ii) obviously holds while (iii) follows from $A \setminus B = A \cap B^c = (A^c \cup B)^c$ which is in R by (ii') and (iii').

Example: (a) Let X be an infinite set. Define F by $A \in F$ if, and only if, either $\| A \| < \infty$ or $\| A^c \| < \infty$. Then F is a field but it is not a σ-field.

(b) Let X be an infinite set. Define F by $A \in F$ if, and only if, either $\| A \|$ is countable or $\| A^c \|$ is countable. Then F is a σ-field.

Theorem: The intersection of any non-empty collection of rings, fields or σ-fields in X is, respectively, a ring, field or σ-field.

Corollary: Given a collection A of subsets of X there exists

(i) a minimal ring containing A, denoted by $R(A)$,

(ii) a minimal σ-field containing A, denoted by $\sigma(A)$.

These are minimal in that if

(i) R is any ring containing A then $R(A) \subseteq R$,

(ii) F is a σ-field containing A then $\sigma(A) \subseteq F$.

Proof: Simply choose $R(A)$ to be the intersection of all rings containing A and $\sigma(A)$ to be the intersection of all σ-fields containing A. These are non-empty intersections since there is always at least one ring or σ-field on a non-empty set X, namely the power set $P(X)$.

Definition: We say that $R(A)$ is the ring *generated* by A. Similarly we say that $\sigma(A)$ is the σ-field *generated* by A, often denoted by $B(A)$, the *Borel* field generated by A.

Theorem: Let $R(C)$ be the ring generated by the semi-ring C in X. Then $R(C)$ is the collection of finite unions of disjoint sets from C, that is

$$R(C)=\left\{A\subseteq X : A=\bigcup_{i=1}^{n} E_i \text{ for some disjoint members of } C\right\}.(*)$$

Proof: Let A denote the right hand side of (*). If $A \in A$ then $A = \bigcup_{i=1}^{n} E_i$ for some disjoint members of C. But the ring $R(C)$ is closed under finite unions and in particular, closed under finite unions of elements of C and so $A \in R(C)$. Hence $A \subseteq R(C)$.

Next we will show that A is a ring which we do by verifying the definition. Take any $A, B \in R(C)$ so

$$A=\bigcup_{i=1}^{m} E_i \text{ and } B=\bigcup_{j=1}^{n} F_j$$

for some finite disjoint collections $\{E_i\}$ and $\{F_j\}$ in C. Then

$$A/B=\left(\bigcup_{i=1}^{m} E_i\right)\cap\left(\bigcup_{j=1}^{n} F_j\right)^c$$

$$=\left(\bigcup_{i=1}^{m} E_i\right)\cap\left(\bigcup_{j=1}^{n} F_j^c\right)$$

$$=\bigcup_{i=1}^{m}\left(E_i\cap\left(\bigcap_{j=1}^{n} F_j^c\right)\right)$$

$$=\bigcup_{i=1}^{m}\left\{\bigcap_{j=1}^{n}(E_i\setminus F_j)\right\}$$

Yet C is a semi-ring so

$$E_i\setminus F_j=\bigcup_{\ell=1}^{L_{ij}} H_{ij\ell},$$

a disjoint union of sets from C. So

$$\bigcap_{j=1}^{n}(E_i\setminus F_j)=\bigcap_{j=1}^{n}\bigcup_{\ell=1}^{L_{ij}} H_{ij\ell},$$

$$= \left(\bigcup_{\ell_1=1}^{L_{i1}} H_{i1\ell_1}\right) \cap \left(\bigcup_{\ell_2=1}^{L_{i2}} H_{i2\ell_2}\right) \cap \ldots \cap \left(\bigcup_{\ell_n=1}^{L_{in}} H_{in\ell_n}\right)$$

$$= \bigcup_{\ell_1=1}^{L_{i1}} \bigcup_{\ell_2=1}^{L_{i2}} \ldots \left(\bigcup_{\ell_2=1}^{L_{in}} H_{i2\ell_2}\right) \cap \ldots \cap \left(\bigcup_{\ell_n=1}^{L_{in}} H_{in\ell_n}\right)$$

a disjoint union. Again since C is a semi-ring we have

$$H_{i1\ell_1} \cap H_{i2\ell_2} \cap \ldots \cap H_{in\ell_n} \in C.$$

So combining (a) and (b) we see that $A \setminus B$ is a disjoint union of sets from C, that is, $A \setminus B \in A$

Similarly $E_i \cap F_j \in C$ and so

$$A \cap B = \left(\bigcup_{i=1}^{m} E_i\right) \cap \left(\bigcup_{j=1}^{n} F_j\right)$$

$$= \bigcup_{i=1}^{m} \bigcup_{j=1}^{n} (E_i \cap F_j)$$

$$\in A.$$

Then $A \cup B = (A \setminus B) \cup (A \cap B) \cup (B \setminus A)$ is a disjoint union of sets that we have seen above are in A and so $A \cup B \in A$. Thus A is a ring.

Trivially $C \subseteq A$. But by definition $R(C)$ is the smallest ring containing C. Hence **R**$(C) \subseteq A$.

Combining we get our result. $R(C) = A$.

Corollary: $\varepsilon = R(P)$.

Proof: This follows immediately from Theorem and the definition of ε as the collection of finite disjoint unions of sets from P, a semi-ring.

Definition: The *Borel sets* in $\mathbb{R}$ are the elements of the σ-field generated by P. It can be denoted by $\sigma(P)$, $B(P)$ or just B.

Theorem: In $\mathbb{R}$ we have

(i) $u \subseteq B(P)$,

(ii) $P \subseteq B(U)$,

(iii) $B(P) = B(u)$.

Proof:

(i) Given any $A \in u$ we can write

$$A = \bigcup_{i=1}^{\infty}(a_i, b_i) \text{ for some } a_i \text{ and } b_i,$$

$$= \bigcup_{i=1}^{\infty}\bigcup_{n=1}^{\infty}\left(a_i, b_i - \frac{1}{n}\right]$$

Each $(a_i, b_i - 1/n] \in P$ and so A is contained in any σ-field containing P, in particular $B(P)$. Hence $u \subseteq B(P)$.

(ii) Given any $A \in P$ we can write

$$A = (a, b] \qquad \text{for some } a \text{ and } b,$$

$$= \bigcap_{n=1}^{\infty}\left(a, b + \frac{1}{n}\right).$$

Each $(a, b + 1 \setminus n) \in u$ and, since σ-fields are closed under countable intersections, A lies in any σ-field containing u, in particular, $B(u)$. Hence $P \subseteq B(u)$.

(iii) Part (i) tells us that $B(P)$ is a σ-field containing u. But $B(u)$ is the minimal σ-field containing u. Hence $B(u) \subseteq B(P)$. Similarly part (ii) implies $B(P) \subseteq B(u)$. Hence $B(P) = B(u)$.

Note: For a collection of subsets A of a set X the σ-field generated by A need not contain the topology generated by A. This is because topologies contain possibly uncountable unions. Exceptionally, the usual topology, u, in $\mathbb{R}$ is made up from countable unions of intervals, (a, b). Hence any σ-field containing all intervals (a, b) must also contain U.

Chapter 9

Differentiation Theory of Functions

In the following sections we discuss the theory of differentiation of measures; but before going into the general theory of differentiation of measure, we review some basic facts about differentiation of functions of a real variable, since these are the ideas constitute the starting point for the theory of differentiation of measures.

Theorem: Every monotonic function f is differentiable almost everywhere.

Lebesgue's original proof assumed f to be continuous, but that assumption may be dropped, since every monotonic function has only jump discontinuities, and is continuous almost everywhere. The theorem is proved by considering the upper and lower, left and right, derived numbers. The upper and lower derived numbers are

$$\lambda^{\pm} = \liminf_{h \to 0^{\pm}} \frac{f(x+h)-f(x)}{h}, \quad \Lambda^{\pm} = \liminf_{h \to 0^{\pm}} \frac{f(x+h)-f(x)}{h}.$$

The derivative of f at x exists if all four derived numbers have the same finite value.

Lebesgue's theorem is easily extended to the larger class of functions of *bounded variation*. Let f be a real-valued function on the interval $[a, b]$, where $-\infty \cdot a < b \leq \infty$. The total variation function of f on $[a, x]$ is defined to be

$$T_f(a,x) = \sup_{P_n} \left\{ \sum_{j=1}^{n} | f(x_j) - f(x_{j-1}) |, n \in \mathbb{N} \right\},$$

$$P_n = \{a = x_0 < x_1 < \ldots < x_n = x\}.$$

Clearly $T_f(a, x)$ is a monotone non-decreasing function of x. If $T_f(a, b)$ is finite, then we say that f is of bounded variation on $[a, b]$. The class of functions of bounded variation on $[a, b]$ is denoted by $BV\,[a, b]$.

Theorem: *The functions of bounded variation on* $[a, b]$ *form a normed linear space* *with norm $||f||_{var} = T_f(a, b)$. The space $BV[a, b]$ is a complete normed linear space with this norm.*

Proof: One must first show that $||.||_{var}$ satisfies the triangle inequality.

There is a minor technical point in that constant functions have total variation zero; so the points in the Banach space are actually equivalence classes of functions which differ by a constant. The proof of completeness requires the use of Helly's theorem.

Jordan Decomposition

Theorem: *Every function of bounded variation can be written as the difference of two non-decreasing functions.*

Proof: Let $T(x) = T(a, x)$. It is clear that T is a non-decreasing function of x, since for $x < y$, $T(y) = T(x) + T(x, y)$.

Noting the identity $f = T - (T - f)$, we see that the theorem will be proved by showing that $T - f$ is non-decreasing. To show that $T(x) - f(x) \leq T(y) - f(y)$, or equivalently that

$$f(y) - f(x) \leq T(y) - T(x),$$

we note that $|f(y) - f(x)|$ is a particular sum of the type given in (5.1); hence $|f(y) - f(x)| \leq T_f(x, y)$, and the result follows.

Corollary: If $f \in BV([a, b])$ then the right and left hand limits of $f(x)$ exist for all x. The only discontinuities of f are jump discontinuities, and there are at most a countable number of them in $[a, b]$. Finally, f is differentiable almost everywhere.

The Cantor function is a continuous, monotonically non-decreasing function whose derivative vanishes almost everywhere.

Such a function is called a *singular function.* The Cantor function cannot be written as the integral of its derivative. Thus, the Cantor function is not an antiderivative.

Definition: A function F on a bounded interval $[a, b]$ is absolutely continuous iff given any $\varepsilon > 0$ and any collection (x_j, y_j), finite or countably infinite, of non-overlapping sub-intervals of (a, b), there exists a $\delta > 0$ such that

$$\sum_j F(y_j) - F(x_j) < \varepsilon \quad \text{whenever} \quad \sum_j (y_j - x_j) < \delta$$

Absolute continuity is a necessary and suffcient condition that a function be differentiable almost everywhere and equal to the integral of its derivative.

The necessity is not hard to establish. First note that the result is a slam dunk if f is bounded on $[a, b]$; for in that case

$$\left|\sum_j F(y_j) - F(x_j)\right| \le \sum_j \int_{x_j}^{y_j} |f| \, dm \le \sum_j \int_{x_j}^{y_j} C dm$$

$$= C\sum_j (y_j - x_j) = C\delta.$$

Hence we may choose $\delta = \varepsilon/C$.

If f is integrable but not bounded, we may decompose f into the sum of a bounded function g and a function h such that $\int |h| \, dm < \varepsilon/2$ where ε is any positive number. This is so because the bounded functions are dense in the integrable functions. It follows that

$$\left|\sum_j F(y_j) - F(x_j)\right| \le \sum_j \int_{x_j}^{y_j} |g| + |h| \, dm \le \sum_j \int_{x_j}^{y_j} C dm + \frac{\varepsilon}{2}$$

$$< C\sum_j (y_j - x_j) + \frac{\varepsilon}{2} < C\delta + \frac{\varepsilon}{2}.$$

The result follows by choosing δ so small that $C\delta < \varepsilon/2$.

The proof that absolute continuity implies that F is an indefinite integral is more involved; the classical proof is given in Riesz & Nagy.

But rather than recount the classical proof here, we shall defer it to the study of differentiation of measures, and obtain the result as a corollary of the Radon-Nikodym theorem.

Theorem: Every continuous monotone function can be decomposed into the sum of an absolutely continuous function and a singular function, both monotone. By the Jordan decomposition theorem, the same applies to continuous functions of bounded variation.

Proof: Let $f(x)$ be continuous and non-decreasing. Then f' exists almost everywhere, by Lebesgue's theorem. Now

$$\frac{f(x+h) - f(x)}{h} \ge 0$$

and this ratio tends to $f'(x)$ a.e. as $h \to 0$. On the other hand, its integral over an interval (a, b) is equal to

$$\frac{1}{h}\int_b^{b+h} f(x)dx - \frac{1}{h}\int_a^{a+h} f(x)dx;$$

hence, since f is continuous, its limit as $h \to 0$ is $f(b) - f(a)$. By Fatou's lemma we then have

$$\int_a^b f'(x)dx \le f(b) - f(a).$$

Since $f' \to 0$, it follows that $f' \in l^1(dx)$. Setting $g(x) = \int_a^x f'(t)dt$ and $h(x) = f(x) - g(x)$, we see that $h'(x) = 0$ *a.e.* Moreover, h is non-decreasing, since, for $y > x$,

$$h(y) - h(x) = f(y) - f(x) - \int_x^y f'(t) \ge 0.$$

Thus h is a non-decreasing singular function.

The result may be extended to general functions of bounded variation, as follows. A *saltus* function is one which has jump discontinuities at a countable set of point $\{x_j\}$. It can be written in the form

$$s(x) = \sum_{x_j \le x} u_j + \sum_{x_j < x} v_j,$$

where the series $\sum u_j$ and $\sum v_j$ are absolutely convergent. When u_j and v_j are positive, the function s is monotone increasing; moreover, by a theorem of Fubini, its derivative is zero almost everywhere.

Every monotone function f can be decomposed into the sum of a continuous function and a saltus function, both of which are monotone. We simply define s to be the saltus function whose jumps coincide in location and size with those of f. Then $g = f - s$ is continuous; and moreover, since $g' \ge 0$ *a.e.*, g is monotone by the argument used in the proof of Theorem.

Theorem: Every function of bounded variation f can be decomposed into the sum of an absolutely continuous function, a saltus function, and a singular function.

SIGNED MEASURES

Till now, we have only considered positive measures; we now turn to more general measures, which can take on both positive and negative values, or, more generally, complex values. A real-valued measure μ, defined on the Borel subsets of a topological space X is called a *signed* (Borel) *measure* if it can be written in the form $v = \mu_1 - \mu_2$, where μ_1, μ_2 are positive Borel measures, at least one of which is finite. A complex Borel measure is one which can be decomposed as a sum $v_1 + iv_2$ where v_1, v_2 are signed measures.

A signed measure vanishes on the empty set, assumes at most one of the values $\pm\infty$, and is countably additive: $v\left(\bigcup_j E_j\right) = \sum_j v(E_j)$, this sum converging absolutely if $v\bigcup_j E_j$ is finite. Given a measurable set E, a *partition* of E is a (finite or countable) disjoint collection $\{E_j\}$ such that $E = \bigcup_j E_j$. The partition is finite if $j = 1, \ldots n$. The collection of finite partitions of a set E is denoted by $P(E)$. If v is a signed measure (real or ocmplex), we define the *total variation* of v, denoted by $|v|$, to be

$$|v|(E) = \sup_{P(E)} \left\{ \sum_{j=1}^{n} |v(E_j)| \right\}.$$

Theorem: The total variation of a real or complex-valued signed measure v is a positive, countably additive measure.

Proof: It is clear from its definition that $|v|(\phi) = 0$; and it is straight for ward to show that $|v|$ is monotone and finitely additive. We thus need to establish its countable additivity. Suppose $\{E_j\}$ is a countable partition of E. By the monotonicity and finite additivity of $\{v\}$,

$$|v|(E) \geq (v)\left(\bigcup_{j=1}^{n} E_j \right) = \sum_{j=1}^{n} |v|(E_j)$$

for all n. If the infinite series $\sum_j |v(E_j)|$ diverges, then $|v|(E) = \infty$.

On the other hand, if the series converges, then $|v|(E) \geq \sum_{j=1}^{\infty} |v|(E_j)$, and we have to establish the opposite inequality. Let $\varepsilon > 0$ and let $\{E_i'\}$, $i = 1, \ldots n$ be a finite partition of E such that

$$|v|(E) < \sum_{i=1}^{n} |v|(E_i') + \varepsilon.$$

Then

$$v(E_i') = \sum_{j=1}^{\infty} v(E_{ij}), \; E_{ij} = E_i' \cap E_j,$$

and it follows that

$$|v|(E) < \sum_{i=1}^{n} |v(E_i')| + \varepsilon = \sum_{i=1}^{n} \sum_{j=1}^{\infty} |v(E_{ij})| + \varepsilon$$

$$\leq \sum_{j=1}^{\infty} \sum_{i=1}^{n} |v(E_{ij}| + \varepsilon \leq \sum_{j=1}^{\infty} \sum_{i=1}^{n} |v|(E_{ij}| + \varepsilon \; = \sum_{j=1}^{\infty} |v|(E_j) + \varepsilon.$$

Since ε is arbitrarily small, the countable additivity of $|v|$ follows.

In order to prove that $|v|$ is a Borel measure, we need to show that $|v|(E)$ is finite if E is compact. However, $|v|(E) \leq \mu_1(E) + \mu_2(E)$; so if E is compact, then $\mu_1(E)$ and $\mu_2(E)$, hence $|v|(E)$, are finite.

If v is a real signed measure, then

$$v^{\pm} = \frac{|v| \pm v}{2}$$

are positive measures, called the *positive* and *negative* variations of v. We thus have the decompositions:

$$v = v^+ - v^-, \ |v| = v^+ + v^-.$$

The first is a *Jordan decomposition* of the signed measure v, corresponding to the Jordan decomposition of a function of bounded variation.

The Jordan decomposition of a measure is not unique, since if $v^{\pm}$ form a Jordan decomposition of v, then so do $v^{\pm} + v'$ for any positive finite measure v'. However, there is a minimal decomposition in the following sense. We introduce an ordering on signed measures by $v_1 \leq v_2$ if $v_1(E) \leq v_2(E)$ for all sets $E \in X$. Then

Proposition: If $v = v_1 - v_2$ is any Jordan decomposition of the signed measure v, then $v^+ \leq v_1$ and $v^- \leq v_2$.

Proof: From the inequality

$$|v|(E) \leq |v(E)| = |(v_1 - v_2)(E)_j \leq (v_1 + v_2)(E), \ E \in M,$$

it follows that $|v| \leq v_1 + v_2$. Hence

$$v^+ = \frac{|v| + v}{2} \leq \frac{(v_1 + v_2) + (v_1 - v_2)}{2} = v_1,$$

and similarly for v^-.

Two measures μ and v on X, M are *mutually singular* if there exist sets $E, F \in M$ such that $X = E \cup F$, $E \cap F = \phi$, and $\mu(E) = 0$, $v(F) = 0$. We write $\mu \perp v$.

Hahn Decomposition

Theorem: Let $v = v^+ - v^-$ be the Jordan decomposition of the signed measure v. Then $v^+ \perp v^-$, and there exist measurable sets P and N such that

$$P \cup N = X, \ P \cap N = \phi, \ v^+(N) = v^-(P) = 0.$$

Moreover, v is positive on every subset of P and negative on every subset of N.

Proof: Assume that $v^+(X) < \infty$ and let $\{E_j\}$ be a sequence of sets such that $v(E_j) > v^+(X) - 2^{-j}$. Now $v^{\pm}$ are monotone, so for $A \in E_j^c$ we have

$$\nu(A) \leq \nu^+(E_j^c) = \nu^+(X) - \nu^+(E_j) \leq \nu^+(X) - \nu(E_j) \leq \frac{1}{2^j}.$$

Since this holds for all $A \in E_j^c$ we have $\nu^+\left(E_j^c\right) \leq 2^{-j}$. By the Jordan decomposition theorem,

$$-\nu^-(E_j) = \nu(E_j) - \nu^+(E_j) \geq \nu(E_j) - \nu^+(X) \geq -\frac{1}{2^j}.$$

Hence $\nu^-(E_j) \leq 2^{-j}$.

Now let

$$P = \limsup_j E_j,\ N = P^C = \liminf_j E_j^c.$$

Then

$$0 \leq \nu^+(N) = \limsup_j \nu^+\left(E_j^c\right) \leq \liminf_j \frac{1}{2^j} = 0;$$

$$0 \leq \nu^-(P) \leq \nu^-\left(\bigcup_{j\geq k} E_j\right) \leq \sum_{j=k}^{\infty} \nu^-(E_j) \leq \sum_{j=k}^{\infty} \frac{1}{2^j} = \frac{1}{2^{k-1}},$$

for any k. Therefore $\nu^-(P) = 0$.

If $E \subset N$ then $\nu(E) = \nu^+(E) - \nu^-(E) = -\nu^-(E) < 0$, hence ν is negative on all subsets of N. Similarly, ν is positive on all subsets of P.

THE LEBESGUE-RADON-NIKODYM THEOREM

In the example above, $\nu_f^{\pm}$ are mutually singular; and, by the Hahn decomposition theorem, the same holds for the components $\mu^{\pm}$ of the Jordan decomposition of a measure μ. The Borel measure generated by the Cantor function and Lebesgue measure are mutually singular with respect to one another.

The measure ν_f has the property that $\nu_f(E) = 0$ whenever $\mu(E) = 0$. The measure ν_f is said to be *absolutely continuous* with respect to μ, and we write $\nu_f \ll \mu$. The measure-theoretic analog of Theorem is the following, known as the Lebesgue decomposition theorem.

Theorem: Let ν be a σ-finite signed measure and μ a positive σ-finite measure on (X, M). Then there exists a unique decomposition

$$\nu = \rho + \lambda,\ \rho \ll \mu\ \lambda \perp \mu,$$

where ρ, λ, are σ-finite. Moreover, there is a function $f \in L^1(X, d\mu)$ such that $d\rho = fd\mu$.

When $d\rho = fd\mu$, the density function f is called the Radon-Nikodym derivative of ρ with respect to λ, is denoted by $d\rho/ d\lambda$. We prove Theorem in a sequence of lemmas.

Lemma: Let ν be finite, and μ a positive measure on (X, M), with ν. Then $\nu \ll \mu$ iff for every $\varepsilon > 0$ there is a $\delta > 0$ such that $|\nu(E)| < \varepsilon$ whenever $\mu(E) < \delta$.

Proof: By the Jordan decomposition theorem, it suffices to prove the result when ν is a positive measure. The sufficiency of the ε, δ condition is immediate; so we have only to establish its necessity. Suppose $\nu \ll \mu$ but that the ε, δ condition fails. Then there is an $\varepsilon > 0$ and a sequence of sets $\{E_j\}$ such that $\nu(E_j) > \varepsilon$ while $\mu(E_j) < 2^{-j}$. Let $E = \limsup_j E_j$. By Fatou's lemma, $\nu(E) \geq \limsup_j \nu(E_j) \geq \varepsilon$. On the other hand,

$$\mu(E) = \mu\left(\bigcap_{k\geq 1}\bigcup_{j\geq k} E_j\right) = \lim_{k\to\infty} \mu\left(\bigcup_{j\geq k} E_j\right)$$

$$\leq \lim_{k\to\infty} \sum_{j=k}^{\infty} \mu(E_j) \leq \lim_{k\to\infty} 2^{1-k} = 0.$$

Thus $\nu(E) > \varepsilon$ while $\mu(E) = 0$, and so ν is not absolutely continuous with respect to μ. Therefore the ε, δ condition is a necessary condition for absolute continuity.

Corollary: The antiderivative of an integrable function is absolutely continuous. More generally, for any positive measure μ and $f \in L^1(X, \mu)$ we have that for every $\varepsilon > 0$ there exists a $\delta > 0$ such that $\left|\int_E fd\mu\right| < \varepsilon$ whenever $\mu(E) < \delta$.

Lemma: Let $\nu \ll \mu$, where ν is positive, finite, and not identically 0, and μ is positive. Then there exist $\varepsilon > 0$ and a measurable set E such that $\mu(E) > 0$ and $\nu(E') \geq \varepsilon\mu(E')$ for all $E' \subset E$.

Proof: Let $\nu_j = \nu - j^{-1}\mu$, $j = 1, 2, \ldots$, and let P_j, N_j be the corresponding Hahn decompositions of ν_j: $\nu_j^-(P_j) = 0, \nu_j^+(N_j) = 0$. The sequence of measures $\{\nu_j\}$ is increasing; hence the P_j are increasing, and their complements therefore decreasing. Setting $P = \bigcup_j P$ and $N = \bigcap_j N_j$ we have $P = N^c$. Since ν_j is negative on N_j, we have

$$0 \leq \nu(N_j) \leq \frac{1}{j}\mu(N_j) \leq \frac{1}{j}\mu(N_1) \to 0,, \text{ as } j \to \infty;$$

and therefore $\nu(N) = 0$. Since ν is not zero, $\nu(P) > 0$; and therefore $\nu(Pj) > 0$ for some j.

The lemma is proved by taking $E = P_j$ and $e = j^{-1}$ for this value of j.

Lemma: Let v be finite and positive, and put

$$F = \left\{ f : f \in L^1(d\mu), v(E) \geq \int_E f d\mu,\ E \in M \right\}.$$

Then there is a unique function $g \in F$, finite a.e., such that

$$\int_E g d\mu = \sup_F \int_X f d\mu.$$

Remark: The class F is nonempty, since one can always take $f = \varepsilon$, the positive number guaranteed in Lemma.

Proof: Let M denote the supremum on the right and let $\{f_j\}$ be a sequence of functions in F such that $\int f_j d\mu \to M$. The sequence $g_n = \max[f_1, f_2, ..., f_n]$, is a monotone increasing sequence of functions; and by the monotone convergence theorem $M = \lim_j \int g_j d\mu = \int \lim_j g_j d\mu$.

Now $g_j \in F$ for all j. In fact, let $E_j = \{x \in E: g_j(x) = f_j(x)g$. Then

$$\int_E g_n d\mu = \sum_{j=1}^{n} \int_{E_j} f_j d\mu \leq \sum_{j=1}^{n} v(E_j) = v(E).$$

By the monotone convergence theorem, F is closed under monotone convergence, so $g = \lim_j g_j \in F$ hence

$$\int_X g d\mu = \lim_n \int_X g_n d\mu = M.$$

The proof of uniqueness is left to the reader.

Radon-Nikodym

Theorem: Let μ and v be σ-finite measures on a set X, and suppose that $v \ll \mu$. Then there is an $f \in L^1(X, M, d\mu)$ such that

$$v(E) = \int_X f d\mu,$$

for all $E \in M$ for which $v(E)$ is finite.

Proof: The proof can be reduced to the case in which v and μ are real, positive and finite. Let f be the maximal function guaranteed by Lemma. Then $v_1(E) = v(E) - \int_E f d\mu$ satisfies $v_1 \geq 0$ and $v_1 \ll \mu$. If v_1 is not identically zero, then by Lemma there is an $\varepsilon > 0$ and a set E' such that

$$\mu(E') > 0,\ v_1(E'') > \varepsilon\mu(E''),\ E'' \subset E'.$$

It follows that

$$v(E) \geq \int_E f + \varepsilon\chi_{E'} d\mu,\ E \in M,$$

contradicting the maximality of f.

The proof that absolute continuity of a function F is sufficient that it be an antiderivative is an immediate corollary of the Radon Nikodym theorem. If F is a monotone increasing real valued function on the line, then the Lebesgue-Stieltjes measure v_F generated by F is the Borel measure generated by the premeasure $v_0((a, b]) = F(b) - F(a)$.

If F is an absolutely continuous function on the line, then v_F is absolutely continuous with respect to Lebesgue measure. Therefore $v_F = f\,dm$, where f is the Radon-Nikodym derivative of v_F with respect to m. Hence for any interval $[a, b]$,

$$v_F((a, b]) = F(b) - F(a) = \int_{[a,b]} fdm = \int_a^b f(x)dx.$$

It follows that $F' = f$ at points of continuity of f.

If F is monotone, then v_F is a positive measure; and if F is of bounded variation, then v_F is a signed measure with total variation $|v_F|(I) = T_F(I)$, where I is an interval in $\mathbb{R}$. On the other hand, the Cantor measure, the Borel measure generated by the Cantor function, is singular with respect to Lebesgue measure; and more generally, singular functions generate Borel measures which are singular with respect to Lebesgue measure.

Corollary: Let μ and v satisfy the hypotheses of the Radon-Nikodym theorem. If $E \in M$ and if $g \in L^1(E, dv)$, then $gf \in L^1(X, d\mu)$ and $\int_E gdv = \int_E fgd\mu$.

Proof: When $g = \chi_E$ the result is an immediate consequence of the Radon-Nikodym theorem. The result extends immediately to simple functions by additivity, and then to any $g \in L_1(dv)$ by approximating g by simple functions and going to the limit.

Lemma: Let μ, v be positive measures and suppose that if f is any function such that

$$\int_E fd\mu \leq v(E),\ E \in M$$

then it follows $f = 0$ a.e. $[\mu]$. Then $\mu \perp v$.

Proof: Consider the measure $\mu \wedge v = (\mu + v - |\mu - v|)/2$, where $|\mu - v|$ denotes the total variation of $\mu - v$. For any set $E \in M$, $|\mu - v|(E) \geq |\mu(E) - v(E)|$, so

$$(\mu \wedge v)(E) \leq \frac{\mu(E) + v(E) - |\mu(E) - v(E)|}{2} = \min(\mu(E), v(E)) \leq \mu(E).$$

Consequently $\mu \wedge \nu \ll \mu$, and so by the Radon-Nikodym theorem, there is a function g such that

$$(\mu \wedge \nu)(E) = \int_E g d\mu$$

for all E for which $(\mu \wedge \nu)(E)$ is finite.

Now also $\nu \geq \mu \wedge \nu$, and this implies, by the hypothesis of the lemma, that $g = 0$. Hence $\mu \wedge \nu = 0$, and it follows that

$$\mu + \nu = |\mu - \nu| = (\mu - \nu)^+ + (\mu - \nu)^-.$$

Letting $X = P \cup N$ be the Hahn decomposition relative to $\mu - \nu$, we see that

$$(\mu + \nu)(P) = (\mu - \nu)^+(P) \leq \mu(P), (\mu + \nu)(N) = (\mu - \nu)^-(N) \leq \nu(N);$$

hence $\nu(P) = \mu(N) = 0$.

We are now ready to prove the Lebesgue decomposition theorem. We may assume that μ and ν are finite. Let f be the maximal element of F as given in Lemma. Then

$$\nu(E) \geq \int_E f d\mu, \ E \in M.$$

If $f = 0$ a.e. then $\nu \perp \mu$ and the Lebesgue decomposition is $\rho = 0$, $\lambda = \nu$. Otherwise, let $\nu_2(E) = \nu(E) - \int_E f d\mu$, and suppose g is any function in $L^1(d\mu)$ such that $\nu_2(E) \geq \int_E g d\mu$ for all $E \in M$. Then $g = 0$ by the maximality of f, hence $\nu_2 \perp \mu$. The Lebesgue decomposition of ν is thus given by

$$\rho(E) = \int_E f d\mu, \ \lambda(E) = \nu(E) - \int_E f d\mu.$$

If ρ', λ' is a second Lebesgue decomposition, then $\rho - \rho' = \lambda' - \lambda$. But the measure on the left side is absolutely continuous while the right side is singular with respect to μ; thus both sides must be zero.

The Dirac measure is the Borel measure generated by the Heaviside step function; it assigns a measure of 1 to the point 0, and zero measure to all sets not containing 0. It is singular with respect to Lebesgue measure, and therefore does not have a Radon-Nikodym derivative with respect to Lebesgue measure. It is the measure-theoretic analog of the saltus functions in the decomposition of real valued functions of bounded variation.

The Lebesgue decomposition theorem has a similar extension in the case of Borel measures on the line. An *atom* of a measure ν is a point x for which $\nu(x) \neq 0$ iff $x \in A$. A measure is called *nonatomic* if it has no atoms. An absolutely continuous function has no atoms; and if F is of bounded variation, then the atoms ν_F are precisely the points of jump discontinuity of F, and so are countable. Let ν_F be the Borel measure generated by F, where F has finite total variation. By Theorem we may write $F = F_1 + F_2 + F_3$, where F_1 is absolutely continuous, F_2 is singular, and F_3 is a saltus function. It follows immediately that $v_F = v_1 + v_2 + v_3$, where $v_1 << m$, $v_2 \perp m$ and v_3 is atomic.

DIFFERENTIATION ON $\mathbb{R}^n$

The function class $L^1_{loc} = L^1_{loc}(\mathbb{R}^n, dm)$ is the class of Borel measurable functions on $\mathbb{R}^n$ which are Lebesgue integrable over any bounded measurable set. The measure $v_f(E) = \int_E f dm$ is absolutely continuous with respect to Lebesgue measure on $\mathbb{R}^n$, and its Radon-Nikodym derivative is f. In one dimension, when f is continuous, f is obtained as the limit

$$\lim_{r\to 0} \frac{v_f(B_r(x))}{m(B_r(x))} = \lim_{h\to 0} \frac{1}{2h} \int_{x-h}^{x+h} f dm = f(x),$$

where $B_r(x)$ is the interval $(x - r, x + r)$.

In this section, we examine this relationship in the more general case when $f \in L^1_{loc}(\mathbb{R}^n, m)$. Let $B_r(x)$ denote the ball of radius r centered at $x \in \mathbb{R}^n$, and consider the Borel measure on v_f on $\mathbb{R}^n$, where $f \in L^1_{loc}(\mathbb{R}^n, dm)$.

It is straightforward to prove that

$$f(x) = \lim_{r\to 0} \frac{v_f(B_r(x))}{m(B_r(x))}$$

for all $x \in \mathbb{R}^n$ belonging to the *Lebesgue set* L_f defined by

$$L_f = \left\{ x : \lim_{r\to 0} \frac{1}{m(B_r(x))} \int_{B_r(x)} | f(y) - f(x) | dy = 0 \right\}.$$

It is not hard to see that Lebesgue set contains the points of continuity of f.

Lebesgue Differentiation Theorem

Theorem: Let $f \in L^1_{loc}$. If $x \in L_f$ then the Radon-Nikodym derivative of v_f is given by for all x in the Lebesgue set of f. The Lebesgue set of f contains the points of continuity of f; its complement has measure 0.

The proof that the complement of the Lebesgue set has measure 0 will be carried out in a series of lemmas.

Lemma: Let C be a collection of open balls in $\mathbb{R}^n$ and let $U = \bigcup_{B\in C} B$. For every $c <$

$m(U)$ there exists a finite collection of disjoint sets $B_j \in C$, $j = 1, \ldots, k$, such that $c < 3^n \sum_{j=1}^{k} m(B_j)$.

Proof: Choose a compact set $K \subset U$ with $m(K) > c$. Since K is compact there is a finite collection of balls $A_1, \ldots, A_m$ which cover K; and we may order them in decreasing radii, $r_1 \geq r_2 \geq \ldots$ Choose $B_1 = A_1$; cast out all the A_i which have a nonempty intersection with B_1; and continue the process until all the $A_i^{'}s$ are exhausted. We have now obtained a finite collection $B_1, \ldots B_k$ of disjoint open balls.

There are two possibilities to consider. If U is originally a disjoint union of open balls, then $B_j = A_j$ and the result is immediate.

Otherwise some of the A_i will be not included in the $A_j^{'}s$, and for each such A_i, there is a smallest integer j such that $A_i \cap B_j \neq \phi$.

By the construction, the radius of A_i will be at most that of B_j; and so $A_i \subset B_j^*$, the ball concentric with B_j with three times the radius. Then $K \subset \bigcup_{j=1}^{k} B_j^*$ and

$$c < m(K) \leq \sum_{j=1}^{k} m(B_j^*) = 3^n \sum_{j=1}^{k} m(B_j).$$

The central method is based on the use of the *Hardy-Littlewood maximal function*, defined by

$$Hf(x) = \sup_{r>0} \frac{1}{m(B_r(x))} \int_{B_r(x)} |f| \, dm, f \in L_{loc}^1.$$

Hardy-Littlewood Maximal Theorem

Theorem: For $f \in L^1(\mathbb{R}^n, m)$ and $\alpha > 0$,

$$m(\{x : Hf(x) > \alpha\}) \leq \frac{3^n}{\alpha} \| f \|_1 .$$

Proof: Let $U_\alpha = \{x: Hf > \alpha\}$; and let $c = m(U_\alpha)$. For each $x \in U_\alpha$ there is a radius r_x such that

$$\alpha m(B_{r_x}(x)) \leq \int_{B_{r_x}(x)} \int |f| \, dm.$$

Since the balls $B_{r_x}(x)$ cover U_α, there is a finite disjoint union of them, $\{B_j\}$, for which

$$c < 3^n \sum_{j=1}^{k} m(B_j) \le \frac{3^n}{\alpha} \sum_{j=1}^{k} \int_{B_j}\int |f|\,dm \le \frac{3^n}{\alpha} \int_{\mathbb{R}^n}\int |f|\,dm.$$

Since this result is true for all $c < m(U_\alpha)$, the theorem follows, with constant $C = 3^n$.

We also need the *Tschebyshev inequality*: Given any $f \in L^1(\mathbb{R}^n, m)$, and all $\alpha > 0$,

$$m(F_\alpha) \le \frac{1}{\alpha} \| f \|_1, F_\alpha = \{x : | f(x) | \ge \alpha\}.$$

This simple inequality is well known in Probability theory; its proof is left to the reader.

Lemma: or $x \in \mathbb{R}^n$ and $r > 0$, let

$$E(f, x, r) = \frac{\int\int_{B_r(x)} f dm}{m\left(B_r^x\right)}.$$

Then if $f \in L^1_{loc}, \lim_{r\to 0} E(f, r, x) = f(x)$ a.e.

Proof: If f is continuous, the proof is well known, and is the standard method of proof of the Fundamental Theorem of Calculus. In the general case, we approximate f by a continuous function g.

We may restrict ourselves to a compact ball $K \subset \mathbb{R}^n$, since the result is purely local; thus we may truncate f by setting it identically equal to 0 outside K. Thus, given any $\varepsilon > 0$ there is a continuous function g with support in K such that $\int_K | f - g | \, dm < \varepsilon$. Since K is compact, g is uniformly continuous on K; and thus for every $\delta > 0$ there is an $r > 0$ such that

$$|g(x) - g(y)| < \delta,\ |x - y| < r,\ x, y \in K.$$

Therefore $|E(g, x, r) - g(x)| < \delta$; and $\lim_{r\to 0} E(g, x, r) = g(x)$.

Therefore

$$\limsup_{r\to 0} | E(f, x, r) - f(x) | =$$

$$\limsup_{r\to 0} | E(f - g, x, r) + (E(g, x, r) - g(x)) + (g(x) - f(x)) |$$

$$\le H(f - g)(x) + | f(x) - g(x) |.$$

Let

$$E_\alpha = \left\{ x : \limsup_{r>0} | E(f,x,r) - f(x) | > \alpha \right\}, F_\alpha = \{x : | f - g | (x) > \alpha\}.$$

Then by the inequality above, $E_\alpha \subset F_{\alpha/2} \cup \{x : H(f-g)(x) > \alpha/2\}$. By the Tchebyshev inequality, and the Maximal Theorem,

$$\frac{\alpha}{2} m(F_{\alpha/2}) \leq \iint_K | f - g | \, dm < \varepsilon,$$

$$\frac{\alpha}{2} m(\{x : H(f-g)(x) > \alpha/2\}) < 3^n \iint_K | f - g | \, dm < 3^n \varepsilon;$$

hence

$$m(E_\alpha) \leq \frac{2\varepsilon}{\alpha}(3^n + 1).$$

Since ε is arbitrary, $m(E_\alpha) = 0$ for all $\alpha > 0$: Since $\lim_{r \to 0} E(f, r, x) = f(x)$ on the complement of $\bigcup_n E_{1/n}$, the result follows.

We may now complete the proof of the Lebesgue di®erentiation theorem. Let

$$E_z = \left\{ x : \lim_{r \to 0} E(| f - z |, x, r) = | f(x) - z | \right\}.$$

Then $m\left(E_z^c\right) = 0$ by the Lemma above. Let D be a countable dense subset of $\mathbb{C}$ and $E = \bigcup_{z \in D} E_z$. Then $m(E^c) = 0$; and for $x \in E$ and any $\varepsilon > 0$ there exists a $z \in D$ such that $| f(x) - z | < \varepsilon/2$. Then

$$\begin{aligned}\lim_{r \to 0} E(| f - f(x) |, x, r) &\leq \lim_{r \to 0} (E(| f - z |, x, r) + E(| z - f(x) |, x, r)) \\ &< \lim_{r \to 0} E(| f - z |, x, r) + \frac{\varepsilon}{2} = | f(x) - z | + \frac{\varepsilon}{2} \\ &< \varepsilon.\end{aligned}$$

Since " is arbitrary, it follows that $\lim_{r \to 0} E(| f - f(x) |, x, r) = 0$ for $x \in E$, and the proof of the theorem is complete.

LEBESGUE APPROACH TO INTEGRATION

The Riemann integral is too restrictive to handle interchange of limits and integration gracefully and is not adequate to the task of building complete metric spaces based on integral norms. The Lebesgue theory proposes an alternate definition that resolves these difficulties, yet coincides with Riemann integral on all Riemann integrable functions. The approach embodied in Lebesgue's theory generalizes easily and usefully beyond the context of integration on $\mathbb{R}^n$, and we intend to present the theory abstractly enough to see its general structure and to pursue several different applications.

The purpose of this lecture is to outline the basic deffinitions and strategy of Lebesgue's approach to integration. To motivate the discussion and to bring out the central ideas, we will ¯rst formulate a list of the minimum prc,,erties we want of an integral operator and its class of integrands. We then deduce some basic consequences of these requirements, which in turn reveal will reveal the basic structure of the theory we want. Although the initial list of properties is based on trying to extend the Riemann integral of a function on $\mathbb{R}^n$, the ideas we abstract from the discussion lead to a general theory of measure and integration.

This lecture will be informal. We will make precise statements, but we will not prove everything; that will be left to future lectures.

Minimal Requirements for a Theory of Integration Generalizing the Riemann Integral

We consider integration over $\mathbb{R}^n$. Given a real-valued function f, f^+ and f^- will denote its positive and negative parts, respectively. Let $V = [a_1, b_1] \subset [a_n, b_n]$ be a rectangle in $\mathbb{R}^n$. The notation $(R)\int_V fdx$ will denote the Riemann integral of f over the rectange V. Given a subset G of $\mathbb{R}^n$, χ_G denotes the indicator function of G.

We want to have an integration theory that is well-behaved with respect to limit operations on integrands, so we will write down the minimal properties that we wish our integration theory to possess. There are two components to consider: a set B of real-valued functions on $\mathbb{R}^n$ we wish to integrate, and an integral operation $f \to \int f$.

Of B we want at the minimum:

F1 B contains all continuous functions.

F2 If $f(x) = \lim_n f_n(x)$ for all x and $f_n \in B$ for every n, then $f \in B$. (B is closed under point-wise limits.)

F3 If $f, g \in B$ and $\chi, \chi \in \mathbb{R}$, then $\chi f + \chi g \in B$. (B is a vector space.)

F4 If $f \in B$, then so are $f^+ = \max\{f, 0\}$ and $f^- = \subset \min\{f, 0\}$.

F5 If $f \in B$ and V is a rectangle, then $f\chi_V \in B$.

The reason for the requirements F1] – [F3] is clear: our integral should be able to handle at least the continuous functions, but, unlike the Riemann integral, it should also handle limits of integrable functions; and of course, it should handle, sums of integrable functions.

Requirements [F4] and [F5] are made for slightly more technical reasons that will become apparent in a moment.

The integration operation will assign to every non-negative f, a value $\int f$ in $\mathbb{R}\cup\{\infty\}$. Thus integrals with in finite values are allowed; this is very convenient.

For an arbitrary function f in B, let

$$\int f \triangleq \int f^{+} \subset \int f^{-}$$

if the right-hand side is the indeterminate form $\infty \subset \infty$. If both integrals $\int f^{+}$ and $\int f^{-}$ are in nite, $\int f$ is left undete ned. Thus, the integral takes values in the extended reals, $\overline{\mathbb{R}} \triangleq \mathbb{R}\cup\{\infty \subset \infty\}$. Of the integral, we require:

1. If f is continuous and $V \subset \mathbb{R}^n$ is a rectangle, $\int f\chi_V = (R)\int_V fdx$. $\int$ coincides (with the Riemann integral on continuous functions on bounded sets.)
2. (Monotone convergence property) If $f_n \in B$ for every n and if $0 \subset f_1 \subset f_2 \subset\subset\subset$, then

$$\int \lim_n f_n = \lim_n \int f_n .$$

3. (Linearity) If f, $g \in B$ and χ, $\chi \in \mathbb{R}$, then

$$\int [\chi f + \chi g] = \chi \int f + \chi \int g,$$

as long as the right-hand side is not the indeterminate form $\infty \subset \infty$.

It is easy to see that there is a smallest set of functions from $\mathbb{R}^n$ to $\mathbb{R}$ satisfying [F1] and [F2]. Just take the intersection of all sets of functions satisfying [F1] and [F2], and verify that this intersection still satisfies [F1] and [F2]. Functions in this smallest set are called the *Baire functions* on $\mathbb{R}^n$, as they were first defined by Rene Baire(1899).

The set of Baire functions can be constructed by trans finite recursion, omitted here. The set of Baire functions on $\mathbb{R}^n$ is in fact a proper subset of the set of all real-valued functions on $\mathbb{R}^n$. Also, it automatically satisfies [F3]||[F5]. The set of continuous functions satisfies [F3] and the Baire functions inherit the vector space property by transfinite reduction, and [F4] and [F5] can also be proved by transfinite induction. But we need not worry about that here.

Just think of the Baire functions as the smallest set of functions closed under [F1]||[F5]. Since we want to be conservative and ask for the minimum that our requirements entail, *we will henceforth take B to be precisely the set of Baire functions.* The Baire functions are also closed under the operation of multiplication; that is, if f and g are Baire functions, then so is fg. This can also be proved by trans nite induction.

For the moment we ask you to admit it as a fact (or you can add it as a required property [F6].) When we get around to full proofs we will avoid all arguments by trans nite induction.

Basic consequences of [F1]-[F5]

We don't know yet whether there exists a theory satisfying the requirements [F1]-[F5]. But this does not prevent us exploring their hypothetical consequences! In particular, what do they imply about measuring volumes of sets? To frame the question mathematically, define the class of sets:

$$B \triangleq \{A \subset \mathbb{R}^n \mid \chi_A \in B\},$$

and, for each set A in B, de ne its volume

$$|A| \triangleq m(A) \triangleq \int \chi_A.$$

Here is what can first be said about B and m, followed by incomplete proofs. It is assumed throughout that $\int$ satisfies.

Lemma: B is the smallest class of subsets of $\mathbb{R}^n$ such that B contains all open sets and

1. $\mathbb{R} \in B$;
2. $A \in B$ implies $A^c \in B$;
3. If $A_1, A_2, \ldots$ are all in B, then $\bigcup_1^\infty A_n \in B$
4. If $A_1, A_2, \ldots$ are all in B, then $\bigcap_1^\infty A_n \in B$.

Lemma: If V is the rectangle $[a_1, b_1) \subset \subset\subset\subset [a_n, b_n)$ in $\mathbb{R}^n$,

$$m(V) = \int \chi_V = \prod_1^n (b_i \subset a_i)$$

Let $A_1, A_2, \ldots$ belong to B and assume they are pairwise disjoint – $A_i \cap A_j = \phi$ for all $i \neq j$. Then

$$m\left(\bigcup_1^\infty A_i\right) = \sum_1^\infty m(A_i).$$

Remarks: In equation of Lemma we could as well have let the intervals in the product defining V be open or closed, or any combination of open, closed, and half open intervals. We chose the form appearing to x a standard choice that will be convenient later on.

Proof of Lemma: Let U be any open set in $\mathbb{R}^n$. Let $\text{dist}(x, U^c) \triangleq \inf\{|x \subset y| \mid y \in U^c\}$ denote the function which to every x assigns its distance to U^c. This is a continuous function as is $\max\{1, n \text{ dist}(x, U^c)\}$, for every positive integer n. Now for every x

$$\chi_U(x) = \lim_x \max\{1, n \text{ dist}(x, U^c)\}$$

exhibi χ_U as a pointwise limit of continuous functions, so χ_U is a Baire function and $U \in B$.

To see that [B1] is true, simply note that $\chi_{\mathbb{R}} \subset 1$ is a continuous function and so $\chi_{\mathbb{R}}$ is in B by [F1].

To see that [B2] is true, note that if $A \in B$, then $\chi_A \in B$. By [F3], $\chi_{U^c} = 1 \subset \chi_A$ is also in B and hence $A^c \in B$. The same reasoning shows

$$\text{if } A, B \in B \text{ and } A \subset B, \text{ then } A \subset B \in B.$$

Recalling that Baire functions are closed under multiplication (here is the only place we use this unproved fact), it follows that

$$A, B \in B \text{ imply } A \cap B \in B,$$

because $\chi_{A \cap B} = \chi_A \chi_B$. Now if A and B are two sets in B, it follows that $A \subset B = A \subset (A \cap B) \in B$. Since $\chi_{A \cup B} = \chi_B + \chi_{A-B}$, [F2] implies, $A \cup B$ is in B also. By iteration B will be closed under finite unions.

Now consider a countable union $A = \bigcup_1^\infty A_i$ of sets in B, since each finite union $\bigcup_1^n A_i$ is in B and since $\chi_A = \lim_n \chi_{\bigcup_1^n A_i}$, it follows again from [F2] that $A \in B$. This proves [B3]. To prove [B4] use [B2], [B3], and the general identity $\bigcap_\lambda C_\lambda = \left(\bigcup_\lambda C_\lambda^c\right)^c$.

We have shown that B satisfies [B1]-[B4], but have yet to show it is the smallest such class containing open sets. We defer this, because it is more naturally proved later.

Proof of Lemma: This is a straightforward consequence of properties. Let $A_1, A_2, \ldots$ be a sequence of pairwise disjoint sets in B, and let $A \triangleq \bigcup_1^\infty A_i$ Let $f_n \triangleq \chi_{\bigcup_1^n A_i}$. By the disjointness of the sets $f_n = \sum_1^n \chi_{A_i}$, and so from linearity of the integration operation, property [I3],

$$\int f_n = \sum_1^n \int \chi_{A_i} = \sum_1^n m(A_i)$$

The functions f_n are non-negative, increase with increasing n, and $\lim_n f_n(x) = \chi_A$. Therefore, the monotone convergence property requires

$$m(A) = \int \chi_{\bigcup^{\infty} A_i} = \lim_n \int f_n$$

$$= \lim_n \sum_1^n m(A_i) = \sum_1^{\infty} m(A_i).$$

MEASURES AND C-ALGEBRAS

At this point we are ready to frame some abstract definitions to reformulate Lemmas. Let S be an arbitrary butfixed nonempty set.

Definition: A collection of subsets of S is called a χ-algebra if

1. $S \in .$;
2. $A \in .$ implies $A^c \in .$;
3. If $A_n \in .$ for every n, $\bigcup_1^{\infty} A_i \in$ ..
4. If $A_n \in .$ for every n, $\bigcap_1^{\infty} A_i \in$ ..

In words, a χ-algebra is a subset of a power set which is closed under complements, countable unions and countable intersections. Note that properties [S1], [S2], [S3] imply [S4], and properties [S1], [S2], [S4] imply [S3], so including both [S3] and [S4] in the definition is redundant; however, both are stated to emphasize that a χ-algebra is closed under both countable unions and countable intersections.

Notice [S1]-[S4] simply generalize to an abstract setting the properties of [B1]-[B4] of B stated in Lemma.

Lemma says that B is the smallest χ-algebra containing the open sets; this will be explained further below.

Definition: Let $(S,.)$ be a non-empty set together with a χ-algebra of its subsets. A function χ which assigns to each set $A \in .$ a number in $[0, \infty]$ is called a measure if

1. $\chi(\phi) = 0$;

2. (Countable additivity) If $A_1, A_2,...$ is a sequence of pairwise disjoint sets in.·, then

$$\chi\left(\bigcup_1^{\infty} A_i\right)=\sum_1^{\infty}\chi(A_i)$$

Observe that [M2] is just a generalization of the property of Lemma, which was a consequence in turn of the requirement of monotone convergence for integration. Notice also, that if $A_1,...., A_n$ is a finite sequence of pair-wise disjoint sets, then application of [M2] with $A_{n+1} = \phi,..., A_{n+2} = \phi$, etc., leads to finite additivity:

$$\chi\left(\bigcup_1^{n} A_i\right)=\sum_1^{n}\chi(A_i).$$

Moreover, if there exists at least one set A with $\chi(A) < \infty$, then application of finite additivity to $\chi(A) = \chi(A) + \chi(\phi)$, yields the property [M1], $\chi(\phi) = 0$.

Let us now summarize the results so far using the language of these new definitions. A theory of integration satisfying properites [F1]-[F3] entails at least the existence of a countably additive measure m on the smallest χ-algebra containing the open sets, such that $m(V)$ is the usual volume of V for all rectangles V.

Before continuing, we give some other examples of our newly defined objects and make some simple observations.

Examples of χ-algebras. Let S be non-empty.

1. The power set $P(S)$ of S is a χ-algebra.
2. $\{S, \phi\}$ is a χ-algebra.
3. Let $S = B_1 \cup B_2 \cup \subset\subset\subset\cup B_m$ and suppose that the B_i's are pairwise disjoint.Consider the collection C of all unions $\bigcup_{i\in K} B_i$, where K ranges over all subsets of $\{1, 2,..., m\}$. This is also a χ-algebra. Since C is finite, any union or intersection of its elements reduces to a finite union or intersection. The sets $B_1,...,$ B_m are called the *atoms* of C.

Examples of measures

1. (Trivial measures) On $(S, P(S))$, let $\chi(A) = 0$ for every $A \in P$. Let $\chi(A) = \infty$ for every $A \in P(S)$. Both χ and χ are measures.
2. (Counting measure on a finite set) Let S be a finite set. On $(S, P(S)$, let $\chi(A)$ be the number of elements in A.
3. (Finite atomic measures) Take a general non-empty set S with an arbitrary χ-algebra of its subsets. Let $x_1,..., x_n$ be distinct points in S. For each $A \in$.·, let $\chi(A)$ be the number of points in the set $\{x_1,..., x_n\} \cap A$.

We have described B as the smallest χ-algebra of subsets of $\mathbb{R}^n$ containing the open sets, but we have yet to show that this description is really meaningful.

This we do next. Only the following observation is needed: if $\{x_1,...., x_n\}$ is a nonempty family of χ-algebras of a fixed set S, then the intersection $\bigcap_{\lambda\in\Lambda} \lambda$ is again a χ-algebra.

Now let $\mathbb{R}$ be a non-empty collection of subsets of S. The family $R \triangleq \{C \mid R \subset C, C$ is a χ-algebra$\}$ is nonempty because it contains $P(S)$. Define

$$\chi(R) \triangleq \bigcap_{C\in\Lambda} C$$

This is the smallest χ-algebra containing R and is called the χ-algebra generated by R. Specifying χ-algebras by generating sets is very convenient.

Beyond simple examples such as those above, it is in general not possible to give explicit constructions of χ-algebras except by transfinite recursion.

Definition: In $\mathbb{R}^n$, the χ-algebra generated by the collection of open sets is called the Borel χ-algebra, and its members are called Borel sets.

Using these definitions, Lemma says that the class of sets whose indicators are Baire functions is the χ-algebra of Borel sets.

More Consequences and Abstract Definitions

In this section we return to considering integrals of functions $f \in B$. To do this, it is useful to give another, simple characterization of B, which will be introduced in an abstract setting. The essential facts behind the theory are the following identities, which say the operations of union and intersection commute with forming inverse images of a function.

Let X and Y be any sets and let f be a function from X into Y. Let $\{A_\lambda \mid \chi \in .\}$ be a family of subsets of Y. Then

$$f^{-1}\left(\bigcup_{\lambda\in\Lambda} A_\lambda\right) = \bigcup_{\lambda\in\Lambda} f^{-1}(A_\lambda)$$

$$f^{-1}\left(\bigcap_{\lambda\in\Lambda} A_\lambda\right) = \bigcap_{\lambda\in\Lambda} f^{-1}(A_\lambda)$$

$$f^{-1}(U^c) = (f^{-1}(U))^c$$

These identities are easy to derive from the definitions of inverse image, union, and intersection.

The following abstract definitions are vitally important.

Definition:

1. Given a nonempty set S and a χ-algebra of its subsets, a function f: $S \rightarrow \mathbb{R}$ is said to be σ-measurable if $f^{-1}(U) = \{x \mid f(x) \in U\} \in$ for every Borel set U contained in $\mathbb{R}$. (When the identity of is clear from the context, one simply says "measurable" instead of "χ-measurable.)
2. A measurable simple function on $(S,\cdot)$ is a function of the form $\sum i = 1^n c_i \chi_{A_i}$, where each c_i is a real number and each A_i belongs to.·.
3. A function f: $\mathbb{R}^n \rightarrow \mathbb{R}$ is said to be Borel measurable if $f^{-1}(U)$ is a Borel subset of $\mathbb{R}^n$ for every Borel set U of $\mathbb{R}$. It is easy to check that a measurable simple function is measurable.

The following lemmas show why these definitions are useful. To avoid repetition S always denotes a general non-empty set and § a χ-algebra of its subsets.

Lemma: (a) If f is a measurable function, there exists a sequence $\{f_n\}$ of measurable simple functions such that limn $f_n(x) = f(x)$ for every $x \in S$.

(b)The set of measurable functions on $(S,\cdot)$ is closed under pointwise limits.

Lemma: The set B of Baire functions on $\mathbb{R}^n$ is equal to the set of Borel measurable functions on $\mathbb{R}^n$.

We shall defer proofs to the formal development of the theory, since that is nice and simple, and important to what we want to see next about the integral. It suces to show this claim for $f \subset 0$ because to prove it for arbitrary, write f as a difference of its positive and negative parts $f = f^+ \subset f^-$, and express each part as a limit of simple measurable functions.

We will show that, in fact, every nonnegative, measurable f is an *increasing limit* of simple measurable functions. Indeed, for every positive integer n, let

$$f_n(x) \triangleq \sum_{k=0}^{n2^n} k2^{-n} \chi_{f^{-1}((k/2^n,(k+1)/2^n])}(x).$$

In other words, if $\frac{k}{2^n} < x \subset \frac{k+1}{2}$ and $0 \subset k \subset n2^n$, set $f(x) = \frac{k}{2^n}$; otherwise, let $f(x) = 0$. Any interval $(k/2^n, (k+1)/2^n]$ is a Borel set in $\mathbb{R}$, and so, since f is measurable by assumption, $f^{-1}((k/2^n, (k+1)/2^n])$ is in.·. Thus f_n is a measurable simple function. Clearly, if $0 \subset x \subset n$, $0 < f(x) \subset f_n(x) < 2^{-n}$, and so f is a pointwise limit of the sequence $\{f_n\}$. The reader should check that $f_n \subset f_{n+1}$ for every n.

Return to our hypothetical integral $\int$ on $\mathbb{R}^n$. Observe first that the integral of any Borel measurable simple function can be expressed in terms of the volume measure m. Indeed, by the linearity property,

$$\int \sum_{1}^{n} c_i \chi_{A_i} = \sum_{1}^{n} c_i \int \chi_{A_i} = \sum_{1}^{n} c_i m(A_i)$$

Now let f be any nonnegative Borel measurable function and let $\{f_n\}$ be the sequence just constructed converging to f. Then we have from the monotone convergence principle that

$$\int f = \lim_{n} \int f = \lim_{n} \sum_{k=0}^{n2^n} \frac{k}{2^n} m\left(\left\{x \left| \frac{k}{2^n} < f(x) \subset \frac{k}{2^n}\right.\right\}\right).$$

This formula expresses $\int f$ in terms of the measure m.

This completes the picture we wanted to draw of the consequences of [F1]-[F5]. We worked from our hypothetical integral to a measure on Borel sets and then, from the monotone convergence property.

The idea of Lebesgue's theory is to work backward from measure to integral. First, construct a measure m on the Borel sets, or even a χ-algebra containing the Borel sets, such that for rectangles V $m(V)$ gives the usual volume.

Once that is accomplished, *define* the integral by a formula like. The theory will be successful, if one can prove properties and from the new definition of integral.

CLASSES OF FUNCTIONS

We first met a sequence as a particularly easy sort of function, defined on $\mathbb{N}$, rather than $\mathbb{R}$. We now move to the more general study of functions. However, our earlier work wasn't a diversion; we shall see that sequences prove a useful tool both to investigate functions, and to give an idea of appropriate methods.

Our main reason for being interested in studying functions is as a model of some behaviour in the real world. Typically a function describes the behaviour of an object over time, or space. In selecting a suitable class of functions to study, we need to balance generality, which has chaotic behaviour, with good behaviour which occurs rarely. If a function has lots of good properties, because there are strong restrictions on it, then it can often be quite hard to show that a given example of such a function has the required properties. Conversely, if it is easy to show that the function belongs to a particular class, it may be because the properties of that class are so weak that belonging may have essentially no useful consequences.

It turns out that there are a number of "good" classes of functions which are worth studying. In this chapter and the next ones, we study functions which have steadily more and more restrictions on them. Which means the behaviour steadily improves; and at the same time, the number of examples steadily decreases. A perfectly general function has essentially nothing useful that can be said about it; so we start by studying continuous functions, the first class that gives us much theory.

In order to discuss functions sensibly, we often insist that we can "get a good look" at the behaviour of the function at a given point, so typically we restrict the domain of the function to be well behaved.

Definition: A subset U of $\mathbb{R}$ is open if given $a \in U$, there is some $\delta > 0$ such that $(a-\delta, a+\delta) \subseteq U$.

In fact this is the same as saying that given $a \in U$, there is some open interval containing a which lies in U. In other words, a set is open if it contains a neighbourhood of each of its points. We saw in that an open interval is an open set. This definition has the effect that if a function is defined on an open set we can look at its behaviour near the point a of interest, from both sides.

Functions And Mapping

Definition: Let A and B be any two non-empty sets. Let A = $\{p, q, r\}$ and $B = \{a, b, c, d\}$. Suppose by some rule or other, we assign to each element of A a 'unique' element of B. Let p be associated to a, q be associated to b, r be associated to c etc. The set $\{(p, a), (q, b), (r, c)\}$ is called a function from set A to set B. If we denote this set by 'f' then we write f: $A \to B$ which is read as "f" is a function of A to B or 'f' is a mapping from A to B.

Definition: Let A and B be two non-empty sets. Suppose there exists a rule of correspondence, denoted by f which associates to each member of A, a unique member of B. Then 'f' is called a function or a mapping from set A to set B.

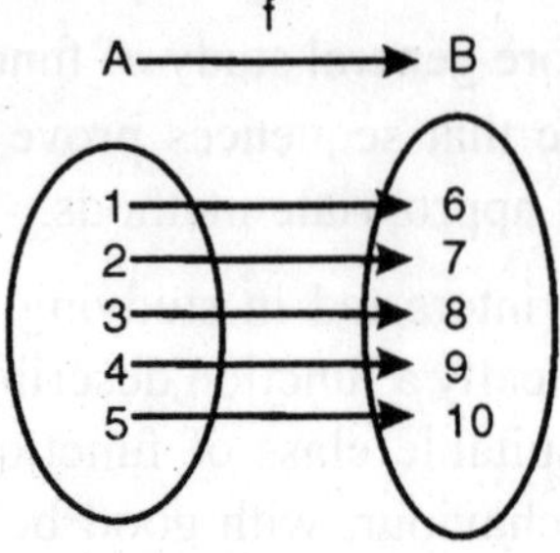

It is denoted by f: $A \to$ B or A $\xrightarrow{f}$ B

Note that:

1. The set A is called "domain" of 'f'
2. The set B is called "codomain" of 'f'
3. The element $y \in B$ which the function 'f' associates to the element $x \in A$ is denoted by $f(x)$ i.e.

   $y = f(x)$ and y is called the 'f' image of x or value of 'f' at x. The element x is also called as pre-image of y. Every element of A has a unique image but each element of B need not have as the image of an element in A. There can be more than one element of A which has the same image in B.

4. The set of these images in B is called the range of 'f' i.e. the range of a function as the images of its domain. We denote the range of $f: A \to B$ by f (A). Thus $f(A) = \{f(x): x \in A\} \Rightarrow f(A) \subseteq B$.

5. If A and B are any two non-empty sets, then a function f from A to B is a subset f of $A \times B$ satisfying the following conditions:

   (a) $\forall\, x \in A, (x, y) \in f$ for some $y \in B$

   (b) $(x, y) \in f$ and $(x, y_1) \in f \Rightarrow y = y1$

The first condition ensures that there is a rule which assigns to each element $x \in A$ some element $y \in B$. Thus each element in A will have an image. The second condition guarantees that the image is unique.

In other words, a function is a set of ordered pair (x, y), such that for each first element 'x, there corresponds a unique element y.

6. The domain variable is often referred as the independent variable and the range variable is referred as the dependent variable.

Formula or Equation:Consider the function 'f' exhibited by the adjoining arrow diagram.

Let A = { 1, 2, 3, 4, 5 }

B = { 5, 7, 9, 11, 13, 15 }

Note that, if we take any element $x \in A$ then the element of the co-domain set B related to x is obtained by adding 3 to twice of x.

Applying this rule we get,

$f(1) = 2(1) + 3 = 5, f(2) = 2(2) + 3 = 7,$

$f(3) = 2(3) + 3 = 9$ etc.

In general we can write $f(x) = 2x + 3$

$x \in A$

This is the formula which exhibits the function 'f'. If we denote the value of f at x by, $y \in B$, we get $y = 2x + 3$

$x \in A$. This becomes an equation which exhibits the function f.

Remarks:

1. If a function is exhibited by a formula, then using this formula we can find the range of the function.

2. Sometimes, the domain is not known or is not given. In such a case, the domain is

taken as that set of elements at which the values of the function can be found. For example

(a) $f(x) = x^2 - 3x + 4$. In this case, the set R of all reals is taken as the domain of 'f'.

(b) $f(x) = \sqrt{16 - x^2}$ In order to have a real valued function, $16 - x^2 \geq 0$ i.e. $-4 \leq x \leq 4$ which the desired domain of 'f'

(c) $f(x) = \dfrac{x+7}{x-7}$ then for $x = 4$, $x + 7$ is not defined. Hence in this case, the domain is the $x - 4$ set R of reals except 4 i.e. R – {4 }

Example: If A is the area of the circle. Describe this function.

Solution: If we call this function f then r is the independent variable and A becomes the dependent variable and we have $A = f(r) = \pi r^2$ for $r \geq 0$. The domain must be stated with the constraint $r \geq 0$ as we can't have a circle with negative radius.

Example: A relation is given as: { (1, 1), $(2, \frac{1}{2})$, $(3, \frac{1}{3})$, (4, ¼), $(5, \frac{1}{5})$ } is it a function ? Justify. If it is a function state its domain and range. Exhibit the relation by a formula.

Solution: Let A = { 1,2,3,4, 5} and $B = \{1, \frac{1}{2}, \frac{1}{3}, \frac{1}{4}, \frac{1}{5}\}$

Every element in *A* occurs once and only once in the ordered pairs and is related to one and only one element of the set *B*, exhibiting a relation.

Hence it is a function from the set *A* to the set *B*. Let us denote it by f. Then domain of *A* = The set A and the range of *f* = The set *B*. Also, the second member in each ordered pair is the reciprocal of the first member.

Thus if we denote any element of the set *A* by *x* and that of *B* by *y* then

The formula is $f(x) = 1/x$ i.e. $y = 1/x \; x \in A$

Example: A function *f* is given as $f(x) = x^2 - 6x + 5$. State its domain. Find $f(-2)$, $f(2)$, $f(1)$, $f(a+h)$, $\dfrac{f(a+h) - f(a)}{h}$, $h \neq 0$

Solution: The domain of '*f*' is the set R of all reals

$$f(x) = x^2 - 6x + 5 \text{ Þ } f(-2) = (-2)^2 - 6(-2) + 5 = 21$$

$$f(2) = (2)^2 - 6(2) + 5 = -3, \; f(1)$$

$$= (1)^2 - 6(1) + 5 = 0$$

$$f(a+h) = (a+h)^2 - 6(a+h) + 5 = a^2 + 2ah + h^2 - 6a - 6h + 5$$

and

$$= \frac{f(a+h)-f(a)}{h}$$

$$= \frac{(a^2+2ah+h^2-6a-6h+5)-(a^2-6h+5)}{h}$$

$$= \frac{h(2a-6+h)}{h}$$

Example: Find the range of each of the following functions

1. $f(x) = 3x + 4$ for $-5 \leq x \leq 8$
2. $g(x) = 2x^2 + 3$ for $-4 \leq x \leq 3$
3. $\phi(x) = x^2 - 6x + 7$ for all $x \in R$

Solution:

1. The domain of ' f ' $= \{ x \mid x \in R, -5 \leq x \leq 8 \}$

   Now $-5 \leq x \leq 8 \therefore 3(-5) \leq 3(x) \leq 3(8)$

   $\therefore -15 \leq 3x \leq 24 \therefore -15 + 4 \leq 3x + 4 \leq 24 + 4$

   $\therefore -11 \leq f(x) \leq 28$

   The range of 'f' $= \{ f(x) \in R, -11 \leq f(x) \leq 28 \}$
2. $-4 \leq x \leq 3 \Rightarrow 0 \leq x^2 \leq 16$ (why ?)

   $\therefore 2(0) \leq 2(x) \leq 2(16) \therefore 0 + 3 \leq 2x^2 + 3 \leq 30 + 3$

   $\therefore 3 \leq 9(x) \leq 35$

   This is the range of 'g '
3. $h(x) = x^2 - 6x + 7 = x^2 - 6x + 9 - 9 + 7 = (x-3)^2 - 2$

   (Note this !)

Since $x \in R$, $(x-3) \in R \Rightarrow (x-3)2 \geq 0 \Rightarrow (x-5)2 - 2 \geq 0 - 2$

$\therefore h(x) \geq -2$. This is the range of 'h'.

Example: Find the largest possible set of real numbers which can be the domain of

$$f(x) = \sqrt{1 - \frac{1}{x}}$$

Solution: Since $\sqrt{1-\frac{1}{x}}$ is defined only when $\sqrt{1-\frac{1}{x}} \geq 0$

i.e. $f(x)$ is defined for $1 - \frac{1}{x} \geq 0$

i.e. $f(x) = \frac{x-1}{x} \geq 0$ Setting numerator and denominator separately equal to zero

and solving, we get $x = 0$ and $x = 1$ (cut points)

Interval	*Test value*	*Value of f (x)*	*Sign of f (x)*
$(-\infty, 0)$	-1	$+2$	+
$(0, 1)$	$\frac{1}{2}$	$-\frac{1}{2}\frac{1}{2}=-1$	–
$(1, \infty)$	2	$\frac{1}{2}$	+

Graph of expression of $f(x) > 0$

Graph of expression of $f(x) \geq 0$

$\therefore$ The domain is the set $= \{ x \mid x < 0 \text{ or } x \geq 1 \} = (-\infty, 0) \cup [1, \infty)$

Example: If $f(x) = \dfrac{x+5}{3x-4}$ and $t = \dfrac{5+4x}{3x-1}$

then show that $f(t) = x$

Solution:

$$f(x) = \frac{x+5}{3x-4} \quad \therefore f(t) = \frac{t+5}{3t-4} \text{ putting } t = \frac{5+4x}{3x-1}$$

$$\therefore f(x) = \frac{\dfrac{5+4x}{3x-1}+5}{3\left(\dfrac{5+4x}{3x-1}\right)}$$

$$= \frac{\dfrac{5+4x+15x-5}{3x-1}}{\dfrac{5+12x-12x+4}{3x-1}}$$

$$\therefore f(t) = \frac{19x}{19} = x \therefore f(t) = x$$

Example: If $f(x) = x^2 - 2x + 3, x \in R$. Find the values of x for which $f(x) = f(3x-1)$

Solution:

$$f(x) = x^2 - 2x + 3 \Rightarrow f(3x-1) = (3x-1)^2 - 2(3x-1) + 3$$

$$= 9x^2 - 6x + 1 - 6x + 2 + 3$$

$$= 9x^2 - 12x + 6$$

$$\therefore f(x) = f(3x-1) \text{ gives us}$$

$$x^2 - 2x + 3 = 9x^2 - 12x + 6$$

$$8x^2 - 10x + 3 = 0$$

$$(4x - 3)(2x - 1) = 0 \quad \text{...(factorizing)}$$

$$4x - 3 = 0 \text{ or } 2x - 1 = 0$$

$$\therefore x = 3/4 \text{ or } x = 1/2$$

$\therefore$ The required values of x are 3/4 and 1/2

LIMIT

In mathematics, the concept of a "limit" is used to describe the behavior of a function as its argument either "gets close" to some point, or as it becomes arbitrarily large; or the behavior of a sequence's elements as their index increases indefinitely. Limits are used in calculus and other branches of mathematical analysis to define derivatives and continuity. The concept of the "limit of a function" is further generalized to the concept of topological net, while the limit of a sequence is closely related to limit and direct limit in category theory.

THE CONCEPT OF THE LIMIT OF A SEQUENCE

Th concept on which the whole of analysis ultimately rests is that of the limit of a sequence. We shall first make the position clear by considering several examples.

$a_n = 1/n$: We consider the sequence

$$a_1 = 1, a_2 = \frac{1}{2}, a_3 = \frac{1}{3}, \ldots, a_n = \frac{1}{n}, \ldots$$

No number of this sequence is zero; but we see that the larger the number n, the closer to zero is the number a_n. Hence, if we mark off around the point 0 an interval as small as we please, then starting with a definite value of the subscript all the numbers a_n will fall into this interval. We express this state of affairs by saying that, as n increases, the numbers a_n tend to 0, or that they possess the limit 0, or that the sequence a_1, a_2, a_3, converges to 0.

If numbers are represented as points on a line, this means that the points $1/n$ crowd closer and closer to the point 0 as n increases.

The situation is similar in the case of the sequence

$$a_1 = 1, a_2 = -\frac{1}{2}, a_3 = \frac{1}{3}, a_4 = -\frac{1}{4}, \ldots, a_n = \frac{(-1)^{n-1}}{n}, \ldots$$

Here too, the numbers a_n tend to zero as n increases; the only difference is that the numbers a_n are sometimes larger and sometimes smaller than the limit 0; as we say, they oscillate about the limit.

The convergence of the sequence to 0 is usually expressed symbolically by the equation

$$\lim_{x\to\infty} a_n = 0,$$

or occasionally by the abbreviation

$$a_n \to 0.$$

a_{2m} = 1/m; a_{2m-1} = 1/2m: In the preceding examples, the absolute value of the difference between a_n and the limit steadily becomes smaller as n increases. This is not necessarily always the case, as is shown by the sequence

$$a_1 = \frac{1}{2}, a_2 = 1, a_3 = \frac{1}{4}, a_4 = \frac{1}{2}, a_5 = \frac{1}{6}, a_6 = \frac{1}{3}, \ldots;$$

that is, in general, for even values $n = 2m$, $a_n = a_{2m}$, $= 1/m$, for odd values $n = 2m - 1$, $a_n = a_{2m-1} = 1$. This sequence also has a limit, namely zero, since every interval about the origin, no matter how small, will contain all the numbers a_n from a certain value of n onwards; but it is not true that every number lies nearer to the limit zero than the preceding one.

$a_n = n/(n + 1)$: We consider the sequence

$$a_1 = \frac{1}{2}, a_2 = \frac{2}{3}, \ldots a_n = \frac{n}{n+1}, \ldots,$$

where the integral subscript n takes all the values 1, 2, 3, ... If we write $a_n = 1 - 1/(n + 1)$, we see at once that, as n increases, the numbers a_n will approach closer and closer to the number 1, in the sense that, if we mark off any interval about the point 1, all the numbers a_n following a certain a_N must fall into that interval. We write

$$\lim_{x\to\infty} a_n = 1.$$

The sequence

$$a_n = \frac{n^2 - 1}{n^2 + n + 1}$$

behaves in a similar manner. This sequence also tends to a limit as n increases to the limit 1, in fact, in symbols, $\lim_{x\to\infty} a_n = 1$. We see this most readily, if we write

$$a_n = 1 - \frac{n+2}{n^2 + n + 1} = 1 - r_n;$$

now we have only to show that the numbers r_n tend to 0 as n increases. For all values of n greater than 2, we have $n + 3 < 2n$ and $n + n + 1 > n$. Hence we have for the remainder r_n

$$0 < r_n < \frac{2n}{n^2} = \frac{2}{n}(n > 2),$$

which shows at once that r_n tends to 0 as n increases. Our discussion yields at the same time an estimate of the amount by which the number a_n (for $n > 2$) can at most differ from the limit 1; this difference certainly cannot exceed 2/n. This example illustrates the fact, which we should naturally expect, that for large values of n the terms with the highest indices in the numerator and denominator of the fraction for a_n predominate and that they determine the limit.

$a_n = \sqrt[n]{p}$ Let p be any fixed positive number. We consider the sequence $a_1, a_2, a_3, \ldots, a_n$, where

$$a_n = \sqrt[n]{p}$$

We assert that

$$\lim_{n\to\infty} a_n = \lim_{n\to\infty} \sqrt[n]{p} = 1$$

We can prove this very easily by using a lemma which we shall find also useful for other purposes.

If $1 + h$ is a positive number (that is, if $h > -1$) and n is an integer greater than 1, then $(1 + h)n > 1 + nh$.......

Assume that Inequality (1) already has been proved for a certain $m > 1$; multiply both sides by $(1+h)$ and obtain

$$(1 + h)^{m+1} > (1 + mh)(1 + h) = 1 + (m + 1)h + mh^2$$

If we omit on the right hand side the positive term mh, the inequality remains valid. We thus obtain

$$(1 + h)^{m+1} > 1+(m + 1)h$$

However, this is our inequality for the index $m + 1$. Hence, if the inequality holds for the index m, is also holds for $m + 1$. Since it holds for m= 2, it also holds for m ˉ 3, whence for $m = 4$, and so on, whence it holds for every index. This is a simple example of a proof by mathematical induction, a type of proof which is often useful.

Returning to our sequence, we distinguish between the case $p > 1$ and the case $p < 1$ (if $p = 1$, then $\sqrt[n]{p}$ is also equal to 1 for every n and our statement becomes trivial).

If $p > 1$, then $\sqrt[n]{p}$ will also be greater than 1; let $\sqrt[n]{p} = 1 + h_n$, where h_n is a positive quantity depending on n and we find by Inequality

$$p = (1+h_n)n > 1+nh_n$$

whence follows at once that

$$0 < h_n < \frac{p-1}{n}$$

Thus, as n increases, the number h_n must tend to 0, which proves that the numbers a_n converge to the limit 1, as stated. At the same time, we have a means for estimating how close any a_n is to the limit 1; the difference between a_n and 1 is certainly not greater than $(p — 1)/n$.

If $p < 1$, then $\sqrt[n]{p}$ will likewise be less than 1 and therefore may be taken equal to $1/(1 + h_n)$, where h_n is a positive number. It follows from this, using Inequality (1), that

$$p = \frac{1}{(1+h_n)^n} < \frac{1}{1+nh_n}$$

(By decreasing the denominator, we increase the fraction. It follows that

$$1 + nh_n < \frac{1}{p},$$

whence

$$h_n < 1\frac{1/p-1}{n}$$

This shows that h_n tends to 0 as n increases. As the reciprocal of a quantity tending to 1, $\sqrt[n]{p}$ it self tends to 1.

$\alpha_n = \alpha^n$:

We consider the sequence $a_n = \alpha^n$, where α is fixed and n runs through the sequence of positive integers.

First, let α be a positive number less than 1. We may then put $\alpha = 1/(1 + h)$, where h is positive and Inequality yields

$$\alpha_n = \frac{1}{(1+h)^n} < \frac{1}{1+nh} < \frac{1}{nh}$$

Since the number h and, consequently, $1/h$ depends only on n and does not change as n increases, we see that as n increases α^n tends to 0:

$$\lim_{n\to\infty} \alpha^n = 0 \quad (0 < \alpha < 1)$$

The same relationship holds when α is zero or negative, but greater than –1. This is immediately obvious, since in any case

$$\lim_{n\to\infty} |\alpha|^n = 0$$

If $\alpha = 1$, then α^n will obviously be always equal to 1 and we shall have to regard the number 1 as the limit of α^n.

If $\alpha > 1$, we set $\alpha = 1 + h$, where h is positive and see at once from our inequality that, as

n increases, α^n does not tend to any definite limit, but increases beyond all bounds. We express this state of affairs by saying that α^n tends to infinity as n increases, or that α^n becomes infinite; in symbols

$$\lim_{n\to\infty} \alpha^n = \infty \qquad (\alpha > 1)$$

Nevertheless, as we must explicitly emphasize, the symbol } does not denote a number with which we can calculate as with any other number; equations or statements which express that a quantity is or becomes infinite never have the same sense as an equation between definite quantities. In spite of this, such modes of expression and the use of the symbol are extremely convenient, as we shall often see in the following pages.

If $\alpha = -1$, the values of α^n will not tend to any limit, but, as n runs through the sequence of positive integers, it will take the values +1 and –1 alternately. Similarly, if $\alpha < -1$, the value of α^n will increase numerically beyond all bounds, but its sign will be in sequence positive and. negative.

GEOMETRICAL ILLUSTRATION OF THE LIMITS OF A^n

If we consider the curves $y = x^n$ and $y = x^{1/n} = \sqrt[n]{x}$ and restrict ourselves, for the sake of convenience, to non-negative values of x, the preceding limits are illustrated by Figures respectively. In the case of the curves $y = x^n$, we see that in the interval 0 to 1 they approach closer and closer to the x-axis as n increases, while outside that interval they climb more and more steeply and draw in closer and closer to a line parallel to the y-axis. All the curves pass through the point with co-ordinates $x=1$, $y=1$ and through the origin.

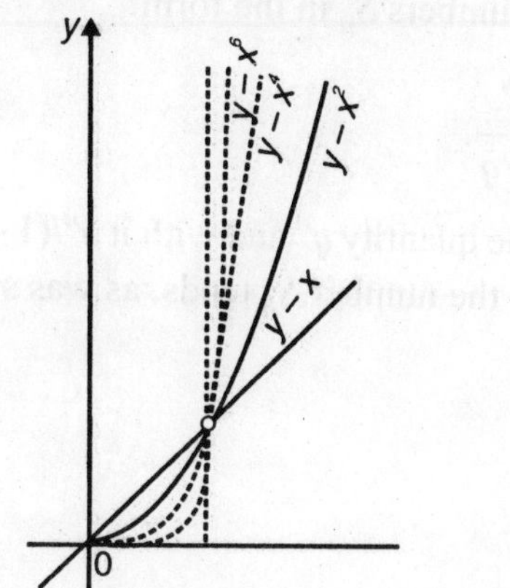

Fig. x^n as n increases

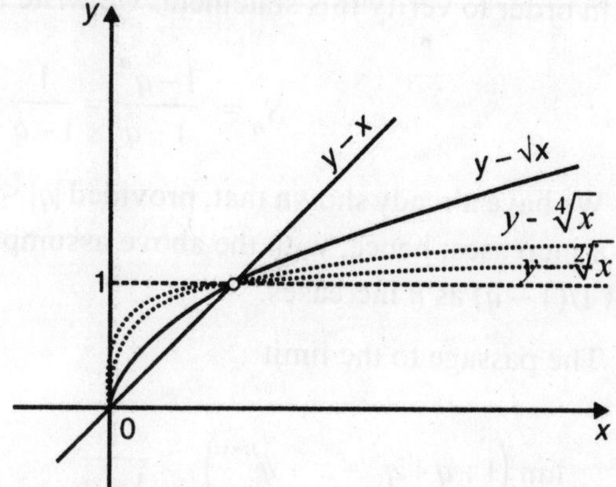

Fig. $x^{1\backslash n}$ as n increases

In the case of the functions $y = x^{1/n} = +\sqrt[n]{x}$, the curves approach closer and closer to the line parallel to the x-axis and at a distance 1 above it. On the other hand, all the curves must pass through the origin. Hence, in the limit, the curves approach the broken line consisting of the part of the y-axis between the points $y = 0$ and $y = 1$ and of the parallel to the x-axis $y = 1$. Moreover, it is clear that the two figures are closely related, as one would expect from the fact that the functions $y = \sqrt[n]{x}$ are actually the inverse functions of the n-th powers, from which we infer that each figure is transformed into the other on reflection in the line $y = x$.

The Geometric Series

An example of a limit which is more or less familiar from elementary mathematics is the geometric series

$$1 + q + q^2 + \ldots + q^{n-1} = S_n;$$

the number q is called the common ratio of the series. The value of this sum may, as is well known, be expressed in. the form

$$S_n = \frac{1 - q^n}{1 - q},$$

provided that $q > 1$; we can derive this expression by multiplying the sum S_n by q and subtracting the equation thus obtained from the original equation, or we may verify the formula by division.

There arises now the question what happens to the sum S_n when n. increases indefinitely? The answer is: The sum S_n has a definite limit S if q lies between -1 and +1, these end values being excluded, and it is then true that

$$S = \lim_{n \to \infty} S_n = \frac{1}{1 - q}.$$

In order to verify this statement, we write the numbers S_n in the form

$$S_n = \frac{1 - q^n}{1 - q} = \frac{1}{1 - q} - \frac{q^n}{1 - q}.$$

We have already shown that, provided $|q| < 1$, the quantity q^n and with it $q^n/(1 - q)$ tends to 0 as n increases; hence, with the above assumption, the number S_n tends, as was stated, to the limit $1/(1 - q)$ as n increases.

The passage to the limit

$$\lim_{n \to \infty} \left(1 + q + q^2 + \ldots + q^{n-1}\right) = \frac{1}{1 - q}$$

is usually expressed by saying that when $|q| < 1$, the geometric series can be extended to infinity and that the sum of the infinite geometric series is the expression $1/(1 - q)$.

The sums S_n of the finite geometric series are also called the partial sums of the infinite geometric series $1 + q + q +$ .... (We must draw a sharp distinction between the sequence of numbers $S_1, S_2, \ldots$, and the geometric series.)

The fact that the partial sums S_n of a geometric series tend to the limit $S = 1/(1-q)$ as n increases may also be expressed by saying that the infinite geometric series $1 + q + q + \ldots$ converges to the sum $S = 1/(1-q)$ when $|q| < 1$.

$a_n = \sqrt[n]{n}$:

We shall show that the sequence of numbers

$a_1 = 1, a_2 = \sqrt{2}, a_3 = \sqrt[3]{3}, \ldots, a_n = \sqrt[n]{n}, \ldots$

tends to 1 as n increases, i.e., that

$$\lim_{n\to\infty} \sqrt[n]{n} = 1.$$

Here we make use of a slight artifice. Instead of the sequence $a_n = \sqrt[n]{n}$, we first consider the sequence. $b_n = \sqrt{a_n} = \sqrt{\sqrt[n]{n}} = \sqrt[n]{n}$. When $n > 1$, the term b_n is also greater than 1. We can therefore set $b_n = 1 + h_n$, where h_n is positive and depends on n. We therefore have

$$\sqrt{n} = (b_n)^n = (1 + h_n)^n \geqq 1 + nh_n,$$

so that

$$h_n \leqq \frac{\sqrt{n}-1}{n} \leqq \frac{\sqrt{n}}{n} = \frac{1}{\sqrt{n}}.$$

We now have

$$1 \leqq a_n = b_n^2 = 1 + 2h_n + h_n^2 \leqq 1 + \frac{2}{\sqrt{n}} + \frac{1}{n}.$$

Obviously, the right hand side of this inequality tends to 1, and so does a_n.

$a_n = \sqrt{n+1} - \sqrt{n}$

We assert that

$$\lim_{n\to\infty} \left(\sqrt{n} + 1 - \sqrt{n}\right) = 0.$$

In order to prove this formula, we need only write this expression in the form

$$\sqrt{n+1} - \sqrt{n} = \frac{\left(\sqrt{n+1} - \sqrt{n}\right)\left(\sqrt{n+1} + \sqrt{n}\right)}{\sqrt{n+1} + \sqrt{n}} = \frac{1}{\sqrt{n+1} + \sqrt{n}};$$

we see at once that this expression tends to 0 as n increases.

$a_n = \dfrac{n}{\alpha^n}$: Let α be a number greater than 1. We assert that as n increases the sequence of the numbers $a_n = n/a_n$ tends to the limit 0.

As in the case of $\sqrt[n]{n}$ above, we consider the sequence

$$\sqrt{a_n} = \frac{\sqrt{n}}{\left(\sqrt{\alpha}\right)^n}$$

We set $\sqrt{\alpha} = 1 + h$ Here $h > 0$, since α and hence α is greater than 1. By Inequality, we have

$$\sqrt{\alpha} = (1+h)^n > 1 + nh,$$

so that

$$\sqrt{\alpha_n} = \frac{\sqrt{n}}{(1+h)^n} \leq \frac{\sqrt{n}}{1+nh} \leq \frac{\sqrt{n}}{nh} = \frac{1}{h\sqrt{n}}$$

Hence

$$a_n \leq \frac{1}{nh^2}$$

Chapter 10

Riemann Integration

In a calculus class integration is introduce d as 'finding the area under a curve'. While this interpretation is certainly useful, we instead want to think of 'integration' as more sophisticated form of summation. Geometric considerations, in our situation, will not be so fruitful, whereas the summation interpretation of integration will make many of its properties easy to remember.

First, as usual, we need to define integration before we can discuss its properties. We will start with defining the Riemann integral, and we will move to the Riemann-Stieltjes and the Lebesgue integral later.

The Riemann integral, created by Bernhard Riemann, was the first rigorous definition of the integral of a function on an interval. While the Riemann integral is unsuitable for many theoretical purposes, it is one of the easiest integrals to define. Some of these technical deficiencies can be remedied by the Riemann-Stieltjes integral, and most of them disappear in the Lebesgue integral.

Riemann Integration

Let $f(x)$ be a non-negative real-valued function of the interval $[a, b]$, and let $S = \{(x, y) \mid 0 < y < f(x)\}$ be the region of the plane under the function $f(x)$ and above the interval $[a, b]$. We are interested in measuring the area of S. Once we have measured it, we will denote the area by:

The basic idea of the Riemann integral is to use very simple approximations for the area of S. By taking better and better approximations, we can say that "in the limit" we get exactly the area of S under the curve.

Note that where f can be both positive and negative, the integral corresponds to *signed area*; that is, the area above the x-axis minus the area below the x-axis.

Partitions of an Interval

A partition of an interval $[a, b]$ is a finite sequence $a = x_0 < x_1 < \ldots < x_n = b$. Each $[x_i, x_{i+1}]$ is called a *subinterval* of the partition. The *mesh* of a partition is defined to be the length of the longest subinterval $[x_i, x_{i+1}]$, that is, it is $\max(x_{i+1} - x_i)$ where $0 \leq i \leq (n-1)$. It is also called the *norm* of the partition.

A *tagged partition of an interval* is a partition of an interval together with a finite sequence of numbers $t_0, \ldots, t_{n-1}$ subject to the conditions that for each i, $x_i \leq t_i \leq x_{i+1}$. In other words, it is a partition together with a distinguished point of every subinterval. The mesh of a tagged partition is defined the same as for an ordinary partition.

Suppose that $x_0, x_1, x_2, \ldots, x_n$ together with are a tagged partition of $[a, b]$, $y_0, \ldots y_m$ and that together with are another tagged partition of $[a, b]$. We say that y_0, y_m and s0. s_{m-1} together are a *refinement* of $x_0, x_1, x_2, \ldots, x_n$ together with $t_0, \ldots, t_{n-1}$ if for each integer i with, $0 \leq i \leq n$, there is an integer $r(i)$ such that $x_i = y_{r(i)}$ and such that $t_i = s_j$ for some j with $r(i) \leq j \leq r(i+1)$. Said more simply, a refinement of a tagged partition takes the starting partition and adds more tags, but does not take any away.

We can define a partial order on the set of all tagged partitions by saying that one tagged partition is bigger than another if the bigger one is a refinement of the smaller one.

A partition P of the closed interval $[a, b]$ is a finite set of points $P = \{x_0, x_1, x_2, \ldots, x_n\}$ such that

$$a = x_0 < x_1 < x_2 < \ldots < x_{n-1} < x_n = b$$

The maximum difference between any two consecutive points of the partition is called the norm or mesh of the partition and denoted as $| P |$, i.e.

$$| P | = max \{x_j - x_{j-1}, j = 1 \ldots n\}$$

A refinement of the partition P is another partition P' that contains all the points from P and some additional points, again sorted by order of magnitude.

Riemann sums

If $P = \{x_0, x_1, x_2, \ldots, x_n\}$ is a partition of the closed interval $[a, b]$ and f is a function defined on that interval, then the n-th Riemann Sum of f with respect to the partition P is defined as:

$$R(f, P) = \sum_{j=1}^{n} f(t_j)\,(x_j - x_{j-1})$$

where t_j is an arbitrary number in the interval $[x_{j-1}, x_j]$.

In other words choose a real-valued function f which is defined on the interval $[a, b]$. The *Riemann sum* of f with respect to the tagged partition $x_0, x_1, x_2, \ldots x_n$ together with $t_0, \ldots, t_{n-1}$ is:

$$\sum_{i=0}^{n-1} f(t_i)(x_{i+1} - x_i)$$

Each term in the sum is the product of the value of the function at a given point and the length of an interval. Consequently, each term represents the area of a rectangle with height $f(t_i)$ and length $x_{i+1} - x_i$. The Riemann sum is the signed area under all the rectangles.

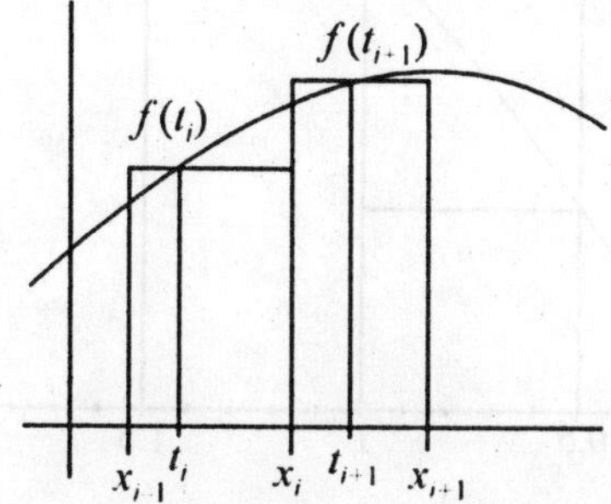

Note: If the function f is positive, a Riemann Sum geometrically corresponds to a summation of areas of rectangles with length $x_j - x_{j-1}$ and height $f(t_j)$. Riemann sums have the practical disadvantage that we do not know which point to take inside each subinterval.

To remedy that one could agree to always take the left endpoint (resulting in what is called the left Riemann sum) or always the right one (resulting in the right Riemann sum). However, it will turn out to be more useful to single out two other close cousins of Riemann sums:

Lower and Upper sum

Let $P = \{x_0, x_1, x_2, \ldots, x_n\}$ be a partition of the closed interval $[a, b]$ and f a bounded function defined on that interval. Then:

1. The upper sum of f with respect to the partition P is defined as:

$$U(f, P) = \sum_{j=1}^{n} c_j (x_j - x_{j-1})$$

where c_j is the supremum of $f(x)$ in the interval $[x_{j-1}, x_j]$.

2. The lower sum of f with respect to the partition P is defined as

$$L(f, P) = \sum_{j=1}^{n} d_j (x_j - x_{j-1})$$

where d_j is the infimum of $f(x)$ in the interval $[x_{j-1}, x_j]$.

Here is an example where the upper sum in displayed in dark brown and the lower sum in light brown.

The partition $P = \{0.5, 1, 1.5, 2\}$, and the numbers for the sums are chosen:

1. For the upper sum: $c_1 = f(1)$, $c_2 = f(1.5)$, and $c_3 = f(1.5)$

for the lower sum: $d_1 = f(0.5)$, $d_2 = f(1)$, and $d_3 = f(2)$

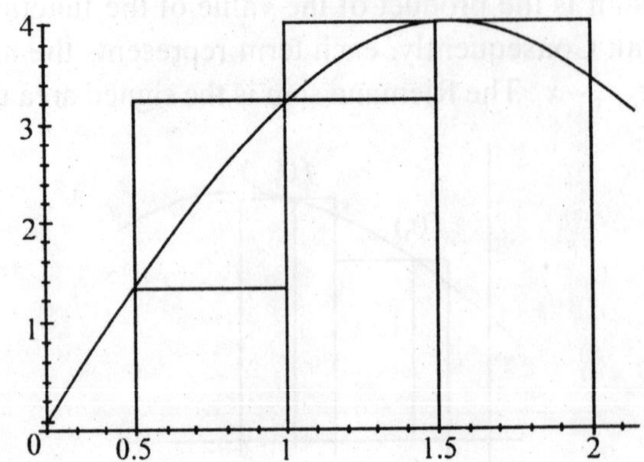

Size of Riemann Sums

Suppose $P = \{x_0, x_1, x_2, ..., x_n\}$ is a partition of the closed interval $[a, b]$, f a bounded function defined on that interval. Then we have:

1. The lower sum is increasing with respect to refinements of partitions, i.e. $L(f, P') \geq L(f, P)$ for every refinement P' of the partition P.
2. The upper sum is decreasing with respect to refinements of partitions, i.e. $U(f, P') \leq U(f, P)$ for every refinement P' of the partition P.
3. $L(f, P) \leq R(f, P)\ U(f, P)$ for every partition P

In other words, the lower sum is always less than or equal to the upper sum, and the upper sum is decreasing with respect to a refinement of the partition while the lower sum is increasing with respect to a refinement of the partition. Hence, a natural question is: will the two quantities ever coincide ?

Riemann Integral

Riemann integration is the formulation of integration most people think of if they ever think about integration. It is the only type of integration considered in most calculus classes; many other forms of integration, notably Lebesgue integrals, are extensions of Riemann integrals to larger classes of functions.

The Riemann integral was developed by Bernhard Riemann in and was, when invented, the first rigorous definition of integration applicable to not necessarily continuous functions.

Suppose f is a bounded function defined on a closed, bounded interval $[a, b]$. Define the *upper* and *lower* Riemann integrals, respectively, as

$I^*(f) = \inf\{U(f, P): P \text{ a partition of } [a, b]\}$

$I_*(f) = \sup\{L(f, P): P \text{ a partition of } [a, b]\}$

Then if $I^*(f) = I_*(f)$ the function f is called Riemann integrable and the Riemann integral of f over the interval $[a, b]$ is denoted by

$$\int_a^b f(x)dx$$

Note that upper and lower sums depend on the particular partition chosen, while the upper and lower integrals are independent of partitions. However, this definition is very difficult for practical applications, since we need to find the *sup* and *inf* over *any* partition.

Riemann Lemma

Suppose f is a bounded function defined on the closed, bounded interval $[a, b]$. Then f is Riemann integrable if and only if for every > 0 there exists at least one partition P such that

$$| U(f, P) - L(f, P) | < \varepsilon$$

Now we can state some easy conditions that the Riemann integral satisfies. All of them are easy to memorize if one thinks of the Riemann integral as a somewhat glorified summation.

Properties of Riemann Integration

Suppose f and g are Riemann integrable functions defined on $[a, b]$. Then

1. $\int_a^b cf(x) + dg(x)dx = \int_a^b f(x)dx + d\int_a^b g(x)dx$

2. If $a < c < b$ then $\int_a^b f(x)dx = \int_a^c f(x)dx + \int_c^b f(x)dx$

3. $\left|\int_a^b f(x)dx\right| \le \int_a^b | f(x) | dx$

4. If g is another function defined on $[a, b]$ such that $g(x) < f(x)$ on $[a, b]$, then $\int_a^b g(x)dx \le \int_a^b f(x)dx$.

5. If g is another Riemann integrable function on $[a, b]$ then $f(x) \cdot g(x)$ is integrable on $[a, b]$.

Now we can illustrate the relation between Riemann integrable and continuous functions.

Theorem: **(Riemann Integrals of Continuous Functions)** Every continuous function on a closed, bounded interval is Riemann integrable. The converse is false.

Note that this theorem does not say anything about the actual value of the Riemann integral. Also, we have as a free extra condition that that f is bounded, since every continuous function on a compact set is automatically bounded.

Since differentiable functions are continuous, this result implies that

{*differentiable functions*} ⊂ {*continuous functions*} ⊂ {*integrable functions*} and each set is in fact a proper subset of the next.

Theorem: Riemann Integral of almost Continuous Function.

If f is a bounded function defined on a closed, bounded interval $[a, b]$ and f is continuous except at countably many points, then f is Riemann integrable.

The converse is also true: If f is a bounded function defined on a closed, bounded interval $[a, b]$ and f is Riemann integrable, then f is continuous on $[a, b]$ except possibly at countably many points.

Now that we have easy conditions to determine which functions are integrable, it would also be convenient to have a nice shortcut to easily compute the actual value of an integral.

Theorem: Fundamental Theorem of Calculus. Suppose f is a bounded, integrable function defined on the closed, bounded interval $[a, b]$, define a new function:

$$F(x) = \int_a^x f(t)dt$$

Then F is continuous in $[a, b]$. Moreover, if f is also continuous, then F is differentiable in (a, b) and

$F'(x) = f(x)$ for all x in (a, b).

In many calculus texts this theorem is called the *Second Fundamental Theorem of Calculus*.

This theorem has an easy corollary that enables us to quickly find the value of an integral in many situations.

Corollary: Integral Evaluation Shortcut. Suppose f is an continuous function defined on the closed, bounded interval $[a, b]$, and G is a function on $[a, b]$ such that $G'(x) = f(x)$ for all x in (a, b). Then

$$\int_a^b f(x)dx = G(b) - G(a)$$

The function G is often called *Antiderivative* of f, and this corollary is called *First Fundamental Theorem of Calculus*

Before we look at several examples, we should rephrase these results in less rigorous notation. The first theorem says, basically:

$$\frac{d}{dx}\int f(x)dx = f(x)$$

while the corollary states, basically, that:

$$\int \frac{d}{dx} f(x)dx = f(b) - f(a)$$

Hence, loosely speaking, integration and differentiation are inverse operations of each other.

Integration Techniques

This section provides integration techniques, i.e., methods for finding the actual value of an integral. We have already found the most basic technique (the *Integral Evaluation Shortcut* or *First Fundamental Theorem of Calculus*): to evaluate an integral over an interval $[a, b]$, find an anti-derivate F of the integrand f and compute $F(b) - F(a)$.

Anti-derivative

For a given function f the function F with the property that $F'(x) = f(x)$ is called the *anti-derivative of f*. There are two standard mechanisms that can be useful: substitution and integration by parts.

Substitution Rule

If f is a continuous function defined on $[a, b]$, and s a continuously differentiable function from $[c, d]$ into $[a, b]$. Then

$$\int_a^b f(x(t))s'(t)dt = \int_{s(a)}^{s(b)} f(x)dx$$

This theorem, in words, says that if you can identify a composition of functions as well as the derivative of one of the composed functions, there's a good chance you can find the anti-derivative and evaluate the corresponding integral.

Think about the substitution rule as a "change of variable": if a particular expression in x makes an integrand difficult, change it to u. Then compute the derivative of u, i.e. $du/$

$dx = u'(x)$ and "solve" it for du: $du = u'(x)\,dx$. If you can use $u'(x)\,dx$ to change the integrand into an expression involving only u and no x, then the substitution worked (and has hopefully simplified the integrand). Otherwise try a different substitution or a different technique altogether.

Another theorem that is commonly used is "integration by parts".

Integration by Parts

Suppose f and g are two continuously differentiable functions. Let $G(x) = f(x)\,g(x)$. Then

$$\int_a^b f(x)g'(x)dx = \{G(b) - G(a)\} - \int_c^b f(x)g(x)dx$$

In order for integration by parts to be useful, three conditions should be satisfied:

1. The integrand must be a product of two expressions (one of which could be l).
2. You must know the anti-derivative of one of the two expressions.
3. The derivative of the other expression should become easier.

Integration by parts can also be used to prove more interesting facts that we might need later.

Integration by Parts and Limits

Suppose f: $[a, b] \to R$ is a continuously differentiable function. Show that:

$$\lim_{x\to\infty} \int_a^b f(x)\sin(nx)dx = 0$$

$$\lim_{x\to\infty} \int_a^b f(x)\cos(nx)dx = 0$$

Integration by parts can be used to provide an effective means to compute the value of an integral numerically. Of course, Riemann sums can be used to approximate an integral, but convergence is usually slow. A much faster convergence scheme is based on the *trapezoid rule*. To prove it, we need the Mean Value Theorem for Integration:

Mean Value Theorem for Integration

If f and g are continuous functions defined on $[a, b]$ so that $g(x) \geq 0$, then there exists a number $c \in [a, b]$ with

$$\int_a^b f(x)g(x)dx = f(c)\int_a^b g(x)dx$$

Now we can state and prove the Trapezoid Rule.

Trapezoid Rule

Let f be a twice continuously differentiable function defined on $[a, b]$ and set

$$K = sup\{| f''(x) |, x \in [a, b]\}$$

If $h = (b - a) / n$, where n is a positive integer, then

$$\int_a^b f(x)dx \left(\frac{1}{2} f(a) + \sum_{j=1}^{n-1} f(a + jh) + \frac{1}{2} f(b) \right) h + R(n)$$

where $| R(n)| < K/12\,(b - a)\, h^2$.

The Trapezoid Rule is useful because the error $R(n)$ depends on the square of h. If h is small, h^2 is a lot smaller so that the Trapezoid Rule provides a good approximation to the numeric value of an integral (as long as f is twice continuously differentiable).

Our final integration technique uses *partial fraction decomposition* of a polynomial to simplify rational functions so that they can be integrated.

Partial Fraction Decomposition

Suppose $p(x)$ is a polynomial of degree n such that

$$p(x) = \prod_{j=0}^{k} q_j(x)$$

where each q_j is polynomial of degree j that is irreducible over R. If $s(x)$ is another polynomial of degree less than n with no factors in common with $p(x)$, then the rational function $s(x) / p(x)$ can be written as a finite sum

$$\frac{s(x)}{p(x)} = \prod_{j=0}^{k} \frac{r_j(x)}{q_j(x)}$$

where each r_j is a polynomial of degree less than j.

This theorem sounds complicated, so we should restate it in a more useful form:

1. $p(x)$ is a polynomial that can be factored into polynomials q_j.
2. each q_j can not be factored any further (with real coefficients)
3. $s(x)$ is another polynomial whose factors, if any, do not cancel any of the q_j, and whose degree is less than the degree of p.
4. Then $s(x)/p(x)$ can be written as a finite sum of simple rational functions whose denominators consist of the q_j's and whose numerators have degree at most $j-1$.

5. finding the r_j amounts to solving a system of linear equations, as the examples will show.

Several of the assumptions are technical:

1. If the degree of s is greater than or equal to the degree of p, then you can split $s(x)/p(x)$ using long division, and the remainder will be of the form where the theorem applies.
2. If s does have factors in common with p, first cancel out those factors, then apply the theorem to the resulting rational function.

Measures

The Riemann integral is certainly useful (and complicated) enough, but it does have a few limitations and oddities:

One of the limitations of the Riemann integral is that it is based on the concept of an "interval", or rather on the length of subintervals $[x_{j-1}, x_j]$.

We therefore need to find a generalization of the "length" concept of a set in the real line. That new "length" concept, which we will call "measure", should satisfy two key conditions:

1. The new "measure" concept should be applicable to intervals, unions of intervals, and to more general sets (such as a Cantor set). Ideally, it should be defined for all sets.
2. The new "measure" concept should share as many properties as possible with the standard length of an interval, such as:
   (a) The 'measure' of a set should be non-negative.
   (b) The 'measure' of an interval should be the length of that interval.
   (c) The 'measure' of countably many disjoint sets should be the sum of the 'measures' of the individual sets.

To define what will eventually be called *Lebesgue Measure*, we follow a two-stage strategy:

Step I: We will define a concept extending length that is defined for all sets (to satisfy condition 1 above).

Step II: We will modify that concept so that it looks as close as possible to the standard length concept (to satisfy condition 2 above).

The Step-I concept is called *outer measure*, defined as follows:

Outer Measure

If A is any subset of R, define the (Lebesgue) outer measure of A as:

$$m^*(A) = inf\{\sum l(A_n)\}$$

where the infimum is taken over all countable collections of open intervals A_n such that $A \subset \cup A_n$ and $l(A_n)$ is the standard length of the interval A_n.

Outer measure is defined for all sets in R and has some of the properties we wanted our concept of measure to have.

Properties of Outer Measure

Outer measure has the following properties:

1. Outer measure m^* is a non-negative set function whose domain is $P(R)$, i.e. the power set of R.
2. The outer measure of an interval is its length.
3. Outer measure is countably subadditive, i.e. if $\{A_n\}$ is a countable collection of sets, then

$$m^*(\cup A_n) \le \sum_{m^*}(A_n)$$

Since outer measure is defined for all sets in R, our first key condition above is satisfied. But outer measure is not quite similar to a length, because it is only sub-additive ($m^*(A \cup B) \le m^*(A) + m^*(B)$), not additive ($m^*(A \cup B) = m^*(A) + m^*(B)$ for disjoint sets A and B). Therefore one of the requirements of the second key condition is not satisfied.

To make our outer measure additive, something has to give. We decide to restrict the domain to gain countable additivity, which will result in our *stage-two* definition.

There are several ways to restrict an outer measure. A somewhat intuitive way, perhaps, is to define an *inner measure* m_* that is similar to the outer measure but involves a *sup* over suitable sets.

Then sets for which $m^*(A) = m_*(A)$ are called *measurable* and the restriction of the outer (or inner) measure to the measurable sets is called a *measure*.

Using an outer and inner measure to define a measure does not have many advantages, and as a definition it is difficult to deal with. We prefer another approach that is due to Caratheodory to define measurable sets.

Measurable Sets and Lebesgue Measure

Set E is (Lebesque) measurable if for every set A we have that

$$m^*(A) = m^*(A \cap E) + m^*(A \cap comp(E))$$

If E is measurable, the non-negative number $m(E) = m^*(E)$ is the (Lebesgue) measure of the set E.

Since outer measure is subadditive, we have

$$m^*(A) = m^*((A \cap E) \cup (A \cap comp\ (E)))$$
$$m^*(A \cap E) + m^*(A \cap comp(E))$$

Therefore, to show that a set E is measurable it is sufficient to prove that

$$m^*(A) \geq m^*(A \cap E) + m^*(A\ comp(E))$$

The motivation for this definition is that it ensures that for two disjoint measurable sets E and F we have that

$$m(E \cup F) = m(E) + m(F)$$

as we will show later, i.e. measure is additive.

The next two results shows that a measure has all of the properties of a length, but is no longer defined for all sets.

Properties of Lebesgue Measure

1. All intervals are measurable and the measure of an interval is its length
2. All open and closed sets are measurable
3. The union and intersection of a finite or countable number of measurable sets is again measurable
4. If A is measurable and A is the union of countable number of measurable sets A_n, then $m(A) \leq \sum_m (A_n)$.
5. If A is measurable and A is the union of countable number of disjoint measurable sets A_n, then $m(A) = \sum_m (A_n)$.

Outer measure is defined for all sets, and according to the above list of properties, most "common" sets such as intervals, closed or open sets, unions and intersections of measurable sets are all measurable. But we have to pay for the property that measure is (countably) additive with the fact that not *every* set is measurable.

Proposition: (Not all Sets are Measurable): There are sets that are not (Lebesgue) measurable, i.e. not every set is (Lebesgue) measurable.

To summarize, we introduced a new concept called measure in two stages, each of which had a typical "good news, bad news" property.

1. In the first stage we defined *outer measure*.
   (a) *Good news:* outer measure is defined for all sets.

(b) *Bad news:* outer measure is not additive, i.e. it is not quite comparable to a length.

2. In the second stage we defined *measure* by restricting outer measure to the measurable sets.

(a) *Good news:* measure is additive, i.e. it is a good generalization of length.

(b) *Bad news:* measure is not defined for all sets.

Before concluding this section, we want to prove one more result that will be important for the next section:

Monotone Sequences of Measurable Sets

If $\{A_n\}$ is a sequence of measurable sets that is decreasing, i.e. $A_j \subset A_{j+1}$ for all j, and $m(A_1)$ is finite, then

$$lim\ m(A_j) = m(\cap A_j)$$

If $\{A_n\}$ is a sequence of measurable sets that is increasing in the sense that $A_{j+1} \subset A_j$ for all j, then

$$lim\ m(A_j) = m(\cup A_j)$$

Lebesgue Integral

We previously defined the Riemann integral roughly as follows:

1. Subdivide the domain of the function (usually a closed, bounded interval) into finitely many subintervals (the partition)
2. Construct a simple function that has a constant value on each of the subintervals of the partition (the Upper and Lower sums)
3. Take the limit of these simple functions as you add more and more points to the partition.

If the limit exists it is called the Riemann integral and the function is called Riemann integrable. Now we will take, in a manner of speaking, the "opposite" approach:

1. Subdivide the *range* of the function into finitely many pieces
2. Construct a simple function by taking a function whose values are those finitely many numbers
3. Take the limit of these simple functions as you add more and more points in the range of the original function

If the limit exists it is called the Lebesgue integral and the function is called Lebesgue integrable. To define this new concept we use several steps:

1. We define the Lebesgue Integral for "simple functions".

2. We define the Lebesgue integral for bounded functions over sets of finite measure.
3. We extend the Lebesgue integral to positive functions (that are not necessarily bounded).
4. We define the general Lebesgue integral.

First, we need to clarify what we mean by "simple function".

Characteristic and Simple Function

For any set A the function

$$X_A(x) = \begin{cases} 1 \text{ if } x \in A \\ 0 \text{ if } x \notin A \end{cases}$$

is called the *characteristic function* of A. A finite linear combination of characteristic functions.

$$s(x) = \sum_{a_i} X_{Ej}(x)$$

is called *simple function* if all sets E_i are measurable.

A function f defined on a measurable set A that takes no more than finitely many distinct values $a_1, a_2, ..., a_n$ can always be written as a simple function

$$f(x) = \sum_{a_i} X_{An}(x)$$

where

$$A_n = \{x \in A: f(x) = a_n\}$$

Therefore simple functions can be thought of as dividing the *range* of f, where the resulting sets A_n may or may not be intervals.

For simple functions we define the Lebesgue integral as follows:

Lebesgue Integral for Simple Function

If $s(x) = \sum_{a_n} X_{An}(x)$ is a simple function and $m(A_n)$ is finite for all n, then the Lebesgue Integral *of* s is defined as

$$\int s(x)dx = \sum_{a_n} m(A_n)$$

If E is a measurable set, we define

$$\int_E S(x)dx = \int X_E s(x)dx$$

Just as step functions were used to define the Riemann integral of a bounded function f over an interval $[a, b]$, simple functions are used to define the Lebesgue integral of f over a set of finite measure.

Lebesgue Integral for Bounded Function

Suppose f is a bounded function defined on a measurable set E with finite measure. Define the *upper* and *lower* Lebesgue integrals, respectively, as

$$I^*(f)_L = inf\left\{\int s(x)dx\right\} : s \text{ is simple and } s \geq f\}$$

$$I_*(f)_L = sup\left\{\int s(x)dx\right\} : s \text{ is simple and } s \leq f\}$$

If $I^*(f)_L = I_*(f)_L$ the function f is called Lebesgue integrable over E and the Lebesgue integral of f over E is denoted by

$$\int_E f(x)dx$$

Now a function f can be integrated (if it is integrable) using either the Riemann or the Lebesgue integral. Fortunately, for many simple functions the two integrals agree and the Lebesgue integral is indeed a generalization of the Riemann integral.

Riemann implies Lebesgue Integrable

If f is a bounded function defined on $[a, b]$ such that f is Riemann integrable, then f is Lebesgue integrable and

$$\int_a^b f(x)dx = \int_{[a,b]} f(x)dx$$

For most practial applications this theorem is all that is necessary: for continuous functions or bounded functions with at most countably many discontinuities over intervals *[a, b]* there is no need to distinguish between the Lebesgue or Riemann integral. All integration techniques we learned apply equally well, using either integral. But for more complicated situations or more theoretical purposes the Lebesgue integral is more useful, but then techniques such as integration by parts or substitution may no longer apply.

Properties of Lebsgue Integral

Theorem (X, B, μ) *Let be a measure space, f: X → [− ∞, ∞] and g: X → [−∞, ∞] be Lebesgue integrable functions, and A, B ∈ B. Then the following properties hold:*

1. $\left|\int_A f d\mu\right| \leq \int_A |f| d\mu$

2. If $f \leq g$, then $\int_A f d\mu \leq \int_A g d\mu$.

3. $\int_A f d\mu = \int_X XAf\,\mu$, where XA denotes the characteristic function of A.

4. If $c \in \mathbb{R}$, then $\int_A cf d\mu = c\int_A f\,d\mu$.

5. If $\mu(A) = 0$, then $\int_A f d\mu = 0$.

6. $\int_A (f + d) d\mu = \int_A f d\mu + \int_A g d\mu$.

7. If $A \cap B = \phi$, then $\int_{A \cup B} f d\mu = \int_A f d\mu + \int_B f d\mu$.

8. If $f = g$ almost everywhere with respect to μ, then $\int_A f d\mu = \int_A g d\mu$.

Proof:

1. $\left|\int_A f d\mu\right| = \int_A f^+ d\mu - \int_A f^- d\mu$ by definition

$\leq \left|\int_A f^+ d\mu\right| + \left|\int_A f^- d\mu\right|$ by the triangle inequality

$= \int_A f^+ d\mu + \int_A f^- d\mu$ by the properties of the Lebesgue integral of nonnegative measurable functions (property 1),

$= \int_A (f^+ + f^-) d\mu$ by the properties of the Lebesgue integral of nonnegative measurable functions (property 7),

$= \int_A |f| d\mu$

2. Since $f \leq g$, the following must hold:

(a) $f^+ = \max\{0, f\} \leq \max\{0, g\} = g^+$;

(b) $-f \geq -g$;

(c) $f^- = \max\{0, -f\} \geq \max\{0, -g\} = g^-$.

Thus, by the properties of the Lebesgue integral of nonnegative measurable functions (property2) $\int_A f^+ d\mu \leq \int_A g^+ d\mu$ and $\int_A f^- d\mu \geq \int_A g^- d\mu$. Therefore, $-\int_A f^- d\mu \leq -\int_A g^- d\mu$.

Hence, $\int_A f^+ d\mu - \int_A f^- d\mu \leq \int_A g^+ d\mu - \int_A f^- d\mu \leq \int_A g^+ d\mu - \int_A g^- d\mu$. It follows that $\int_A f d\mu \leq \int_A g d\mu$.

3. $\int_A f d\mu = \int_A f^+ d\mu - \int_A f^- d\mu$ by definition

$= \int_X XAf^+ d\mu - \int_X XAf^- d\mu$ by the properties of the Lebesgue integral of nonnegative measurable functions (property 3),

$= \int_X (XAf)^+ d\mu - \int_X (XAf)^- d\mu$

$= \int_X XAf d\mu$ by definition

4. If $c \geq 0$, then

$\int_A cf d\mu = \int_A (cf)^+ d\mu - \int_A (cf)^- d\mu$ by definition

$= \int_A cf^+ d\mu - \int_A cf^- d\mu$

$= c\int_A f^+ d\mu - c\int_A f^- d\mu$ by theproperties of the Lebesgue integral of nonnegative measurable functions (property 5)

$= c\left(\int_A f^+ d\mu - \int_A f^- d\mu\right)$

$= c\int_A f d\mu$ by definition

5. If $c < 0$, then

$= \int_A cf d\mu = \int_A (cf)^+ d\mu - \int_A (cf)^- d\mu$ by definition

$= \int_A (-c) f^- d\mu - \int_A (-c) f^+ d\mu$

$= -c\int_A f^- d\mu + c\int_A f^+ d\mu$ by the properties of the Lebesgue integral of nonnegative measurable functions (property 5)

$= c\left(-\int_A f^- d\mu + \int_A f^+ d\mu\right)$

$= c\int_A f d\mu$ by definition.

2. Note that $\int_A f^+ d\mu = 0$ and $\int_A f^- d\mu = 0$ by the properties of the Lebesgue integral of nonnegative measurable functions (property 6). It follows that $\int_A f d\mu = 0$.

7. Let $\{S_n\}$ be a nondecreasing sequence of nonnegative simple functions converging pointwise to $f^+ + g^+$ and $\{t_n\}$ be a nondecreasing sequence of nonnegative simple functions converging pointwise to $f^- + g^-$. Note that, for every n, $\int_A S_n d\mu - \int_A t_n du = \int_A (S_n - t_n) d\mu$.

Since f and g are integrable and $|f+g| \leq |f|+|g|$, $f+g$ is integrable. Thus,

$\int_A f d\mu + \int_A g d\mu = \int_A f^+ d\mu - \int_A f^- d\mu + \int_A g^+ d\mu - \int_A g^- d\mu$ by definition

$= \int_A f^+ d\mu + \int_A g^+ d\mu - \left(\int_A f^- d\mu + \int_A g^- d\mu \right)$

$= \int_A (f^+ + g^+) d\mu - \left(\int_A (f^- + g^-) d\mu \right)$ by the properties of the Lebesgue integral of nonnegative measurable functions (property7)

$= \lim_{n\to\infty} \int_A S_n d\mu - \left(\lim_{n\to\infty} \int_A t_n d\mu \right)$ by Lebesgue's monotone convergence theorem

$= \lim_{n\to\infty} \left(\int_A S_n d\mu - \int_A t_n d\mu \right)$

$= \lim_{n\to\infty} \int_A (S_n - t_n) d\mu$

$= \int_A (f^+ + g^+ - (f^- + g^-)) d\mu$ by Lebesgue's dominated convergence theorem

$= \int_A (f^+ - f^- + g^+ - g^-) d\mu$

$= \int_A (f+g) d\mu$ by definition.

8. $\int_{A\cup B} f d\mu = \int_{A\cup B} f^+ d\mu - \int_{A\cup B} f^- d\mu$ by definition

$= \int_A f^+ d\mu + \int_B f^+ d\mu - \left(\int_A f^- d\mu + \int_B f^- d\mu \right)$ by the properties of the Lebesgue integral of nonnegative measurable functions (property 8),

$= \int_A f^+ d\mu - \int_A f^- d\mu + \int_B f^+ d\mu - \int_B f^- d\mu$

$= \int_A f d\mu + \int_B f d\mu$ by definition

5. Let E = $\{x \in A: f(x) = g(x)\}$. Since f and g are measurable functions and $A \in B$, it must be the case that $E \in B$. Thus, $A - E \in B$. By hypothesis, $\mu(A \setminus E) = 0$. Note that $E \cap (E \setminus A) = \phi$ and $E \cup (A \setminus E) = A$.

Thus,

$$= \int_A f d\mu = \int_E f d\mu + \int_{A\setminus B} f d\mu = \int_E f d\mu + 0$$

$$= \int_E g d\mu + 0 = \int_E g d\mu + \int_{A\setminus E} g d\mu = \int_A g d\mu.$$

At this point we could stop: we have extended the concept of integration to (bounded) functions defined on general sets (measurable sets with finite measure) without using partitions (subintervals).

The new concept, the Lebesgue integral, agrees with the old one, Riemann integral, when both apply and removes some of the oddities mentioned before.

But we can continue to extend the Lebesgue integral to functions that are unbounded, including functions that may occasionally be *equal* to infinity. To do that, we first need to define the concept of a measurable function.

Measurable Function

Let f be a function from $E \subset R$ into $R\ \{-\infty, \infty\}$. The function f is called (Lebesgue) measurable if

1. The domain E of the function is a measurable set.
2. For every real number a the set $f^{-1}(-\infty, a)$ is a measurable set.

In other words, functions whose values are real numbers or possibly plus or minus infinity are measurable if the inverse image of every interval $(-\infty, a)$ is measurable. That is somewhat comparable to one of the equivalent definitions of continuous functions: a function f is continuous if the inverse image of every open interval is open. However, not every measurable function is continuous, while every continuous function is clearly measurable.

Measurable functions that are bounded are equivalent to Lebesgue integrable functions.

Proposition: Bounded Measurable Functions are Integrable. If f is a bounded function defined on a measurable set $\boldsymbol{E}$ with finite measure. Then f is measurable if and only if f is Lebesgue integrable.

Measurable functions do not have to be continuous, they may be unbounded and they can, in particular, be equal to plus or minus infinity. On the other hand, measurable functions are "almost" continous.

Measurable Functions are Almost Continuous

Suppose f is a measurable function defined on an interval $[a, b]$ such that the set where f is plus or minus infinity has measure zero. Then, for any > 0 we can find a step function g and a continuous function h such that

$$| f(x) - g(x) | < \varepsilon$$

$$| f(x) - h(x) | < \varepsilon$$

except on a set of measure less than ε.

Using measurable functions allows us to extend the Lebesgue integral first to non-negative functions that are not necessarily bounded and then to general measurable functions.

Lebesgue Integral of Non-Negative Functions

If f is a measurable function defined on E and h is a bounded measurable function such that m ({x: $h(x)$ # 0}) is finite, then we define

$$\int_E f(x)dx = \sup\left\{\int_E h(x)dx, h \le f\right\}$$

If $\int_E f(x)dx$ is finite, then f is called Lebesgue integrable over E.

The final step to define the Lebesgue integral of a general function is now easy.

The General Lebesgue Integral

Let f be a measurable function and define the positive and negative parts of f, respectively, as:

$$f^+(x) = \max(f(x), 0)$$

$$f^-(x) = \max(-f(x), 0)\,.$$

so that $f = f^+ - f^-$. Then f is Lebesgue integral if f^+ and f^- are Lebesgue integrable and

$$\int_E f(x)dx = \int_E f^+(x)dx - \int_E f^-(x)dx$$

Riemann Versus Lebesgue

If f is L-integrable, so is $|f|$, but the converse is not true (I think).

There is a function f that is R-integrable but $|f|$ is not R-integrable

Riemann integral for bounded functions only, Lebesgue for bounded and unbounded functions.

Riemann integral can be extended to improper Riemann integrals, but can not allow functions that are extended real valued.

An improper R-integral may exist without the function being L-integrable $f(x) = \sin(x) / x$ for $x > 0$.

If f is L-integrable and the improper R-integral exists, then both agree.

Mention sequences of R-integrable function vs. L-integrable.

Lebesgue's Theorem: a bounded function f on $[a, b]$ is Riemann integrable if and only if the set of points where f is not continuous has measure zero.

Definition: A collection of $n + 1$ distinct points of the interval $[a, b]$

$$P := \{x_0 = a < x_1 < \ldots < x_{i-1} < x_i < \ldots < b =: x_n\}$$

is called a *partition* of the interval. In this case, we define the *norm* of the partition by

$$\| P \| := \max_{1 \le i \le n} \Delta x_i .$$

where $\Delta x_i := x_i - x_{i-1}$ is the *length* of the *i*-th subinterval $\lfloor x_{i-1}, x_i \rfloor$.

Definition: For a given partition *P*, we define the *Riemann upper sum* of a function *f* by

$$U(P, f) := \sum_{i=1}^{n} M_i \Delta x_i$$

where M_i denotes the supremum of *f* over each of the subintervals $\lfloor x_{i-1}, x_i \rfloor$. Similarly, we define the *Riemann lower sum* of a function *f* by

$$L(P, f) := \sum_{i=1}^{n} m_i \Delta x_i$$

where m_i denotes the infimum of *f* over each of the subintervals $\lfloor x_{i-1}, x_i \rfloor$. Since $m_i \le M_i$, we note that

$$L(P, f) \le U(P, f)$$

for any partition *P*.

Definition: Suppose P_1, P_2 are both partitions of [*a*, *b*], then P_2 is called a *refinement of* P_1, denoted by

$$P_1 \prec P_2 ,$$

if as sets $P_1 \subseteq P_2$.

Note: *If* $P_1 \prec P_2'$, *it follows that* $\| P_2 \| \le \| P_1 \|$ *since each of the subintervals formed by* P_2 *is contained in a subinterval which arises from* P_1.

Lemma: If $P_1 \prec P_2$, then

$$L(P_1, f) \le L(P_2, f).$$

and

$$U(P_2, f) \le U(P_1, f).$$

Proof: Suppose first that P_1 is a partition of [*a*, *b*] and that P_2 is the partition obtained from P_1 by adding an additional point z. The general case follows by induction, adding one point at at time. In particular, we let

$$P_1 := \{x_0 = a < x_1 < \dots < x_{i-1} < x_i < \dots < b =: x_n\}$$

and

$$P_2 := \{x_0 = a < x_1 < \dots < x_{i-1} < x_i < \dots < b =: x_n\}$$

for some fixed i. We focus on the upper Riemann sum for these two partitions, noting that the inequality for the lower sums follows similarly. Observe that

$$U(P_1, f) := \sum_{j=1}^{n} M_j \Delta_j$$

and

$$U(P_2, f) := \sum_{j=1}^{i-1} M_j \Delta_j + M(z - x_{i-1}) + \tilde{M}(x_i - z) + \sum_{j=k+1}^{n} M_j \Delta x_j$$

where $M := \sup_{[x_{i-1}, z]} f(x)$ and $\tilde{M} := \sup_{[z, x_i]} f(x)$. It then follows that $U(P_2, f) \le U(P_1, f)$ since

$$M, \tilde{M} \le M_i.$$

Defination: If P_1 and P_2 are arbitrary partitions of $[a, b]$, then the common refinement of P_1 and P_2 is the formal union of the two.

Corollary: Suppose P_1 and P_2 are arbitrary partitions of $[a, b]$, then

$$L(P_1, f) \le U(P_2, f).$$

Proof: Let P be the common refinement of P_1 and P_2, then

$$L(P_1, f) \le L(P, f) \le U(P, f) \le U(P_2, f).$$

Definition: The *lower Riemann integral* of f over $[a, b]$ is defined to be

$$\underline{\int_a^b} f(x)dx := \sup_{\substack{\text{all partitions} \\ P \text{ of } [a,\, b]}} L(P, f)$$

Similarly, the *upper Riemann integral* of f over $[a, b]$ is defined to be

$$\overline{\int_a^b} f(x)dx := \inf_{\substack{\text{all partitions} \\ P \text{ of } [a,\, b]}} U(P, f)$$

By the definitions of least upper bound and greatest lower bound, it is evident that for any function f there holds

$$\underline{\int_a^b} f(x)dx \le \overline{\int_a^b} f(x)dx.$$

Definition: A function f is *Riemann integrable over* $[a, b]$ if the upper and lower Riemann integrals coincide. We denote this common value by $\int_a^b f(x)dx$.

Examples:

1. $\int_a^b kdx = k(b-a)$.

2. $\int_a^b xdx = \frac{1}{2}(b^2 - a^2)$.

[***Hint:*** Use $\sum_{i=1}^{n} i = n(n+1)/2$.]

Theorem: A necessary and sufficient condition for f to be Riemann integrable is given $\in > 0$, there exists a partition P of $[a, b]$ such that

$$(*)\ U(P, f) - L(P, f) < \in.$$

Proof: First we show that (*) is a sufficient condition. This follows immediately, since for each $\in > 0$ that there is a partition P such that (*) holds,

$$\sum_{i=1}^{n} i = n(n+1)/2 \int_a^{-b} f(x)dx - \int_{-a}^{b} f(x)dx \le U(P,f) - L(P,f) < \in.$$

Since $\in > 0$ was arbitrary, then the upper and lower Riemann integrals of f must coincide. To prove that (*) is a necessary condition for f to be Riemann integrable, we let By the definition of the upper Riemann integral as a infimum of upper sums, we can find a partition P_1 of $[a, b]$ such that

$$\int_a^b f(x)dx \le U(P_1, f) < \int_a^b f(x)dx + \in/2$$

Similarly, we have

$$\int_a^b f(x)dx - \in/2 < L(P_2, f) \le \int_a^b f(x)dx.$$

Let P be a common refinement of P_1 and P_2, then subtracting the two previous inequalities implies,

$$U(P, f) - L(P, f) \le U(P_1, f) - L(P_2, f) < \text{Î}.$$

Definition: A *Riemann sum* for f for a partition P of an interval $[a, b]$ is defined by

$$R(P, f, \xi) := \sum_{j=1}^{n} f(\xi_j)\Delta x_i$$

where the ξ_j, satisfying $x_{j-1} \le \xi_j \le x_j (1 \le j \le n)$, are arbitrary.

Corollary: Suppose that f is Riemann integrable on $[a, b]$, then there is a unique number ($\gamma = \int_a^b f(x)dx$) such that for every $\in > 0$ there exists a partition P of $[a, b]$ such that if $P \prec P_1, P_2$, then

(i) $0 \le U(P_1, f) - \gamma < \in$

(ii) $0 \le \gamma - L(P_2, f) < \in$

(iii) $|\gamma - R(P_1, f, \xi)| < \in$ where $R(P_1, f, \xi)$ is any Riemann sum of f for the partition P_1. In this sense, we can interpret

$$\int_a^b f(x)dx) = \lim_{\|P\| \to 0} R(P, f, \xi).$$

although we would actually need to show a little more to be entirely correct.

Proof. Since $L(P_2, f) \le \gamma \le U(P_1, f)$ for all partitions, we see that parts i.) and ii.) follow from the definition of the Riemann integral. To see part iii.), we observe that $m_j \le f(\xi_j) \le M_j$ and hence that

$$L(P_1, f) \le R(P_1, f, \xi) \le U(P_1, f).$$

But we also know that both

$$L(P_1, f) \le \gamma \le U(P_1, f)$$

and condition (*) hold, from which part iii.) follows.

Theorem: If f is continuous on $[a, b]$, then f is Riemann-integrable on $[a, b]$.

Proof: We use the condition (*) to prove that f is Riemann-integrable. If $\in > 0$, we set $\in_{0:} = \in/(b - a)$.

Since f is continuous on $[a, b]$, f is uniformly continuous. Hence there is a $\delta > 0$ such that $| f(y) - f(x) | < \in_0$ if $| y - x | < \delta$. Suppose that $\| P \| < \delta$, then it follows that $|M_i - m_i| \le \in_0$ $(1 \le i \le n)$. Hence

$$U(P, f) - L(P, f) = \sum_{i=1}^{n} (M_i - m_i)\Delta x_i \le \in_0 (b - a) = \in.$$

Theorem: If f is monotone on $[a, b]$, then f is Riemann-integrable on $[a, b]$.

Proof: If f is constant, then we are done. We prove the case for f monotone increasing. The case for monotone decreasing is similiar.

We again use the condition (*) to prove that f is Riemann-integrable. If $\in > 0$, we set $\delta := \in/(f(b) - f(a))$ and consider any partition P with $\| P \| < \delta$. Since f is monotone increasing on [a, b], then $M_i = f(x_i)$ and mi $= f(x_{i-1})$. Hence

$$U(P, f) - L(P, f) = \sum_{i=1}^{n} (M_i - m_i)\Delta x_i$$

$$= \sum_{i=1}^{n} (f(x_i) - f(x_{i-1}) \Delta x_i$$

$$\leq \| P \| \sum_{i=1}^{n} (f(x_i) - f(x_{i-1})$$

$$< \delta(f(b) - f(a)) = \in.$$

Theorem: (Properties of the Riemann Integral) Suppose that f and g are Riemann integrable and k is a real number, then

(i) $\int_a^b kf(x)dx = k \int_a^b f(x)dx$

(ii) $\int_a^b f + gdx = \int_a^b fdx + \int_a^b gdx$

(iii) $g \leq f$ implies $\int_a^b gdx \leq \int_a^b fdx$.

(iv) $\left| \int_a^b fdx \right| \leq \int_a^b | f | dx$

Proof: To prove part *i.*), we observe that in case $k \geq 0$, then $\sup_{[x_{i-1}, x_i]} kf(x) = kM_i$ and $\inf_{[x_{i-1}, x_i]} kf(x) = km_i$.

Hence $U(P, kf) = kU(P, f)$ and $L(P, kf) = kL(P, f)$. In the case that *k; SPMlt;*0, then $\sup_{[x_{i-1}, x_i]} kf(x) = km_i$ and $\inf_{[x_{i-1}, x_i]} kf(x) = kM_i$. It follows in this case that $U(P, kf) = kL(P, f)$ and $L(P, kf) = kU(P, f)$ and so

$$\int_a^{-b} kf(x)dx = k \int_{-a}^{b} f(x)dx$$

$$\int_{-a}^{b} kf(x)dx = k \int_a^{-b} f(x)dx.$$

To prove property ii.)

we notice that $\sup_I (f + g) \leq \sup_I f + \sup_I g$ and $\inf_I f + \inf_I g \leq \inf_I (f + g)$ for any interval I (for example, $I = [x_{i-1}, x_i]$). Hence,

1. $L(P, f) + L(P, g) \leq L(P, f + g) \leq U(P, f + g) \leq U(P, f) + U(P, g)$

   Let $\in > 0$, then since f and g are Riemann integrable, there exist partitions P_1, P_2 such that

2. $U(P_1, f) - L(P_1, f) < \in/2$, $U(P_2, g) - L(P_2, g) < \in/2$.

If we let P be a common refinement of P_1 and P_2, then by combining inequalities (1) and (2), we see that see that

$$U(P, f+g) - L(P, f+g) \leq U(P, f) - L(P, f) + U(P, g) - L(P, g)$$
$$\leq U(P_1, f) - L(P_1, f) + U(P_2, g) - L(P_2, g)$$
$$\leq \in/2 + \in/2 = \in.$$

Property iii.) follows directly from the definition of the upper and lower integrals using, for example, the inequality $\sup_I g(x) \leq \sup_I f(x)$.

Property iv.) is proved by applying property iii.) to the inequality

$$-|f| \leq f \leq |f|,$$

from which it follows that $-\int_a^b |f|\,dx \leq \int_a^b f dx \leq \int_a^b |f|\,dx$. But this inequality implies property iv.).

Definition: We extend the definition of the integral to include general limits of integration which are consistent with our earlier definition.

1. $\int_a^a f(x)dx = 0$.

2. $\int_b^a f(x)dx = -\int_a^b f(x)dx$.

Theorem: If f is Riemann integrable on $[a, b]$, then it is Riemann integrable on each subinterval $\lfloor c,d \rfloor \subseteq \lfloor a,b \rfloor$. Moreover, if $c \in \lfloor a,b \rfloor$, then

3. $\int_a^b f(x)dx = \int_a^c f(x)dx + \int_c^b f(x)dx$

Proof: We show first that condition (*) holds for the interval $[c, d]$. Suppose $\in > 0$, then by $(_*)$ applied to f over the interval $[a, b]$, we have that there exists a partition P of $[a,b]$ such that condition $(_*)$ holds.

Let $\tilde{p}$ be the refinement obtained from P which contains the points c and d. Let P be the partition obtained by restricting the partition $\tilde{p}$ to the interval $[c, d]$, then

$$U(P^*, f) - L(P^*, f) \leq U(\tilde{P}, f) - L(\tilde{P}, f) \leq U(P, f) - L(P, f) < \in$$

and so f is Riemann integrable over $[c, d]$.

To prove the identity (3), we use the fact that condition $(_*)$ holds when f is Riemann integrable. Let $\in > 0$, then for $\in/3 = 0$, we may apply $(_*)$ to each of the intervals $I = [a, b]$, $[a, c]$ and $[c,b]$, respectively, to obtain partitions P_I which satisfy

4. $0 \le U_I(P_I, f) - \int_I f dx \le U_I(P_I, f) - L_I(P_I, f) < \in /3$.

We let P be the partition of $[a, b]$ formed by the union of the two partitions $P_{[a,c]}$, $P_{[c,b]}$ and $\tilde{p}$ be the common refinement of P and $P_{[a,b]}$. Observing that

(5) $U_{[a,b]}(\tilde{P}, f) = U_{[a,c]}(\tilde{P}_1, f) + U_{[c,b]}(\tilde{P}_2, f)$,

we can combine with inequality (4) to obtain

$$\left| \int_a^c f dx + \int_c^b f dx - \int_a^b f dx \right| \le \left| U_{[a,c]}(\tilde{P}, f) - \int_a^c f dx \right| + \left| U_{[c,b]}(\tilde{P}, f) - \int_c^b f dx \right|$$

$$+ \left| U_{[a,b]}(\tilde{P}, f) - \int_c^b f dx \right|$$

$$< 3\in_0 = \in.$$

Since $\in > 0$ was arbitrary, then equality (3) must hold.

Theorem: (Intermediate Value Theorem for Integrals) If f is continuous on $[a, b]$, then there exists ξ between a and b such that

$$\int_a^b f(x) dx = f(\xi)(b - a).$$

Proof: Since f is continuous on $[a, b]$ and for $\eta := \dfrac{\int_a^b f dx}{b - 1}$ there holds

$$\min_{[a,b]} f(x) \le \eta \le \max_{[a,b]} f(x),$$

then by the Intermediate Value Theorem for continuous functions, there exists a $\xi \in [a, b]$ such that $f(\xi) = \eta$.

Theorem: (Fundamental Theorem of Calculus, Part I. Derivative of an Integral) Suppose that f is continuous on $[a, b]$ and set $F(x) := \int_a^x f(y) dy$, then F is differentiable and $F'(x) = f(x)$ for $a < x < b$.

Proof: Notice that

$$\frac{F(x_0 + h) - F(x_0)}{h} = \frac{\int_{x_0}^{x_0 + h} f dx}{h} = f(\xi)$$

for some ξ between x_0 and $x_0 + h$. Hence, as $h \to 0$, then $\xi = \xi_h$ converges to x_0 and so the displayed difference quotient has a limit of $f(x_0)$ as $h \to 0$.

Theorem:(Fundamental Theorem of Calculus) Suppose that F is function with a continuous derivative on $[a, b]$, then

$$\int_a^b F'(y)dy = F(x)|_{x=a}^{x=b} := F(b) - F(a).$$

Proof: Define $G(x) := \int_a^x F'(y)dy$, and set $H := F - G$. Since the derivative of H is identically zero by Part *I* of the Fundamental Theorem of Calculus, then the Mean Value Theorem implies that $H(b) = H(a)$. Expressing this in terms of F and G gives

$$F(b) - \int_a^b F'(y)dy = F(a),$$

which establishes the theorem.

Chapter 11

Shortcomings of Riemann Integration

RIEMANN REVISITED

The development of the integral in most introductory analysis courses is centered almost exclusively on the Riemann integral. This relatively intuitive approach begins by taking a partition $P = \{x_0, ..., x_n\}$ of the domain of the real-valued function f in question. Given P, the Riemann sum of f is simply

$$\sum_{i=1}^{n}(x_i - x_{i-1}).f(x_i) \text{ where } x_{i-1} < x_i < x_i.$$

The integral of f, if it exists, is the limit of the Riemann sum as $n \to \infty$.

Shortcomings of Riemann Integration

Although the Riemann integral suffices in most daily situations, it fails to meet our needs in several important ways. First, the class of Riemann integrable functions is relatively small.

Second and related to the first, the Riemann integral does not have satisfactory limit properties. That is, given a sequence of Riemann integrable functions $\{f_n\}$ with a limit function $f = \lim_{n\to\infty} f_n$, it does not necessarily follow that the limit function f is Riemann integrable.

Third, all L_p spaces except for L_∞ fail to be complete under the Riemann integral.

Example: Consider the sequence of functions $\{f_n\}$ over the interval $E = [0, 1]$.

$$f_n(x) = \begin{cases} 2^n & \text{if } \dfrac{1}{2^n} \le x \le \dfrac{1}{2^{n-1}} \\ 0 & \text{otherwise} \end{cases}$$

The limit function of this sequence is simply $f = 0$. In this example, each function in the sequence is integrable as is the limit function. However, the limit of the sequence of integrals is not equal to the integral of the limit of the sequence. That is,

$$\lim_{n\to\infty} \int_{f_n}^{1} f_n(x)dx = 1 \neq 0 = \int_0^1 \lim_{n\to\infty} f_n(x)dx.$$

We will return to this function as an example of how the Lebesgue integral fails as well.

Example: Consider the sequence of functions $\{d_n\}$ over the interval $E = [0, 1]$.

$$d_n(n) = \begin{cases} 1 & \text{if } x \in (r_n) \\ 0 & \text{otherwise} \end{cases}$$

where $\{r_n\}$ is the set of the first n elements of some decided upon enumeration of the rational numbers. Each function d_n is Riemann integrable since it is discontinuous only at n points. The limit function $D = \lim_{n\to\infty} d_n$ is given by

$$D(x) = \begin{cases} 1 & \text{if } x \text{ is rational} \\ 0 & \text{if } x \text{ is irrational} \end{cases}$$

This function, known as the Dirichlet function, is discontinuous everywhere and therefore not Riemann integrable. Another way of showing that $D(x)$ is not Riemann integrable is to take upper and lower sums, which result in 1 and 0, respectively.

A NEW WAY TO COUNT RECTANGLES: LEBESGUE INTEGRATION

An equally intuitive, but long in coming method of integration, was presented by Lebesgue in 1902. Rather than partitioning the domain of the function, as in the Riemann integral, Lebesgue chose to partition the range. Thus, for each interval in the partition, rather than asking for the value of the function between the end points of the interval in the domain, he asked how much of the domain is mapped by the function to some value between two end points in the range.

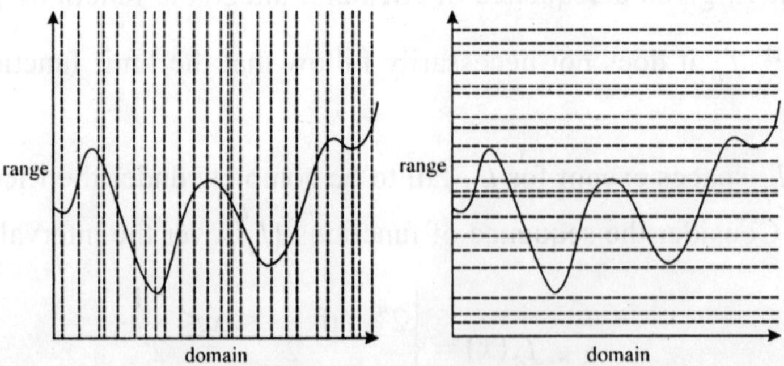

Fig. Two Ways to Count Rectangles – Partitioning the Range as Opposed to Partitioning the Domain of a Function.

Partitioning the range of a function and counting the resultant rectangles becomes tricky since we must employ some way of determining (or measuring) how much of the domain is sent to a particular portion of a partition of the range. Measure theory addresses just this problem.

As it turns out, the Lebesgue integral solves many of the problems left by the Riemann integral. With this in mind, we now turn to measure theory.

MEASURE THEORY

Measure theory is a rich subject in and of itself. However, we present it here expressly for the purpose proposed to define the "length" of an arbitrary set so as to formalize the idea of the Lebesgue integral. As such, only the very basics of measure theory are presented here and many of the rote proofs are left to the reader.

MEASURE

Given an interval $E = [a, b]$ and a set S of subsets of E which is closed under countable unions, we define the following.

Definition: A set function on S is a function which assigns to each set $A \in S$ a real number.

Definition: A set function μ on S is called a measure if the following properties hold.

1. Semi-Positive-Definite: $0 \leq \mu(A) \leq b - a$ for all $A \in S$.
2. Trivial Case: $\mu(\phi) = 0$.
3. Monotonicity: $\mu(A) \leq \mu(B)$ for all $A, B \in S, A \subset B$
4. Countable Additivity: if $A = \bigcup_{n=1}^{\infty} A_n$, then $\mu(A) = \bigcup_{n=1}^{\infty} \mu(A_n)$, where $A_n \in S$ for $n = 1, 2,...$ and $A_n \bigcap A_m = \phi$; for $n \neq m$.

As can be seen from the definition, the concept of measure is a general one indeed.

Example: Examples of measure abound. To begin at the beginning, consider the trivial measure: $\mu(A) = 0$ for all $A \in S$.

Example: Let $\{x_1,..., x_n\} \subset E$ be a finite set of points and $E = [a, b]$. Let $a < x_i < x_j < b$ for all $i < j$. Consider the set $A = E / \{x_1,..., x_n\}$, the interval E without the points $x_1,..., x_n$. Let our measure be such that the measure of any interval with endpoints $x < y$ is $y - x$. Then,

$$
\begin{aligned}
m(A) &= m([a, x_1) \cup (x_1, x_2) \cup ... \cup (x_{n-1}, x_n) \cup (x_n, b]) \\
&= m([a, x_1)) + m((x_1, x_2)) + ... + m((x_{n-1}, x_n)) + m((x_n, b]) \\
&= (x_1 - a) + (x_2 - x_1) + ... + (x_n - x_{n-1}) + (b - x_n)
\end{aligned}
$$

$= b - a$

$= m(E).$

Notice that the only complication which may arise in this example is how we define our set of subsets, *S*.

OUTER AND INNER MEASURE

To keep us grounded in reality, we would like to use a measure which puts our intuitive notion of length at ease. Thus we define the following.

Definition: The outer measure of any interval I on the real number line with endpoints $a < b$ is $b - a$ and is denoted as $m^*(I)$.

We would like to generalize this definition of outer measure to all sets of real numbers, not just intervals. To do this, we use the following theorem, upon which the remainder of our work depends heavily.

Theorem: Every non-empty open set $G \subset R$ can be uniquely expressed as a finite or countably infinite union of pairwise disjoint open intervals.

Proof Idea: Let $x \in G$ and consider the open interval $I_x = (a_x, b_x)$ constructed from

$$a_x = glb\{y \mid (y, x) \subset G\}$$

and

$$b_x = lub\{z \mid (x, z) \subset G\}.$$

I_x is called the component of x in G. $G = \bigcup_{x \in G} I_x$ is exactly the decomposition of G we are looking for. It is straight-forward to show that, given $x_1, x_2 \in G$, either I_{x_1} and I_{x_2} are disjoint or they are equal. The countability of this collection of open intervals follows easily from their disjointness; simply pick one rational number from each interval and use it to label that interval. Since the intervals are disjoint, no two will have the same label and since the rationals are countable, so will be the intervals which they label. Uniqueness of the decomposition is, as usual, a straight-forward proof by contradiction.

With this theorem, it is possible to extend the definition of outer measure to open sets of real numbers.

Definition: The outer measure $m^*(G)$ of an open set $G \subset E$ is given by $\sum_i m^*(I_i)$ where the I_i form the unique decomposition of G into a finite or a countably infinite union of pairwise disjoint open intervals. From this directly follows a definition of outer measure for any set.

Definition: The outer measure $m^*(A)$ of any set $A \subset R$ is given by $glb\{m(G) \mid A \subset G$ and G open in $E\}$.

Now that outer measure is well-defined for arbitrary subsets of E, we turn to a closely related measure, inner measure.

Definition: The inner measure of any set $A \subset E$, denoted $m_*(A)$, is defined as $m^*(E) - m^*(E/A)$, where E/A is the compliment of A with respect to E.

Intuitively, the inner measure is in some ways "measuring" the same thing as the outer measure, only in a more roundabout way.

We cannot take it for granted, however, that the inner and outer measures of any given set are the same, although we are very interested in the cases when they are, as we shall soon see. For now, though, let us state without proof some simple observations regarding inner and outer measure.

Lemma: The measures m_* and m^* both exhibit monotonicity. That is, given $A \subset B \subset E$, it follows that,

1. $m^*(A) \leq m^*(B)$
2. $m^*(A) \leq m_*(B)$

Lemma: Given a set $A \subset E$, it follows that $m_*(A) \leq m^*(A)$.

Lemma: The outer measure m^* exhibits subadditivity.

That is, whenever $\{A_n \mid n = 1, 2, ...\}$ is a set of subsets of E, then

$$m^*\left(\bigcup_{n=1}^{\infty} A_n\right) \leq \sum_{n=1}^{\infty} m^*(A_n).$$

SHORTCOMINGS OF OUTER AND INNER MEASURE

Now that we have some concrete measures to work with, we must examine whether they satisfy our needs. The first shortcoming we discover is presented below.

Theorem: The outer measure, m^*, is not countably additive on the set of all subsets of E.

Proof Idea: By construction. It is possible to construct pairwise disjoint sets A_n such that $E = \bigcup_{n=1}^{\infty} A_n$ and $m^*(E) \neq \bigcup_{n=1}^{\infty} m^*(A_n)$.

This is rather disappointing. It would be nice to have a measure which is defined on all subsets of E and still satisfies our intuition enough to use it as the basis for the development of the Lebesgue integral.

However, it turns out that the set of subsets of E on which the outer measure m is countably additive is large enough to make it worth our while to continue using the outer measure.

MEASURABLE SETS AND MEASURABLE FUNCTIONS

MEASURABLE SETS

Let's look at a certain subset of all the sets on which inner and outer measure are well-defined – the set of Lebesgue measurable sets.

Definition: A set $A \subset E$ is Lebesgue measurable, or measurable, if $m^*(A) = m_*(A)$, in which case the measure of A is denoted simply by $m(A)$ and is given by $m(A) = m^*(A) = m_*(A)$. θ

A straight-forward extension of this definition applies to unbounded sets,

Definition: The measure for an unbounded set A is defined simply as,

$$m(A) = \lim_{n\to\infty} m\left(A \bigcap [-n,n]\right)$$

Lemma: Measurable sets have the following properties.

1. A set $A \subset E$ is measurable if and only if

$$m^*(A) + m^*(E/A) = b - a,$$

where a and b are the endpoints of the interval E.

2. A set $A \subset E$ is measurable if and only if its complement E/A is measurable.

Proof: The lemma follows directly from the definition of measurable (3.1).

Example: We see that a set consisting of an interval which is missing a finite number of points has the same outer measure as the interval itself. That is,

$$m^*(E/\{x_i,\dots, x_n\}) = m^*(E).$$

Thus, by Lemma, $E/\{x_i,\dots, x_n\}$ is measurable if and only if $m^*(\{x_i,\dots, x_n\}) = 0$.

Theorem: The outer measure, m^*, is countably additive on the set of all measurable subsets of E. That is, whenever $\{A_n \mid n = 1, 2,\dots\}$ is a set of measurable subsets of E, then

$$m^*\left(\bigcup_{n=1}^{\infty} A_n\right) = \sum_{n=1}^{\infty} m^*(A_n)$$

Proof Idea: Although the proof of this theorem is not difficult, it is somewhat lengthy in its technical detail and is therefore left for the enrichment of the reader.

So, knowing that we intend to define the Lebesgue integral in terms of the outer measure, it seems that the size of the class of "Lebesgue integrable" functions may somehow be limited by the fact that, as far as we know, the outer measure is only defined on measurable set.

MEASURABLE FUNCTIONS

Definition: Let A be a bounded measurable subset of R and $f: A \to R$ a function. Then f is said to be measurable on A if $\{x \in A \mid f(x) > r\}$ is measurable (as a set) for every real number r.

There are many equivalent definitions of measurable which follow similar lines. In this definition we see the beginnings of being able to "count the rectangles" created by partitioning the range of a function rather than the domain.

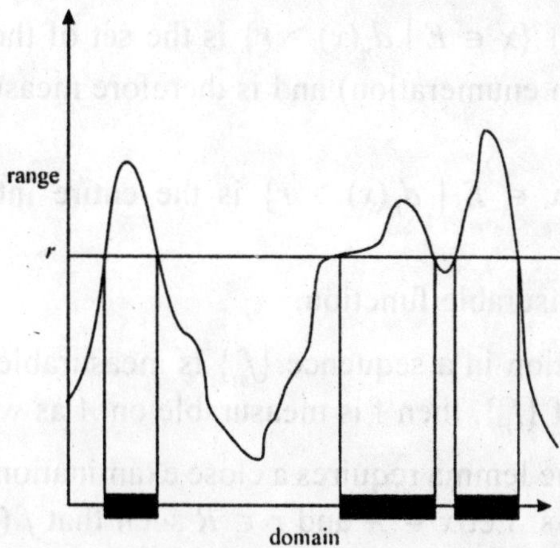

Fig. The Function f is Measurable if the Shaded Region of the Domain is Measurable as a set for all Choices of the Real Number r.

Example: Consider again the sequence of functions over $E = [0, 1]$,

$$f_n(x) = \begin{cases} 2^n & if \dfrac{1}{2^n} \leq x \leq \dfrac{1}{2^{n-1}} \\ 0 & otherwise \end{cases}$$

We want to show that each f_n is a measurable function. There are three cases to consider, corresponding to three possible choices for the real number r. They are,

1. $r \geq 2n$: The set $\{x \in E \mid f_n(x) > r\}$ is the null set and is therefore measurable.
2. $0 \leq r < 2^n$: The set $\{x \in E \mid f_n(x) > r\}$ is the closed interval $\left[\dfrac{1}{2^n}, \dfrac{1}{2^{n-1}}\right]$ and is therefore measurable (with a measure of $\dfrac{1}{2^n}$).
3. $r < 0$: The set $\{x \in E \mid f_n(x) > r\}$ is the entire interval E and is therefore measurable.

Thus, each f_n is a measurable function.

Example: Consider again the sequence of functions over $E = [0, 1]$

$$d_n(x) = \begin{cases} 1 & if\ x \in (r_n) \\ 0 & otherwise \end{cases}$$

We want to show that each d_n is a measurable function. There are three cases to consider, corresponding to three possible choices for the real number r. They are,

1. $r \geq 1$: The set $\{x \in E \mid d_n(x) > r\}$ is the null set and is therefore measurable.
2. $0 \leq r < 1$: The set $\{x \in E \mid d_n(x) > r\}$ is the set of the first n rational numbers (in a decided upon enumeration) and is therefore measurable (with a measure of 0).
3. $r < 0$: The set $\{x \in E \mid d_n(x) > r\}$ is the entire interval E and is therefore measurable.

Thus, each d_n is a measurable function.

Lemma: If each function in a sequence $\{f_n\}$ is measurable on a set A and if f is the pointwise limit function of $\{f_n\}$, then f is measurable on A as well.

Proof: The proof of the lemma requires a close examination of the definitions of limit and measurable, as follows. Let $x \in A$ and $r \in R$ such that $f(x) > r$. Let p be a natural number such that $f(x) > r + 1/p$. Then, by definition of limit, there exists a natural number N such that for all $n > N$, $f_n(x) > r + 1/p$. Thus,

$$f(x) = \lim_{n \to \infty} f_n(x) > r + 1/p > r.$$

This implies that

$$\{x \in A \mid f(x) > r\} = \bigcup_{p=1}^{\infty} \bigcup_{N=1}^{\infty} \bigcap_{n=N+1}^{\infty} f_n(x) > r + 1/p > r.$$

Since this set is measurable and r was arbitrary, it follows that f is measurable.

SIMPLE FUNCTIONS

Similar to the way which step functions are put to use in the development the Riemann integral, our development of the Lebesgue integral will make use of a pedestrian class of measurable functions, aptly named simple functions.

Definition: A simple function $f: A \to R$ is a measurable function which takes on finitely many values.

Theorem: A function $f: A \to R$ is measurable if and only if it is the pointwise limit of a sequence of simple functions.

THE LEBESGUE INTEGRAL

INTEGRATING BOUNDED MEASURABLE FUNCTIONS

We introduce the Lebesgue integral by first restricting our attention to bounded measurable functions. The approach here is almost identical to that used in constructing the Riemann integral except that here we partition the range rather than the domain of the function.

Let $f: A \to R$ be a bounded measurable function on a bounded measurable subset A of R. Let $l = glb\{f(x) \mid x \in A\}$ and $u > lub\{f(x) \mid x \in A\}$ where u is arbitrary insofar as it is greater than the least upper bound of f on A.

As with the Riemann integral, we'll define the Lebesgue integral of f over an interval A as the limit of some "Lebesgue sum".

Definition: The Lebesgue sum of f with respect to a partition $P = \{y_0, ..., y_n\}$ of the interval $[l, u]$ is given as

$$L(f,P) = \sum_{i=1}^{n} y_i^* m(\{x \in A \mid y_{i-1} \le f(x) < y_i\})$$

where $y_i^* \in [y_{i-1}, y_i]$ for $i = 1, ..., n$ and f is a bounded measurable function over a bounded measurable set $A \subset R$.

This is the new way to count rectangles; the y_i^* is the height of the rectangle and the $m(\{x \in A \mid y_{i-1} \le f(x) < y_i\})$ serves as the base of the rectangle. The definition of the actual Lebesgue integral is virtually identical to that of the Riemann integral.

Definition: A bounded measurable function $f: A \to R$ is Lebesgue integrable on A if there is a number $L \in R$ such that, given $\varepsilon > 0$, there exists a $\delta > 0$ such that $| L(f, P) - L | < \varepsilon$ whenever $\| P \| < \delta$. L is known as the Lebesgue integral of f on A and is denoted by $\int_A f dm$.

CRITERIA FOR INTEGRABILITY

As with the definition of the Riemann integral, the definition of the Lebesgue integral does not illuminate us as to whether a given function is Lebesgue integrable and, if it is, what the Lebesgue integral is. The following theorem, presented without proof, fills this gap.

Theorem: A bounded measurable function f is Lebesgue integrable on a bounded measurable set A if and only if, given $\varepsilon > 0$, there exist simple functions $\underline{f}$ and $\overline{f}$ such that

$$\underline{f} \le f \le \overline{f}$$

and

$$\int_A \overline{f}dm - \int_A \underline{f}dm < \varepsilon.$$

Corollary: If f is a bounded measurable function on a bounded measurable set A, then f is Lebesgue integrable on A. Furthermore,

$$\int_A fdm = \text{lub}\left\{\int_A \underline{f}dm \mid \underline{f} \textit{ is simple and } \underline{f} \le f\right\}$$

$$= \text{glb}\left\{\int_A \overline{f}dm \mid \overline{f} \textit{ is simple and } \overline{f} \ge f\right\}$$

Example: Let's find the Lebesgue integral of the Dirichlet function. We know by Example that the Dirichlet function is measurable. Thus, by Corollary, it is Lebesgue integrable. Specifically, we claim that $\int_{[0,1]} Ddm = 0$. Looking back to the definition of the Lebesgue integral, we must show that, given $\varepsilon > 0$, there exists a $\delta > 0$ such that $|L(D, P)| < \varepsilon$ whenever the width of the partition P is less than δ. For $|L(D, P)|$, this means that we must find δ such that

$$|L(D,P)| = \sum_{i=1}^{n} y_i^* m(\{x \in [0,1] \mid y_{i-1} \le D(x) < y_i\}) < \varepsilon,$$

where $P = \{y_0 = 0, ..., y_n = 2\}$ is a partition of the interval $[0, 2]$ in the range, $y_i \in [y^*_{i-1}, y_i]$ for $i = 1, ..., n$, and the absolute value signs have been omitted since all quantities here are inherently positive. We accomplish this by picking $\delta = \varepsilon/3$. Then

$$|L(D,P)| = \sum_{i=1}^{n} y_i^* m(\{x \in [0,1] \mid y_{i-1} \le D(x) < y_i\})$$

$$= y_1^* m(\{x \in [0,1] + 0 \le D(x) < \varepsilon/3\})$$

$$+ y_j^* m(\{x \in [0,1] y_{i-1} \le D(x) < y_j\}),$$

where $y_{j-1} \le 1 < y_j$. All other terms in the sum vanish because $D(x)$ only takes on two values, 0 and 1. The first term achieves its maximum when $y_1^* = \varepsilon/3$ and when $m(\{x \in [0, 1] \mid 0 \le D(x) < \varepsilon/3\} = 1$. Thus,

$$|L(D, P)| = \sum_{i=1}^{n} y_i^* m(\{x \in [0,1] \mid y_{i-1} \le D(x) < y_j\})$$

$$\le [\varepsilon/3] + [y_j^* m(\{x \in [0,1] \mid y_{j-1} \le D(x) < y_j\})].$$

Also, we know that $y_j^* < 1 + \varepsilon/3$. Finding an upper bound on

$$m(\{x \in [0, 1] \mid y_{j-1} \leq D(x) < y_j\})$$

is equivalent to asking for an upper bound on the measure of the rational numbers in the interval [0, 1]. Since any non-empty open interval centered on a rational number contains other rational numbers, an upper bound on the measure of the rational numbers is simply the sum of the measures of all the intervals surrounding the rationals. That is, given an enumeration of the rationals, let the *n*th rational number be contained in an interval of measure $\varepsilon/3^{n+1}$. In this way,

$$m(\{x \in [0, 1] \mid y_{j-1} \leq D(x) < y_j\}) < \sum_{n=1}^{\infty} \frac{\varepsilon}{3^{n+1}} = \frac{\varepsilon}{6},$$

and we finally get that

$$\begin{aligned} | L(D, P)| &\leq \varepsilon/3 + y_j^* \, m(\{x \in [0, 1] \mid y_{j-1} \leq D(x) < y_j\}) \\ &< \varepsilon/3 + (1 + \varepsilon/3)(\varepsilon/6) \\ &= \varepsilon/3 + \varepsilon/3 + \varepsilon^2/18 \\ &< \varepsilon, \end{aligned}$$

where we have assumed without loss of generality that $\varepsilon < 1$. This completes the proof that $\int_{[0,1]} Ddm = 0$. Combined with Lemma, we also know that the irrationals must have measure 1 on the interval [0, 1]. This result makes intuitive sense since the rationals are countably infinite whereas the irrationals are uncountably infinite.

PROPERTIES OF THE LEBESGUE INTEGRAL

It is comforting to know that the Lebesgue integral shares many properties with the Riemann integral.

Theorem: Let *f* and *g* be bounded measurable functions on a bounded measurable set *A*. Then,

1. Monotonicity: If $f \leq g$, then $\int_A fdm \leq \int_A gdm$.
2. Linearity:

$$\int_A (f+g)dm = \int_A fdm + \int_A gdm$$

and

$$\int_A cfdm = d\int_A fdm \text{ for } c \in R.$$

- For any numbers l, $u \in R$ such that $l \leq f \leq u$, it follows that $l \cdot m(A) \leq \int_A fdm \leq u \cdot m(A)$.
- $\left|\int_A fdm\right| \leq \int_A |f| \, dm$.
- If A and B are disjoint bounded measurable sets and $f: A \cup B \to R$ is a bounded measurable function, then

$$\int_{A \cup B} fdm = \int_A fdm + \int_B fdm.$$

- Countable Additivity: If $A = \bigcup_{i=1}^{\infty} A_i$ where the A_i are pairwise disjoint bounded measurable sets, then

$$\int_A fdm = \sum_{i=1}^{\infty} \int_A fdm.$$

INTEGRATING UNBOUNDED MEASURABLE FUNCTIONS

Thus far, we have restricted ourselves to integrating bounded measurable functions. The generalization of the Lebesgue integral to unbounded measurable functions is straightforward but will not be given in full detail here. In general, though, it involves the introduction of $\pm\infty$ as possible values of the integral.

COMPARISON OF LEBESGUE AND RIEMANN INTEGRALS

Having achieved what we first set out to do – to define an integral based on a new way of counting rectangles, let us look back at the motivation behind this endeavor. We are striving to define an integral such that the class of integrable functions is large and well-behaved. Ideally, we would like the class of integrable functions to have nice limit properties; that the integral of the limit is the limit of the integral. Aside from examining the convergence properties of the Lebesgue integral, we are also interested in how it behaves relative to the Riemann integral. Is a Riemann integrable function Lebesgue integrable and, if so, what are the values of the respective integrals?

CONVERGENCE OF THE LEBESGUE INTEGRAL

The following lemmas and theorems build upon each other and end with a broad statement which identifies when the limit and Lebesgue integral operations are interchangeable.

Lemma: Let g be a non-negative measurable function on a bounded measurable set A. If $\{A_n\}$ are measurable subsets of A such that

$$A_1 \subset A_2 \subset A_3 \infty...,$$

and if $L \in R$ is such that $L \geq \int_{A_n} gdm$ for all n, then $L \geq \int_{\cup A_n} gdm$.

Proof Idea: Given that the Lebesgue integral depends on the measure of subsets of the domain of the function, and that measure follows our intuition, this lemma is easy to believe without proof.

Monotone Convergence Theorem

Theorem: Let A be a bounded measurable subset of R and $\{f_n\}$ be a sequence of measurable functions on A such that $0 \leq f_1 \leq f_2 \leq \cdots$. Let f be the pointwise limit of $\{f_n\}$. That is,

$$\lim_{n \to \infty} f_n(x) = f(x)$$

for all $x \in E$. Then f is integrable and

$$\lim_{n \to \infty} \int_A f_n dm = \int_A fdm..$$

Proof: By Lemma, f is measurable since each f_n is measurable. Monotonicity of the Lebesgue integral implies that $\left\{\int_A f_n dm\right\}$ is an increasing sequence. So, not discounting ∞ as a possibility, this sequence has a limit L. Since $f \geq f_n$ implies that $\int_A fdm \geq \int_A f_n dm$ for all n, it follows that $\int_A fdm \geq L$

On the other hand, we can also show that $\int_A fdm \leq L$. Let $c \in (0, 1)$ and let $g: A \to R$ be a simple function such that $0 \leq g \leq f$. Consider the sets $A_n = \{x \mid f_n(x) \geq cg(x)\}$, which happen to satisfy the hypotheses of Lemma. The reason for including c is so that $cg < f$, which implies that $\bigcup_{n=1}^{\infty} A_n = A$. For any $n = 1, 2, \cdots$, we have

$$L = \lim_{n \to \infty} \int_A f_n dm \geq \int_{A_n} f_n dm \geq c \int_{A_n} gdm.$$

By Lemma, $L \geq c \int_{\cup A_n} gdm$. Since $\bigcup_{n=1}^{\infty} A_n = A$ and $c \in (0, 1)$ was arbitrary, it follows that $L \geq \int_{\cup A_n} gdm$. By Corollary,

$$\int_A f dm = \text{lub}\left\{\int_A \underline{f} dm \mid \underline{f} \text{ is simple and } \underline{f} \leq f\right\},$$

we get $L \geq \int_A f dm$.

Combining the above two inequalities gives $L = \int_A f dm$, as desired.

Fatou's Lemma

Lemma: Let $A \subset R$ be a bounded measurable set. If $\{f_n\}$ is a sequence of non-negative measurable functions on A and

$$f(x) = \varliminf_{n\to\infty} f_n(x) \equiv \lim_{n\to\infty} \{\text{glb } f_k(x) \mid k \geq n\}$$

for every $x \in A$, then $\int_A f dm \leq \varliminf_{n\to\infty} \int_A f_n dm$

Proof: Let

$$g_n(x) = glb\{ f_k(x) \mid k \geq n\}$$

for each $n = 1, 2, \cdots$ and for each $x \in A$. Each g_n is a measurable function since the set

$$\{x \mid g_n(x) > r\} = \bigcup_{k=n}^{\infty} \{x \mid f_k(x) > r\}$$

is measurable. Notice that $g_n \leq f_n$ for all n and that $\{g_n\}$ is a monotonically increasing sequence of non-negative functions.

By definition,

$$\lim_{n\to\infty} g_n(x) = \lim_{n\to\infty} \{\text{glb } f_k(x) \mid k \geq n\} \equiv \varliminf_{n\to\infty} f_n(x) = f(x).$$

Therefore, by the Monotone Convergence Theorem,

$$\lim_{n\to\infty} \int_A g_n dm = \int_A f dm.$$

This, combined with

$$\int_A g_n dm \leq \int_A f_n(x) dm \text{ for each } n \text{ and each } x \in A$$

due to the monotonicity of the Lebesgue integral, leads to

$$\lim_{n\to\infty}\int_A g_n dm = \int_A f dm \le \underline{\lim}_{n\to\infty}\int_A f_n dm,$$

as desired.

Lebesgue Dominated Convergence Theorem

Theorem: Let $A \in R$ be a bounded measurable set and let $\{f_n\}$ be a sequence of measurable functions on A such that $\lim_{n\to\infty} f_n(x) = f(x)$ for all $x \in A$. If there exists a function g whose Lebesgue integral is finite such that $|f_n(x)| \le g(x)$ for all n and all $x \in A$, then

$$\lim_{n\to\infty}\int_A f_n dm = \int_A f dm.$$

Proof: To begin, the functions f_n and f have finite Lebesgue integrals since g does (this fact does take some proof, but is omitted here). By our choice of g we have $f_n + g \ge 0$ on the set A. So,

$$\int_A (f+g)dm \le \underline{\lim}_{n\to\infty}\int_A (f_n+g)dm$$

by Fatou's Lemma and hence,

$$\int_A f dm \le \underline{\lim}_{n\to\infty}\int_A f_n dm$$

by the linearity of the Lebesgue integral.

The same argument can be used with the function $g - f_n$ to obtain

$$-\int_A f dm \le \underline{\lim}_{n\to\infty}\left(-\int_A f_n dm\right).$$

But since

$$\underline{\lim}_{n\to\infty}\left(-\int_A f_n dm\right) = \lim_{n\to\infty}\left\{\text{glb}\left(-\int_A f_k dm\right) \mid k \ge n\right\}$$

$$= \lim_{n\to\infty}\left\{\text{glb}\int_A f_k dm \mid k \ge n\right\}$$

$$\equiv \overline{\lim}_{n\to\infty}\int_A f_n dm$$

it follows that

$$\int_A fdm \geq \overline{\lim_{n\to\infty}} \int_A f_n dm .$$

Finally, combining the above two inequalities,

$$\int_A fdm \leq \underline{\lim_{n\to\infty}} \int_A f_n dm \leq \overline{\lim_{n\to\infty}} \int_A f_n dm \leq \int_A fdm,$$

shows that $\lim_{n\to\infty} \int_A f_n dm$ exists and is equal to $\int_A fdm$, as desired.

Convergence of the Riemann Integral

Now that we have a grasp of the convergence properties of the Lebesgue integral, let's compare it with those of the Riemann integral. The Riemann equivalent to the Lebesgue Dominated Convergence Theorem.

Theorem: Let $a, b \in R, a < b$, and let $\{f_n\}$ be a uniformly convergent sequence of Riemann integrable functions on $[a, b]$ such that $\lim_{n\to\infty} f_n(x) = f(x)$ for all $x \in [a, b]$. Then

$$\lim_{n\to\infty} \int_a^b f_n(x)dx = \int_a^b f(x)dx.$$

Recall that $\{f_n\}$ is uniformly convergent if, given $\varepsilon > 0$, there exists a natural number N such that $| f(x) - f_n(x) | <$ whenever $n > N$, for all $x \in [a, b]$. Clearly, the hypotheses placed on $\{f_n\}$ in order for the Lebesgue Dominated Convergence Theorem to hold are much less stringent than requiring $\{f_n\}$ to converge uniformly. Thus, we can at least expect that the class of Lebesgue integrable functions has somewhat better limit properties than those of the class of Riemann integrable functions.

A FINAL COMPARISON

The following theorem, stated without proof, explicitly relates the Riemann and Lebesgue integrals.

Theorem: If f is Riemann integrable on $[a, b]$, then f is Lebesgue integrable on $[a, b]$, and

$$\int_a^b f(x)dx = \int_{[a,b]} fdm .$$

What's more, we've already seen that the converse of this theorem isn't true. Thus, not only does the class of Lebesgue integrable functions have better limit properties, but it is also larger than the class of Riemann integrable functions.

Example: Consider the sequence of functions $\{f_n\}$ over the interval $E = [0, 1]$.

$$f_n(x) = \begin{cases} 2^n & if \dfrac{1}{2^n} \le x \le \dfrac{1}{2^{n-1}} \\ 0 & otherwise \end{cases}$$

The limit function of this sequence is simply $f = 0$. In this example, each function in the sequence is Riemann integrable, as is the limit function. However, the limit of the sequence of Riemann integrals is not equal to the Riemann integral of the limit of the sequence. That is,

$$\int_0^1 f_n(x)dx = 1 \neq 0 = \int_0^1 \lim_{n\to\infty} f_n(x)dx.$$

By Theorem it is also the case that

$$\int_{[0,1]} f_n dm = 1 \neq 0 = \int_{[0,1]} \lim_{n\to\infty} f_n(x)dx\,.$$

This example goes to show that, although the Lebesgue integral is superior to the Riemann integral insofar as the size of the class of integrable functions and the limit properties of these functions, there are still functions which defy Lebesgue integration.

One point we have not yet touched on is the effect of the Lebesgue integral on the L_p spaces. It turns out that L_2 is complete under the Lebesgue integral. This is of considerable importance, especially when dealing with the theory behind Fourier series. However interesting that topic may be, though, it will have to wait for another time.